$1 + 2 + \cdots + N$

$\sum_{i=1}^{N} i$

Introduction to STATISTICAL INFERENCE

This book is in the
ADDISON-WESLEY SERIES IN STATISTICS

Z. W. Birnbaum
Consulting Editor

Introduction to
STATISTICAL INFERENCE

HAROLD FREEMAN
Massachusetts Institute of Technology

ADDISON-WESLEY PUBLISHING COMPANY, INC.
READING, MASSACHUSETTS · PALO ALTO · LONDON

Printed in the United States of America

Library of Congress Catalog Card No. 62-19140

PREFACE

This book is an introduction to statistical inference. It is for the student who has a modest knowledge of calculus, but who has not been trained in matrix algebra, and the mathematical exposition and rigor are, I believe, appropriate to this background. Occasionally the student will find such words as *transcendental, identically,* . . . , not met in his training; they may not be defined here and at the level of our exposition they will not be critical to the point at hand. Discussions of these terms will be found in elementary textbooks in analysis.

The writing is sometimes brief and the reading therefore hard. In part this was meant to provide an alternative choice. But it does seem to me well for a student to settle fewer main questions by reading and more with his own hands. Instead of working only on peripheral home problems decorating the text (there are about ninety, some are quite difficult, and all are carefully discussed at the end of the text), he is here sometimes asked to regard certain details of the central theorems of the text as home problems and wherever necessary to spell out the proofs, possibly on the pages of the book itself. In this plan, as I imagine it, the function of the teacher would be to explain whatever after reasonable effort remains unclear to the student, and to introduce material not at all in the text.

I have confined the text to fairly basic matters in probability and statistics. Even so, the choice of topics is occasionally curious, reflecting the weaknesses of the author. Some aging but interesting mathematical material like Hermite polynomials is present, some higher-regarded topics hardly mentioned. There is, unfortunately, nothing on decision theory or on sequential analysis. Some topics are present because they involve mathematical concepts and methods of value to a student outside the areas of probability and statistics.

This text can be covered in two terms. A one-term course in statistical inference could consist of

Part I: All chapters
Part II: Chapters 9, 10, 11, 16, 17
Part III: Chapters 19, 21 (Section 1 only), 23
Part IV: Chapters 25, parts of 28, 29

A one-term course in elementary probability could consist of

Part I: All chapters
Part II: All chapters

Challenging a modern trend, the symbolism and numbering system of this book are simple but, I believe, systematic and helpful. Numbers in parentheses, e.g., (1–1), always refer to a displayed mathematical expression in the text. These are consecutively numbered within chapters. Numbers in brackets, e.g., [10], always refer to an entry among the complete list of references on pages 388–397. Problem 1–1, etc., always refers to a home problem; these problems are given at the ends of chapters.

The Appendix contains some major tables and figures which may be useful. These are referred to in the text as Table or Fig. A–1, etc. A complete paged list of these tables and figures is given on pages 398–399. If by their nature complete tables are brief, complete tables are shown; if lengthy, brief excerpts. Although few numerical examples are encountered in this text, these tables and excerpts will, I believe, be found useful to understanding and checking one's understanding of the quantities involved.

In addition to the complete list of references on pages 388–397, references are included at the end of each chapter. These are not routine; they have been carefully selected. A special effort has been made to refer to relevant journal literature which the student may be able, in part, to follow. As there appeared to be no gain in numbering references at the end of chapters, they are not numbered. All are included in the complete list of references on pages 388–397.

It is better, in fact almost imperative, to learn a subject from more than one source. The student should read as his taste, mathematical preparation, and library resources suggest, in several of the books listed on page 382. Particularly, he should try to solve some of the problems he will encounter in them.

As I read over the manuscript and note its faults, all of them mine, it occurs to me that it may be more thoughtful to omit mention of the many persons whose help I have had. At whatever risk, I thank Dr. Jack Kiefer, Dr. Norman Levinson, and Dr. Paul Samuelson for valuable suggestions, Dr. Phoebus Dhrymes and Dr. Edward Kane for careful reading, many improvements, and, along with Drs. Marian Krzyzaniak, Ronald Howard, George Perry, Georges Magaud, for solutions to problems. I thank Dr. William Jackson Hall for his admirable Chapel Hill lectures from which I have too often borrowed ideas and content. I am indebted to an anonymous reader of an earlier draft for sharp criticism, particularly with respect to motivation and accuracy of statement, which forced me to make a thorough revision; I believe the current text is improved in these important respects. I owe much to Dr. Robert Solow for improving an earlier draft and my heaviest debts are to Dr. David Durand and Dr. John Pratt whose constructive detailed criticisms have saved the reader countless poor moments. I thank Beatrice Rogers whose intelligent and unfailing help at a distance of 800 miles was critical to getting this job done at all.

I am indebted to many persons for typing services. I would particularly thank Tomlin Brown and Grace Locke for their excellent typescript of the final draft and Inez Crandall for painstaking work on all drafts.

I am indebted to the John Simon Guggenheim Memorial Foundation for an earlier grant during which some of this was planned and to the National Science Foundation for a sabbatical year at the University of North Carolina during which most of this was written. In appropriate places I acknowledge my debt to those who have given me permission to reproduce tables and to use material and problems from their own work.

I will warmly welcome corrections, criticism, and suggestions, minor and major, from readers.

Cambridge, Massachusetts Harold Freeman
December 1962.

CONTENTS

PART I PROBABILITY AND RANDOM VARIABLES

ile recognizing that for M finite, r/M is for almost all observable omena unstable, we require that the *limit* (1–1) exist; for many omena it may not exist. Ultimate stability of relative frequency is cit in our definition of probability. It is, therefore, in those fields re M can be large and where, for M large, stability in the relative uencies of events is an empirical fact that probabilities are most legiti- ely defined and most profitably used. As examples, we may name h diverse fields as gambling, genetics, mass production, mortality, and clear physics. Note that the "probability" that men are living on Mars s outside the range of our definition.

It is not quite implied that one actually makes $M \to \infty$ trials, but that $\to \infty$ trials are conceivable. With this in mind, it becomes evident that umerical values of $p(E)$ (for example, that the probability of a coin alling head is $\frac{1}{2}$) are somewhere between conventions and truths; they an never be *proved* (in fact, experience with most coins does not support p (head) $= \frac{1}{2}$). In card playing, where the frequencies of critical events are low, it would literally take centuries of playing to check the conventional numerical values of probabilities (reached by a combination of finite experience, common sense, and deductive argument) on which (good) card-playing ability rests.

1–2 Simple atomic and molecular events. We write (1–1) briefly, remembering that probability is defined only for large M,

$$p(E) \approx \frac{r}{M}. \tag{1–2}$$

Now consider the two possible outcomes of a single trial, the events E and $\overline{E}$. They are *mutually exclusive;* if one occurs (and one must occur) on a single trial, the other cannot occur on that trial. These two mutually exclusive events are (or better, this set of mutually exclusive events is) also *exhaustive;* on a single trial *only* E or *only* $\overline{E}$ can occur. Each of the examples given in Section 1–1 falls into this category; i.e., a coin cannot fall head *and* tail on a single trial; moreover (since we exclude the possibilities of falling on edge or being lost) it must fall *either* head or tail.

By (1–2) we have, entirely heuristically for we replace $\approx$ by $=$,

$$p(\overline{E}) = \frac{M - r}{M} = 1 - \frac{r}{M} = 1 - p(E),$$

$$p(E) + p(\overline{E}) = 1. \tag{1–3}$$

But two mutually exclusive events may not be exhaustive, that is, the outcome of a single trial may not be dichotomous. If there are three or

Part I

Probability and Random Variables

In Part I the elements of probability and the concept of random variables are introduced. These elements and concepts underlie all of the probability calculus as well as all of statistical inference. The probability structure of one, two, and several random variables, in fairly general form, is discussed. Part I closes with a discussion of the problem of transforming from one set of random variables to another.

CHAPTER 1

ELEMENTS OF PROBABILITY

1–1 Probability as an extension of relative frequenc ments of an axiomatic approach to probability are intro 1–6 and are further developed in the later chapters of Pa duction to probability will be primarily heuristic; a serio the foundations of probability belongs in a separate course.

Our early discussion will regard the probability of an relative frequency of its occurrence in a large number of tria discussion will view probability as a function satisfying ax themselves are motivated by elementary properties of relative There will therefore be no serious difference between the tw view.

On a single trial or experiment an event E may occur; for head on the toss of a coin, blindness on examination of a person a wrong number on a phone call. In each of these examples the of the trial or experiment is, or is imagined to be, *dichotomous:* head blindness or sight, wrong or right number (these are simple ex to begin with; an experiment generally has several possible outc If the outcome is dichotomous, there exists on such a trial or experi an alternative event which we label $\overline{E}$: tail, sight, the right num We shall sometimes call any possible outcome of a trial or experiment *atomic* event.

Now imagine M, the number of trials or experiments (the number tosses of a coin, the number of persons examined for sight, the number phone calls) where, roughly speaking, these trials or experiments are performed under constant conditions. The probability $p(E)$ of the event E stands for

$$p(E) = \lim_{M \to \infty} \frac{\text{number of occurrences } (r) \text{ of } E}{\text{number of trials } (M)}. \qquad (1\text{–}1)$$

Probability, as defined by (1–1), is evidently motivated by, and is an extension of, the familiar concept of relative frequency r/M, where M is *finite.* It will be evident throughout Part I that both the axioms to which we later subject probabilities and the properties we find for probabilities are, with few exceptions, satisfied by (finite) relative frequencies.

more possible outcomes of a trial, we label them $E, F, G, \ldots$. Six mutually exclusive outcomes are associated with a trial consisting of a throw of a die; blindness in a person can be and generally is classified by degree of blindness into a number of categories; an individual's vote may be cast for one of several candidates. Our definition of probability (1–2) becomes, with obvious symbolism,

$$p(E) \approx \frac{r}{M}, \qquad p(F) \approx \frac{s}{M}, \qquad p(G) \approx \frac{t}{M}, \ldots \qquad (1\text{–}4)$$

Now we introduce an event which itself consists of one or more possible outcomes of a trial, that is, one or more atomic events. We shall sometimes call such an event a *molecular* event. Let $p(E + F)$ stand for the probability of the atomic event E *or* the atomic event F *or both*. (The last is impossible here, for the two events E and F, while not exhaustive, are mutually exclusive.) We have, heuristically, replacing $\approx$ by $=$,

Formula when E and F cannot both occur

$$p(E + F) = \frac{r + s}{M} = \frac{r}{M} + \frac{s}{M} = p(E) + p(F), \qquad \text{*only if mutually exclusive*} \qquad (1\text{–}5)$$

which, noting (1–3) and (1–4), is 1 only if E and F are also exhaustive. We also have

$$p(E + F + G) = p(E) + p(F) + p(G), \qquad \text{*only if mutually exclusive*} \qquad (1\text{–}6)$$

the expression on the left of (1–6) standing for the probability of E or F or G, *or any two* (impossible here), *or all three* (impossible here) events. Keep in mind that (1–5) and (1–6) are valid only if E and F, and E, F, and G are mutually exclusive. If the set of events E, F, and G is also exhaustive, the sum in (1–6) is 1. Note that while our definition of probability and the heuristic consequences of this definition given in (1–3), (1–5) and (1–6) are valid only for $M \to \infty$, corresponding *exact* statements could be made for relative frequencies; the motivational role of relative frequency is evident. Finally, the extension of (1–6) to k mutually exclusive events is straightforward.

So far we have considered what we might call *simple* atomic events. Only one fact, so to speak, is asked of a trial or experiment: on which side does the coin fall? for which candidate do you vote? is a person blind or not? Now we turn to *compound* atomic events, that is, to events which may occur on a single trial or experiment, but which themselves take the form of combinations of simple events. Two or more facts, so to speak, are asked of a trial or experiment; for example, is a person blind or not *and* can he hear or not? (Our use of the words "simple" and "compound" is peculiar to this text.)

1–3 Compound atomic events. Again, with M large so that all probabilities are properly defined, consider a trial in which (E or $\overline{E}$) *and* (F or $\overline{F}$) must occur, where, as before, $\overline{E}$ and $\overline{F}$ are the complementary or alternative or opposite events to E and F, respectively. Clearly E and $\overline{E}$ exclude each other, as do F and $\overline{F}$. But now let us consider a compound event such as EF, where EF (sometimes written $E \times F$) stands for the event E *and* F. The compound events EF, $E\overline{F}$, $\overline{E}F$, $\overline{E}\overline{F}$, may, and one must, occur. For example, consider a trial which consists of tossing a coin, which may fall head or tail, *and* drawing a card, which may be black or red. On each trial a compound atomic event will occur, and now it is the four compound events EF (head, black card), $E\overline{F}$ (head, red card), $\overline{E}F$ (tail, black card), $\overline{E}\overline{F}$ (tail, red card) which are mutually exclusive (for only one of them can occur during a single trial) and exhaustive (for one of them must occur). In Fig. 1–1 we show the frequencies, in a large number M of trials, of these four compound atomic events.

Note that while we may choose to call such an event as EF compound, EF is an atomic event; it is no more decomposable than the simple atomic event E of Section 1–2. On any trial the "whole" compound event EF occurs or fails to occur. Whether the outcome of a trial is a simple atomic or a compound atomic event depends merely on the nature of the trial.

	F	$\overline{F}$
E	r_{11}	r_{12}
$\overline{E}$	r_{21}	r_{22}

FIG. 1–1. Frequencies of four mutually exclusive and exhaustive compound atomic events.

We have, by definition,

$$p(EF) = \frac{r_{11}}{M}, \qquad p(E\overline{F}) = \frac{r_{12}}{M},$$

$$p(\overline{E}F) = \frac{r_{21}}{M}, \qquad p(\overline{E}\overline{F}) = \frac{r_{22}}{M},$$

and, from (1–5), for what is here the *molecular* event E and the *molecular* event F,

$$p(E) = \frac{r_{11} + r_{12}}{M} = p(EF) + p(E\overline{F}),$$

$$p(F) = \frac{r_{11} + r_{21}}{M} = p(EF) + p(\overline{E}F), \tag{1–7}$$

and for $p(E + F)$, by which we mean, as before, the probability of E or F or both (the last of which now *can* occur), the important result

$$p(E + F) = \frac{r_{11} + r_{12} + r_{21}}{M} = \frac{(r_{11} + r_{12}) + (r_{11} + r_{21}) - r_{11}}{M}$$

$$= p(E) + p(F) - p(EF), \tag{1–8}$$

of which (1–5) is a special case in which the event E and F *cannot* occur.

If on each trial three, rather than two, dichotomous outcomes (E or $\overline{E}$) *and* (F or $\overline{F}$) *and* (G or $\overline{G}$) must occur, then the possible outcomes of a trial are the compound atomic events

$$EFG,\quad EF\overline{G},\quad E\overline{F}G,\quad E\overline{F}\overline{G},\quad \overline{E}FG,\quad \overline{E}F\overline{G},\quad \overline{E}\overline{F}G,\quad \overline{E}\overline{F}\overline{G}.$$

We have, corresponding to (1–8), that the probability of E or F or G or any two (which now can occur) or all three (which now can occur) is

$$p(E + F + G)$$
$$= p(E) + p(F) + p(G) - p(EF) - p(EG) - p(FG) + p(EFG), \quad (1\text{–}9)$$

of which (1–6) is a special case in which E, F, and G *cannot* occur together or by pairs; that is, in (1–6), E, F, and G are mutually exclusive.

Equation (1–9) can be obtained either by direct consideration of the eight-compartment frequency table shown in Fig. 1–2 (which is entirely analogous to Fig. 1–1), or it can be obtained from (1–8),

$$p(E + H) = p(E) + p(H) - p(EH),$$

where now $H = F + G$.

Similar results can be obtained when the outcome of a trial or experiment are compounded of finitely many dichotomous events (E or $\overline{E}$) *and* (F or $\overline{F}$) *and* (G or $\overline{G}$) *and* . . . . For outcomes which are compounded of finitely many simple events each of which includes *more* than two categories

$$(E_1 \text{ or } E_2 \text{ or } E_3 \text{ or } \ldots)$$

and

$$(F_1 \text{ or } F_2 \text{ or } F_3 \text{ or } \ldots)$$

and

$$(G_1 \text{ or } G_2 \text{ or } G_3 \text{ or } \ldots),$$

that is, compound atomic events of the form

$$E_1F_1G_1, \qquad E_1F_1G_2,$$
$$E_1F_1G_3, \qquad E_1F_2G_1, \text{ etc.,}$$

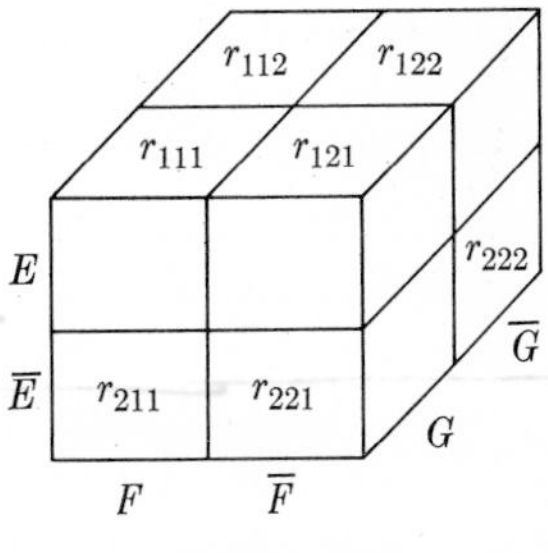

FIG. 1–2. Frequencies of eight mutually exclusive and exhaustive compound atomic events.

we need more elaborate counting methods to determine the probabilities of various molecular events; some will be provided in Chapter 9.

These and earlier consequences of the definition of probability are readily, if not rigorously, obtained with the aid of Venn diagrams, which afford pictorial representation of events by allocating regions to (simple or compound, atomic or molecular) events. We can check certain ele-

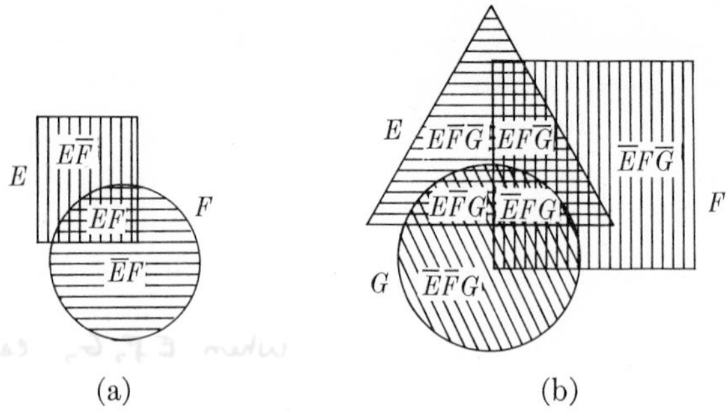

FIG. 1–3. Venn diagrams showing atomic and molecular events.

mentary results in probability by noting a few elementary logical properties of such regions. Thus noting from Fig. 1–3(a) the equivalence of the molecular events (shaded regions)

$$\{E \text{ or } F \text{ or both}\} \qquad \text{and} \qquad \{E \text{ plus } F \text{ minus } EF\},$$

one readily gets (1–8); similarly (1–9) is obtained from Fig. 1–3(b). Squares, circles, and triangles are used merely for neatness. Note that the regions *outside* the enclosed spaces (and these outside regions may be thought of as enclosed) are $\bar{E}\bar{F}$ and $\bar{E}\bar{F}\bar{G}$ respectively. Thus all atomic events, as well as whatever molecular events are of interest, can be shown in Venn diagrams.

1–4 Conditional probabilities. In a wide class of problems in the real world, the events of greatest interest are those whose occurrence is conditional on the occurrence of other events. For this reason *conditional* probability, that is, probability conditional on certain events having occurred, is now introduced. By the symbol $p(E|F)$ we mean the probability of the event E given that the event F has occurred. First, restricting ourselves to the mutually exclusive and exhaustive compound atomic events EF, $E\bar{F}$, $\bar{E}F$, and $\bar{E}\bar{F}$ for which probabilities are assumed to exist, and with relative frequency again as motivation, $p(E|F)$ is naturally defined, using the notation of Fig. 1–1, by

$$p(E|F) = \frac{r_{11}}{r_{11} + r_{21}}, \tag{1–10}$$

the denominator containing only those frequencies associated with the unconditional occurrence of F, and the numerator the subset of those frequencies in which E also occurs. By specifying the previous occurrence of F, we have merely introduced a new (and restricted) space of possible events (here reduced from four to two compound atomic events) and have defined probabilities on all events in this restricted space. The total frequency $r_{11} + r_{21}$ associated with the event F must, of course, be large for the conditional probability of E to be defined by (1–10) at all.

A heuristic algebraic operation on (1–10) produces

$$p(E|F) = \frac{r_{11}}{r_{11} + r_{21}} = \frac{r_{11}}{M} \cdot \frac{M}{r_{11} + r_{21}} = p(EF) \cdot \frac{1}{p(F)} \cdot \tag{1–11}$$

Now, excluding $p(F) = 0$, we may write (1–11)

$$p(EF) = p(E|F) \cdot p(F), \tag{1–12}$$

which we take as the definition of $p(E|F)$. We may likewise obtain

$$p(EF) = p(F|E) \cdot p(E), \tag{1–13}$$

by which we define $p(F|E)$ for all E except $p(E) = 0$. Expressions similar to (1–12) and (1–13) for the probabilities of the remaining three compound atomic events $p(E\bar{F})$, $p(\bar{E}F)$, and $p(\bar{E}\bar{F})$ defining the remaining conditional probabilities $p(E|\bar{F})$, $p(\bar{F}|E)$, etc., may be written.

1–5 Independence. A natural definition of the *independence* of event E from event F would be

$$p(E|F) = p(E); \tag{1–14}$$

the probability of the event E is unaffected by the occurrence of the event F. Equally naturally, $p(F|E) = p(F)$ could serve as definition of the independence of F from E. Noting (1–11), such a definition as (1–14) would be restricted to $p(F) > 0$, since for $p(F) = 0$, $p(E|F)$ is not defined. The restriction is trivial, but for convenience we adopt the symmetrical definition of independence of E and F

$$p(EF) = p(E)p(F), \tag{1–15}$$

which, noting (1–12), is equivalent to (1–14). We will regard (1–15) as a special case of (1–12) and (1–13), and we will adopt (1–15) as our definition of independence, even if $p(E) = 0$ or $p(F) = 0$.

If E or F are independent, so are E and $\bar{F}$, $\bar{E}$ and $\bar{F}$, etc., one of which the student is asked to show in problem 1–4.

The mathematical statement (1–15) of independence is simple and natural, but which events in the world are in fact independent in the sense of (1–15) and which are not is harder to settle. Roughly, we seem to find more independence among phenomena in physical science, more dependence among phenomena in social and biological science. But the question is at all times difficult to settle. Assume that we can extend the concept of independence to particular trials. Then, even in coin-tossing where independence in the outcomes of successive tosses appears to be well established, both on the basis of evidence over centuries as well as our appraisal of the act of tossing a coin, there are those who believe, for a curious variety of reasons, that after six straight heads a head is *less* likely than a tail, as well as those who believe, for perhaps better reasons which will be considered in Chapter 24, that a head is *more* likely than a tail.

The natural extension of (1–12) is

$$p(EFG) = p(G|EF)p(EF) = p(G|EF)p(E|F)p(F), \qquad (1\text{–}16)$$

which can readily be motivated by consideration of the eight-compartment frequency table shown in Fig. 1–2:

$$\begin{aligned} p(EFG) &= \frac{r_{111}}{M} = \frac{r_{111}}{M} \cdot \frac{M}{r_{111} + r_{112}} \cdot \frac{r_{111} + r_{112}}{M} \\ &= \frac{r_{111}}{r_{111} + r_{112}} \cdot \frac{r_{111} + r_{112}}{M} \\ &= p(G|EF)p(EF), \text{ etc.} \end{aligned}$$

The extension of (1–15), which we describe as the conditions for *mutual independence* of E, F, and G, is

$$p(EFG) = p(E)p(F)p(G), \qquad (1\text{–}17)$$

as well as pairwise independence of all possible subsets of events, here

$$p(EF) = p(E)p(F), \quad p(EG) = p(E)p(G), \quad p(FG) = p(F)p(G). \qquad (1\text{–}18)$$

Mutual independence appears often to be practically assured (that is to say, it is difficult to produce practically significant examples* contradicting the following) if events are *pairwise* independent, though pairwise independence is not sufficient to guarantee mutual independence, as shown in the following (simple but contrived) example by Bernstein (Kolmogorov [108]). Let E, F, G, and H be mutually exclusive and exhaustive atomic events with *equal* (this is critical to the example) probabilities $\frac{1}{4}$. Let A, B, and C be the molecular events $E + F$, $E + G$, and $E + H$ respectively. From earlier results of this chapter, or guided by Fig. 1–4,

$$\begin{aligned} p(A) = p(B) &= p(C) = \tfrac{1}{2}, \\ p(AB) = p(AC) &= p(BC) = p(E) = \tfrac{1}{4} = (\tfrac{1}{2})^2 \\ &= p(A)p(B) = p(A)p(C) = p(B)p(C), \end{aligned}$$

* The force of this remark is reduced by the following noncontrived example: let A be head on the first toss, B be head on the second toss, C a match.

that is, A and B are independent, A and C are independent, and B and C are independent. But E is also the common element of A, B, and C, so

$$p(ABC) = p(E)$$
$$= \tfrac{1}{4} \neq p(A)p(B)p(C);$$

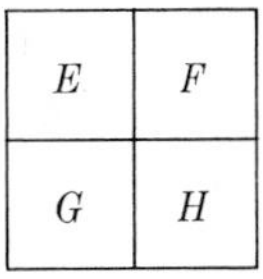

FIG. 1-4. Venn diagram showing four mutually exclusive and exhaustive events.

A, B, and C, though pairwise independent, are not mutually independent.

Events may satisfy our conditions for independence, and therefore be independent, without being intuitively independent; we give an example due to Feller [57]. Let male and female children be equally likely. Consider families with three children. Consider the molecular events

A both sexes represented among the children of a family,

B at most one girl among the children of a family.

We regard the mutually exclusive and exhaustive outcomes of a trial (here a sexed *and* ordered census of the children in a family with three children) as the sexed and ordered compound atomic events

$$MMM, \quad MMF, \quad MFM, \quad MFF, \quad FMM, \quad FMF, \quad FFM, \quad FFF,$$

and not,* say, the sexed outcomes without regard to order, MMM, MMF, MFF, FFF. If now we argue (doubtfully) that the sex of a child in a family is independent of the sex of other children in the family, we have, from (1-17), that the probability of each of these eight compound atomic events is $\frac{1}{8}$. Now the number of these events favorable to the molecular event A is 6, the number favorable to the molecular event B is 4, and the number favorable to the molecular event AB is 3. As the eight atomic events are mutually exclusive, we have, extending (1-6),

$$p(A) = \tfrac{6}{8}, \qquad p(B) = \tfrac{4}{8}, \qquad p(AB) = \tfrac{3}{8},$$

and (1-15) is satisfied; A and B are independent.

1-6 Elements of an axiomatic approach to probability. To this point the argument has rested, somewhat loosely, on relative frequency. The definition of the probability of an event was a loose extrapolation of relative frequency, for M large. The definition of conditional probability was

* As might some who, considering the experiment of tossing a coin twice, view the three *unordered* outcomes HH, HT, TT rather than the four *ordered* outcomes HH, HT, TH, TT as equally likely.

similarly motivated by the notion of relative frequency within a restricted space of events. Moreover, our results were obtained by heuristic manipulation of the numerators r, s, ... and denominator M of probabilities which, once defined for $M \to \infty$, are no longer decomposable into numerators and denominators.*

Better (by which we mean in the rigorous manner of almost all branches of modern mathematics, of which probability is one), we may begin with axioms. In this approach probabilities are merely functions which satisfy the axioms. We prefer as few axioms as possible, and, roughly speaking, we chose those likely to yield the largest number of theorems valuable in the real world. Since it is the relative frequency of events in the real world which we hope to describe and better understand, the axioms usually selected are suggested by elementary properties of relative frequencies. There is, therefore, little conflict in fact between the consequences of the axiomatic approach (formalized by Kolmogorov [108] in 1933) and our loose earlier definition of probability as the limit of an observable relative frequency (formalized, not loosely but controversially, by von Mises in 1919; see [124]).

Let S be a well-defined collection of distinct objects called elements. With respect to a single trial or experiment, let S be an enumerable set of points, each point representing a possible outcome of a trial; S is the set of all possible outcomes of a trial, i.e., the set of all possible atomic events. The points of S may represent simple or compound atomic events, depending on the nature of the trial. Let A be any subset of points in the set S; A is a molecular event, i.e., a collection of one or more possible outcomes of a trial, each outcome favorable to the event A.

Write p for a function which assigns to each molecular event A a real number $p(A)$, called the probability of A. We do not concern ourselves with the possibility (of much interest in a rigorous approach to probability) that certain molecular events A may be such that probabilities $p(A)$ cannot be defined on them (if the set S contains infinitely many points, not every subset may be *measurable*.) *Practically*, this possibility does not exist.

In modern probability theory, the following minimal set of axioms is generally met:

(a) $p(A) \geq 0$,

(b) $p(S) = 1$,

---

* This is true, although many of our results could be obtained using limits. For example, for A and B mutually exclusive,

$$p(A + B) = \lim_{M\to\infty} \frac{r + s}{M} = \lim_{M\to\infty} \frac{r}{M} + \lim_{M\to\infty} \frac{S}{M} = p(A) + p(B).$$

(c) $p(A + B) = p(A) + p(B)$ for A and B any pair of mutually exclusive molecular events; an extension of the last is to hold for a finite or infinite number of mutually exclusive molecular events (which can here be interpreted as sets of points with no points in common).

With these axioms, we can immediately obtain such simple results as

$$\begin{aligned} p(A + \bar{A}) &= p(A) + p(\bar{A}) && \text{(using axiom c),} \\ &= 1 && \text{(now using axiom b),} \qquad (1\text{–}19) \\ p(A) &\leq 1 && \text{(now using axiom a),} \end{aligned}$$

where the complement $\bar{A}$ consists of those points in S which are not in A. Thus the natural (1–19) is not axiomatized.

To deal with events A and B which are not mutually exclusive, we must manage to express them in terms of mutually exclusive events, since it is only on such events that we have an axiom. No appeal to any nonaxiomatized properties of relative frequencies can be made here. (We might, of course, have begun by axiomatizing properties of events which were not mutually exclusive.)

For example, let us produce (1–8) from our axioms. Note from (a relettering of) the Venn diagram in Fig. 1–3(a) that the event $A + B$ is identical with the logical sum of the mutually exclusive events A and $\bar{A}B$. Using axiom (c), we have

$$p(A + B) = p(A) + p(\bar{A}B).$$

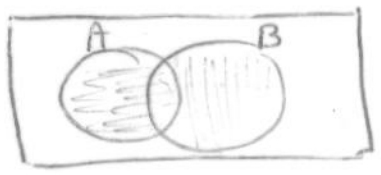

But events $\bar{A}B$ and AB are mutually exclusive and their logical sum is B; therefore

$$p(B) = p(\bar{A}B) + p(AB),$$

which, combined with the preceding equation, yields (1–8).

Note that, unlike definition (1–1), the axioms (a), (b), and (c) do not suggest *how* to estimate $p(A)$. They are, for example, compatible with probability equal to $\frac{1}{11}$ for each of the eleven possible sums of faces, the compound atomic events 2, 3, . . . , 12, in a trial consisting of throwing two dice (probabilities hardly supported by experience with ordinary dice).* Many numerical estimates of probabilities will satisfy these axioms; we must learn how to select those which are best for the particular problem at hand.

---

* As an anonymous reader corrected an earlier edition, it is *not* within the dicemaker's art to create two (independent) dice which can produce such (equal) probabilities! Questions of this sort will be considered in Chapter 20 on convolution.

Collateral Reading

H. Cramér, *The Elements of Probability Theory and Some of Its Applications.* New York, John Wiley and Sons, 1955. Contains a valuable historical introduction to probability.

W. Feller, *An Introduction to Probability Theory and Its Applications.* 2nd ed., New York, John Wiley and Sons, 1957 (first published in 1950). The best reference.

S. Goldberg, *Probability, An Introduction.* Englewood Cliffs (N.J.), Prentice-Hall, Inc., 1960. A good introduction to the theory of sets.

A. A. Kolmogorov, *Foundations of the Theory of Probability.* 2nd English ed., New York, Chelsea Publishing Co., 1954 (first published in 1933). The original axiomatic formulation.

R. von Mises, *Wahrscheinlichkeitsrechnung.* Leipzig and Vienna, Fr. Deuticke, 1931. A popular account of the frequency theory appears in the author's *Probability, Statistics and Truth.* New York, Macmillan Co., 1939 (2nd revised English ed., 1957).

E. Parzen, *Modern Probability Theory and Its Applications.* New York, John Wiley and Sons, 1960. A careful reference on the material of this chapter.

H. Richter, *Wahrscheinlichkeitstheorie.* Berlin, Springer-Verlag, 1956.

Problems

1–1. From the three axioms, show that the probability of an impossible event (a subset containing no points) is zero. (Accordingly, this natural result is not axiomatized.)

1–2. (Brown [21], resurrected by Parzen [149]). Prove that the 13th day of the month is more likely to be Friday than any other day of the week! A somewhat laborious but not difficult problem.

1–3. Fifty-two persons stand in line, each to draw, without replacement, a card from an ordinary deck. The ace of spades is the winning card. What place in the line is best? Any psychological aspects, such as the peculiar pains of being first or last in a line, are to be neglected.

1–4. If E and F are independent, $\overline{E}$ and $\overline{F}$ are independent.

1–5. (Cramér [37]). Prove that $p(AB) \leq p(A) \leq p(A + B) \leq [P(A) + p(B)]$.

1–6. If $E_1, E_2, \ldots, E_n$ are mutually independent,

$$p(E_1 + E_2 + \cdots + E_n) = 1 - p(\overline{E}_1)p(\overline{E}_2) \ldots p(\overline{E}_n).$$

1–7. (Solow). Can you show (perhaps by an example) that if A and C are independent, they may not be independent given B?

1–8. (Hoel [89]). A, B, and C toss a coin, in that order, until a (winning) head appears. What are their respective chances of winning?

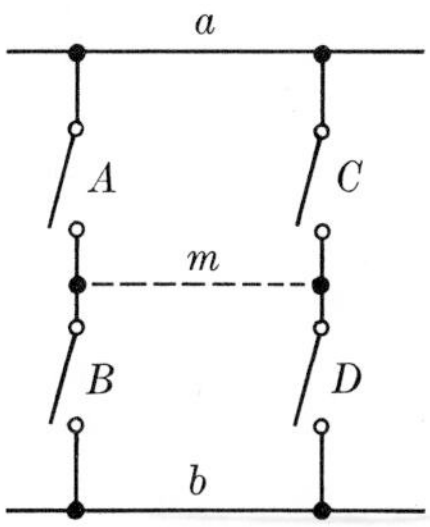

Fig. 1–5. A network of four switches across a power line.

1–9. (Fox). A network of switches, A, B, C, D, are connected across the power lines a and b, shown in Fig. 1–5. The switches operate electrically. Each switch has an independent operating mechanism, and all are controlled simultaneously on the same impulse; that is, it is intended that on an impulse all switches shall close (or open) simultaneously. But each switch has a probability p of failure. What is the probability that the circuit from a to b will fail to close? Will fail to open? Originally bus m was not in the circuit. Would the addition of the bus improve the operation?

CHAPTER 2

RANDOM VARIABLES

2–1 Random variables and a new event space. The events to which we have already attached probabilities belong to an elementary class. In the earlier examples of Chapter 1, they were simple and dichotomous; for example, head or tail, success or failure. In the later examples of Chapter 1, the number of categories was somewhat extended; for example, to the simple but not dichotomous six faces of a die, to the four and eight compound atomic events of Section 1–3. We now want to introduce a new way of referring to some of the events to which probabilities can be attached.

By the introduction of a function $\mathbf{x}(E)$ we associate with each point E in S a real number x; that is, we construct a correspondence between the atomic events $E_1, E_2, \ldots$, the points in the event space S, and the real numbers $x_1, x_2, \ldots$. The correspondence need not be one to one; we may speak of *mapping* several points of S into one real number x. The original atomic event space S may then be viewed as having been mapped into a new event space consisting of points on the real line, and molecular events A of interest in S, that is, particular subsets of points of S, now become particular collections of points on the real line. In a trial or experiment consisting of a toss of a coin, the outcomes head and tail, which in Chapter 1 were discussed as points E and $\overline{E}$ in S, are in Chapter 2 transformed by the function $\mathbf{x}$ into the numbers $\mathbf{x}(E) = x_1$, $\mathbf{x}(\overline{E}) = x_2$ on the real line. For example, we can put $x_1 = 1$ and $x_2 = 0$, which can then be usefully interpreted (as, say, 17 and 3 cannot) as the possible number of heads in one trial; we have chosen origin and numbers so that the values of x have significant physical meaning. The six faces of a die may be transformed into the real numbers 1, 2, 3, 4, 5, 6 with immediate interpretations for all ordinary dice games. The molecular event "an odd face" simply becomes the subset of numbers 1, 3, 5 in the new event space.

Transformations from S to the real line x may not be effective for all *compound* atomic events. The compound atomic events of Section 1–3,

EF	head, black card,	$E\overline{F}$	head, red card,
$\overline{E}F$	tail, black card,	$\overline{E}\overline{F}$	tail, red card,

may be replaced by four numbers on the real line; but whatever these numbers, it is difficult to imagine a useful interpretation of them. For here *two* facts are asked of a trial (which side does the coin show, which color is the card) and a transformation of the four *joint* outcomes onto the *one-*

dimensional real line may not be effective. We return to and solve this problem in Chapter 4.

If, however, the four compound atomic events were the outcomes of tossing a coin (head or tail) and tossing a second coin (head or tail),

EF	head, head,	$E\bar{F}$	head, tail,
$\bar{E}F$	tail, head,	$\bar{E}\bar{F}$	tail, tail,

a transformation of *these* four points of S into the real numbers 2, 1, 1, 0 might be useful, for the numbers can be interpreted as the "number of heads in one trial." The molecular event "at least one head" would consist of the subset of numbers 2, 1. Note that here the two different events $E\bar{F}$ and $\bar{E}F$ in the original event space S are mapped into the same number 1 in the new event space; some information (the order in which head appears) is lost in this mapping. But if only the number of heads and not the order in which they appear is of interest, the present mapping of these four compound atomic events of S onto the real line is adequate. Similarly, for the common American two-dice game, the mapping into the sum of the top faces (2, 3, . . . , 12) is adequate.

Most variables with which we are familiar (length, weight, price, income, strength, density) already fit a mapping on real numbers; others, such as intelligence, preference, taste, and color, have been or can often be quantified to fit.

The function $\mathbf{x}(E)$ which effects this shift from the old to the new event space is called a *random function*, or more commonly, a *random variable*, the adjective "random" reminding us that the numerical outcome x of a trial or experiment on $\mathbf{x}$ is uncertain, and therefore that the *value* of the function (not the function itself) is uncertain.* A random variable is simply a real-valued function defined on S.

Note that no particular probabilities are implied by any mapping. In the new event space of the real numbers x, a space which is more convenient for analysis than the original event space of points, we now define probabilities which satisfy equivalents of the three axioms (a), (b), and (c) of Chapter 1 and which without *practical* exception [that is, for all molecular events (all points and all intervals on the real line) that can be practically constructed] can be assigned to every point and every interval on the real line. The axioms can be introduced in several ways. In Sections 2–2 and 2–3, the method of introduction has immediate intuitive support. In Section 2–4 we will introduce the axioms in an alternative manner which has certain mathematical advantages.

---

* In this text random functions are generally shown boldface ($\mathbf{x}$) and their numerical values lightface (x); on occasion, this convention is abandoned.

Same axioms for new space

2–2 Discrete random variables.

To motivate the reintroduction of the axioms of Chapter 1 for use here, we return to the concept of relative frequency. First, consider the *discrete* random variable **x** of which, on any one trial, the possible outcomes (now the numerical values of **x**) are the real numbers $x_1, x_2, \ldots$. If a large number M of trials on **x** performed under constant conditions yield a stable pattern of relative frequencies, $r_1/M, r_2/M, \ldots$, of the real numbers $x_1, x_2, \ldots$, respectively, then, similarly to (1–1), we could define the probability of the event $\mathbf{x} = x_i$ by

$X(E) = P(E) = X_i = \frac{r}{M}$
$$p(\mathbf{x} = x_i) = \lim_{M \to \infty} \frac{r_i}{M}, \quad i = 1, 2, \ldots,$$

and proceed heuristically as in the first five sections of Chapter 1. Instead, but with this motivation in mind, we postulate, in the axiomatic spirit of Section 1–6, the existence of a function $f(x_i)$ which is connected to probability statements on the random variable **x** by

$$p(\mathbf{x} = x_i) = f(x_i), \tag{2–1}$$

where $f(x_i)$ is here called (somewhat unusually*) the *density function* of the discrete random variable **x**. The behavior of the random variable **x**, as described by $f(x_i)$, is evidently determined by the probabilities on the real line x on which the random variable **x** is defined. As may be evident in Fig. 2–1, complete information on the probability behavior of **x** is contained in $f(x_i)$, $i = 1, 2, \ldots$.

With relative frequency in mind, we will now subject $f(x_i)$ to axioms exactly analogous to (a), (b), and (c) of Chapter 1. In the light of (2–1) it is a matter of taste as to whether axioms are placed, as in Chapter 1, on the basic probabilities [the left side of (2–1)], or on the density function [the right side of (2–1)], which in fact carries the major burden in the forthcoming mathematical analysis. Here the axioms are placed (somewhat uncommonly) on the latter.

The equivalents here of axioms (a) and (b) of Chapter 1 are

Axioms (a) $f(x_i) \geq 0,$ (b) $\sum_i f(x_i) = 1, \quad i = 1, 2, \ldots.$

$P(A) \geq 0$ $P(S) = 1$

As for the addition axiom (c), we have

(c) $p(a \leq \mathbf{x} \leq c) = \sum_{a \leq x_i \leq c} f(x_i)$ $P(A \text{ or } B)$

which, for $a < b < c$,

$$= \sum_{a \leq x_i \leq b} f(x_i) + \sum_{b < x_i \leq c} f(x_i) = p(a \leq \mathbf{x} \leq b) + p(b < \mathbf{x} \leq c),$$

---

* Some defense of this unusual convention (of speaking of density functions of *discrete* random variables) is offered on page 23.

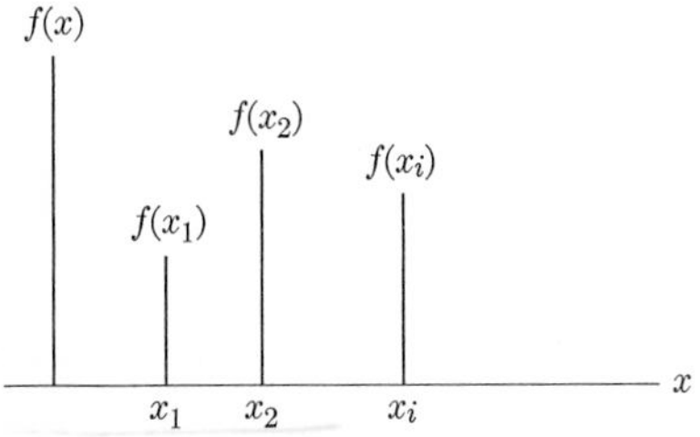

FIG. 2–1. Density function of a discrete random variable.

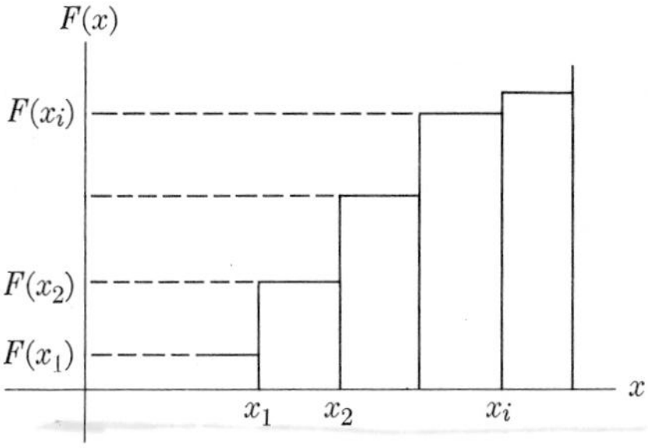

FIG. 2–2. Distribution function of a discrete random variable.

where $b < x_i \leq c$ indicates that the sum is taken over all values of $\mathbf{x}$ between b and c (but to avoid overlap not including b) which the random variable $\mathbf{x}$ may possibly assume.

As we are often interested in the molecular event $\mathbf{x} \leq x_i$, we now introduce $F(x_i)$ the *distribution function* of the discrete random variable $\mathbf{x}$ (Fig. 2–2) and define it by

$$p(\mathbf{x} \leq x_i) = F(x_i) = \sum_{x \leq x_i} f(x_i). \tag{2–2}$$

The student may find such properties of (the monotonic nondecreasing) function $F(x_i)$ as

$$F(-\infty) = 0, \qquad F(\infty) = 1, \qquad 0 \leq F(x_i) \leq 1,$$

$$p(a < \mathbf{x} \leq b) = \sum_{a < x_i \leq b} f(x_i) = F(b) - F(a). \tag{2–3}$$

To illustrate: for a single die, the appropriate mapping would often be

E: 

$$\begin{array}{cccccc} \downarrow & \downarrow & \downarrow & \downarrow & \downarrow & \downarrow \end{array}$$

$$x: \quad 1 \quad 2 \quad 3 \quad 4 \quad 5 \quad 6$$

and, for most dice, we write

$$f(1) = f(2) = f(3) = f(4) = f(5) = f(6) = \tfrac{1}{6}.$$

Axioms (a) and (b) are evidently satisfied. As one example of the satisfaction of axiom (c),

$$p(1 \leq \mathbf{x} \leq 5) = p(1 \leq \mathbf{x} \leq 3) + p(3 < \mathbf{x} \leq 5) = \tfrac{3}{6} + \tfrac{2}{6} = \tfrac{5}{6}.$$

Note that

$$F(0) = 0, \qquad F(6) = 1, \qquad 0 \le F(x) \le 1,$$

and, for example,

$$p(1 < \mathbf{x} \le 5) = F(5) - F(1) = \tfrac{5}{6} - \tfrac{1}{6} = \tfrac{4}{6}. \tag{2–4}$$

2–3 Continuous random variables. For a continuous random variable $\mathbf{x}$ we assume that all molecular events of *practical* interest will be represented by intervals on the real line x, and to each of these intervals a probability may be assigned. The probability at a point is 0. Once again, we do not concern ourselves with the possibility (of much interest in a strict exposition) that to certain (unusual) sets of points on the real line, probabilities may not be assigned. Again idealizing our experience with relative frequencies and now using the calculus of continuous variables, we define probability over the interval $x, x + dx$ in terms of "area,"

$$p(x \le \mathbf{x} \le x + dx) = f(x)\,dx, \tag{2–5}$$

where $f(x)$ is the *density function* of the continuous random variable $\mathbf{x}$ (Fig. 2–3) and $f(x)\,dx$ is the *probability element* of $\mathbf{x}$. We replace axioms (a) and (b) by

$$\text{(a)}\quad f(x) \ge 0,$$

$$\text{(b)}\quad \int_{-\infty}^{\infty} f(x)\,dx = 1.$$

As for the addition axiom (c), we have

$$p(a \le \mathbf{x} \le c) = \int_a^c f(x)\,dx$$

which, for $a < b < c$,

$$= \int_a^b f(x)\,dx + \int_b^c f(x)\,dx = p(a \le \mathbf{x} \le b) + p(b \le \mathbf{x} \le c).$$

Note also that for a continuous random variable, any overlap of the integrals [in (c) at the point b] is of no consequence since $\int_b^b f(x)\,dx = 0$.

Again, out of interest in the event $\mathbf{x} \le x$, we introduce $F(x)$, the *distribution function* of the continuous random variable $\mathbf{x}$ (Fig. 2–4) and define it by

$$F(x) = \int_{-\infty}^{x} f(t)\,dt\,, \tag{2–6}$$

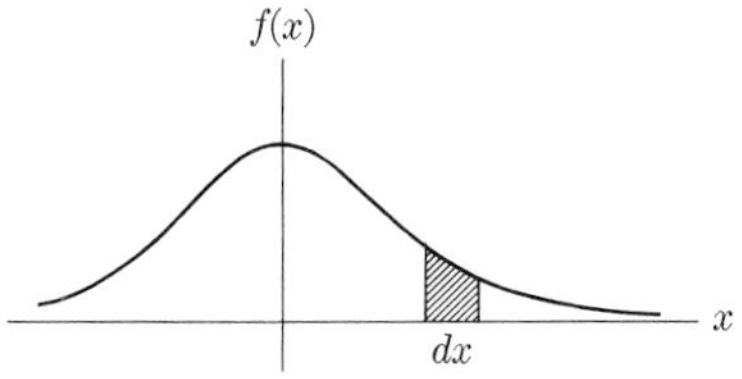

FIG. 2–3. Density function of a continuous random variable.

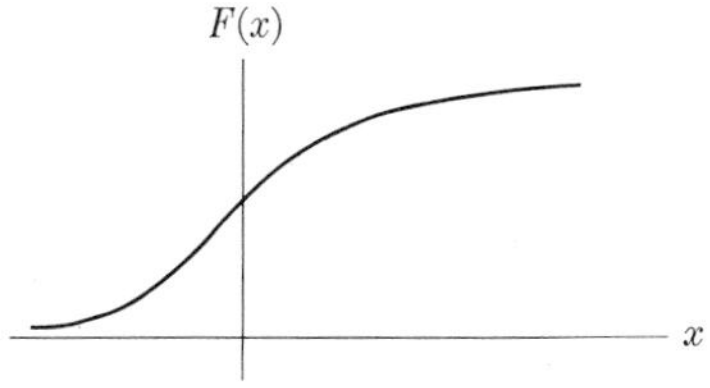

FIG. 2–4. Distribution function of a continuous random variable.

where $f(x)$ is the density function of $\mathbf{x}$. We can obtain such results as

$$F(-\infty) = 0, \qquad F(\infty) = 1,$$

and, from the elementary calculus, the fundamental result, at points of continuity of f,

$$\frac{dF(x)}{dx} = f(x), \tag{2–7}$$

relating continuous distribution and density functions. From (2–5) and (2–6), we also have

$$p(a \leq \mathbf{x} \leq b) = \int_a^b f(x)\,dx = F(b) - F(a), \tag{2–8}$$

and from axioms (a) and (b), we find

$$0 \leq F(x) \leq 1,$$

and F is evidently monotonic nondecreasing.

2–4 Alternative axiomatic formulation. The alternative (perhaps less intuitive but from a mathematical point of view superior) approach is to *begin* with the distribution function

$$p(\mathbf{x} \leq x) = F(x) \tag{2–9}$$

rather than the density function (2–1). In this approach certain *theorems* of Sections 2–2 and 2–3 now become *axioms*, while certain axioms of Sections 2–2 and 2–3 now become theorems. If we omit a customary axiom on the continuity of F, for this will not be critical to our discussion, the proposed axioms are

(a) $F(x)$ is monotonic nondecreasing, that is, if $b > a$, $F(b) \geq F(a)$,

(b) $F(-\infty) = 0$ and $F(\infty) = 1$,

and finally, an additivity axiom for mutually exclusive events; for example, for the mutually exclusive events $\mathbf{x} \leq a$ and $a < \mathbf{x} \leq b$ (disjoint intervals on the real line), we require

$$p(\mathbf{x} \leq a) + p(a < \mathbf{x} \leq b) = p(\mathbf{x} \leq b),$$

or generally

$$\text{(c)}\quad p(\mathbf{x} \in E_1 + E_2 + \ldots) = p(\mathbf{x} \in E_1) + p(\mathbf{x} \in E_2) + \ldots$$

($\in$ means "belongs to"), for an enumerable number of disjoint intervals $E_1, E_2, \ldots$ on the real line.

Evidently the distribution function $F(x)$ is a mathematical idealization of finite cumulative relative frequencies,

$$\frac{r_1}{M}, \frac{r_1 + r_2}{M}, \ldots, \frac{r_1 + r_2 + \cdots}{M}$$

(which themselves satisfy the axioms). With this axiomatic structure, the following results, among others, are immediately obtained from the definition (2–9),

$$p(a < \mathbf{x} \leq b) = F(b) - F(a), \quad b > a,$$

$$p(-\infty < \mathbf{x} < +\infty) = 1 \tag{2–10}$$

and

$$p(a < \mathbf{x} \leq b) = [F(b) - F(a)] \geq 0.$$

Also, for A and B not necessarily mutually exclusive events (not necessarily disjoint intervals on the real line), we find

$$p(\mathbf{x} \in A + B) = p(\mathbf{x} \in A) + p(\mathbf{x} \in B) - p(\mathbf{x} \in AB), \tag{2–11}$$

which follows from

$$p(A + B) = p(A\overline{B}) + p(AB) + p(\overline{A}B)$$

by inserting

$$p(A\overline{B}) + p(AB) = p(A), \qquad p(\overline{A}B) + p(AB) = p(B)$$

for $p(A\overline{B})$ and $p(\overline{A}B)$; (2–11) corresponds exactly to (1–8).

For $F(x)$ discrete, by which we mean that there exists a finite or enumerable set of points $x_1, x_2, \ldots$ on the real line with corresponding positive numbers, $p_1, p_2, \ldots$ such that $\sum_i p_i = 1$, with $F(x)$ defined by

$$F(x) = \sum_i p_i \tag{2–12}$$

over all i for which $x_i \leq x$ (illustrated in Fig. 2–2), we can obtain the important result for a discrete random variable,

$$p(\mathbf{x} = x_i) = p_i \quad \text{for} \quad \mathbf{x} = x_i, \quad i = 1, 2, \ldots \tag{2–13}$$

and 0 for $\mathbf{x} \neq x_i$. This is a result with which, as (2–1), we *began* the discussion of discrete random variables in Section 2–2.

For $F(x)$ absolutely continuous, by which we mean that a function $f(t) \geq 0$ exists such that

$$F(x) = \int_{-\infty}^{x} f(t) \, dt, \tag{2–14}$$

we readily obtain the following important result:

$$p(a \leq \mathbf{x} \leq b) = \int_{a}^{b} f(t) \, dt, \tag{2–15}$$

as a consequence of (2–9) and (2–14); $f(x)$, as before, is the density function of $\mathbf{x}$. The equivalence of probability and "area" or, for a discrete random variable, probability and "height," are here consequences of the axioms and not axioms themselves.

If $f(x)$ itself is continuous near x [it need not be even if $F(x)$ is], we have, except for differentials of higher order,

$$p(x \leq \mathbf{x} \leq x + dx) = f(x) \, dx, \quad \text{prob. in} \tag{2–16}$$
neighborhood of x

which provides us with an interpretation of the quantity $f(x)$ which we have called the density function. We see that the density function $f(x)$ multiplied by the small interval dx on the real line gives us the probability in the neighborhood of the point x. This is an important interpretation, since for continuous $\mathbf{x}$ the probability *at* the point x is 0. Thus density has meaning in the vicinity of a point, not at it. To justify our earlier (unusual) description of $f(x_i)$ as the density function of a *discrete* random variable [for a discrete random variable, $f(x_i)$ gives the probability *at* a point], we may think of (2–16), for the discrete case, with dx representing not the length of the small interval x to $x + dx$ but the number of points of increase in the interval; in the limit, this number will be 1 or 0. Thus the language of density attached to discrete $f(x_i)$ is not unreasonable, and it simplifies and unifies without loss the discussion of the discrete and continuous cases.

2–5 Summary and concluding remarks. In this chapter we shifted from S, the space of events represented by points and subsets of points, to x, the space of events represented by points and intervals on the real line, with probabilities of points and subsets in S equivalent, without *practical* exception, to probabilities of corresponding points and intervals on the real line. Moreover, as certain points and intervals on the real line are of particular interest (for example, $\mathbf{x} = x$, $x \leq \mathbf{x} \leq x + dx$, $\mathbf{x} \leq x$), we introduced for mathematical convenience certain functions $f(x)$ and $F(x)$ which neatly describe probabilities at these points and over these intervals. The necessary axioms were placed on these functions, on the density function $f(x)$ in Sections 2–2 and 2–3, and, alternatively, on the distribution function $F(x)$ in Section 2–4. From here on, much of the analysis is in terms of $f(x)$ and $F(x)$.

We may note that certain definitions of the integral permit discrete and continuous random variables to be treated simultaneously; such discussion may be of interest to students. See, for example, Cramér [36].

We may ask here what the general natures of problems remaining to be considered for a single random variable are. They include (1) determination of the probabilities of various molecular events in which mutual exclusion is not present, particularly for certain special density functions $f(x)$; (2) estimation from observable data of the form of $f(x)$ and its principal characteristics; (3) formulation of hypotheses on certain characteristics of $f(x)$ and testing of these hypotheses by appeal to data (in the main, to observable data). These problems will occupy many of the following chapters.

Collateral Reading

H. D. Brunk, *An Introduction to Mathematical Statistics.* Boston, Ginn and Co., 1960.

H. Cramér, *The Elements of Probability Theory and Some of Its Applications.* New York, John Wiley and Sons, 1955.

———, *Mathematical Methods of Statistics.* Princeton, Princeton University Press, 1946.

W. Feller, *An Introduction to Probability Theory and Its Applications.* 2nd ed., New York, John Wiley and Sons, 1957.

R. V. Hogg and A. T. Craig, *Introduction to Mathematical Statistics.* New York, Macmillan Co., 1959.

M. E. Munroe, *Theory of Probability.* New York, McGraw-Hill Book Co., 1951. Unusual and valuable.

S. S. Wilks, *Mathematical Statistics.* Princeton, Princeton University Press, 1943. An excellent discussion. More advanced in the 1962 Wiley volume.

CHAPTER 3

RANDOM VARIABLES; EXPECTED VALUES

3–1 Definition of expectation. In this chapter we are interested in characteristics of probability distributions. A *probability distribution*, which is a generic term used to describe the probability behavior of a random variable (distribution and density functions are probability distributions) contains information in diffuse form. We may want to focus on one or a few characteristics which summarize the information in the distribution relevant to our interest. Of the many characteristics of a probability distribution which might be described here, the most valuable, both with respect to containing high and frequently desired summary information and to mathematical tractability, are certain quantities, generally called parameters since they vary in value from one probability distribution to another, which take the form of weighted averages of certain functions $g(\mathbf{x})$ of the random variable $\mathbf{x}$, the weights being the densities $f(x)$ of $\mathbf{x}$.

For brevity, we shall generally consider continuous random variables; the corresponding definitions and results for discrete random variables are readily written down.

The mathematical expectation or the expected value or the mean of $\mathbf{x}$ is written $E(\mathbf{x})$ and is defined by

$$E(\mathbf{x}) = \int_{-\infty}^{\infty} xf(x)\,dx. \tag{3–1}$$

Now consider a second random variable $\mathbf{y}$ with density function $h(y)$, and where

$$\mathbf{y} = g(\mathbf{x}).$$

We have

$$E(\mathbf{y}) = \int_{-\infty}^{\infty} yh(y)\,dy = E[g(\mathbf{x})] = \int_{-\infty}^{\infty} g(x)h\{g(x)\}\,d\{g(x)\},$$

and a most important theorem is that this expression for $E[g(\mathbf{x})]$ may be reduced to

$$E[g(\mathbf{x})] = \int_{-\infty}^{\infty} g(x)f(x)\,dx. \tag{3–2}$$

We call $E[g(\mathbf{x})]$ the mathematical expectation or the expected value or the mean of $g(\mathbf{x})$.

A proof of this important result for continuous random variables requires the change-of-variable technique to be discussed in Chapter 8. Therefore we confine the demonstration here to the discrete case. Given the discrete random variable $\mathbf{x}$ with values $x_1, x_2, \ldots, x_n$, and respective probabilities $f(x_1), f(x_2), \ldots, f(x_n)$. The mean of $\mathbf{x}$ is defined by

$$E(\mathbf{x}) = \sum_{i=1}^{i=n} x_i f(x_i).$$

Consider the relationship $y = g(x)$. We wish to prove that

$$E(\mathbf{y}) = E[g(\mathbf{x})] = \sum_{i=1}^{n} g(x_i) f(x_i). \tag{3-3}$$

By definition we have

$$E(\mathbf{y}) = \sum_{i=1}^{m} y_i h(y_i),$$

where the random variable $\mathbf{y}$ has values $y_1, y_2, \ldots, y_m$ with respective probabilities $p_1, p_2, \ldots, p_m$.

If the relationship between $\mathbf{y}$ and $\mathbf{x}$ is one to one ($m = n$), we have

$$\begin{aligned} E(\mathbf{y}) = \sum_{i=1}^{n} y_i h(y_i) &= \sum_{i=1}^{n} y_i p(\mathbf{y} = y_i) \\ &= \sum_{i=1}^{n} g(x_i) p(\mathbf{y} = y_i) \\ &= \sum_{i=1}^{n} g(x_i) p(\mathbf{x} = x_i) \\ &= \sum_{i=1}^{n} g(x_i) f(x_i), \end{aligned}$$

as was to be shown. But (as suggested in Fig. 3-1) several values of $\mathbf{x}$

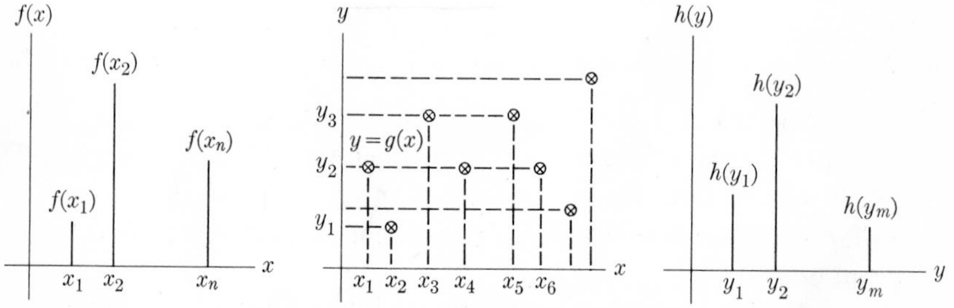

FIG. 3-1. Density functions $f(x)$ and $h(y)$ of the discrete random variables $\mathbf{x}$ and $\mathbf{y}$, and the relationship $y = g(x)$ between $\mathbf{x}$ and $\mathbf{y}$.

may map into one value of $\mathbf{y}$ $(m < n)$; in Fig. 3–1, x_1, x_4, and x_6 map into y_2. For these points we would have

$$E(\mathbf{y}) = \sum_i y_i p(\mathbf{y} = y_i) = g(x_1)f(x_1) + g(x_4)f(x_4) + g(x_6)f(x_6),$$

and it is evident that the theorem (3–3) still follows.

Had we been able to begin with a more general definition of an integral, these results would have been reached with greater logic and rigor. We would have described $E\{g(\mathbf{x})\}$ as the integral of $g(\mathbf{x})$ with respect to the distribution function $F(x)$ and would have written

$$E\{g(\mathbf{x})\} = \int_{-\infty}^{\infty} g(x)\, dF(x).$$

We would then have proceeded to show that for a discrete random variable, this integral reduces to (3–3), while for a continuous random variable, this integral reduces to (3–2). (See, for example, Wadsworth and Bryan [192].)

For $E[g(\mathbf{x})]$ to exist, the integral or sum defining $E[g(\mathbf{x})]$ must converge absolutely, that is,

$$\sum_i |g(x_i)|f(x_i) < \infty \quad \text{or} \quad \int |g(x)|f(x)\, dx < \infty, \tag{3–4}$$

and for most $g(\mathbf{x})$ and $f(x)$ met in statistics this occurs. When $f(x)$ vanishes outside a finite interval or when the number of terms in the sum is finite, this condition is surely satisfied (provided $g = \infty$ is excluded); otherwise, convergence need be checked.

3–2 Examples and discussion of expectation. There are particular $g(\mathbf{x})$ whose expectations, when they exist, effectively describe important characteristics of the probability distribution of $\mathbf{x}$. These expectations contain much interesting information on $f(x)$ and are therefore met often enough to be given special names and symbols. We consider these particular cases now. Setting $g(\mathbf{x}) = \mathbf{x}$ gives

$$E(\mathbf{x}) = \int_{-\infty}^{\infty} xf(x)\, dx,$$

already introduced. If possible without confusion, $E(\mathbf{x})$ is sometimes written λ_1 or λ; it is called the arithmetic mean of $\mathbf{x}$. In structure, $E(\mathbf{x})$ is similar to the common average or mean $\bar{x}$ often computed from observed numbers $x_1, x_2, \ldots, x_n$ occurring with respective frequencies $f_1, f_2, \ldots, f_n$,

$$\bar{x} = \frac{x_1 f_1 + x_2 f_2 + \cdots + x_n f_n}{N} = x_1 \frac{f_1}{N} + x_2 \frac{f_2}{N} + \cdots + x_n \frac{f_n}{N},$$

$f_1 + f_2 + \cdots + f_n = N$. The quantity $\bar{x}$ may be regarded as the historical motivation of the definition (3–1) and will later reappear when we consider the problem of estimating λ_1 from observed data. Geometrically, λ_1 is one of a number of possible devices for locating the "center" of the probability distribution of $\mathbf{x}$; in mechanics, λ_1 is the abscissa of the center of gravity of a unit of mass distributed along the x-axis with amount of mass $f(x)\,dx$ lying between x and $x + dx$.

Setting $g(\mathbf{x}) = \mathbf{x}^k$ gives

$$E(\mathbf{x}^k) = \int_{-\infty}^{\infty} x^k f(x)\,dx, \tag{3–5}$$

often written λ_k, the kth moment of $\mathbf{x}$ about the origin. The arithmetic mean of $\mathbf{x}$, λ_1, is of greatest interest; it is followed in interest by λ_2, λ_3, and λ_4. Setting $g(\mathbf{x}) = (\mathbf{x} - \lambda_1)^k$ gives

$$E(\mathbf{x} - \lambda_1)^k = \int_{-\infty}^{\infty} (x - \lambda_1)^k f(x)\,dx, \tag{3–6}$$

often written μ_k, the kth moment of $\mathbf{x}$ about the arithmetic mean λ_1. In particular, for $k = 2$,

$$E(\mathbf{x} - \lambda_1)^2 = \int_{-\infty}^{\infty} (x - \lambda_1)^2 f(x)\,dx, \tag{3–7}$$

variously written σ^2 or $\sigma^2(\mathbf{x})$ or V or μ_2, and variously called the variance of $\mathbf{x}$, the variance of the probability distribution of $\mathbf{x}$, the second moment of $\mathbf{x}$ about the mean. Geometrically, σ^2 is one of a number of devices for measuring the variability of $\mathbf{x}$; in mechanics, σ^2 is the moment of inertia (with respect to an axis perpendicular to the x-axis and through the center of gravity) of the unit mass referred to in connection with $E(\mathbf{x})$. Further, σ, the positive square root of σ^2, is called the standard deviation of $\mathbf{x}$. In one respect σ may be a more intuitive measure of variability (its dimensions are those of $\mathbf{x}$), but the more convenient mathematical properties of the variance lead to its use in a wide variety of applications.

Setting $g(\mathbf{x}) = e^{\theta \mathbf{x}}$, θ a dummy real variable, gives

$$E(e^{\theta \mathbf{x}}) = \int_{-\infty}^{\infty} e^{\theta x} f(x)\,dx, \tag{3–8}$$

written $M_{\mathbf{x}}(\theta)$ and called, for reasons which will be made clear in Section 3–4, the moment generating function of $\mathbf{x}$. It is due to Laplace.

A few useful results follow readily from (3–2):

$$E(a + b\mathbf{x}) = a + bE(\mathbf{x}),$$

$$\mu_1 = 0 \text{ (and therefore a useless moment)},$$

$$\mu_{2k+1} = 0 \text{ (not meaning that } \textit{all } 2k + 1 \text{ moments vanish)}$$

the last not true in general, but true if $\mathbf{x}$ is symmetrical about its mean (and if $2k + 1$ moments exist). The converse of the last result, that $\mathbf{x}$ is symmetrical about its mean if $\mu_{2k+1} = 0$, is not true.

We also have

$$\mu_2 = \lambda_2 - \lambda_1^2, \tag{3–9}$$

a result used occasionally in the text, showing the relationship between the often-wanted variance of $\mathbf{x}$ and the sometimes more readily found first and second moments of $\mathbf{x}$ about the origin. Results similar to (3–9) can readily be obtained for higher values of k.

As already noted, for certain $g(\mathbf{x})$ and $f(x)$, $E[g(\mathbf{x})]$ may be undefined. Also, for many $g(\mathbf{x})$, $E[g(\mathbf{x})]$ may be intractable and can be evaluated in closed form not for all $f(x)$ as implied so far in this chapter [for, to this point, no restrictions have been placed on $f(x)$], but only for certain obliging $f(x)$. For example, $g(\mathbf{x}) = \sqrt{\mathbf{x}}$ (a function which will arise later in sampling theory) is often intractable. Results obtained for particular $f(x)$ are of course less general; they hold only for the density function actually employed.

Geometrical interpretation of the lower moments, particularly of λ and σ^2, is helpful but not decisive to the importance of moments, and, when applied to third and higher moments, can even be treacherous.

Note also that despite the heavy (and proper) preoccupation by statisticians with variability as measured by the variance, the fact is that even $\sigma^2(\mathbf{x}) < \sigma^2(\mathbf{y})$ does not give lucid information as to the nature of the smaller variability of $\mathbf{x}$ about its mean as against that of $\mathbf{y}$ about its mean. It says that the probability distribution of $\mathbf{x}$ is more concentrated near $E(\mathbf{x})$, in a certain sense, than is the probability distribution of $\mathbf{y}$ near $E(\mathbf{y})$. If $\sigma^2(\mathbf{x})$ is small, we can say that most deviations $x - E(\mathbf{x})$ are small; if $\sigma^2(\mathbf{x})$ is large, we can only say that not all deviations $x - E(\mathbf{x})$ are small.

Example. An interesting example combining the elementary probability results of Chapter 1 and the notion of expectation of this chapter has been given by Dorfman [49]. The problem is to detect every diseased member of a population of N persons. The blood test available for detection is expensive; it is also so sensitive that it will detect the presence of unhealthy blood in a pooled sample taken from n persons, all but one of whom are healthy. Let the probability of the disease be p. We may test all N persons individually (an expensive procedure) or make analyses on N/n pooled blood-sample groups of n persons each, analyzing the individuals in a group only if the group blood analysis is positive. Intuition suggests that for p small, grouping should reduce the number of analyses.

The probability of a diseased individual is p, of one not diseased is $1 - p$. The probability of all in a group of size n not diseased is $(1 - p)^n$, of at least one in a group of size n diseased is $1 - (1 - p)^n$. If all N individuals are tested, the number of analyses is N. If grouping is used, the expected number of analyses is N/n (the number of groups) plus $n[1 - (1 - p)^n](N/n)$ (the expected number of individual tests, assuming all members of a diseased group must be tested, i.e., only two rounds of testing are allowed). The ratio

$$\frac{N/n + n[1 - (1 - p)^n](N/n)}{N} = \frac{1}{n} + [1 - (1 - p)^n]$$

is small for small p.

3–3 The Bienaymé-Chebychev theorem. The following theorem, showing a relationship between the standard deviation and the probability distribution of $\mathbf{x}$, is remarkable in that it holds for all discrete and continuous random variables. It is due to Bienaymé, 1853, and in more general form, to Chebychev, 1866. We have, for a continuous random variable $\mathbf{x}$,

$$\begin{aligned}\sigma^2 = E(\mathbf{x} - \lambda)^2 &= \int_{-\infty}^{\infty} (x - \lambda)^2 f(x)\,dx \\ &= \int_{-\infty}^{\lambda - k\sigma} (x - \lambda)^2 f(x)\,dx + \int_{\lambda - k\sigma}^{\lambda + k\sigma} (x - \lambda)^2 f(x)\,dx \\ &\qquad + \int_{\lambda + k\sigma}^{\infty} (x - \lambda)^2 f(x)\,dx. \qquad (3\text{–}10)\end{aligned}$$

The argument involves replacing the equality in (3–10) by the inequality equal to or greater than ($\geq$). This is accomplished by replacing the first and third integrals by smaller quantities and dropping the (positive) middle integral altogether.

In the first of the three integrals of (3–10), with $k > 0$,

$$x \leq \lambda - k\sigma \quad \text{implies} \quad x - \lambda \leq -k\sigma \quad \text{implies} \quad (x - \lambda)^2 \geq k^2\sigma^2,$$

while in the third integral of (3–10),

$$x \geq \lambda + k\sigma \quad \text{implies} \quad x - \lambda \geq k\sigma \quad \text{implies} \quad (x - \lambda)^2 \geq k^2\sigma^2.$$

If therefore in the first and third integrals we replace the present variable integrands $(x - \lambda)^2$ by the constant integrand $k^2\sigma^2$ (leaving the second integral alone), the equality (3–10) will be replaced by the inequality

$$\sigma^2 \geq \left[\int_{-\infty}^{\lambda - k\sigma} k^2\sigma^2 f(x)\,dx + \int_{\lambda - k\sigma}^{\lambda + k\sigma} (x - \lambda)^2 f(x)\,dx + \int_{\lambda + k\sigma}^{\infty} k^2\sigma^2 f(x)\,dx\right]. \qquad (3\text{–}11)$$

But if the variance σ^2 is equal to or greater than the sum of three *positive* terms it is surely equal to or greater than the sum of the first and third. Omitting the second integral in (3–11) and expressing the first and third integrals in terms of probabilities, we obtain

$$\sigma^2 \geq \{k^2\sigma^2 p[\mathbf{x} \leq (\lambda - k\sigma)] + k^2\sigma^2 p[\mathbf{x} \geq (\lambda + k\sigma)]\}$$

$$\geq k^2\sigma^2 p(|\mathbf{x} - \lambda| \geq k\sigma)$$

or, finally,

$$p(|\mathbf{x} - \lambda| \geq k\sigma) \leq \frac{1}{k^2}\cdot \tag{3–12}$$

This result connects probabilities of deviations of $\mathbf{x}$ with the standard deviation (if both exist), holds for all $k > 0$ though of principal interest for $k > 1$, and shows (remarkably) that no matter what the form of $f(x)$, the probability outside $\lambda \pm k\sigma$ is limited (less than or equal to $1/k^2$). For example, not more than 25 percent of the probability can lie outside plus or minus two standard deviation units of the mean of a probability distribution.

If something is known of $f(x)$, sharper results can be obtained. For example, if $f(x)$ has a single peak and has high-order contact with the x-axis at $x = \pm\infty$, k^2 may be replaced by $2.25k^2$, a result due in 1922 to Camp and, separately, Meidell. If, of course, $f(x)$ is *exactly* known, we can *exactly* determine $p|\mathbf{x} - \lambda|$. This is a familiar situation in probability; we often have the choice between weaker theorems [an inequality on $p(|\mathbf{x} - \lambda|)$] of greater generality [$f(x)$ unspecified] and sharper theorems [$p|\mathbf{x} - \lambda|$ exactly determined] of lesser generality [the particular form of $f(x)$ specified].

Because of its generality, the Bienaymé-Chebychev inequality has much utility in modern probability theory. It, as well as the Camp-Meidell variant, is useful in modern industrial quality control, where the proportion of output outside $\pm k\sigma$ of the mean quality λ is of interest.

3–4 Moment generating function. We return to the expectation (3–8), the particular case of

$$E[g(\mathbf{x})] = \int_{-\infty}^{\infty} g(x)f(x)\,dx,$$

in which

$$g(\mathbf{x}) = e^{\theta\mathbf{x}},$$

where θ is a dummy real variable of little interest itself, but necessary (as the reader can best discover by deleting θ) to the following argument. Since

$$e^{\theta x} = 1 + \frac{\theta x}{1!} + \frac{(\theta x)^2}{2!} + \cdots \tag{3–13}$$

(the Taylor series expansion of $e^{\theta x}$), we find, whenever the integral (3–8) exists (and if $\int_{-\infty}^{\infty} \sum_{j=1}^{\infty}$ may be replaced by $\sum_{j=1}^{\infty} \int_{-\infty}^{\infty}$),

$$E(e^{\theta \mathbf{x}}) = \lambda_0 + \frac{\theta}{1!}\lambda_1 + \frac{\theta^2}{2!}\lambda_2 + \cdots. \tag{3–14}$$

Thus $E(e^{\theta \mathbf{x}})$ may be said to generate the moments $\lambda_0, \lambda_1, \lambda_2, \ldots$ of the random variable $\mathbf{x}$, and herein lies its importance in statistical theory. It is called the moment generating function of $\mathbf{x}$ and is written $M_{\mathbf{x}}(\theta)$.

Note that

$$\left.\frac{\partial^k M_{\mathbf{x}}(\theta)}{\partial \theta^k}\right|_{\theta=0} = \lambda_k; \tag{3–15}$$

any specific moment λ_k of the random variable $\mathbf{x}$ can be readily obtained by differentiation of the moment generating function of $\mathbf{x}$.

We shall now accept that the moment generating function of $h(\mathbf{x})$, an arbitrary continuous function of the random variable $\mathbf{x}$,

$$M_{h(\mathbf{x})}(\theta) = \int_{-\infty}^{\infty} e^{\theta h(x)} f(x)\, dx, \tag{3–16}$$

generates the moments of $h(\mathbf{x})$, with special cases

$$h(\mathbf{x}) = a + b\mathbf{x}, \qquad M_{h(\mathbf{x})}(\theta) = e^{\theta a} M_{\mathbf{x}}(\theta b), \tag{3–17}$$

$$h(\mathbf{x}) = a + bu(\mathbf{x}), \qquad M_{h(\mathbf{x})}(\theta) = e^{\theta a} M_{u(\mathbf{x})}(\theta b), \tag{3–18}$$

and of particular interest,

$$h(\mathbf{x}) = \mathbf{x} - \lambda_1, \qquad M_{h(\mathbf{x})}(\theta) = e^{-\theta \lambda_1} M_{\mathbf{x}}(\theta). \tag{3–19}$$

The final example shows the relationship between the function generating moments of a random variable about its mean and the function generating moments of that random variable about the origin.

As already indicated, $g(\mathbf{x})$ and $f(x)$ must be such that the integral (or, in the discrete case, the sum) in (3–2) converges absolutely, that is, satisfies (3–4). If in place of $e^{\theta \mathbf{x}}$ in (3–8) we use $e^{i\theta \mathbf{x}}$, where $i = \sqrt{-1}$, the absolute convergence of $E(e^{i\theta \mathbf{x}})$ is assured; the corresponding integral, called the *characteristic function* of $\mathbf{x}$, opens up broader and deeper techniques for identifying probability distributions and dealing with their moments. But this method is not available at our level of exposition.

3–5 Relationship between moments and probability distributions. Relationships among probability distributions, moment generating functions, and moments form a profound and difficult area of probability to which we can give little attention; it is here that the characteristic function is a particularly effective tool. But, however roughly and conditionally, a few things had better be said. Given a probability distribution, we can get, uniquely, its moment generating function and its moments provided they exist; the process has been described in Section 3–4. But the critical converse theorem, which we shall often use, that the moment generating function (or the moments) *uniquely* determines the probability distribution, is difficult to prove and can be proved at all only under certain mild conditions. Here we are able to show only a particularly modest result (suggested by Mood [129]) that if the moments associated with two density functions $f(x)$ and $u(x)$ are the same (or if the moment generating functions are the same, since the latter are uniquely related to the moments) and if the difference $f(x) - u(x)$ has a power series expansion about the origin (a fairly nonrestrictive condition), then $f(x) = u(x)$, identically. We have

$$f(x) - u(x) = a + bx + cx^2 + \cdots, \tag{3–20}$$

and we form

$$\int_{-\infty}^{\infty} [f(x) - u(x)]^2\, dx = \int_{-\infty}^{\infty} (a + bx + cx^2 + \cdots)[f(x) - u(x)]\, dx,$$

replacing one $[f(x) - u(x)]$ by (3–20). On integration, the right-hand side reduces to 0, for the corresponding moments of the two random variables are given to be the same. As the left-hand integrand is nonnegative, it is evident that for the left-hand integral to be 0, we must have $f(x) = u(x)$, identically. Results of a similar character can be obtained with weaker conditions on $f(x)$ and $u(x)$.

Collateral Reading

H. D. Brunk, *An Introduction to Mathematical Statistics*. Boston, Ginn and Co., 1960.

H. Cramér, *The Elements of Probability Theory and Some of Its Applications*. New York, John Wiley and Sons, 1955.

W. Feller, *An Introduction to Probability Theory and Its Applications*. 2nd ed., New York, John Wiley and Sons, 1957.

A. M. Mood, *Introduction to the Theory of Statistics*. New York, McGraw-Hill Book Co., 1950.

Problems

3–1. For A constant, show that

$$\int_{-\infty}^{\infty} (x - A)^2 f(x)\, dx$$

is minimum when $A = E(\mathbf{x})$.

3-2. The median $M(\mathbf{x})$ of a continuous random variable $\mathbf{x}$ is defined by

$$\int_{-\infty}^{M(\mathbf{x})} f(x)\, dx = \int_{M(\mathbf{x})}^{\infty} f(x)\, dx = \tfrac{1}{2}.$$

For A constant, show that

$$\int_{-\infty}^{\infty} |x - A| f(x)\, dx$$

is minimum when $A = M(\mathbf{x})$. The median $M(\mathbf{x})$ is evidently a characteristic of $f(x)$ which is *not* definable in terms of moments λ_k or μ_k. For certain $f(x)$, $M(\mathbf{x})$ contains much information.

3–3. Let

$$\mathbf{u} = \frac{\mathbf{x} - E(\mathbf{x})}{\sigma(\mathbf{x})};$$

then show that $E(\mathbf{u}) = 0$ and $\sigma^2(\mathbf{u}) = 1$. The useful random variable $\mathbf{u}$ is called a standardized (or normalized or dimensionless) random variable. This simple but valuable result will be frequently used in later chapters.

3–4. (Markov's inequality.) Let $\mathbf{x}$ be a nonnegative random variable. Let $c > 0$. Show that $p(\mathbf{x} \geq c) \leq E(\mathbf{x})/c$. Note that the variance need not be known (or exist). Unlike the random variable of the Bienaymé-Chebychev inequality, $\mathbf{x}$ is here restricted to nonnegative values [often a realistic (or unimportant) restriction]. Substituting $(\mathbf{x} - \lambda)^2$ for $\mathbf{x}$, derive the Bienaymé-Chebychev inequality.

CHAPTER 4

TWO RANDOM VARIABLES

4–1 Axioms for two random variables. Of the two approaches to an axiomatic structure for one random variable described in Chapter 2, we confine ourselves here to the second, given in Section 2–4. Following the argument introduced there for one random variable, we now postulate the existence of a two-variable distribution function $F(x, y)$ which is connected by definition to a *joint* probability statement on the random variables $\mathbf{x}$, $\mathbf{y}$ by

$$p(\mathbf{x} \leq x, \mathbf{y} \leq y) = F(x, y), \tag{4–1}$$

where the left-hand side of (4–1) stands for the probability that $\mathbf{x}$ is less than or equal to x *and* that $\mathbf{y}$ is less than or equal to y. A two-variable distribution function for discrete random variables is shown in Fig. 4–1.

The function $F(x, y)$ will now be required to satisfy three axioms entirely similar to those introduced in Section 2–4 for one random variable. The first difference in two dimensions of $F(x, y)$ shall be nonnegative; corresponding to axiom (a) of page 21 for a single random variable which required a nonnegative first difference of $F(x)$ in order to ensure nonnegative probabilities, we require here for a pair of random variables and, as we shall soon verify, for the same reason, the more cumbersome condition

$$\text{(a)} \quad [F(b, d) - F(a, d) - F(b, c) + F(a, c)] \geq 0, \quad a < b, c < d.$$

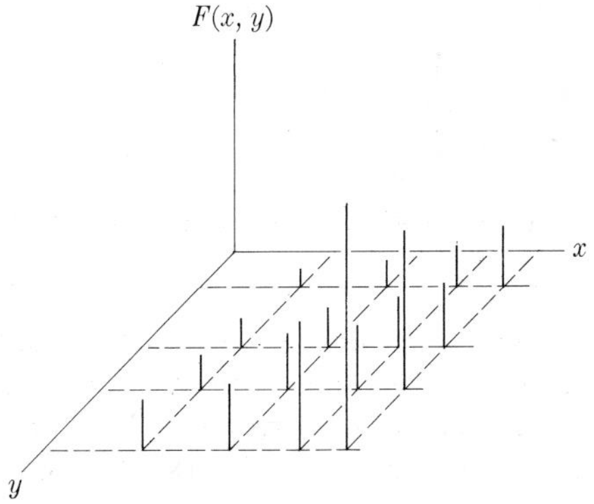

FIG. 4–1. Joint distribution function for discrete random variables.

As before, we also require

(b) $F(-\infty, y) = F(x, -\infty) = 0, \qquad F(\infty, \infty) = 1,$

and, representing mutually exclusive events by the disjoint sets of points $E_1, E_2, \ldots$ in the two-dimensional space of x, y, we may write the addition axiom for a pair of random variables **x**, **y**,

(c) $p(\mathbf{x}, \mathbf{y} \in E_1 + E_2 + \cdots) = p(\mathbf{x}, \mathbf{y} \in E_1) + p(\mathbf{x}, \mathbf{y} \in E_2) + \cdots.$

Note that no axioms are placed on $F(-\infty, -\infty)$, for we can *prove* this to be 0, and that no axioms are placed on $F(x, \infty)$ or on $F(\infty, y)$ which will turn out to be descriptions of the probability behavior of **x** and **y** *separately*.

The motivation underlying the introduction of two-variable probabilities and the two-variable distribution function $F(x, y)$ lies in the fact that a trial or experiment often produces *two* numbers (e.g., a person may be examined for height *and* weight) and, as we saw in Section 2–1, the probability structure associated with one random variable is generally not adequate to handle such problems. For example, the four compound atomic events of page 16 each of which was composed of answers to two questions (i.e., the face of a coin and the color of a card), found no satisfactory representation on the real line. (Such numbers as 2, 1, 1, 0 could be used, with the random variable representing the number of heads *plus* the number of black cards in a single trial, but such a random variable would seldom be of practical interest.) But the four events do find satisfactory representation as points in the real *plane*, as illustrated in Fig. 4–2, with the coordinate x usefully interpreted as the number of heads in one trial (0 or 1) and the coordinate y as the number of black cards in one trial (0 or 1).

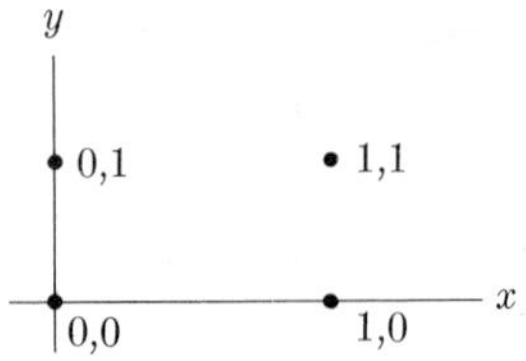

FIG. 4–2. Representation of the compound atomic events of page 16 in the plane.

Generally, the random functions or, in more common language, the random variables **x**, **y** map the compound atomic events of S involving *two* characteristics into points in the real *plane*. Sets of points in S, that is, molecular events in S, are transformed by the random variables **x**, **y** into regions in the real plane. Following the discussion of Chapter 2, it should be reasonably clear that $F(x, y)$, aided by theorems following from (4–1) and the axioms on pages 35–36 when questions involving complicated molecular events are asked, will fully describe the joint probability behavior of $\mathbf{x}$, $\mathbf{y}$. As earlier for one random variable, we again take it for granted that, practically speaking, *all* regions in the plane however complex in structure have probabilities attached to them and therefore, that the distribution function $F(x, y)$ will describe the probability behavior of $\mathbf{x}$, $\mathbf{y}$ over all regions that can be constructed in the plane.

From the nature of (4–1) and the axioms to which $F(x, y)$ is subjected, it is evident that $F(x, y)$ is a mathematical idealization of joint cumulative relative frequencies of two variables; such (finite) joint cumulative relative frequencies satisfy our axioms for $F(x, y)$. Also note that, similarly to Sections 2–2 and 2–3, we might have created an axiomatic structure for the random variable pair $\mathbf{x}$, $\mathbf{y}$ by equating probability and volume.

4–2 Elements of distribution theory for two random variables. Corresponding to one of (2–10) which stated that for one random variable

$$p(a < \mathbf{x} \leq b) = F(b) - F(a), \quad b > a,$$

and which from axiom (a) of Chapter 2 is nonnegative, we now have, for a pair of random variables,

$$p(a < \mathbf{x} \leq b, c < \mathbf{y} \leq d) = F(b, d) - F(a, d) - F(b, c) + F(a, c) \quad (4\text{–}2)$$

[and not $F(b, d) - F(a, c)$, as is readily seen in Fig. 4–3]; and we now see that axiom (a) of this chapter was chosen to ensure that the joint probability on the left side of (4–2) be nonnegative.

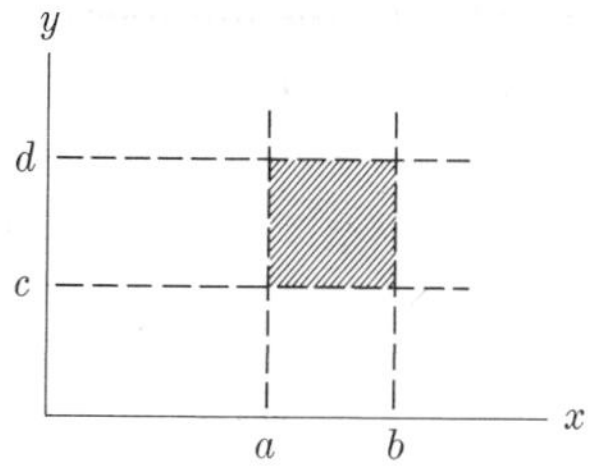

FIG. 4–3. The event $(a < \mathbf{x} \leq b, \quad c < \mathbf{y} \leq d)$.

Other theorems include

$$p(\mathbf{x}, \mathbf{y} \in \text{any region } R) \leq 1, \qquad p(-\infty < \mathbf{x} < \infty, -\infty < \mathbf{y} < \infty) = 1,$$

and for E_1 and E_2 not necessarily disjoint regions in the space of x, y,

$$p(\mathbf{x}, \mathbf{y} \in E_1 + E_2) = p(\mathbf{x}, \mathbf{y} \in E_1) + p(\mathbf{x}, \mathbf{y} \in E_2) - p(\mathbf{x}, \mathbf{y} \in E_1E_2), \quad (4\text{–}3)$$

corresponding exactly to (2–11) and (1–8), and proved by arguments entirely similar to those supporting (2–11) and (1–8).

For $F(x, y)$ *discrete*, by which we mean that there exists an enumerable set of points (x_1, y_1), (x_1, y_2), (x_2, y_1), (x_2, y_2), ..., (x_i, y_j), ... with associated positive numbers p_{11}, p_{12}, p_{21}, p_{22}, ..., p_{ij}, ... satisfying $\sum_i \sum_j p_{ij} = 1$, with

$$F(x_h, y_k) = \sum_i \sum_j p_{ij} \tag{4-4}$$

summed over all i and j for which $x_i \leq x_h$ and $y_j \leq y_k$, we can obtain, similarly to (2–13) for one discrete random variable,

$$p(\mathbf{x} = x_i, \mathbf{y} = y_j) = p_{ij} \quad \text{discreet} \tag{4-5}$$

for all i and j for which values of the random variables $\mathbf{x}$, $\mathbf{y}$ exist, and 0 otherwise. We shall sometimes replace p_{ij} by the symbol $f(x_i, y_j)$ and shall generally call both of these the joint density function of the discrete random variables $\mathbf{x}$, $\mathbf{y}$.

For $F(x, y)$ *continuous*, by which we mean that a function $f(x, y)$ exists such that

$$F(x, y) = \int_{-\infty}^{y} \int_{-\infty}^{x} f(u, v) \, du \, dv, \quad \text{continuous}$$

or the rough equivalent

$$f(x, y) = \frac{\partial}{\partial x} \frac{\partial}{\partial y} F(x, y), \quad \text{rough equivalent}$$

we immediately have

$$p(a \leq \mathbf{x} \leq b, c \leq \mathbf{y} \leq d) = \int_c^d \int_a^b f(u, v) \, du \, dv, \tag{4-6}$$

an important relationship between joint probability and the joint density function $f(x, y)$, or

$$p(x \leq \mathbf{x} \leq x + dx, y \leq \mathbf{y} \leq y + dy) = f(x, y) \, dx \, dy, \tag{4-7}$$

with an interpretation of (4–6) and (4–7) analogous to that given for (2–15) and (2–16) for one random variable.

Terminology To summarize the terminology for two random variables, $F(x, y)$ is the joint distribution function of the random variables $\mathbf{x}$, $\mathbf{y}$; p_{ij} or $f(x_i, y_j)$ is the joint density function of discrete $\mathbf{x}$, $\mathbf{y}$; $f(x, y)$ is the joint density function of continuous $\mathbf{x}$, $\mathbf{y}$ and $f(x, y) \, dx \, dy$ is the joint probability element of continuous $\mathbf{x}$, $\mathbf{y}$.

$F(x, y)$
$p_{ij} = f(x_i, y_j)$
$f(x, y)$

$f(x, y) \, dx \, dy$

4–3 Marginal distributions. We often want to determine the probability behavior of one variable, given the joint probability behavior of two. We interpret this to be equivalent to

$$p(\mathbf{x} \le x,\ \text{no condition on } \mathbf{y}), \tag{4–8}$$

which is equivalent to $F(x, \infty)$. Note that $F(x, \infty)$ is itself a *bona fide* distribution function, that is, $F(x, \infty)$ satisfies the axioms; specifically, $F(x, \infty)$ is monotonic nondecreasing (as the student is asked later to show) $F(-\infty, +\infty) = 0$, $F(\infty, \infty) = 1$, and the additivity axiom is satisfied. In fact, $F(x, \infty)$ is the distribution function of $\mathbf{x}$, for this is what (4–8) means.

For the discrete case

$$F(x_h, \infty) = \sum_i \sum_j p_{ij} \tag{4–9}$$

summed over all j, and over i such that $x_i \le x_h$; $F(x_h, \infty)$ is the distribution function of $\mathbf{x}$. The result is similar for $\mathbf{y}$.

For the continuous case

$$F(x, \infty) = \int_{-\infty}^{x} \int_{-\infty}^{\infty} f(u, v)\, du\, dv = \int_{-\infty}^{x} \left[du \int_{-\infty}^{\infty} f(u, v)\, dv\right] = \int_{-\infty}^{x} g(u)\, du, \tag{4–10}$$

where, noting the right-hand integral of (4–10),

$$g(x) = \int_{-\infty}^{\infty} f(x, y)\, dy \tag{4–11}$$

is evidently the density function of $\mathbf{x}$. When so obtained (or in the discrete case above, by summation) from the density function of two (or more) random variables, $g(x)$ is sometimes called the *marginal* distribution of $\mathbf{x}$. Similarly,

$$F(\infty, y) = \int_{-\infty}^{y} \int_{-\infty}^{\infty} f(u, v)\, du\, dv = \int_{-\infty}^{y} \left[dv \int_{-\infty}^{\infty} f(u, v)\, du\right] = \int_{-\infty}^{y} h(v)\, dv, \tag{4–12}$$

where, noting the right-hand integral of (4–12),

$$h(y) = \int_{-\infty}^{\infty} f(x, y)\, dx \tag{4–13}$$

is evidently the density function or the marginal distribution of $\mathbf{y}$. The probability behavior of one variable is obtained from the joint probability behavior of two variables by integrating (or in the discrete case which we omitted, summing) over the unwanted variable. The reverse process is *generally* impossible; we cannot *generally* obtain $F(x, y)$ from the separate distribution functions $G(x) = F(x, \infty)$ and $H(y) = F(\infty, y)$.

4–4 Independence. In Chapter 1 we defined independence of events A and B by $p(AB) = p(A)p(B)$. For the extended class of random variables, we define independence of $\mathbf{x}$ and $\mathbf{y}$ by

$$f(x, y) = g(x)h(y), \tag{4–14}$$

where $g(x)$ and $h(y)$ are the marginal distributions of $\mathbf{x}$ and $\mathbf{y}$ respectively; this is equivalent to $p\{\mathbf{x} \in A, \mathbf{y} \in B\} = p\{\mathbf{x} \in A\}p\{\mathbf{y} \in B\}$ for all A, B.

A necessary and sufficient condition for (4–14) to hold is that $f(x, y)$ can be factored into the product of two functions, one a function only of x, the other a function only of y, say

$$f(x, y) = r(x)s(y). \tag{4–15}$$

For (4–14) to be satisfied, (4–15) is evidently necessary. As for sufficiency, if (4–15) is true, we have

$$g(x) = f(x, -\infty < y < \infty) = \int_{-\infty}^{\infty} r(x)s(y)\,dy$$

$$= r(x)\int_{-\infty}^{\infty} s(y)\,dy = r(x)\cdot A$$

$$h(y) = f(-\infty < x < \infty, y) = \int_{-\infty}^{\infty} r(x)s(y)\,dx$$

$$= s(y)\int_{-\infty}^{\infty} r(x)\,dx = s(y)\cdot B,$$

from which

$$r(x) = \frac{1}{A}g(x), \qquad s(y) = \frac{1}{B}h(y), \qquad \text{and} \qquad r(x)s(y) = \frac{1}{AB}g(x)h(y).$$

But

$$1 = \int_{-\infty}^{\infty}\int_{-\infty}^{\infty} f(x, y)\,dx\,dy = \int_{-\infty}^{\infty}\int_{-\infty}^{\infty} r(x)s(y)\,dx\,dy$$

$$= \int_{-\infty}^{\infty} r(x)\,dx \int_{-\infty}^{\infty} s(y)\,dy = AB,$$

from which $AB = 1$ and the theorem follows. If $f(x, y)$ is factorable, the factorization is unique (except for multiplication of one factor by an arbitrary constant and division of the other factor by the same), and the factors are the marginal distributions.

Finally, note that beginning with definition (4–14), theorems on *distribution* functions corresponding in form to (4–14) can be found. Alternatively, independence can be *defined* in terms of distribution functions by

$$F(x, y) = G(x)H(y), \tag{4–16}$$

and our present definition (4–14) can be obtained as a *theorem.*

4–5 Conditional probability. Following the definitions in Chapter 1 of conditional probabilities, $p(A|B)$ and $p(B|A)$ for the simple events A, B,

$$p(A, B) = p(A|B)p(B) = p(B|A)p(A), \qquad (4\text{–}17)$$

$$P(A|B) = \frac{P(A,B)}{P(B)}$$

we want to introduce, first for the extended class of a pair of discrete random variables $\mathbf{x}$, $\mathbf{y}$, conditional probabilities $p(x_i|y_j)$, $p(y_j|x_i)$. Let A represent the event $\mathbf{x} = x_i$ and B the event $\mathbf{y} = y_j$. We have, following (4–17),

$$p(\mathbf{x} = x_i|\mathbf{y} = y_j) = \frac{p(\mathbf{x} = x_i, \mathbf{y} = y_j)}{p(\mathbf{y} = y_j)} = \frac{f(x_i, y_j)}{h(y_j)} \; ; \; g(x_i) \quad (4\text{–}18)$$

discrete
case

for $h(y_j) \neq 0$ the ratio is called $g(x_i|y_j)$, the conditional density function of $\mathbf{x}$, given $\mathbf{y} = y_j$. Similarly

$$p(\mathbf{y} = y_j|\mathbf{x} = x_i) \qquad \text{leads to} \qquad h(y_j|x_i), \qquad (4\text{–}19)$$

the conditional density function of $\mathbf{y}$, given x_i. We have defined two new discrete conditional density functions $g(x_i|y_j)$ and $h(y_j|x_i)$ in terms of the previously established joint density function $f(x_i, y_j)$ and the marginal distributions $h(y_j)$ and $g(x_i)$ by

$$g(x_i|y_j) = \frac{f(x_i, y_j)}{h(y_j)} \quad \text{and} \quad h(y_j|x_i) = \frac{f(x_i, y_j)}{g(x_i)} ,$$

= joint density funct.
= marginal dist.

or by

$$f(x_i, y_j) = g(x_i|y_j)h(y_j) = h(y_j|x_i)g(x_i), \quad g(x_i) > 0, h(y_j) > 0. \quad (4\text{–}20)$$

The connection between these newly defined conditional density functions and conditional probabilities is postulated to be

$$p(\mathbf{x} = x_i|\mathbf{y} = y_j) = g(x_i|y_j), \qquad p(\mathbf{y} = y_j|\mathbf{x} = x_i) = h(y_j|x_i). \quad (4\text{–}21)$$

We seek the same definition (4–20) for continuous random variables $\mathbf{x}$, $\mathbf{y}$, with $f(x, y)$, $g(x)$, and $h(y)$ viewed as the respective density functions. But for continuous random variables, probabilities at a point vanish. Definition (4–20) can, however, be heuristically justified as follows. In the spirit of (4–17), we write

$$p(x \leq \mathbf{x} \leq x + dx|y \leq \mathbf{y} \leq y + dy) = \frac{p(x \leq \mathbf{x} \leq x + dx, y \leq \mathbf{y} \leq y + dy)}{p(y \leq \mathbf{y} \leq y + dy)}$$

$$= \frac{f(x, y)\, dx\, dy}{h(y)\, dy} , \qquad (4\text{–}22)$$

which ratio we call $g(x|y)\,dx$, the conditional probability element of $\mathbf{x}$ given y. Similarly,

$$p(y \le \mathbf{y} \le y + dy | x \le \mathbf{x} \le x + dx) \qquad \text{leads to} \qquad h(y|x)\,dy, \tag{4–23}$$

the conditional probability element of $\mathbf{y}$ given x. Cancelling $dx\,dy$, we define the two continuous conditional density functions $g(x|y)$ and $h(y|x)$ by

$$g(x|y) = \frac{f(x, y)}{h(y)}, \qquad h(y|x) = \frac{f(x, y)}{g(x)},$$

or by

$$f(x, y) = g(x|y)h(y) = h(y|x)g(x). \tag{4–24}$$

The connection between these newly defined conditional density functions and conditional probabilities is postulated to be

$$p(\mathbf{x} = x|\mathbf{y} = y) = g(x|y), \qquad p(\mathbf{y} = y|\mathbf{x} = x) = h(y|x), \tag{4–25}$$

where again, as discussed on page 23, density should be thought of in the vicinity of a point.

Beginning with definition (4–24), we may find relations among *distribution* functions corresponding to (4–24) as theorems. Alternatively, conditional probability can be defined in terms of distribution functions as follows:

$$F(x, y) = G(x|y)H(y) = H(y|x)G(x), \tag{4–26}$$

and our present definitions (4–24) can be obtained as *theorems.*

For economy in writing, we now consider continuous $\mathbf{x}$, $\mathbf{y}$ only. The functions $g(x|y)$ and $h(y|x)$ are proper density functions; for example, $g(x|y)$ is the ratio of two density functions and is therefore nonnegative. Moreover,

$$\int_{\infty}^{\infty} g(x|y)\,dx = \int_{\infty}^{\infty} \frac{f(x, y)}{h(y)}\,dx = \frac{1}{h(y)} \int_{\infty}^{\infty} f(x, y)\,dx = \frac{h(y)}{h(y)} = 1.$$

Such results as the following are immediate consequences of (4–24) and (4–25):

$$p(a < \mathbf{x} < b|y) = \int_a^b g(x|y)\,dx = \frac{\int_a^b f(x, y)\,dx}{h(y)},$$

$$p(c < \mathbf{y} < d|x) = \int_c^d h(y|x)\,dy = \frac{\int_c^d f(x, y)\,dy}{g(x)}. \tag{4–27}$$

Using (4–24), our definition (4–14) of **x**, **y** independent leads, for **x**, **y** independent, to

$$g(x|y) = \frac{f(x, y)}{h(y)} = \frac{g(x)h(y)}{h(y)} = g(x),$$

and similarly for **y**. In fact

$$g(x|y) = g(x) \qquad \text{and} \qquad h(y|x) = h(y) \tag{4–28}$$

are essentially equivalent and alternative definitions of independence to

$$f(x, y) = g(x)h(y),$$

just as

$$p(A|B) = p(A) \qquad \text{and} \qquad p(B|A) = p(B)$$

are essentially equivalent and alternative definitions of independence to

$$p(AB) = p(A)p(B).$$

Collateral Reading

H. D. Brunk, *An Introduction to Mathematical Statistics.* Boston, Ginn and Co., 1960.

W. Feller, *An Introduction to Probability Theory and Its Applications.* 2nd ed. New York, John Wiley and Sons, 1957.

R. V. Hogg and A. T. Craig, *Introduction to Mathematical Statistics.* New York, Macmillan Co., 1959.

S. S. Wilks, *Mathematical Statistics.* Princeton, Princeton University Press, 1943.

Problems

4–1. From axioms (a) and (b), show that $F(x, y)$ is monotonic nondecreasing in each variable separately.

4–2. Given 7 units, 5 of them good, 2 bad; a sample of two units is drawn without replacement. Determine all possible values of the joint density function, the marginal distributions, and the conditional density function for the random variable pair $\mathbf{x}$, $\mathbf{y}$, where $\mathbf{x}$ is the number of good units in the first unit drawn and $\mathbf{y}$ is the number of good units in the second unit drawn.

4–3. If the random variables $\mathbf{x}$ and $\mathbf{y}$ are independent, any functions $\mathbf{u} = g(\mathbf{x})$ and $\mathbf{v} = h(\mathbf{y})$ of them are independent (an important theorem).

CHAPTER 5

TWO RANDOM VARIABLES; EXPECTED VALUES

5–1 Definitions and examples. Following the argument of Chapter 3 for one random variable, the mathematical expectation or expected value or mean $E[\phi(\mathbf{x}, \mathbf{y})]$ of an arbitrary function $\phi(\mathbf{x}, \mathbf{y})$ of two continuous random variables $\mathbf{x}$, $\mathbf{y}$ is, by definition,

$$E[\phi(\mathbf{x}, \mathbf{y})] = \int_{-\infty}^{\infty} \int_{-\infty}^{\infty} \phi(x, y) f(x, y)\, dx\, dy, \tag{5–1}$$

where $f(x, y)$ is the joint density function of $\mathbf{x}$, $\mathbf{y}$. For two discrete random variables, the corresponding definition is

$$E[\phi(\mathbf{x}, \mathbf{y})] = \sum_i \sum_j \phi(x_i, y_j) f(x_i, y_j) \tag{5–2}$$

summed over all values of i and j for which $\mathbf{x}$ and $\mathbf{y}$ exist. For brevity, we confine the discussion to continuous random variables. As earlier with one random variable, certain expectations may not exist; that is, the integrals (5–1) or the sums (5–2) defining them may not converge absolutely.

As with one random variable, the mathematical expectations of certain functions $\phi(\mathbf{x}, \mathbf{y})$ identify characteristics, high in information and interest, of the joint probability distribution of two random variables; this is the motivation of the present chapter. Such expectations include the following. Setting

$$\phi(\mathbf{x}, \mathbf{y}) = \mathbf{x}^l \mathbf{y}^m$$

gives

$$E[\mathbf{x}^l \mathbf{y}^m] = \int_{-\infty}^{\infty} \int_{-\infty}^{\infty} x^l y^m f(x, y)\, dx\, dy, \tag{5–3}$$

written λ_{lm}, the lmth moment of $\mathbf{x}$, $\mathbf{y}$ about the origin. Setting

$$\phi(\mathbf{x}, \mathbf{y}) = [\mathbf{x} - \lambda_{10}]^l [\mathbf{y} - \lambda_{01}]^m$$

gives

$$E[\mathbf{x} - \lambda_{10}]^l [\mathbf{y} - \lambda_{01}]^m = \int_{-\infty}^{\infty} \int_{-\infty}^{\infty} [x - \lambda_{10}]^l [y - \lambda_{01}]^m f(x, y)\, dx\, dy, \tag{5–4}$$

written μ_{lm}, the lmth moment of $\mathbf{x}$, $\mathbf{y}$ about the respective moments λ_{10} and λ_{01}.

A relationship can be readily found connecting (5–3) and (5–4); it is a simple generalization of (3–9)

$$\mu_2 = \lambda_2 - \lambda_1^2,$$

connecting, for a single random variable, the second moment about the mean and the first and second moments about the origin.

Particular cases of λ_{lm} and μ_{lm} often met in statistical theory and the expressions to which they lead are shown below:

$l = 0, m = 0$ $\quad \lambda_{00} = 1,$

$l = 1, m = 0$ $\quad \lambda_{10} = E(\mathbf{x})$, the mean of $\mathbf{x}$,

$l = 0, m = 1$ $\quad \lambda_{01} = E(\mathbf{y})$, the mean of $\mathbf{y}$,

$l = 1, m = 1$ $\quad \mu_{11} = E[\mathbf{x} - E(\mathbf{x})][\mathbf{y} - E(\mathbf{y})]$, the covariance of $\mathbf{x}$ and $\mathbf{y}$, (5–5)

$l = 2, m = 0$ $\quad \mu_{20} = \sigma^2(\mathbf{x})$, the variance of $\mathbf{x}$,

$l = 0, m = 2$ $\quad \mu_{02} = \sigma^2(\mathbf{y})$, the variance of $\mathbf{y}$.

Note that, just as we determined the probability behavior of either one of two random variables from the joint probability distribution of both, so certain parameters of the probability distributions of either one of the two random variables $\mathbf{x}$, $\mathbf{y}$ are particular cases, obtained by integrating over the unwanted variable, of two-dimensional expectations. We prove one such result:

$$\lambda_{10} = \int_{-\infty}^{\infty}\int_{-\infty}^{\infty} xf(x, y)\,dx\,dy = \int_{-\infty}^{\infty} [x\,dx \int_{-\infty}^{\infty} f(x, y)\,dy]$$
$$= \int_{-\infty}^{\infty} xg(x)\,dx = E(\mathbf{x}).$$

Two further expectations, important in statistical theory, follow. Setting

$$\phi(\mathbf{x}, \mathbf{y}) = e^{\alpha\mathbf{x}+\beta\mathbf{y}}$$

gives

$$E[e^{\alpha\mathbf{x}+\beta\mathbf{y}}] = \int_{-\infty}^{\infty}\int_{-\infty}^{\infty} e^{\alpha x+\beta y} f(x, y)\,dx\,dy, \qquad (5\text{–}6)$$

written $M_{\mathbf{x},\mathbf{y}}(\alpha, \beta)$, the moment generating function of $\mathbf{x}$, $\mathbf{y}$, with generators α and β. Setting

$$\phi(\mathbf{x}, \mathbf{y}) = e^{\beta G(\mathbf{x},\mathbf{y})}$$

gives

$$E[e^{\beta G(\mathbf{x},\mathbf{y})}] = \int_{-\infty}^{\infty}\int_{-\infty}^{\infty} e^{\beta G(x,y)} f(x, y)\,dx\,dy, \qquad (5\text{–}7)$$

written $M_{G(\mathbf{x},\mathbf{y})}(\beta)$, the moment generating function of an arbitrary function $G(\mathbf{x}, \mathbf{y})$ with β as generator.

If, similar to the argument for one random variable, we expand $e^{\alpha x+\beta y}$ in a Taylor series, its expectation reveals all the moments of **x**, **y** about the origin. If, as an alternative to so noting the moments, we differentiate partially this expectation, we find

$$\left.\frac{\partial^{l+m}M(\alpha, \beta)}{\partial\alpha^l\partial\beta^m}\right|_{\alpha=\beta=0} = \lambda_{lm}; \tag{5–8}$$

the lmth moment of **x**, **y** about the origin is obtained. This is the motivation for introducing (5–6) and (5–7). The student should carry out the expansion and differentiation in detail. For example,

$$e^{\alpha x+\beta y} = 1 + \frac{\alpha x + \beta y}{1!} + \frac{(\alpha x + \beta y)^2}{2!} + \cdots,$$

$$E(e^{\alpha \mathbf{x}+\beta \mathbf{y}}) = 1 + \frac{\alpha}{1}\lambda_1(x) + \frac{\beta}{1}\lambda_1(y) + \frac{\alpha^2}{2}\lambda_2(x) + \frac{\beta^2}{2}\lambda_2(y) + \alpha\beta\lambda_{11}(x, y) + \cdots.$$

5–2 Expected values of certain functions of random variables. Later we shall consider the probability behavior of certain simple functions of two random variables, in particular, $\mathbf{x} + \mathbf{y}$ and $\mathbf{xy}$. Here we consider the more limited problem of the expected values of such functions of two random variables.

The following elementary but important theorems follow directly from the definitions of expectation (5–1) and (5–2). For a and b constant,

$$E(a\mathbf{x} + b\mathbf{y}) = aE(\mathbf{x}) + bE(\mathbf{y}), \tag{5–9}$$

$$\begin{aligned} E[(a\mathbf{x} + b\mathbf{y}) - E(a\mathbf{x} + b\mathbf{y})]^2 &= \sigma^2(a\mathbf{x} + b\mathbf{y}) \\ &= a^2\sigma^2(\mathbf{x}) + b^2\sigma^2(\mathbf{y}) + 2ab\rho(\mathbf{x}, \mathbf{y})\sigma(\mathbf{x})\sigma(\mathbf{y}), \end{aligned} \tag{5–10}$$

where

$$\rho(\mathbf{x}, \mathbf{y}) = \frac{\text{cov}(\mathbf{x}, \mathbf{y})}{\sigma(\mathbf{x})\sigma(\mathbf{y})} \tag{5–11}$$

is a function of **x** and **y** which will be considered in detail in the next section. Note that (5–9) and (5–10) hold for all $f(x, y)$.

We prove (5–10) in detail:

$$
\begin{aligned}
E[(a\mathbf{x} + b\mathbf{y}) &- E(a\mathbf{x} + b\mathbf{y})]^2 \\
&= E[a\mathbf{x} + b\mathbf{y} - aE(\mathbf{x}) - bE(\mathbf{y})]^2 \\
&= E\{[a\mathbf{x} - aE(\mathbf{x})] + [b\mathbf{y} - bE(\mathbf{y})]\}^2 \\
&= E\{a[\mathbf{x} - E(\mathbf{x})] + b[\mathbf{y} - E(\mathbf{y})]\}^2 \\
&= a^2E[\mathbf{x} - E(\mathbf{x})]^2 + 2abE[\mathbf{x} - E(\mathbf{x})][\mathbf{y} - E(\mathbf{y})] + b^2E[\mathbf{y} - E(\mathbf{y})]^2 \\
&= a^2\sigma^2(\mathbf{x}) + 2ab\rho(\mathbf{x}, \mathbf{y})\sigma(\mathbf{x})\sigma(\mathbf{y}) + b^2\sigma^2(\mathbf{y}).
\end{aligned}
$$

Even this modest result has utility in application, for the additive function $\mathbf{x} + \mathbf{y}$ is of interest in many areas of science. For example, in the theory of assembly of mass-produced parts, the dimension $x + y$ is often of primary interest. In particular, in *selective* assembly (in which parts with high x are assembled with parts of low y, etc., in order to bring the assembly within engineering tolerances), (5–10) is particularly relevant, with the correlation coefficient $\rho(\mathbf{x}, \mathbf{y})$ playing the key role in describing the selective assembly process.

Note that in the proofs of such theorems as (5–10), or even of the simple but important result (5–9), terms in the integrals (or for discrete random variables, sums) are rearranged. If the integrals or sums defining the expectations are absolutely convergent, the terms in such functions as sums or products may be freely rearranged.

The addition theorem (5–9) involving $\phi(\mathbf{x}, \mathbf{y}) = a\mathbf{x} + b\mathbf{y}$ is a general theorem; in the proof, particular forms of the joint probability distribution of $\mathbf{x}$ and $\mathbf{y}$ need not be specified. For almost equally important multiplicative functions like $\phi(\mathbf{x}, \mathbf{y}) = \mathbf{xy}$ (e.g., the area of a rectangular sheet of steel), the situation is different. Though we will be involved with $\mathbf{xy}$, we cannot generally express $E(\mathbf{xy})$ in terms of (often the more readily determined) $E(\mathbf{x})$ and $E(\mathbf{y})$; the expectation $E(\mathbf{xy})$ can be evaluated in terms of $E(\mathbf{x})$ and $E(\mathbf{y})$ only for a *particular* probability distribution of $\mathbf{x}$, $\mathbf{y}$, or, what amounts to the same restriction, only when certain conditions (such as independence or zero covariance) are placed on the random variables $\mathbf{x}$, $\mathbf{y}$. As with one random variable, such results are, of course, less general; they are specific to the joint probability distribution used or to the conditions imposed. Similar reservations extend to such functions as $\mathbf{x}/\mathbf{y}$ which arise often in practical application. We have, for $\mathbf{x}$, $\mathbf{y}$ independent,

$$E\left(\frac{\mathbf{x}}{\mathbf{y}}\right) = E(\mathbf{x})E\left(\frac{1}{\mathbf{y}}\right),$$

which, however, even for $\mathbf{x}$, $\mathbf{y}$ independent is not equal to $E(\mathbf{x})/E(\mathbf{y})$. Without any conditions on $\mathbf{x}$, $\mathbf{y}$, $E(\mathbf{x}/\mathbf{y})$ is even more intractable.

5-3 Covariance and correlation. The mathematical expectation of the products of random variables is of much interest in statistical theory and statistical inference. The motivation underlying much of this interest is the fact that the expectations of the products of certain random variables throw light on the relationship between them. We have already introduced in (5-5) one such expectation, μ_{11}, the covariance of $\mathbf{x}$ and $\mathbf{y}$; the motivation was to measure relationship between $\mathbf{x}$ and $\mathbf{y}$. From the definition of μ_{11}, it is evident that if large values of the random variable $\mathbf{x}$ are found paired generally with large values of the random variable $\mathbf{y}$ in the function $F(x, y)$ or $f(x, y)$, and if small values of $\mathbf{x}$ are found paired generally with small values of $\mathbf{y}$, μ_{11} will be positive. Human height and weight, regarded as random variables, is an example. Also, if large values of $\mathbf{x}$ are found paired generally with small values of $\mathbf{y}$ in $F(x, y)$ or $f(x, y)$, and if small values of $\mathbf{x}$ are found paired generally with large values of $\mathbf{y}$, μ_{11} will be negative. Market price and quantity bought is an example. Finally, if when values of $\mathbf{x}$ are large, *some* of the paired values of $\mathbf{y}$ in $F(x, y)$ or $f(x, y)$ are large and *some* are small, and similarly if for small values of $\mathbf{x}$, *some* of the paired values of $\mathbf{y}$ are large and *some* are small μ_{11} will be near 0. Human height and human intelligence is a (tentative) example. These situations are roughly described in Fig. 5-1.

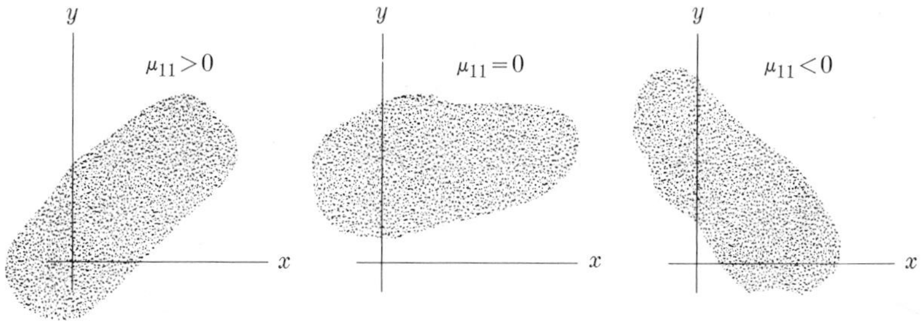

FIG. 5-1. Positive, zero, and negative covariance between $\mathbf{x}$ and $\mathbf{y}$.

The concept of correlation was introduced by Galton in 1885. The crude measure of correlation μ_{11}, also due to Galton, is merely a beginning; it will be refined immediately, and to some extent replaced by other measures in Chapter 6.

The immediate (and important) refinement of μ_{11} is the *correlation coefficient*, a dimensionless measure already introduced in (5-11) and now written in two equivalent forms (it is due to K. Pearson, 1895):

$$\rho(\mathbf{x}, \mathbf{y}) = \frac{\text{cov}\,(\mathbf{x}, \mathbf{y})}{\sigma(\mathbf{x})\sigma(\mathbf{y})} = \frac{\mu_{11}}{(+)\sqrt{\mu_{20}\mu_{02}}}. \tag{5-12}$$

One effect of the denominator of $\rho(\mathbf{x}, \mathbf{y})$ is to eliminate, as μ_{11} itself evidently does not, any dependence of the magnitude of the measure of relationship on the units in which **x** and **y** happen to be expressed; any change in the units of **x** and **y** affect numerator and denominator of (5–12) equally and cancel out. The correlation coefficient $\rho(\mathbf{x}, \mathbf{y})$ is readily seen to be the covariance between the *standardized* random variables,

$$\mathbf{u} = \frac{\mathbf{x} - E(\mathbf{x})}{\sigma(\mathbf{x})}, \qquad \mathbf{v} = \frac{\mathbf{y} - E(\mathbf{y})}{\sigma(\mathbf{y})};$$

that is,

$$\rho(\mathbf{x}, \mathbf{y}) = \text{cov}(\mathbf{u}, \mathbf{v}) = E\left[\frac{\mathbf{x} - E(\mathbf{x})}{\sigma(\mathbf{x})}\right]\left[\frac{\mathbf{y} - E(\mathbf{y})}{\sigma(\mathbf{y})}\right].$$

Note that division of μ_{11} by $\sigma(\mathbf{x})$ and $\sigma(\mathbf{y})$ has no effect on the *degree* of correlation between **x** and **y**, for $\sigma(\mathbf{x})$ and $\sigma(\mathbf{y})$ are functions of **x** and **y** *separately*. Division merely affects the size of the measure of correlation; as we shall soon see, it reduces the range of the measure of correlation from an unwieldy $\pm\infty$ (for μ_{11}) to a convenient ± 1 [for $\rho(\mathbf{x}, \mathbf{y})$].

As well as μ_{11}, $\rho(\mathbf{x}, \mathbf{y})$ will turn out to be relevant only to *linear* relationships, if any, between **x** and **y**, and all of our discussion of the conditions under which μ_{11} tends to be positive, negative, or zero must be restricted to such a relationship. For example, $\rho = 0 = \mu_{11}$ is compatible with *functional* or *perfect nonlinear* relationship between **x** and **y**. Consider any standardized random variable **x** which is symmetrical about its mean. We have $E(\mathbf{x}) = 0$ and $\sigma^2(\mathbf{x}) = 1$. Let **x** and **y** be *functionally* related by $y = x^2$. The numerator of ρ is the covariance

$$\mu_{11} = E[\mathbf{x} - E(\mathbf{x})][\mathbf{y} - E(\mathbf{y})].$$

But

$$E(\mathbf{y}) = E(\mathbf{x}^2) = \sigma^2(\mathbf{x}) = 1,$$

so

$$\mu_{11} = E(\mathbf{x} - 0)(\mathbf{x}^2 - 1) = E(\mathbf{x}^3) - E(\mathbf{x}) = 0,$$

since the expectation of any odd μ-moment of a random variable which is symmetrical about its mean is 0. Although **x** and **y** are functionally related (quadratically), the correlation coefficient ρ is 0.

The assertion of independence of **x** and **y** or lack of linear correlation between **x** and **y**, that is $\rho(\mathbf{x}, \mathbf{y}) = 0$, often have common motivation [this is particularly evident if independence is defined by $g(x|y) = g(x)$, $h(y|x) = h(y)$] but they are not equivalent. From $\rho = 0$ one cannot generally establish independence; the single example above is evidence enough. But if **x** and **y** are independent, $\rho = 0$ follows, as the student is asked to show in problem 5–4. For a good discussion of this relationship see Lancaster [109].

5–4 Properties of the correlation coefficient. We now explore certain properties of ρ; these are more readily found if we consider the standardized random variables

$$\mathbf{u} = \frac{\mathbf{x} - E(\mathbf{x})}{\sigma(\mathbf{x})}, \qquad \mathbf{v} = \frac{\mathbf{y} - E(\mathbf{y})}{\sigma(\mathbf{y})}. \tag{5–13}$$

From (5–10) we have

$$\sigma^2(\mathbf{u} + \mathbf{v}) = \sigma^2(\mathbf{u}) + 2\rho(\mathbf{u}, \mathbf{v})\sigma(\mathbf{u})\sigma(\mathbf{v}) + \sigma^2(\mathbf{v}),$$

$$\sigma^2(\mathbf{u} - \mathbf{v}) = \sigma^2(\mathbf{u}) - 2\rho(\mathbf{u}, \mathbf{v})\sigma(\mathbf{u})\sigma(\mathbf{v}) + \sigma^2(\mathbf{v}).$$

But from problem 3–3

$$\sigma^2(\mathbf{u}) = \sigma^2(\mathbf{v}) = 1,$$

so

$$\sigma^2(\mathbf{u} \pm \mathbf{v}) = 2[1 \pm \rho(\mathbf{u}, \mathbf{v})]. \tag{5–14}$$

Since any variance is nonnegative, it follows from

$$[1 + \rho(\mathbf{u}, \mathbf{v})] \geq 0, \qquad [1 - \rho(\mathbf{u}, \mathbf{v})] \geq 0,$$

that

$$-1 \leq \rho(\mathbf{u}, \mathbf{v}) \leq 1, \tag{5–15}$$

as was asserted in Section 5–3. The lower limit follows from the left and the upper limit from the right inequality directly above. Finally, making use of the answer to problem 5–5, we have

$$-1 \leq \rho(\mathbf{x}, \mathbf{y}) \leq 1.$$

We now show that when $\rho(\mathbf{x}, \mathbf{y})$ assumes its extreme values $+1$ and -1, the relationship between $\mathbf{x}$ and $\mathbf{y}$ is perfect linear *positive*, and perfect linear *negative*, respectively; all values of the random variable pair $\mathbf{x}$, $\mathbf{y}$ lie on a straight line of positive or of negative slope.

Returning to (5–14), we note for $\rho(\mathbf{u}, \mathbf{v}) = +1$,

$$\sigma^2(\mathbf{u} - \mathbf{v}) = 0;$$

the random variable $\mathbf{u} - \mathbf{v}$ is a constant, with all probability concentrated at the constant. That is, returning by way of (5–13) to the original random variables $\mathbf{x}$ and $\mathbf{y}$, we have

$$\frac{\mathbf{x} - E(\mathbf{x})}{\sigma(\mathbf{x})} - \frac{\mathbf{y} - E(\mathbf{y})}{\sigma(\mathbf{y})} = \text{constant}, \tag{5–16}$$

an equation of the form $\mathbf{y} = a + b\mathbf{x}$, b positive, a positive or negative, perfect positive linear relationship between $\mathbf{x}$ and $\mathbf{y}$ [every point (x, y) on the line]. For $\rho(\mathbf{u}, \mathbf{v}) = -1$, again from (5–14),

$$\sigma^2(\mathbf{u} + \mathbf{v}) = 0;$$

the random variable $\mathbf{u} + \mathbf{v}$ is a constant with all probability concentrated at the constant. In terms of $\mathbf{x}$, $\mathbf{y}$,

$$\frac{\mathbf{x} - E(\mathbf{x})}{\sigma(\mathbf{x})} + \frac{\mathbf{y} - E(\mathbf{y})}{\sigma(\mathbf{y})} = \text{constant}, \tag{5–17}$$

an equation of the form $\mathbf{y} = a - b\mathbf{x}$, b positive, a positive or negative, perfect negative linear relationship between $\mathbf{x}$ and $\mathbf{y}$.

Moreover, the converse holds. Suppose the relationship between $\mathbf{x}$ and $\mathbf{y}$ is perfect linear, that is, every point (x, y) lies on the line

$$\mathbf{y} = a \pm b\mathbf{x}. \tag{5–18}$$

We have

$$\begin{aligned} \mu_{11} &= E[\mathbf{x} - E(\mathbf{x})][\mathbf{y} - E(\mathbf{y})] \\ &= E[\mathbf{x} - E(\mathbf{x})][(a \pm b\mathbf{x}) - (a \pm bE(\mathbf{x}))] \\ &= \pm bE[\mathbf{x} - E(\mathbf{x})]^2 = \pm b\sigma^2(\mathbf{x}), \end{aligned}$$

and

$$\rho(\mathbf{x}, \mathbf{y}) = \frac{\mu_{11}}{\sigma(\mathbf{x})\sigma(\mathbf{y})} = \frac{\pm b\sigma^2(\mathbf{x})}{b\sigma^2(\mathbf{x})} = \pm 1. \tag{5–19}$$

This result will reappear in the following chapter, in our discussion of regression.

Collateral Reading

H. D. Brunk, *An Introduction to Mathematical Statistics.* Boston, Ginn and Co., 1960.

W. Feller, *An Introduction to Probability Theory and Its Applications.* 2nd ed., New York, John Wiley and Sons, 1957.

H. O. Lancaster, "Zero correlation and independence," *The Australian Jour. of Stat.*, Vol. 1 (1959), pp. 53–56.

Problems

5–1. Show that the moment generating function of the sum of two independent random variables is the product of the moment generating functions of each.

5–2. For $\mathbf{u}$, $\mathbf{v}$ in $\mathbf{u} = a\mathbf{x} + b\mathbf{y}$ and $\mathbf{v} = c\mathbf{x} + d\mathbf{y}$ to be uncorrelated, what conditions must be imposed on a, b, c, d?

5–3. From the fact that $E(a\mathbf{x} + \mathbf{y})^2 \geq 0$, prove the (famous) Schwarz inequality (met often in mathematical analysis, particularly in the theory of orthogonal functions),

$$E^2(\mathbf{xy}) \leq E(\mathbf{x}^2)E(\mathbf{y}^2)$$

for random variables $\mathbf{x}$, $\mathbf{y}$ with finite variances. From this result, show that $-1 \leq \rho \leq 1$.

5–4. If $\mathbf{x}$ and $\mathbf{y}$ are independent, $\rho(\mathbf{x}, \mathbf{y}) = 0$. The converse is not necessarily true.

5–5. For

$$\mathbf{u} = \frac{\mathbf{x} - E(\mathbf{x})}{\sigma(\mathbf{x})}, \qquad \mathbf{v} = \frac{\mathbf{y} - E(\mathbf{y})}{\sigma(\mathbf{y})},$$

show that $\rho(\mathbf{u}, \mathbf{v}) = \rho(\mathbf{x}, \mathbf{y})$; the correlation coefficient is undisturbed by a linear change of random variables. The result also holds for the more general linear change of variable, $\mathbf{u} = a + b\mathbf{x}$, $\mathbf{v} = c + d\mathbf{y}$.

5–6. If $\mathbf{x}$, $\mathbf{y}$ are independent random variables,

$$\sigma^2(\mathbf{xy}) = \sigma^2(\mathbf{x})\sigma^2(\mathbf{y}) + \lambda^2(\mathbf{x})\sigma^2(\mathbf{y}) + \lambda^2(\mathbf{y})\sigma^2(\mathbf{x}).$$

CHAPTER 6

TWO RANDOM VARIABLES; CONDITIONAL EXPECTATION AND PREDICTION

6–1 Conditional expectation. In Sections 5–3 and 5–4, we began a discussion of the relationship between two random variables $\mathbf{x}$ and $\mathbf{y}$. To measure their *linear* relationship, we introduced μ_{11}, the covariance of $\mathbf{x}$ and $\mathbf{y}$, and the more refined linear measure ρ, the correlation coefficient of $\mathbf{x}$ and $\mathbf{y}$. Let us now look further into the relationship between two random variables $\mathbf{x}$ and $\mathbf{y}$ with distribution function $F(x, y)$ and density function $f(x, y)$. We restrict ourselves to $\mathbf{x}$, $\mathbf{y}$ continuous.

The central quantity will be the conditional distribution function $H(y|x)$ or, better in this elementary exposition, the conditional density function $h(y|x)$. Both completely describe the probability behavior of $\mathbf{y}$ at a fixed value x, for every value x. Similarly, $G(x|y)$ or, better $g(x|y)$, describes the probability behavior of $\mathbf{x}$ at a fixed value y for every value y. As most of the arguments of this chapter are symmetrical, *though inferences from them may not be*, we often confine ourselves to $\mathbf{y}|x$, but the student should at all times be aware of the parallel mathematical structure centering on $\mathbf{x}|y$.

At a fixed value x, $\mathbf{y}$ is a random variable with conditional density function $h(y|x)$. At the fixed value x, this random variable $\mathbf{y}$ has mean and variance which we write $E(\mathbf{y}|x)$ and $V(\mathbf{y}|x)$, respectively. They are naturally defined by

$$E(\mathbf{y}|x) = \int_{-\infty}^{\infty} y h(y|x)\, dy, \quad V(\mathbf{y}|x) = \int_{-\infty}^{\infty} [y - E(\mathbf{y}|x)]^2 h(y|x)\, dy. \tag{6–1}$$

We also have

$$E(\mathbf{x}|y) = \int_{-\infty}^{\infty} x g(x|y)\, dx, \quad V(\mathbf{x}|y) = \int_{-\infty}^{\infty} [x - E(\mathbf{x}|y)]^2 g(x|y)\, dx.$$

The quantities in (6–1), along with $h(y|x)$, are illustrated in Fig. 6–1. In the illustrations, $E(\mathbf{y}|x)$ increases, then decreases as x moves from x' to x'' to x'''; $V(y|x)$ decreases as x moves from x' to x'' to x'''. The unconditional density function $h(y)$, the unconditional mean $E(\mathbf{y})$, and the unconditional variance $V(\mathbf{y})$ are illustrated at the left of Fig. 6–1.

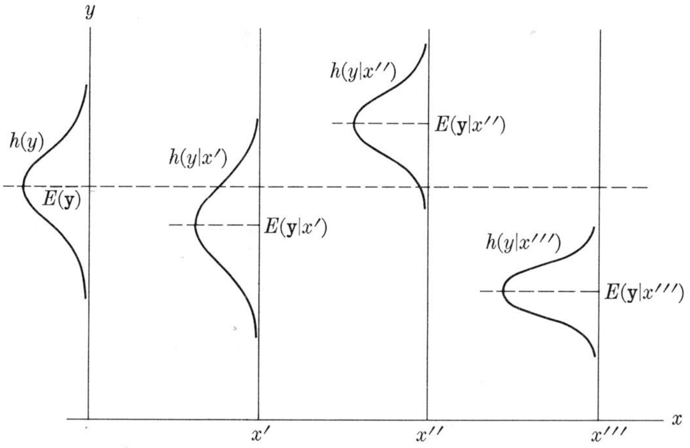

FIG. 6–1. Conditional and unconditional density functions, means, and variances.

While $E(\mathbf{y}|x)$ and $V(\mathbf{y}|x)$ are, for fixed x, parameters of the conditional distribution of y at the fixed x, they are, for x permitted to vary, properly regarded as random variables, for they become functions of the random variable $\mathbf{x}$. The mean and variance may then be written $E(\mathbf{y}|\mathbf{x})$ and $V(\mathbf{y}|\mathbf{x})$; both are functions of the random variable $\mathbf{x}$ and are, therefore (though we do not show this formally), themselves random variables. Thus $E(\mathbf{y}|\mathbf{x})$, regarded as a function of $\mathbf{x}$, is a random variable with its own probability distribution, with mean and variance

$$E[E(\mathbf{y}|\mathbf{x})], \qquad V[E(\mathbf{y}|\mathbf{x})]. \tag{6–2}$$

The same is true of $V(\mathbf{y}|\mathbf{x})$ regarded as a function of $\mathbf{x}$; $V(\mathbf{y}|\mathbf{x})$ becomes a random variable with its own probability distribution, with mean and variance

$$E[V(\mathbf{y}|\mathbf{x})], \qquad V[V(\mathbf{y}|\mathbf{x})]. \tag{6–3}$$

The inner expectations (integrals) of (6–2) and (6–3) are over $\mathbf{y}$ at fixed x; the outer expectations (integrals) are over $\mathbf{x}$. There are, of course, four parallel expectations for $\mathbf{x}|\mathbf{y}$.

The random variable $E(\mathbf{y}|\mathbf{x})$, which describes the *mean* behavior of $\mathbf{y}$ as a function of $\mathbf{x}$, will be of basic interest in our discussion of the relationship between $\mathbf{x}$ and $\mathbf{y}$, for, given a certain principle of goodness of prediction, $E(\mathbf{y}|\mathbf{x})$ turns out to be a "best" predictor of $\mathbf{y}$. The random variable $V(\mathbf{y}|\mathbf{x})$ which describes the *variability* of $\mathbf{y}$ as a function of $\mathbf{x}$ is also of interest and will appear in the discussion.

We now investigate (6–2), the mean and variance of $E(\mathbf{y}|\mathbf{x})$. First, we find the mean of $E(\mathbf{y}|\mathbf{x})$, with subscripts indicating the variable of integration

$$E_x[E_y(\mathbf{y}|\mathbf{x})] = E_x\int_{-\infty}^{\infty} yh(y|x)\,dy = \int_{-\infty}^{\infty}\left[\int_{-\infty}^{\infty} yh(y|x)\,dy\right]g(x)\,dx$$
$$= \int_{-\infty}^{\infty}\int_{-\infty}^{\infty} yf(x, y)\,dx\,dy = E(\mathbf{y}). \tag{6–4}$$

The mean over all x of the conditional mean of $\mathbf{y}$ at fixed x is the unconditional mean of $\mathbf{y}$.

Note $E(\mathbf{y})$ in Fig. 6–1 as the weighted mean of

$$E(\mathbf{y}|x'), \qquad E(\mathbf{y}|x''), \qquad E(\mathbf{y}|x'''),$$

with these conditional means (naturally) weighted by the probabilities of occurrence of x', x'', x''', respectively, as indicated in the development of (6–4).

Second, we find the variance of $E(\mathbf{y}|\mathbf{x})$, later to be considered as a measure of the goodness of the "best" predictor $E(\mathbf{y}|\mathbf{x})$. By definition, and using (6–4), we have

$$V[E(\mathbf{y}|\mathbf{x})] = E_x[E_y(\mathbf{y}|\mathbf{x}) - E(\mathbf{y})]^2, \tag{6–5}$$

again integrating over x, y by first integrating over y with x fixed, then over x.

We will evaluate $V[E(\mathbf{y}|\mathbf{x})]$ in (6–5) by showing that the unconditional variance $V(\mathbf{y})$ is the sum of two variances, one of which is $V[E(\mathbf{y}|\mathbf{x})]$, thereby finding the desired expression for $V[E(\mathbf{y}|\mathbf{x})]$ in terms of $V(\mathbf{y})$:

$$V(\mathbf{y}) = E[\mathbf{y} - E(\mathbf{y})]^2 = E_x[E_y\{[\mathbf{y} - E(\mathbf{y})]^2|\mathbf{x}\}]. \tag{6–6}$$

But the inner expectation of the right-hand term of (6–6) is

$$E_y\{[\mathbf{y} - E(\mathbf{y})]^2|x\} = E_y\{[\overbrace{\mathbf{y} - E(\mathbf{y}|x)} + \overbrace{E_y(\mathbf{y}|x) - E(\mathbf{y})}]^2|x\}.$$

Since $E_y(\mathbf{y}|x) - E(\mathbf{y})$ is a function only of x, the right-hand side becomes

$$V(\mathbf{y}|x) + [E_y(\mathbf{y}|x) - E(\mathbf{y})]^2, \tag{6–7}$$

the cross-product term vanishing, for $E[\mathbf{y} - E(\mathbf{y}|x] = 0$. Finally, taking the outer expectation of (6–7) over x, we have

$$V(\mathbf{y}) = EV(\mathbf{y}|\mathbf{x}) + V[E(\mathbf{y}|\mathbf{x})]. \tag{6–8}$$

The unconditional variance of $\mathbf{y}$ is the sum of two (entirely different) variances, the first being the mean of the conditional variance of $\mathbf{y}$ over all

x, and the second the variance of the conditional mean of **y** over all x. The first is a measure of the average variability of **y** about its conditional mean as **x** moves over its range of values; the second, which is the quantity sought in (6–5), is a measure of the average variability of the *mean* of **y** as **x** moves over its range of values. The latter variability is sometimes said to be explained by or associated with changes in $E(\mathbf{y}|x)$; the former variability is said to be unexplained by or not associated with such changes. Both should be identified in Fig. 6–1. The second describes the average of the evident differences in the conditional means

$$E(\mathbf{y}|x') \text{ (medium)}, \qquad E(\mathbf{y}|x'') \text{ (high)}, \qquad E(\mathbf{y}|x''') \text{ (low)}, \ldots,$$

while the first describes the average of the variability of **y** about each of these means (large, medium, small), variability obviously unaffected by the differences among the conditional means.

We have not obtained numerical values for the means and the variances in (6–2); we have only expressed (6–2) in terms of $E(\mathbf{y})$ and $V(\mathbf{y})$, the mean and the variance of **y**, and $EV(\mathbf{y}|\mathbf{x})$. For the arguments leading up to (6–4) and (6–8) to yield numerical values, the form of $f(x, y)$ would have to be known and the integration in (6–4) and (6–8) carried out in detail. Nonetheless, (6–4), and particularly (6–8), are valuable results as they stand. The latter describes an *orthogonal* partition or breakdown of $V(\mathbf{y})$ (a square which can be expressed as sums of squares, with no cross-product terms). Such partitions play central roles in the statistical theory underlying modern experimental designs, and they lend strong support to the use of variance (rather than say standard deviation) as a measure of variability.

Though our interest has been in obtaining expressions for the mean and variance of $E(\mathbf{y}|\mathbf{x})$, the expressions (6–4) and (6–8) are often used to obtain the unconditional parameters $E(\mathbf{y})$ and $V(\mathbf{y})$ in two additive stages, via the random variable **x**. In such an operation, **x** may itself have important physical meaning or **x** may be a dummy variable introduced merely to permit use of (6–4) and (6–8).

Also note that as all variances are nonnegative, the variance $V[E(\mathbf{y}|\mathbf{x})]$, which we set out to find in (6–5), is seen from (6–8) never to be greater than $V(\mathbf{y})$, the unconditional variance of **y**. The same is true of $EV(\mathbf{y}|\mathbf{x})$, a quantity of particular interest when $V(\mathbf{y}|\mathbf{x})$ is constant over x.

6–2 Linear least-squares prediction. To motivate the discussion of Section 6–1, we must now show the peculiar importance of $E(\mathbf{y}|\mathbf{x})$ in studying relationships between **x** and **y**, or better, in predicting values of **y** from knowledge of **x**. With this as the objective, we begin with a special problem in which $E(\mathbf{y}|\mathbf{x})$, as such, does not appear.

Let us predict values of the random variable $\mathbf{y}$ from values of the random variable $\mathbf{x}$ by constructing a *linear* function

$$\mathbf{y}' = \alpha + \beta\mathbf{x}, \tag{6–9}$$

where α and β are constants, and $\mathbf{y}'$ is the predicted value of $\mathbf{y}$.

It should not be inferred that the conditional expectation $E(\mathbf{y}|\mathbf{x})$ has significance only in linear relationships between $\mathbf{x}$ and $\mathbf{y}$. We begin with (6–9) for two reasons: first, it permits ready discussion of certain problems encountered in prediction; second, linear prediction functions are frequently met in applied statistics, particularly in experimentation. But a general argument in support of $E(\mathbf{y}|\mathbf{x})$ will be given in Section 6–3.

Let α and β be determined so that the (nonnegative) mean error of prediction

$$E_y(\mathbf{y} - \mathbf{y}')^2 = E_{x,y}(\mathbf{y} - \alpha - \beta\mathbf{x})^2 \tag{6–10}$$

is minimum. This is the famous and reasonable criterion of *least-squares* to which we return at length in Chapter 27 on best linear unbiased estimates. Either by expansion and algebraic manipulation of (6–10) along the line

$$E_{x,y}(\mathbf{y} - \alpha - \beta\mathbf{x})^2 = E_{x,y}\{\mathbf{y} - E(\mathbf{y}) - \beta[\mathbf{x} - E(\mathbf{x})] + (E\mathbf{y} - \alpha - \beta E\mathbf{x})\}^2$$

or simply by differentiation of (6–10) with respect to the (continuous) parameters α and β, we find that the values of α and β which minimize (6–10) are

$$\alpha = E(\mathbf{y}) - \rho\frac{\sigma(\mathbf{y})}{\sigma(\mathbf{x})}E(\mathbf{x}), \qquad \beta = \rho\frac{\sigma(\mathbf{y})}{\sigma(\mathbf{x})}, \tag{6–11}$$

which, inserted in (6–9), yield the least-squares linear prediction function

$$\frac{y' - E(\mathbf{y})}{\sigma(\mathbf{y})} = \rho\,\frac{x - E(\mathbf{x})}{\sigma(\mathbf{x})}, \tag{6–12}$$

which, incidentally, also provides us with a new interpretation of ρ^2 [the ratio of the variability of the linear predictor $\mathbf{y}'$ about the mean $E(\mathbf{y})$ to the variability of $\mathbf{x}$ about the mean $E(\mathbf{x})$] which we shall exploit later.

We introduced in (6–9) a linear function of $\mathbf{x}$ to predict $\mathbf{y}$, with the parameters α and β determined by the criterion of minimum mean square error. It develops in (6–12) that both means $E(\mathbf{x})$ and $E(\mathbf{y})$, both standard deviations $\sigma(\mathbf{x})$ and $\sigma(\mathbf{y})$, and the correlation coefficient $\rho(\mathbf{x}, \mathbf{y})$ are involved in the determination of α and β. Thus five parameters, a heavy requirement for a restricted (linear) prediction process, are required in the prediction function (6–12).

6–3 Least-squares prediction, generally. Let us approach the problem of prediction more generally. Let us predict $\mathbf{y}$ by $R(\mathbf{x})$, an unspecified function of $\mathbf{x}$. We use the same least-squares criterion of goodness of prediction, that is, $R(\mathbf{x})$ is determined so that

$$E[\mathbf{y} - R(\mathbf{x})]^2 \tag{6–13}$$

is minimum. If $R(\mathbf{x})$ is a *linear* function of $\mathbf{x}$, the answer has been given by (6–12). But generally

$$E_{x,y}[\mathbf{y} - R(\mathbf{x})]^2 = E_x[E_y\{[\mathbf{y} - R(\mathbf{x})]^2|\mathbf{x}\}] \tag{6–14}$$

is to be minimized. First consider the inner expectation of (6–14); it may be written

$$\begin{aligned} &E_y\{[\overbrace{\mathbf{y} - E(\mathbf{y}|x)} + \overbrace{E(\mathbf{y}|x) - R(x)}]^2|x\} \\ &= E_y\{[\mathbf{y} - E(\mathbf{y}|x)]^2|x\} + E_y\{[E(\mathbf{y}|x) - R(x)]^2|x\}, \end{aligned} \tag{6–15}$$

the argument identical with that preceding (6–8). Again the cross-product vanishes, since $E[\mathbf{y} - E(\mathbf{y}|x)] = 0$. Now take the expectation of (6–15) over x. The first term of (6–15) will lead, by definition, to $E[V(\mathbf{y}|\mathbf{x})]$, a nonnegative quantity. Whatever the expectation of the second term, it is clearly nonnegative. Therefore (6–15) is minimum when

$$R(\mathbf{x}) = E(\mathbf{y}|\mathbf{x}), \tag{6–16}$$

a result which motivates our interest in $E(\mathbf{y}|\mathbf{x})$. The function $R(\mathbf{x})$ which minimizes the mean square error of prediction of $\mathbf{y}$ is the conditional expectation $E(\mathbf{y}|\mathbf{x})$. This important result can also be obtained from (6–8), and indeed is a special case of problem 3–1.

The function $E(\mathbf{y}|\mathbf{x})$ is called the *regression function of* $\mathbf{y}$ *on* $\mathbf{x}$. No restriction to linearity is imposed here although, for a particular $h(y|x)$, $E(\mathbf{y}|\mathbf{x})$ may in fact integrate out to be a linear function of $\mathbf{x}$, and therefore be equal to $\mathbf{y}'$ of (6–12). In such a case, $E(\mathbf{y}|\mathbf{x})$ is known as the *linear regression function of* $\mathbf{y}$ *on* $\mathbf{x}$; an example will be given in Chapter 17.

In least-squares linear prediction (6–12), five parameters were required. In general least-squares prediction (6–16), in return for greater generality, more is required; we need

$$E(\mathbf{y}|\mathbf{x}) = \int_{-\infty}^{\infty} yh(y|x)\,dy$$

and, therefore, $h(y|x)$ at *every* x.

In the linear function (6–12), the five parameters would likely be known only if the joint density function $f(x, y)$, in terms of which all five parameters are defined, were itself known. In the more general prediction process

(6–16), knowledge of $f(x, y)$ [or of $h(y|x)$] is explicitly required. Note that even if $h(y|x)$ were fully known, a prediction problem would remain, for at fixed x, $\mathbf{y}$ is not determined by $h(y|x)$; $\mathbf{y}$ has a *probability distribution* at fixed x. Some principle is needed to select one value of y from this distribution; the principle of this chapter was least-squares and the selected value of y at fixed x was $E(\mathbf{y}|x)$.

Generally, then, we need to know $h(y|x)$ at each x, from which by integration we obtain $E(\mathbf{y}|x)$, the best predictor of $\mathbf{y}$ in the least-squares sense. In the absence of knowledge of $h(y|x)$, we may sometimes flatly propose the linear prediction process (6–12), arguing that over a short range all curves are straight lines. For certain random variables $\mathbf{x}$, $\mathbf{y}$ whose probability behavior we will study in Chapter 17, the linear process (6–12) is best, in the sense that for these random variables we actually find $E(\mathbf{y}|\mathbf{x}) = \mathbf{y}'$. Generally, however, there will be a discrepancy between $\mathbf{y}'$ and the true regression function $E(\mathbf{y}|\mathbf{x})$. Note in Fig. 6–1 the discrepancies that would exist between *any* linear function $y'(= \alpha + \beta x)$ and the three indicated values of $E(\mathbf{y}|x)$. Finally, note that the best predictor may lead to poor predictions; the person whose weight is $x = 150$ pounds and whose height we wish to predict may in fact be $y = 75$ inches tall, while $E(\mathbf{y}|150) = 69$ inches.

6–4 Goodness of prediction. *The linear case.* How good is prediction by the *linear* function (6–12)? The answer depends, of course, on the discrepancy between values of the random variable $\mathbf{y}$ and predictions y' of those values. The obvious measure of average goodness is the magnitude of the mean square error (6–10).

First consider the special case $\beta = 0$. No attention is paid to $\mathbf{x}$; we predict $\mathbf{y}$ without recourse to $\mathbf{x}$. The linear prediction function (6–12) reduces to

$$y' = \alpha = E(\mathbf{y}).$$

At every x, the predicted value of y is the mean $E(\mathbf{y})$. The mean square error of this best (neglecting x) predictor is

$$E[\mathbf{y} - E(\mathbf{y})]^2 = \sigma^2(\mathbf{y}), \tag{6–17}$$

which agrees with problem 3–1; for y' constant, $E[\mathbf{y} - y']^2$ is minimum about $y' = E(\mathbf{y})$.

If we predict $\mathbf{y}$ *with* recourse to $\mathbf{x}$, the linear prediction function is (6–12) as it stands, and by an easy algebraic argument which the student should do in detail, we obtain for the (minimum) mean square error of prediction

$$E(\mathbf{y} - \alpha - \beta\mathbf{x})^2 = (1 - \rho^2)\sigma^2(\mathbf{y}), \tag{6–18}$$

which is never larger than $\sigma^2(\mathbf{y})$. Unless $\rho = 0$, the introduction of the random variable $\mathbf{x}$ reduces the error of prediction of $\mathbf{y}$.

Equation (6–18) can be written

$$\sigma^2(\mathbf{y}) = \rho^2\sigma^2(\mathbf{y}) + E(\mathbf{y} - \alpha - \beta\mathbf{x})^2, \tag{6–19}$$

and in this form can be recognized, as we shall now see, as a special case of the basic orthogonal breakdown (6–8):

$$V(\mathbf{y}) = E[V(\mathbf{y}|\mathbf{x})] + V[E(\mathbf{y}|\mathbf{x})],$$

with the terms on the right in (6–19) in reverse order to those on the right in (6–8). In (6–19) the total unconditional variance $\sigma^2(\mathbf{y})$ is the sum of two variances. The first, $\rho^2\sigma^2(\mathbf{y})$, may be regarded as that part of $\sigma^2(\mathbf{y})$ explained by or attributed to or associated with the linear regression function (6–12), namely $E(\mathbf{y}|\mathbf{x}) = \alpha + \beta\mathbf{x}$. For, going back to (6–8),

$$V[E(\mathbf{y}|\mathbf{x})] = V(\alpha + \beta\mathbf{x}) = \beta^2 V(\mathbf{x}),$$

which, using (6–11), is equal to $\rho^2\sigma^2(\mathbf{y})$. The second term on the right in (6–19) evidently describes that part of $\sigma^2(\mathbf{y})$ which cannot be explained by or attributed to or associated with the linear regression function (6–12). It is easily shown to be the special case, for linear regression, of the general term $EV(\mathbf{y}|\mathbf{x})$ in (6–8).

Note that the basic orthogonal partition (6–8) may now be usefully further partitioned. We write

$$E(\mathbf{y}|\mathbf{x}) = \mathbf{y}' + [E(\mathbf{y}|\mathbf{x}) - \mathbf{y}'], \tag{6–20}$$

where $\mathbf{y}' = \alpha + \beta\mathbf{x}$ is the linear regression function and the term in square brackets is evidently that part of $E(\mathbf{y}|\mathbf{x})$ which is not described by $\mathbf{y}'$.

$V(\mathbf{y})$	$=$	$E[V(\mathbf{y}|\mathbf{x})]$	$+$	$V(\mathbf{y}')$	$+$	$V[E(\mathbf{y}|\mathbf{x}) - \mathbf{y}']$
		part of the variance $V(\mathbf{y})$ which cannot be explained by $\mathbf{x}$		part of the variance $V(\mathbf{y})$ which can be explained linearly by $\mathbf{x}$		part of the variance $V(\mathbf{y})$ which can be explained nonlinearly by $\mathbf{x}$

We can also write (6–19) suggestively,

$$\rho^2(\mathbf{x}, \mathbf{y}) = \frac{\sigma^2(\mathbf{y}) - E(\mathbf{y} - \alpha - \beta\mathbf{x})^2}{\sigma^2(\mathbf{y})}. \tag{6–21}$$

The square of the Pearson correlation coefficient is identical with the ratio of the variance of $\mathbf{y}$ associated with linear regression on $\mathbf{x}$ to total variance of $\mathbf{y}$. This is an expression for (the square of) ρ $(\mathbf{x}, \mathbf{y})$ alternative to (5–11) and more immediately a measure of the *partitioning of variance* than is the form (5–11); the latter is more immediately a measure of the *relationship between* **x** *and* **y**. As the student is asked to show in problem 6–1, (6–21) is of course the square of (5–11).

From (6–18) or (6–19) it is evident that the goodness of prediction of (6–12) is clearly measured by ρ. If $\rho = +1$, that is, if **x** and **y** are in fact perfectly linearly correlated, the error of linear prediction is 0; linear prediction of **y** from **x** is perfect. If $\rho = 0$, the error of prediction of y remains $\sigma^2(\mathbf{y})$; **x** contributes nothing to prediction of **y**.

Alternatively, we might have calculated the correlation coefficient of **y** and **y′** as a measure of the goodness of prediction of **y** by **y′**. As the correlation coefficient was seen in Chapter 5 to be a measure of linear relationship, which is what we have in (6–12), we consider it here. For

$$\mathbf{y}' = \alpha + \beta\mathbf{x},$$

we find

$$\rho(\mathbf{y}', \mathbf{y}) = \rho(\mathbf{x}, \mathbf{y}) \tag{6–22}$$

for *any* α, β. For

$$\text{cov}\,(\alpha + \beta\mathbf{x}, \mathbf{y}) = \beta\,\text{cov}\,(\mathbf{x}, \mathbf{y}), \qquad V(\alpha + \beta\mathbf{x}) = \beta^2 V(\mathbf{x})$$

and (6–22) follows. The goodness of our linear predictor is immediately measured by the degree of linear correlation between **x** and **y**.

The general case. How good a prediction function is $E(\mathbf{y}|\mathbf{x})$ of (6–16)? We might again consider the correlation coefficient of **y** and $E(\mathbf{y}|\mathbf{x})$. By an easy calculation we find

$$\rho^2[\mathbf{y}, E(\mathbf{y}|\mathbf{x})] = \frac{VE(\mathbf{y}|\mathbf{x})}{V(\mathbf{y})}, \tag{6–23}$$

which is not equal to $\rho^2(\mathbf{x}, \mathbf{y})$ unless $E(\mathbf{y}|\mathbf{x})$ is linear. But *quite apart from any relationship to correlation*, the quantity on the right-hand side of (6–23), showing the ratio to the total variability of the random variable **y** of variability explained by or associated with the predicting regression function, parallels (6–21) and is therefore suggestive. Some adopt

$$\eta^2 = \frac{V[E(\mathbf{y}|\mathbf{x})]}{V(\mathbf{y})} \tag{6–24}$$

as a measure of goodness of prediction of **y**. From (6–8) it is evident that

$$0 \le \eta^2 \le 1.$$

Also, from (6–8), it is evident that $\eta^2 = 1$ only if $EV(\mathbf{y}|\mathbf{x}) = 0$, that is, only if **x** and **y** are perfectly related. Note that η is the correlation coefficient of $\mathbf{y}$ and $E(\mathbf{y}|\mathbf{x})$, not of $\mathbf{y}$ and $\mathbf{y}'$ where $\mathbf{y}' = \alpha + \beta\mathbf{x}$. If, for example, $\mathbf{y}$ and $\mathbf{x}$ are perfectly related quadratically, by $y = x^2$, we have $\eta^2 = 1$, while for the same situation, ρ^2 was shown to be 0. From (6–24), $\eta^2 = 0$ only when $VE(\mathbf{y}|\mathbf{x}) = 0$, that is, only when the regression function of $\mathbf{y}$ on $\mathbf{x}$ is constant (horizontal; $\mathbf{x}$ is of no value in predicting $\mathbf{y}$).

6–5 Concluding remarks. The analogous but different structure for $\mathbf{x}|\mathbf{y}$ must be kept in mind. It is, for example, instructive to show that the linear least-squares prediction function

$$\mathbf{x}' = \gamma + \delta\mathbf{y},$$

in which γ and δ are determined by the criterion $E(\mathbf{x} - \mathbf{x}')^2$ minimum, is by no means identical with the line (6–12); the latter always lies nearer the x-axis. We may also find $\sqrt{\beta\delta} = \rho(\mathbf{x}, \mathbf{y})$ as well as other interrelationships between the regression functions of $\mathbf{x}$ on $\mathbf{y}$ and $\mathbf{y}$ on $\mathbf{x}$.

Finally, note that this chapter is concerned with mathematical structure relating to two random variables. It would leave much to be desired as a practical prediction process; no data are involved anywhere. Moreover, it is a specialized mathematical structure; the situation in which values of x are not random but are (carefully) selected by the experimenter, a situation of much interest which will be the subject of Chapter 30, has not been explored. In this connection, the student may want to note if, in the derivation leading to (6–16), $\mathbf{x}$ *need* be a random variable.

In Chapter 30, we will see the relevance of some of the mathematical functions of this chapter to practical prediction. There the central problem will be to *estimate from observed data* certain of these mathematical functions.

Collateral Reading

H. D. Brunk, *An Introduction to Mathematical Statistics.* Boston, Ginn and Co., 1960. Very good discussion.

R. V. Hogg and A. T. Craig, *Introduction to Mathematical Statistics.* New York, Macmillan Co., 1959.

M. G. Kendall, "Regression, structure and functional relationship, Part I," *Biometrika*, Vol. 38 (1951), pp. 11–25.

Problem

6–1. Show that ρ as given in (6–21) is identical with ρ as given in (5–11).

CHAPTER 7

SEVERAL RANDOM VARIABLES

7–1 Axioms and elementary distribution theory. The extension to several random variables of the axioms, definitions, and results for one and two random variables is generally straightforward. In this elementary text we utilize only a few results from the statistical theory of several random variables; these are now written down only for the continuous case. A change in notation from $\mathbf{x}, \mathbf{y}, \mathbf{z}, \ldots$ to $\mathbf{x}_1, \mathbf{x}_2, \ldots, \mathbf{x}_n$ is imperative.

The immediate motivation of this chapter is a natural extension of the motivation underlying one and two random variables. Here the outcome of a trial is imagined to consist of n numbers (a person may be examined for height, weight, . . . , intelligence). Each of the points of the original event space S has n attributes attached to it, and the random variables $\mathbf{x}_1, \ldots, \mathbf{x}_n$ map the points and subsets of points of S into points and regions in the n-dimensional Euclidean space of $x_1, \ldots, x_n$. As earlier, we state that, *practically* speaking, every constructable region in the n-dimensional space has a probability attached to it.

A second motivation, and a more important one, will appear when we consider the probability behavior of a sample of n observations drawn from a population.

Consider n random variables $\mathbf{x}_1, \ldots, \mathbf{x}_n$ with distribution function $F(x_1, \ldots, x_n)$ connected by definition to a joint probability statement on the random variables $\mathbf{x}_1, \ldots, \mathbf{x}_n$ by

$$p(\mathbf{x}_1 \le x_1, \ldots, \mathbf{x}_n \le x_n) = F(x_1, \ldots, x_n). \tag{7–1}$$

As earlier, the distribution function is basic; this function describes the probability of the important event $\mathbf{x}_1 \le x_1, \ldots, \mathbf{x}_n \le x_n$, and the probabilities of all other events of interest can be expressed in terms of the distribution function. Our three axioms are:

(a) The first difference in n dimensions of $F(x_1, \ldots, x_n)$ must be nonnegative, in order to insure nonnegative probabilities. This corresponds to the simple axiom for one random variable (page 21) and to the more cumbersome axiom for two random variables (page 35).

(b) $$F(x_1, \ldots, -\infty, \ldots, x_n) = 0 \text{ for every } \mathbf{x}_i,$$
$$F(\infty, \ldots, \infty) = 1.$$

(c) The additivity axiom for mutually exclusive events $E_1, E_2, \ldots$ (nonoverlapping regions $E_1, E_2, \ldots$ in the n-dimensional space)

$$p(\mathbf{x}, \ldots, \mathbf{x}_n \in E_1 + E_2 + \cdots)$$
$$= p(\mathbf{x}_1, \ldots, \mathbf{x}_n \in E_1) + p(\mathbf{x}_1, \ldots, \mathbf{x}_n \in E_2) + \cdots.$$

For $F(x_1, \ldots, x_n)$ absolutely continuous, that is, if a function $f(x_1, \ldots, x_n)$ exists such that

$$F(x_1, \ldots, x_n) = \int_{-\infty}^{x_n} \cdots \int_{-\infty}^{x_1} f(u_1, \ldots, u_n)\, du_1 \ldots du_n, \tag{7–2}$$

we find, directly from (7–2) and the axioms

$$p[x_1 \leq \mathbf{x}_1 \leq x_1 + dx_1, \ldots, x_n \leq \mathbf{x}_n \leq x_n + dx_n]$$
$$= f(x_1, \ldots, x_n)\, dx_1 \ldots dx_n, \tag{7–3}$$

where $f(x_1, \ldots, x_n)$ is the joint density function of $\mathbf{x}_1, \ldots, \mathbf{x}_n$, as well as the result

$$\int_{-\infty}^{\infty} \cdots \int_{-\infty}^{\infty} f(x_1, \ldots, x_n)\, dx_1 \ldots dx_n = 1.$$

7–2 Mathematical expectation. By definition, the mathematical expectation of the arbitrary function $\phi(\mathbf{x}_1, \ldots, \mathbf{x}_n)$ is, when the expectation exists,

$$E[\phi(\mathbf{x}_1, \ldots, \mathbf{x}_n)] = \int_{-\infty}^{\infty} \cdots \int_{-\infty}^{\infty} \phi(x_1, \ldots, x_n) f(x_1, \ldots, x_n)\, dx_1 \ldots dx_n. \tag{7–4}$$

The following special cases are particularly important for us:

$$E[e^{\theta_1 \mathbf{x}_1 + \cdots + \theta_n \mathbf{x}_n}] = \int_{-\infty}^{\infty} \cdots \int_{-\infty}^{\infty} e^{\theta_1 x_1 + \cdots + \theta_n x_n} f(x_1, \ldots, x_n)\, dx_1 \ldots dx_n, \tag{7–5}$$

written $M_{\mathbf{x}_1, \ldots, \mathbf{x}_n}(\theta_1, \ldots, \theta_n)$, the moment generating function of $\mathbf{x}_1, \ldots, \mathbf{x}_n$, with n generators $\theta_1, \ldots, \theta_n$, and

$$E[e^{\theta G(\mathbf{x}_1, \ldots, \mathbf{x}_n)}] = \int_{-\infty}^{\infty} \cdots \int_{-\infty}^{\infty} e^{\theta G(x_1, \ldots, x_n)} f(x_1, \ldots, x_n)\, dx_1 \ldots dx_n, \tag{7–6}$$

written $M_{G(\mathbf{x}_1, \ldots, \mathbf{x}_n)}(\theta)$, the moment generating function of the arbitrary function $G(\mathbf{x}_1, \ldots, \mathbf{x}_n)$ with a single generator θ. An important special case of (7–6) is

$$G(\mathbf{x}_1, \ldots, \mathbf{x}_n) = \mathbf{x}_1 + \cdots + \mathbf{x}_n, \quad \mathbf{x}_1, \ldots, \mathbf{x}_n \text{ mutually independent,} \tag{7–7}$$

for which (7–6) reduces to

$$\begin{aligned} M_{\mathbf{x}_1+\cdots+\mathbf{x}_n}(\theta) &= \int_{-\infty}^{\infty} \cdots \int_{-\infty}^{\infty} e^{\theta(x_1+\cdots+x_n)} f(x_1, \ldots, x_n)\, dx_1 \ldots dx_n \\ &= \int_{-\infty}^{\infty} e^{\theta_1 x_1} f_1(x_1)\, dx_1 \ldots \int_{-\infty}^{\infty} e^{\theta_n x_n} f_n(x_n)\, dx_n \\ &= M_{\mathbf{x}_1}(\theta_1) \cdot \ldots \cdot M_{\mathbf{x}_n}(\theta_n); \end{aligned} \tag{7–8}$$

the moment generating function of the sum of independent random variables is the product of their moment generating functions, a result we shall often use.

If, in addition to mutual independence, the random variables $\mathbf{x}_1, \ldots, \mathbf{x}_n$ have the same probability behavior [the same density function $f(x)$], (7–8) further reduces to

$$M_{\mathbf{x}_1+\cdots+\mathbf{x}_n}(\theta) = M_{\mathbf{x}}^n(\theta), \tag{7–9}$$

a result which will play a basic role in random sampling theory in Part II.

The definitions of conditional probabilities in terms of conditional density (or distribution) functions are natural extensions of those for two random variables. For example, we shall use the definition

$$g(x_1|x_2, \ldots, x_n) = \frac{f(x_1, x_2, \ldots, x_n)}{g(x_2, \ldots, x_n)}. \tag{7–10}$$

Corresponding to the univariate linear prediction function (6–9), we will have the *multivariate linear regression* or *prediction function*

$$\mathbf{x}_1' = \alpha_0 + \alpha_2\mathbf{x}_2 + \cdots + \alpha_n\mathbf{x}_n, \tag{7–11}$$

with the constants $\alpha_0, \alpha_2, \ldots, \alpha_n$ determined by the least-squares criterion

$$E(\mathbf{x}_1 - \mathbf{x}_1')^2 \text{ minimum,} \tag{7–12}$$

with, as earlier, error of prediction often measured by the magnitude of (7–12). Or better, error of predictions is measured by $\rho^2(\mathbf{x}_1', \mathbf{x}_1)$, the square of the linear correlation coefficient of $\mathbf{x}_1'$ and $\mathbf{x}_1$, which turns out, entirely similar to (6–21) for univariate linear regression, to be the ratio of variability (variance) of $\mathbf{x}_1$ "explained" by the multiple linear regression function (7–11) to the total unconditional variability (variance) of $\mathbf{x}_1$. The function $\rho(\mathbf{x}_1', \mathbf{x}_1)$ is called the *multiple correlation coefficient.* Corresponding to $R(\mathbf{x})$ of (6–16), we have here the general $n - 1$ variable prediction function

$$R(\mathbf{x}_2, \ldots, \mathbf{x}_n), \tag{7–13}$$

which, if the form of R is determined by the least-squares criterion, leads, by an argument entirely similar to that leading to (6–16), to

$$R(\mathbf{x}_2, \ldots, \mathbf{x}_n) = E(\mathbf{x}_1|\mathbf{x}_2, \ldots, \mathbf{x}_n), \tag{7–14}$$

the regression function of $\mathbf{x}_1$ on $\mathbf{x}_2, \ldots, \mathbf{x}_n$, a result entirely similar to (6–16) for two random variables. The error-of-prediction structure of (7–14) is similar in spirit to, but more complex in detail than, that for two random variables. Multivariate regression functions, particularly the multivariate linear function (7–11), play central roles in the statistical theory of modern experimentation.

Finally, we give two important results which the student should derive:

$$E(\mathbf{x}_1 + \mathbf{x}_2 + \cdots + \mathbf{x}_n) = E(\mathbf{x}_1) + E(\mathbf{x}_2) + \cdots + E(\mathbf{x}_n); \qquad (7\text{–}15)$$

$$\begin{aligned}\sigma^2(\mathbf{x}_1 + \mathbf{x}_2 + \cdots + \mathbf{x}_n) &= \sigma_1^2 + \sigma_2^2 + \cdots + \sigma_n^2 \\ &\quad + 2\rho_{12}\sigma_1\sigma_2 + \cdots + 2\rho_{1n}\sigma_1\sigma_n \\ &\quad + \cdots + 2\rho_{n-1,n}\sigma_{n-1}\sigma_n, \end{aligned} \qquad (7\text{–}16)$$

the latter depending only on correlations between *pairs* $(\rho_{12}, \rho_{13}, \ldots)$ of the n random variables, and not on correlations among three or more random variables.

Note that (7–15) and (7–16) are obtained without conditions on $\mathbf{x}_1, \mathbf{x}_2, \ldots, \mathbf{x}_n$. On the contrary

$$E(\mathbf{x}_1 \cdot \mathbf{x}_2 \cdot \ldots \cdot \mathbf{x}_n) = E(\mathbf{x}_1)E(\mathbf{x}_2) \cdot \ldots \cdot E(\mathbf{x}_n) \qquad (7\text{–}17)$$

cannot be reached without conditions on $\mathbf{x}_1, \mathbf{x}_2, \ldots, \mathbf{x}_n$; a sufficient though not necessary condition for (7–17) to hold is $\mathbf{x}_1, \mathbf{x}_2, \ldots, \mathbf{x}_n$ mutually independent, as the student should show.

Mathematical expectations of such linear sums as (7–15) will appear in the following pages. Here we note in passing that such expectations have been extensively applied to returns from games of chance. If, with probabilities $p_1, p_2, \ldots, p_k$, gains $g_1, g_2, \ldots, g_k$ are realized on the occurrences, respectively, of the mutually exclusive and exhaustive events $E_1, E_2, \ldots, E_k$, the expected gain $E(\mathbf{g})$ is

$$E(\mathbf{g}) = p_1g_1 + p_2g_2 + \cdots + p_kg_k;$$

a fair game (and as Feller [56] has shown, it may be unfair indeed) is *defined* by $E(\mathbf{g}) = 0$.

Collateral Reading

A. M. Mood, *Introduction to the Theory of Statistics.* New York, McGraw-Hill Book Co., 1950.

S. S. Wilks, *Mathematical Statistics.* Princeton, Princeton University Press, 1943. (Or the 1962 Wiley volume.) The best reference.

Problems

7–1. Using independent random variables $\mathbf{x}_1, \mathbf{x}_2, \ldots, \mathbf{x}_n$, of identical variances, form the linear functions

$$\mathbf{l} = \sum_{i=1}^{n} a_i \mathbf{x}_i, \qquad \mathbf{m} = \sum_{i=1}^{n} b_i \mathbf{x}_i.$$

Show that the condition for

$$\rho(\mathbf{l}, \mathbf{m}) = 0$$

is

$$\sum_{i=1}^{n} a_i b_i = 0.$$

7–2. The St. Petersburg paradox (D. Bernoulli). A coin is repeatedly tossed, and A is paid when (and only when) a head first appears. The game ends on A being paid. If head first appears on the first toss, A receives one dollar, if on the second toss two dollars, if on the third toss four dollars, . . . , if on the kth toss 2^{k-1} dollars, ... What fixed entrance fee makes the game fair?

7–3. Cramér [37]. Given that a, b, c are randomly chosen between 0 and 1; form

$$ax^2 + 2bx + c = 0.$$

Find the probability that the roots are real.

CHAPTER 8

CHANGE OF VARIABLE

8–1 The nature of the problem. A number of important problems in statistical theory can be approached and solved, in whole or in part, by change of variable. These range from simple changes in location and scale to orthogonal and other more general transformations which we consider in Part III.

We often have a random variable $\mathbf{x}$ with known probability behavior [known density function $f(x)$] and we need to determine the probability behavior [the density function $g(y)$] of a (random) variable $\mathbf{y}$ which is related to $\mathbf{x}$ by a known function $y = \phi(x)$. Even more often we have a set of random variables $\mathbf{x}_1, \ldots, \mathbf{x}_n$ with known probability behavior [known joint density function $f(x_1, \ldots, x_n)$] and we need to determine the probability behavior [the joint density function $g(y_1, \ldots, y_m)$] of a set of (random) variables $\mathbf{y}_1, \ldots, \mathbf{y}_m$ which are related to $\mathbf{x}_1, \ldots, \mathbf{x}_n$ by m known functions $y_i = \phi_i(x_1, \ldots, x_n)$, $i = 1, \ldots, m$. A common and important special case of this last is $\mathbf{y}_1, \ldots, \mathbf{y}_m$ replaced by *one* random variable $\mathbf{y}$ which is related to $\mathbf{x}_1, \ldots, \mathbf{x}_n$ by *one* known function $y = \phi(x_1, \ldots, x_n)$.

We have already encountered this problem and have solved special cases by the use of moment generating functions. For example, for the linear change of variable

$$\mathbf{y} = a + b\mathbf{x},$$

we found

$$M_{a+b\mathbf{x}}(\theta) = \int_{-\infty}^{\infty} e^{(a+bx)\theta} f(x)\, dx = e^{a\theta} M_{\mathbf{x}}(b\theta). \tag{8–1}$$

Thus the moment generating function of $\mathbf{y}$ is determined in terms of the moment generating function of $\mathbf{x}$, and to the extent that moment generating functions provide knowledge of the probability behavior of a random variable, the problem of the probability behavior of $\mathbf{y}$ is solved. Unfortunately this method seldom leads to results as immediate as in this example, and a direct method of change of variable is required.

We will restrict the discussion to random variables which have a one-to-one relationship to each other, that is, for each possible value of $\mathbf{x}$ (or of $\mathbf{x}_1, \ldots, \mathbf{x}_n$) there is *one* value of $\mathbf{y}$ (or of $\mathbf{y}_1, \ldots, \mathbf{y}_m$), and for each possible value of $\mathbf{y}$ (or of $\mathbf{y}_1, \ldots, \mathbf{y}_m$) there is *one* value of $\mathbf{x}$ (or of $\mathbf{x}_1, \ldots, \mathbf{x}_n$). Thus $\mathbf{y}$ is a single-valued function of $\mathbf{x}$, and $\mathbf{x}$ is a single-valued function of $\mathbf{y}$. The one-to-one restriction does not significantly reduce the range of problems in elementary statistical theory to which our results can be applied.

8–2 Discrete random variables. With discrete random variables, the problem of change of variable is readily solved. For example, given a discrete random variable **x** with density function (one with which we shall deal at length in Chapter 10)

$$f(x) = \frac{4!}{x!(4-x)!}\left(\frac{1}{3}\right)^x\left(\frac{2}{3}\right)^{4-x}, \quad x = 0, 1, 2, 3, 4, \qquad (8\text{–}2)$$

consider the change of variable $y = x^2$. As the inverse function $x = \pm\sqrt{y}$ is not single-valued, we restrict the inverse function to the *positive* root. Such restriction is often reasonable in the light of the physical nature of the new variable **y**, if it is such as length, velocity, weight, etc. With this restriction the relationship between **x** and **y** is one-to-one.

We are interested in the probability behavior of **y**. But there is evidently a one-to-one relationship between occurrence of the y-events

$$y\text{:}\quad 0 \quad 1 \quad 4 \quad 9 \quad 16,$$

and the corresponding possible x-events

$$x\text{:}\quad 0 \quad 1 \quad 2 \quad 3 \quad 4,$$

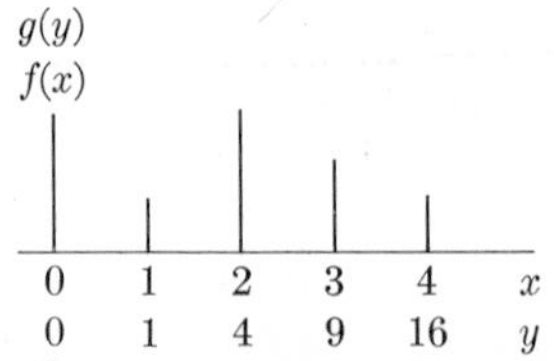

FIG. 8–1. Density functions of the discrete random variables of (8–2) and (8–3).

and it is evident that $g(y) = f(x)$ where $y = x^2$. As f is known and g is not, we must express $g(y)$ in terms of $f(x)$ with x replaced by $+\sqrt{y}$. We have

$$g(y) = f(x) = f\{\sqrt{y}\} = \frac{4!}{\sqrt{y}!(4-\sqrt{y})!}\left(\frac{1}{3}\right)^{\sqrt{y}}\left(\frac{2}{3}\right)^{4-\sqrt{y}},$$

$$y = 0, 1, 4, 9, 16, \qquad (8\text{–}3)$$

an expression more ponderous than, but identical with, (8–2) for $x = +\sqrt{y}$. For example, $g(y)$ for $y = 9$, is equal to $f(x)$ for $x = 3$. The density functions of **x** and of **y** can be shown on one diagram as in Fig. 8–1.

As one-to-one transformations of discrete random variables seldom involve anything more than simple substitution of the new variable for the old, via a transformation $x = \psi(y)$, the remainder of this chapter is restricted to continuous random variables. It will be assumed that derivatives of the transformation functions exist.

8–3 One continuous random variable; increasing functions. For continuous random variables, one-to-one relationship implies that $y = \phi(x)$ is either an increasing or a decreasing function of x over the whole range

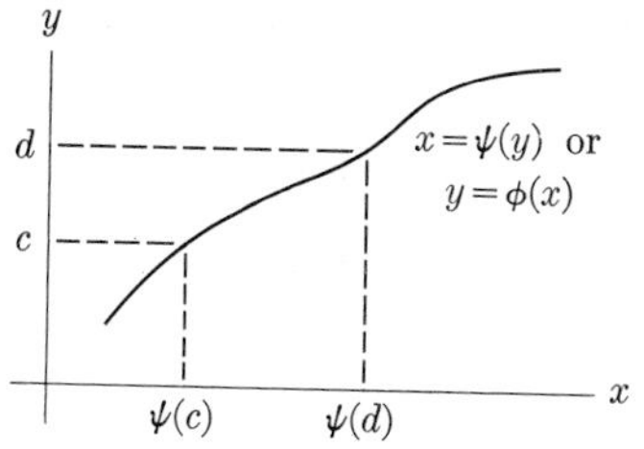

FIG. 8–2. y an increasing function of x.

of x. Either case implies that $y = \phi(x)$ has a unique inverse function, which we write $x = \psi(y)$, and which itself is increasing or decreasing with y as $\phi(x)$ increases or decreases with x.

The case in which y is an increasing function of x is illustrated in Fig. 8–2. It is evident that the events

$$c \leq \mathbf{y} \leq d \quad \text{and} \quad \psi(c) \leq \mathbf{x} \leq \psi(d)$$

are equivalent, that is, must occur simultaneously for arbitrary c, d, $c < d$. For these equivalent events, we have

$$p(c \leq \mathbf{y} \leq d) = p[\psi(c) \leq \mathbf{x} \leq \psi(d)] \tag{8–4}$$

or, for continuous $\mathbf{x}$, $\mathbf{y}$,

$$\int_c^d g(y)\, dy = \int_{\psi(c)}^{\psi(d)} f(x)\, dx, \tag{8–5}$$

where $f(x)$ is the density function of $\mathbf{x}$ and $g(y)$ is the density function of $\mathbf{y}$.

Using (8–5), we can readily calculate probabilities over ranges of the new random variable $\mathbf{y}$ [the left-hand side of (8–5)] from known probabilities over corresponding ranges of the old random variable $\mathbf{x}$ [the right-hand side of (8–5)]. For example, consider

$$f(x) = e^{-x}, \quad x \geq 0,$$

where $x = y^2$ with unique inverse $y = +\sqrt{x}$. Equation (8–5) yields

$$\int_1^{1.414} g(y)\, dy = \int_1^2 e^{-x}\, dx = 0.2325.$$

Thus we can find areas under the density function $g(y)$, although we have not yet explicitly determined the form of $g(y)$ itself. To determine the latter, we transform all x-quantities in the right-hand integral of (8–5) to y-quantities by

$$f(x) = f[\psi(y)] \qquad \text{and} \qquad dx = \frac{d\psi(y)}{dy}\,dy. \tag{8–6}$$

Thus (8–5) becomes

$$\int_c^d g(y)\,dy = \int_c^d f[\psi(y)]\,\frac{d\psi(y)}{dy}\,dy, \tag{8–7}$$

revealing that

$$f[\psi(y)]\,\frac{d\psi(y)}{dy} \tag{8–8}$$

is the density function $g(y)$ of **y**. We see that if y is an increasing function $\phi(x)$ with (unique) inverse $x = \psi(y)$, the density function $f(x)$ of **x** transforms into the density function $f[\psi(y)]\,d\psi(y)/dy$ of **y**. The result is a familiar one in elementary calculus, the only point special to probability here being the identification of $f[\psi(y)]\,d\psi(y)/dy$ as a density function.

In our example, we find

$$g(y) = f\{\psi(y)\}\,\frac{d\psi(y)}{dy} = e^{-y^2}\cdot 2y$$

with graphical description shown in Fig. 8–3.

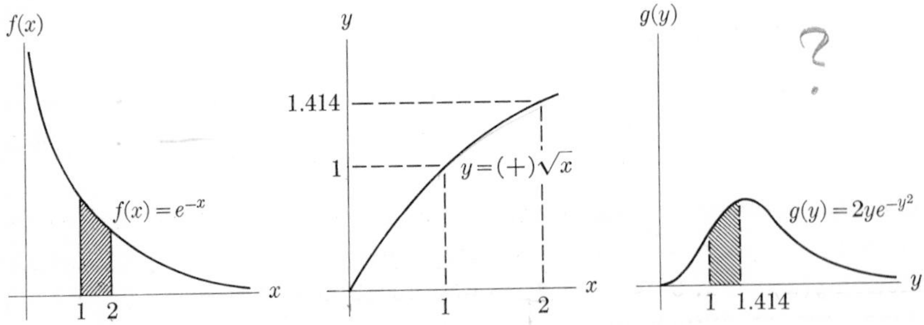

FIG. 8–3. Change of variable from **x** to **y**, with $y = +\sqrt{x}$.

We consider two further examples. Let **v** be the velocity of a gas molecule of mass m. Under quite general conditions, the probability behavior of **v**, originally found by Clerk Maxwell, is described by the density function

$$f(v) = av^2e^{-bv^2}, \quad v > 0,$$

where a and b are constants depending on m. Let **E** be the kinetic energy of a molecule. We have

$$E = mv^2/2, \qquad \text{with unique inverse } v = (+)\sqrt{2E/m},$$

and we seek the probability behavior of **E**. We find

$$g(E) = f\left\{\sqrt{\frac{2E}{m}}\right\} \cdot \sqrt{\frac{2}{m}}\,\frac{1}{2}\,E^{-1/2}$$

$$= \left\{a\,\frac{2E}{m}\,e^{-2bE/m}\right\}\sqrt{\frac{2}{m}}\,\frac{1}{2\sqrt{E}}$$

$$= \alpha E^{1/2}e^{-\beta E}, \quad E > 0,$$

where α and β are constants which depend on a, b, and m.

Our second example is of general interest. Let **x** be a *continuous* random variable with *any* density function $f(x)$. Let the transformation from **x** to **w** be described by

$$w = \varphi(x) = \int_{-\infty}^{x} f(x)\,dx\,[=F(x)], \quad -\infty < x < \infty, \tag{8–9}$$

a one-to-one integral transformation which is an increasing function of x; note that w here is the distribution function of x. Differentiating the inverse function, we have

$$\frac{d\psi(w)}{dw} = \frac{dx}{dw} = \frac{1}{dw/dx} = \frac{1}{f(x)} = \frac{1}{f\{\psi(w)\}},$$

and

$$g(w) = f\{\psi(w)\} \cdot \frac{d\psi(w)}{dw} = \frac{f\{\psi(w)\}}{f\{\psi(w)\}} = 1, \tag{8–10}$$

with $0 \le w \le 1$, for when $x = -\infty$, $w = 0$, and when $x = +\infty$, $w = 1$. Thus (8–9) transforms a continuous random variable **x** with *any* density function $f(x)$ into the continuous random variable **w** with *uniform* density function $g(w) = 1$, $0 \le w \le 1$, as illustrated in Fig. 8–4.

One area of application of the integral transformation (8–9) is the following. We may wish to shift from a random variable **x** with known

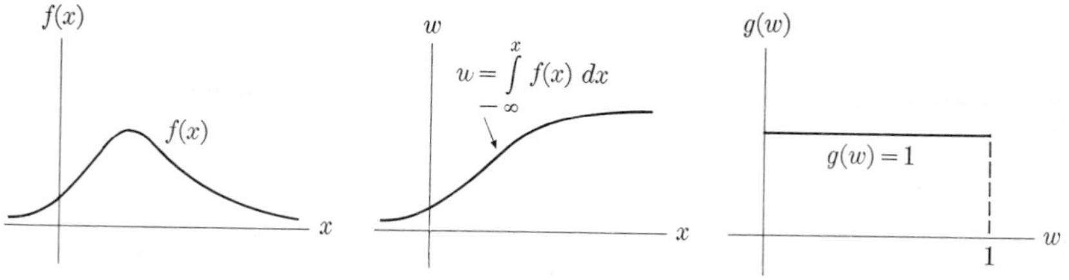

FIG. 8–4. The effect of the integral transformation (8–9) on the random variable **x**.

density function $f(x)$ (which may be difficult to handle or table) to a random variable $\mathbf{y}$ with known density function $g(y)$ (which may be easy to handle or table). We seek the transformation $y = \phi(x)$, $x = \psi(y)$ which will accomplish this. We find it by transforming $\mathbf{x}$ to the uniform random variable $\mathbf{w}$ via the integral transformation (8–9), then transforming the uniform random variable $\mathbf{w}$ to the random variable $\mathbf{y}$ via the *inverse* of the integral transformation (8–9). Practically, this is simply accomplished by transforming *both* $\mathbf{x}$ and $\mathbf{y}$ to uniform $\mathbf{w}$ by

$$w = \int_{-\infty}^{x} f(x)\,dx \qquad \text{and} \qquad w = \int_{-\infty}^{y} g(y)\,dy$$

and equating the integrals, thereby obtaining the desired relationship $\mathbf{y} = \phi(x)$ or $\mathbf{x} = \psi(y)$ between the upper limits of the two integrals. For example (Wadsworth and Bryan [192]), consider

$$f(x) = 6x(1 - x), \quad 0 \le x \le 1,$$
$$g(y) = 3(1 - \sqrt{y}), \quad 0 \le y \le 1.$$

The transformation which converts $\mathbf{x}$ to $\mathbf{y}$ is found from

$$\int_0^x 6x(1 - x)\,dx = \int_0^y 3(1 - \sqrt{y})\,dy,$$

which integrates out to

$$3x^2 - 2x^3 = 3y - 2y^{3/2},$$

with the required transformation being the solution (found by inspection)

$$y = x^2.$$

8–4 One continuous random variable; decreasing functions. If $y = \phi(x)$ is a decreasing function of x (Fig. 8–5), the events $c \le \mathbf{y} \le d$ and $\psi(d) \le \mathbf{x} \le \psi(c)$ are equivalent for arbitrary c, d, $c < d$, and we have

$$p(c \le \mathbf{y} \le d) = p[\psi(d) \le \mathbf{x} \le \psi(c)] \tag{8–11}$$

or, for continuous $\mathbf{x}$, $\mathbf{y}$,

$$\int_c^d g(y)\,dy = \int_{\psi(d)}^{\psi(c)} f(x)\,dx, \tag{8–12}$$

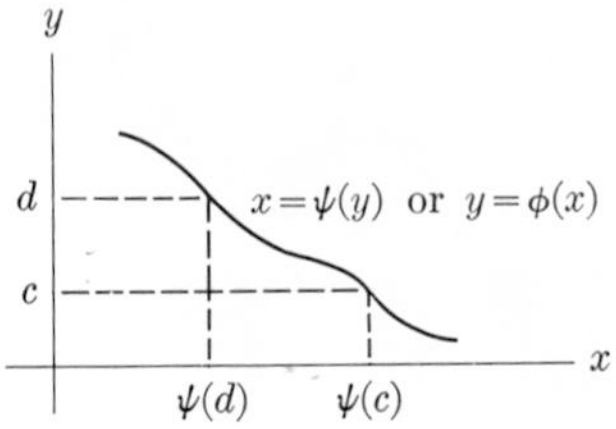

FIG. 8–5. y a decreasing function of x.

where again $f(x)$ is the density function of $\mathbf{x}$ and $g(y)$ is the density function of $\mathbf{y}$. As before, areas under the density function $g(y)$ can be obtained by evaluating the (known) right-hand side of (8–12). Also as before, to determine the density function $g(y)$, we transform all x-quantities in the right-hand integral of (8–12) to y-quantities:

$$\int_c^d g(y)\,dy = \int_c^d -f[\psi(y)]\,\frac{d\psi(y)}{dy}\,dy, \tag{8–13}$$

revealing that

$$-f[\psi(y)]\,\frac{d\psi(y)}{dy} \tag{8–14}$$

(which is positive since $d\psi(y)/dy$ is negative for decreasing functions) is the density function $g(y)$ of **y**.

Example. Given

$$f(x) = e^{-x}, \quad x \geq 0, \qquad y = -3x + 7,$$

the latter a decreasing function with unique inverse (decreasing) function $x = (7 - y)/3$; (8–14) yields

$$g(y) = -e^{-(7-y)/3}(-\tfrac{1}{3}), \quad -\infty < y \leq 7,$$

since $d\psi(y)/dy = -\frac{1}{3}$. The example is illustrated in Fig. 8–6.

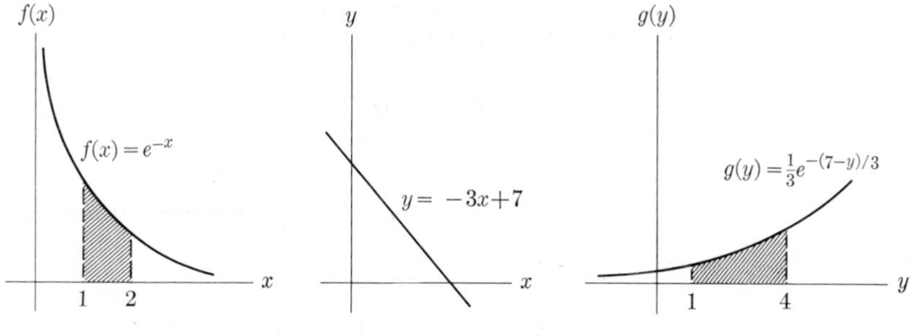

FIG. 8–6. Change of variable from **x** to **y**, with $y = -3x + 7$.

As $d\psi(y)/dy > 0$ when y is an increasing function of x and $d\psi(y)/dy < 0$ when y is a decreasing function of x, (8–8) and (8–14) may be combined to form

$$g(y) = f[\psi(y)]\left|\frac{d\psi(y)}{dy}\right| \tag{8–15}$$

for any monotonic continuous transformation, increasing or decreasing.

Example. Given $\mathbf{x}$ with (general) density function $f(x)$, and the linear transformation $y = ax + b$, find the density function $g(y)$ of $\mathbf{y}$. We find

$$g(y) = f\left(\frac{y-b}{a}\right)\left|\frac{1}{a}\right| \tag{8–16}$$

with the absolute value of $1/a$ covering both possibilities $a < 0$ and $a > 0$. We shall return to this example.

8–5 Change of variable via distribution functions. It is possible and perhaps preferable to consider change of variable via distribution rather than density functions. The answers are, of course, the same, but the argument is instructive and therefore we include it here. First, let us consider increasing functions (Fig. 8–7).

The events

$$\mathbf{y} \leq c \qquad \text{and} \qquad \mathbf{x} \leq a$$

are evidently equivalent, from which

$$p(\mathbf{y} \leq c) = p(\mathbf{x} \leq a),$$

or

$$G(c) = F(a), \tag{8–17}$$

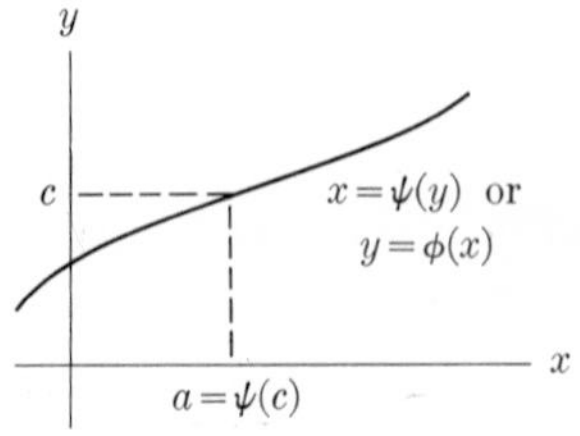

FIG. 8–7. y an increasing function of x.

where G and F are the distribution functions of $\mathbf{y}$ and $\mathbf{x}$, respectively. As c is any value of $\mathbf{y}$, we have, differentiating (8–17) with respect to c,

$$\frac{dG(c)}{dc} = \frac{dF(a)}{dc} = \frac{dF(a)}{da}\,\frac{da}{dc},$$

or

$$g(c) = f(a) \cdot \frac{da}{dc} = f\{\psi(c)\}\,\frac{d\{\psi(c)\}}{dc}.$$

As c is any value of $\mathbf{y}$, we replace c by y and obtain

$$g(y) = f\{\psi(y)\}\,\frac{d\psi(y)}{dy}$$

as before.

For decreasing functions (Fig. 8–8), the events

$$\mathbf{y} \leq c \qquad \text{and} \qquad \mathbf{x} > a$$

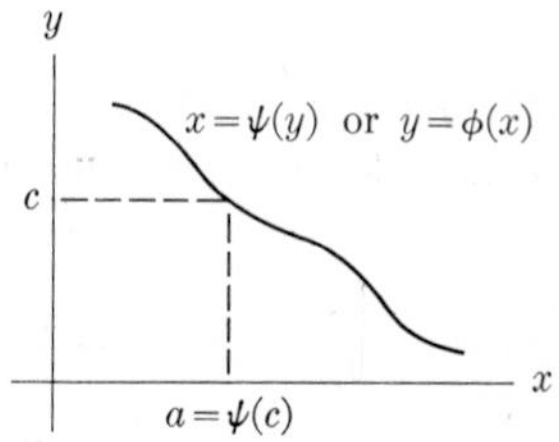

FIG. 8–8. y a decreasing function of x.

are equivalent, from which

$$p(\mathbf{y} \leq c) = p(\mathbf{x} > a)$$

or

$$G(c) = 1 - F(a). \tag{8–18}$$

By the argument used for increasing functions, we are led to

$$g(y) = -f\{\psi(y)\} \frac{d\psi(y)}{dy},$$

which is positive since the derivative is negative. As earlier, the absolute-value convention is generally introduced to cover both increasing and decreasing functions.

For the example on page 76 we would have

$$G(y) = p(\mathbf{y} \leq y) = p(a\mathbf{x} + b \leq y);$$

if $a > 0$, the event $(a\mathbf{x} + b) \leq y$ is equivalent to the event $\mathbf{x} \leq (y - b)/a$; if $a < 0$, the event $(a\mathbf{x} + b) \leq y$ is equivalent to the event $\mathbf{x} \geq (y - b)/a$. This leads to

$$G(y) = F\left(\frac{y - b}{a}\right) \quad \text{for } a > 0$$

and

$$G(y) = 1 - F\left(\frac{y - b}{a}\right) \quad \text{for } a < 0,$$

and, differentiating with respect to y,

$$g(y) = \frac{1}{|a|} f\left(\frac{y - b}{a}\right)$$

for all a, as in (8–16).

8–6 One variable; alternative discussion. The change of variable from $\mathbf{x}$ to $\mathbf{y}$ can be profitably viewed somewhat differently, although the probability aspects of change of variables are not evident in this argument. Writing $F(x)$ for the distribution function and $f(x)$ for the density function of $\mathbf{x}$, we have

$$\frac{d}{dx} F(x) = f(x),$$

and by the chain rule for differentiation

$$\frac{d}{dy} F(x) \frac{dy}{dx} = f(x). \tag{8–19}$$

If the relationship between **x** and **y** is $x = \psi(y)$, (8–19) becomes

$$\frac{dF[\psi(y)]}{dy}\frac{dy}{dx} = f(x),$$

or

$$\frac{dF[\psi(y)]}{dy} = f(x)\frac{dx}{dy}.$$

Now integrating with respect to y, we obtain

$$F[\psi(y)] = \int f[\psi(y)]\frac{dx}{dy}\,dy. \tag{8–20}$$

As

$$F[\psi(y)] = F(x) = \int f(x)\,dx, \tag{8–21}$$

we have, from (8–20) and (8–21),

$$\int f(x)\,dx = \int f[\psi(y)]\frac{dx}{dy}\,dy,$$

as before.

If when $x = a$, $y = c$ and when $x = b$, $y = d$, we have

$$\int_a^b f(x)\,dx = \int_c^d f[\psi(y)]\frac{dx}{dy}\,dy$$

as before.

8–7 Two continuous random variables. We now consider, heuristically and in simpler notation, the problem of transforming from two random variables to two new random variables. We are given two continuous random variables **x** and **y** whose known joint density function is $f(x, y)$. They are related to two (random) variables **u** and **v** by known functions,

$$x = x(u, v) \quad \text{and} \quad y = y(u, v), \tag{8–22}$$

which have unique inverses

$$u = u(x, y) \quad \text{and} \quad v = v(x, y); \tag{8–23}$$

the relationship is one-to-one, that is, there is a one-to-one correspondence between values of the random variable pairs **u**, **v** and **x**, **y**.

As we shall see, the procedure is best routinized if we change from the two original random variables **x**, **y** to *two* new random variables **u**, **v**, even if our interest lies in only one new random variable, say **u**. In the latter case, the second variable $\mathbf{v}$ is introduced as a dummy variable (as simply as possible by, say, $\mathbf{v} = \mathbf{x}$, or $\mathbf{v} = \mathbf{y}$) and the joint distribution of $\mathbf{u}$ and $\mathbf{v}$ is determined; the desired probability behavior of $\mathbf{u}$ is then obtained by integrating out $\mathbf{v}$.

We want to determine

$$p\{u \leq \mathbf{u} \leq u + du, v \leq \mathbf{v} \leq v + dv\} \tag{8–24}$$

which must, of course, be evaluated in terms of the known joint density function $f(x, y)$. We need, therefore, to determine in the x, y-plane the region dA (expressed in terms of u and v) such that $f(x, y)\,dA$ will be equal to (8–24). Then $f(x, y)$ can be written in terms of u and v, using (8–22), thus completing the solution.

As suggested in Fig. 8–9, the infinitesimal element of area dA in the x, y-plane may not be a rectangle. Moreover, the transformation functions $u = u(x, y)$ and $v = v(x, y)$ are often curves in the plane of x, y, and the element of area dA will have curved boundaries.

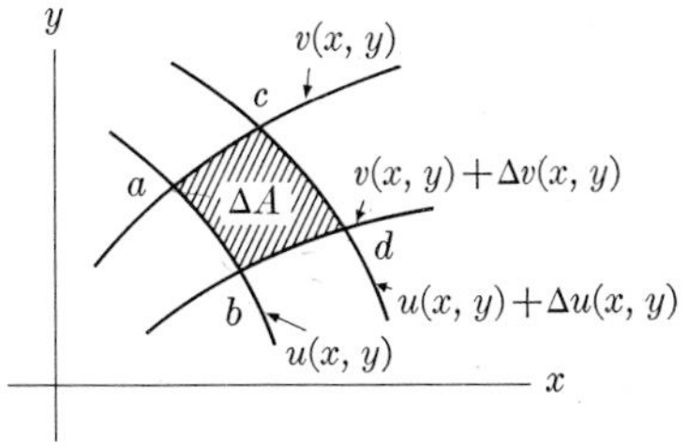

FIG. 8–9. The functions $u = u(x, y)$ and $v = v(x, y)$, and the region ΔA in the x, y-plane.

With $\Delta u \to 0$ and $\Delta v \to 0$, we argue that the quadrilateral region dA may be approximated by a parallelogram with the same vertices. For the area of this parallelogram we will need the x and y coordinates of three vertices, say a, b, c. Let

$$x = x(u, v), \qquad y = y(u, v) \tag{8–25}$$

be the x, y coordinates of the vertex a. Moving from a to b, u remains unchanged and we have for the coordinates of the vertex b,

$$x = x(u, v + \Delta v), \qquad y = y(u, v + \Delta v). \tag{8–26}$$

Similarly, moving from a to c, v remains unchanged and we have for the coordinates of the vertex c,

$$x = x(u + \Delta u, v), \qquad y = y(u + \Delta u, v). \tag{8–27}$$

The Taylor series expansions of (8–26) and (8–27) are

$$x(u, v + \Delta v) = x(u, v) + \frac{\partial x}{\partial v}\Delta v + \cdots,$$

$$y(u, v + \Delta v) = y(u, v) + \frac{\partial y}{\partial v}\Delta v + \cdots,$$

$$x(u + \Delta u, v) = x(u, v) + \frac{\partial x}{\partial u}\Delta u + \cdots,$$

$$y(u + \Delta u, v) = y(u, v) + \frac{\partial y}{\partial u}\Delta u + \cdots,$$

the expansions shown to terms of order Δu, Δv, equivalent, geometrically, to approximating the quadrilateral area dA by a parallelogram. From elementary analytic geometry, the area of the parallelogram approximating dA is the (absolute) value of the determinant

$$\begin{vmatrix} 1 & x \text{ coor. of } a & y \text{ coor. of } a \\ 1 & x \text{ coor. of } b & y \text{ coor. of } b \\ 1 & x \text{ coor. of } c & y \text{ coor. of } c \end{vmatrix} = \begin{vmatrix} 1 & x(u, v) & y(u, v) \\ 1 & x(u, v) + \dfrac{\partial x}{\partial v}\,dv & y(u, v) + \dfrac{\partial y}{\partial v}\,dv \\ 1 & x(u, v) + \dfrac{\partial x}{\partial u}\,du & y(u, v) + \dfrac{\partial y}{\partial u}\,du \end{vmatrix}, \tag{8–28}$$

which reduces to

$$\frac{\partial x}{\partial u}\,du\,\frac{\partial y}{\partial v}\,dv - \frac{\partial y}{\partial u}\,du\,\frac{\partial x}{\partial v}\,dv = \begin{vmatrix} \dfrac{\partial x}{\partial u} & \dfrac{\partial y}{\partial u} \\ \dfrac{\partial x}{\partial v} & \dfrac{\partial y}{\partial v} \end{vmatrix} du\,dv = \frac{\partial(x, y)}{\partial(u, v)}\,du\,dv. \tag{8–29}$$

Thus the joint probability element $f(x, y)\,dA$ [which is equal to the desired joint probability element of **u**, **v**, that is, $g(u, v)\,du\,dv$] becomes

$$f\{x(u, v), y(u, v)\}\left|\frac{\partial(x, y)}{\partial(u, v)}\right| du\,dv, \tag{8–30}$$

writing **x** and **y** in terms of **u** and **v** in the arguments of the known density function f.

The sign of the determinant (8–28) depends merely on the order in which we insert the rows of (8–28), that is, on the order of designating the vertices a, b, c of the parallelogram. As we seek the (nonnegative) area of dA, we have taken the absolute value of the determinant (8–28) in (8–30).

As with one random variable, the *Jacobian* J on the transformation from the random variable pair $\mathbf{x}$, $\mathbf{y}$ to the random variable pair $\mathbf{u}$, $\mathbf{v}$,

$$J = \frac{\partial(x, y)}{\partial(u, v)},$$

must be incorporated as part of the desired density function $g(u, v)$. Thus the joint density function of $\mathbf{u}$, $\mathbf{v}$ is

$$f\{x(u, v), y(u, v)\} \left|\frac{\partial(x, y)}{\partial(u, v)}\right|.$$

This is perhaps the one unique feature of change of variable in statistical theory as against change of variable in general mathematics.

Example (Anderson and Bancroft [2]). Given

$$f(x, y) = e^{-(x+y)}, \quad 0 \le x, y < \infty,$$

$$u = x + y, \qquad v = x/y,$$

we find

$$J = \begin{vmatrix} \dfrac{\partial x}{\partial u} & \dfrac{\partial y}{\partial u} \\[2ex] \dfrac{\partial x}{\partial v} & \dfrac{\partial y}{\partial v} \end{vmatrix},$$

or better, using problem 8–2,

$$\frac{1}{J} = \begin{vmatrix} \dfrac{\partial u}{\partial x} & \dfrac{\partial u}{\partial y} \\[2ex] \dfrac{\partial v}{\partial x} & \dfrac{\partial v}{\partial y} \end{vmatrix} = \begin{vmatrix} 1 & 1 \\[1ex] \dfrac{1}{y} & -\dfrac{x}{y^2} \end{vmatrix} = -\frac{x + y}{y^2} = -\frac{(1 + v)^2}{u},$$

$$g(u, v)\,du\,dv = e^{-u} \cdot [u/(1 + v)^2]\,du\,dv, \qquad 0 \le u, v < \infty.$$

The density function $r(u)$ of $\mathbf{u}$ is given by

$$r(u) = ue^{-u} \int_0^\infty \frac{1}{(1 + v)^2}\,dv = ue^{-u}, \qquad 0 \le u < \infty$$

and the density function $s(v)$ of $\mathbf{v}$ is given by

$$s(v) = \frac{1}{(1 + v)^2} \int_0^\infty ue^{-u}\,du = \frac{1}{(1 + v)^2}, \qquad 0 \le v < \infty$$

These results also follow from the general theorem that if $\mathbf{x}$ and $\mathbf{y}$ are independent (and they are, for $e^{-(x+y)} = e^{-x}e^{-y}$) then any functions $\mathbf{u}$ and $\mathbf{v}$ of $\mathbf{x}$ and $\mathbf{y}$ are independent, as they are.

In this example, the ranges of the original variables **x** and **y** do not depend on each other and, therefore, the ranges of **u** and **v** are independent. In some problems, the range of one of the original variables depends on the other (for example, $1 \leq y \leq x$), and the ranges of **u** and **v** are correspondingly interrelated. The determination of such interrelated ranges of **u**, **v** is a subject in the area of conformal mapping; for most problems, however, a graph or careful replacement of **x**, **y** by the corresponding values of **u**, **v** in the original ranges of **x**, **y** will suffice. We give an example (Anderson-Bancroft [2]). Let

$$0 \leq x \leq \mathbf{y} \qquad 0 \leq y \leq 1;$$

the x, y region is evidently bounded by the three straight lines $x = 0$, $x = y$, $y = 1$. Now consider

$$u = x + y, \qquad v = x - y, \qquad \text{that is,} \qquad x = \frac{u+v}{2}, \qquad y = \frac{u-v}{2}.$$

The x, y region transforms into the u, v region as follows (see Fig. 8–10):

$$x = 0 \rightarrow \frac{u+v}{2} = 0, \qquad u = -v;$$

$$x = y \rightarrow v = 0;$$

$$y = 1 \rightarrow u - v = 2, \qquad u = 2 - v;$$

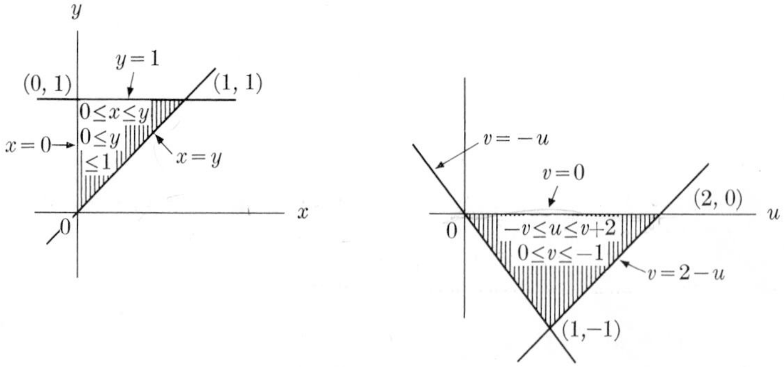

FIG. 8–10. x, y and corresponding u, v regions.

8–8 Several continuous random variables. We now turn briefly to a multivariate change of variable. Again we require that the relationships

$$y_i = \phi_i(x_1, \ldots, x_n), \quad i = 1, \ldots, n,$$

$$x_i = \psi_i(y_1, \ldots, y_n), \quad i = 1, \ldots, n, \qquad (8\text{–}31)$$

between the sets of random variables $\mathbf{x}_1, \ldots, \mathbf{x}_n$ and $\mathbf{y}_1, \ldots, \mathbf{y}_n$ be one-to-one, that is, for each value of the set $\mathbf{x}_1, \ldots, \mathbf{x}_n$ there is one value of the set $\mathbf{y}_1, \ldots, \mathbf{y}_n$, and for each value of the set $\mathbf{y}_1, \ldots, \mathbf{y}_n$, there is one value of the set $\mathbf{x}_1, \ldots, \mathbf{x}_n$. This one-to-one property is guaranteed *locally** if the determinant

$$\frac{\partial(x_1, \ldots, x_n)}{\partial(y_1, \ldots, y_n)} \equiv \begin{vmatrix} \frac{\partial\psi_1}{\partial y_1} & \cdots & \frac{\partial\psi_1}{\partial y_n} \\ \vdots & & \vdots \\ \frac{\partial\psi_n}{\partial y_1} & \cdots & \frac{\partial\psi_n}{\partial y_n} \end{vmatrix}, \tag{8–32}$$

known as the Jacobian of the transformation (8–31), does not change sign. This is an extension, though hardly an obvious one, of the situation for one variable in which, for $\mathbf{y}$ and $\mathbf{x}$ to be functionally related one-to-one, $d\psi(y)/dy$ was required not to change sign. The required property on the Jacobian may hold only for particular values of $\mathbf{x}_1, \ldots, \mathbf{x}_n$, in which case the multivariate change of variable is restricted to those values.

Once again, in the transformation from $\mathbf{x}_1, \ldots, \mathbf{x}_n$ to $\mathbf{y}_1, \ldots, \mathbf{y}_n$, the appropriate change in $f(x_1, \ldots, x_n)$ is simple. As with one variable where we replaced $f(x)$ by $f[\psi(y)]$, here we replace

$$f(x_1, \ldots, x_n) \quad \text{by} \quad f[\psi_1(y_1, \ldots, y_n), \ldots, \psi_n(y_1, \ldots, y_n)]. \tag{8–33}$$

With one variable the transformation was completed by replacing dx by

---

* The guarantee is local but not in the large. For example, for

$$x_1 = e^{y_1} \sin y_2, \qquad x_2 = e^{y_1} \cos y_2$$

the Jacobian J of the transformation is $J = -e^{y_1}$ which does not change sign (J is negative) for all values of y_1. But

$$x_1(y_1, y_2 + 2n\pi) = x_1(y_1, y_2), \qquad x_2(y_1, y_2 + 2n\pi) = x_2(y_1, y_2),$$

and the transformation is evidently not one-to-one. For *local* one-to-one (small deviations, $J \neq 0$ is sufficient. But to guarantee one-to-one in the *large*, various weak sufficient conditions have been proposed; the most convenient, by Samuelson [172], requires that successive principle minors be bounded away from 0 for all y's,

$$\left|\frac{\partial x_1}{\partial y_1}\right| = \epsilon \neq 0, \qquad \begin{vmatrix} \frac{\partial x_1}{\partial y_1} & \frac{\partial x_1}{\partial y_2} \\ \frac{\partial x_2}{\partial y_1} & \frac{\partial x_2}{\partial y_2} \end{vmatrix} = \epsilon \neq 0, \text{ etc.}$$

See also Tukey [190].

$[dx/dy]\, dy$ where the Jacobian

$$\frac{dx}{dy}$$

of the transformation may be regarded as the ratio of the small element of length dx in the space of x to the small element of length dy in the space of y, (and therefore nonnegative). With n variables, the corresponding step is the replacement of

$$dx_1 \ldots dx_n \quad \text{by} \quad \frac{\partial(x_1, \ldots, x_n)}{\partial(y_1, \ldots, y_n)}\, dy_1 \ldots dy_n, \tag{8–34}$$

where the Jacobian

$$\frac{\partial(x_1, \ldots, x_n)}{\partial(y_1, \ldots, y_n)}$$

may be regarded as the ratio of the small element of hypervolume $dx_1 \ldots dx_n$ in the space of $x_1, \ldots, x_n$ to the corresponding small element of hypervolume $dy_1 \ldots dy_n$ in the space of $y_1, \ldots, y_n$ (and therefore nonnegative).

Since the transformation (8–31) is one-to-one, there exists a region S_y in the space of $y_1, \ldots, y_n$ corresponding to a region R_x in the space of $x_1, \ldots, x_n$ such that

$$\int \cdots \int_{S_y} g(y_1, \ldots, y_n)\, dy_1 \ldots dy_n = \int \cdots \int_{R_x} f(x_1, \ldots, x_n)\, dx_1 \ldots dx_n. \tag{8–35}$$

As earlier, changing all x-quantities in the right-hand multiple integral to y quantities via (8–33) and (8–34), we have

$$\int \cdots \int_{S_y} g(y_1, \ldots, y_n)\, dy_1 \ldots dy_n$$
$$= \int \cdots \int_{S_y} f[\psi_1(y_1, \ldots, y_n), \ldots, \psi_n(y_1, \ldots, y_n)]\, \frac{\partial(x_1, \ldots, x_n)}{\partial(y_1, \ldots, y_n)}\, dy_1 \ldots dy_n, \tag{8–36}$$

and the joint multivariate density function of $\mathbf{y}_1, \ldots, \mathbf{y}_n$ is evidently

$$f[\psi_1(y_1, \ldots, y_n), \ldots, \psi_n(y_1, \ldots, y_n)]\, \frac{\partial(x_1, \ldots, x_n)}{\partial(y_1, \ldots, y_n)}. \tag{8–37}$$

As indicated at the beginning of this chapter, we shall often want the density function of *one* new variable $\mathbf{y}_1 = \phi_1(\mathbf{x}_1, \ldots, \mathbf{x}_n)$, and not the multivariate density function of n new variables $\mathbf{y}_1, \ldots, \mathbf{y}_n$. Integration of (8–37) over $y_2, \ldots, y_n$ will produce the density function of $\mathbf{y}_1$.

One of the principal examples in statistical theory of the change from n random variables $\mathbf{x}_1, \ldots, \mathbf{x}_n$ to n (random) variables $\mathbf{y}_1, \ldots, \mathbf{y}_n$ is the set of linear transformations

$$\begin{aligned} y_1 &= a_{11}x_1 + \cdots + a_{1n}x_n \\ &\vdots \\ y_n &= a_{n1}x_1 + \cdots + a_{nn}x_n. \end{aligned}$$

The Jacobian J of this transformation is the determinant

$$J = \begin{vmatrix} a_{11} & \cdots & a_{1n} \\ \vdots & & \vdots \\ a_{n1} & \cdots & a_{nn} \end{vmatrix},$$

and, for a class of linear transformations of much interest in statistical theory, $J = \pm 1$.

Collateral Reading

T. C. Fry, *Probability and Its Engineering Uses.* New York, D. Van Nostrand Co., 1928.

R. P. Gillespie, *Integration.* London, Oliver and Boyd, 1951.

R. V. Hogg and A. T. Craig, *Introduction to Mathematical Statistics.* New York, Macmillan Co., 1959.

N. L. Johnson and H. Tetley, *Statistics, Volume II.* Cambridge (Eng.) Cambridge University Press, 1950. One of the best discussions.

J. W. Tukey, "A smooth invertibility theorem," *Ann. Math. Stat.*, Vol. 29 (1958), pp. 581–584.

G. P. Wadsworth and J. Bryan, *An Introduction to Probability and Random Variables.* New York, McGraw-Hill Book Co., 1960. Very good discussion.

Problems

8–1. Kendall [103]. The random variable $\mathbf{t}$ has density function $f(t)$ in the range $0, \infty$; $\mathbf{z} = \mathbf{t} - T_0$ where $t > T_0$. Find the density function of $\mathbf{z}$.

8–2. Show that

$$\frac{\partial(x, y)}{\partial(u, v)} = \left[\frac{\partial(u, v)}{\partial(x, y)}\right]^{-1}.$$

8–3. Given that $\mathbf{x}$ has density function $f(x)$, $\mathbf{y}$ has density function $g(y)$. Find the density function of $\mathbf{z} = \mathbf{x}/\mathbf{y}$.

Part II

Specific Probability Distributions

In Part II we consider ten specific probability distributions, beginning with a brief preliminary chapter on permutations and combinations and ending with a chapter on a general system embracing many probability distributions. The specific distributions considered may be regarded as examples of the more general probability distributions of Part I. We have selected probability distributions which are mathematically interesting, which are important in statistical theory, and which have extensive histories of successful application to real phenomena.

CHAPTER 9

PERMUTATIONS AND COMBINATIONS

9–1 Permutations and combinations. In this section we consider permutations and combinations themselves. In Section 9–3 we consider the statistical motivation of our interest in these quantities.

We are given n different (distinguishable) objects which are to be arranged in sets of m objects. The first of the m places in the set of m objects can be occupied by any one of the n objects, the second by any one of the $n - 1$ objects not already used, . . . , the mth by any one of the $[n - (m - 1)]$ objects not already used. The number of such sets, that is, the number of different ordered sets of m objects from n different objects, is written P^n_m and is evidently given by (Pascal, 1654)

$$P^n_m = n(n-1)\ldots[n-(m-1)] = \frac{n!}{(n-m)!}, \qquad (9\text{–}1)$$

where

$$n! = n(n-1)(n-2)\ldots 3\cdot 2\cdot 1,$$

$$(n-m)! = (n-m)[n-(m+1)][n-(m+2)]\ldots 3\cdot 2\cdot 1.$$

The quantity P^n_m is often called the *number of permutations* of n different objects taken m at a time. Note that by the basic argument of the second paragraph by which P^n_m was reached, P^n_n must be $n!$, which suggests that if (9–1) is to apply to $m = n$, we must have $0! = 1$.

Example. Four different objects a, b, c, d are to be taken 2 at a time:

$$P^4_2 = \frac{4!}{2!} = 12.$$

The twelve permutations are

$$\begin{array}{cccccc} ab & ac & ad & bc & bd & cd \\ ba & ca & da & cb & db & dc. \end{array}$$

Now exclude from (9–1) those sets which are merely different orders of the *same* objects. *Each* set of m different objects is included $m!$ times in (9–1), for $m!$ is the number of permutations of m different objects taken m at a time. The number of sets, excluding such differences in order, of n different objects taken m at a time is written C^n_m and is evidently given by (Pascal, 1654)

$$C^n_m = \frac{P^n_m}{m!} = \frac{n!}{m!(n-m)!}. \qquad (9\text{–}2)$$

The quantity C^n_m is often called the *number of combinations* of n different objects taken m at a time.

Example. Four different objects a, b, c, d are to be taken 2 at a time:

$$C^4_2 = \frac{4!}{2!2!} = 6.$$

The six combinations are

ab (or ba) ac (or ca) ad (or da) bc (or cb) bd (or db) cd (or dc).

The student may prefer to reason in the reverse direction; C^n_m combinations produce $C^n_m \cdot m!$ permutations, since each combination of m objects produces $m!$ permutations.

The quantities C^n_m were utilized by Pascal to simplify solutions of problems in probability. He computed them with the aid of the recursion formula shown in problem 9–2 and arranged them in the triangle shown in Fig. 9–1. Several properties of this triangle (which assumes $C^n_0 = 1$, to be discussed later), some of which the student can discover by careful examination, were known to the Pythagoreans in the sixth century B.C. Small tables of $n!$, P^n_m, and C^n_m are shown on pages 400–403.

n	C^n_m										
0	1										
1	1	1									
2	1	2	1								
3	1	3	3	1							
4	1	4	6	4	1						
5	1	5	10	10	5	1					
6	1	6	15	20	15	6	1				
7	1	7	21	35	35	21	7	1			
8	1	8	28	56	70	56	28	8	1		
9	1	9	36	84	126	126	84	36	9	1	
10	1	10	45	120	210	252	210	120	45	10	1

FIG. 9–1. Pascal triangle. The index m runs horizontally.

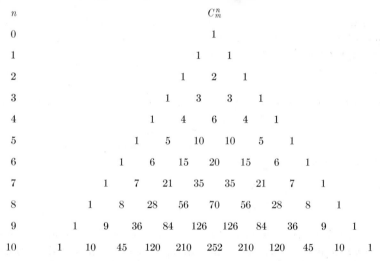

Note that the m objects may be drawn from the n objects at once or one at a time, but if the latter, *without replacement* of any object drawn; no one of the n objects is used more than once in a set of m objects. If replacement is permitted, *each* place in the ordered set of m objects can be filled by any one of n objects, and the number of ordered sets of n different objects taken m at a time is n^m.

Example. Four different objects a, b, c, d are to be taken 2 at a time, with replacement. The number of ordered pairs is $4^2 = 16$. The sixteen ordered pairs are

aa	*ba*	*ca*	*da*
ab	*bb*	*cb*	*db*
ac	*bc*	*cc*	*dc*
ad	*bd*	*cd*	*dd.*

It is not implied that the P_m^n permutations or the C_m^n combinations or the n^m arrangements with replacement are equally likely to occur. In a particular problem, this will or will not be true, but no particular probabilities are implied by the enumerations of the chapter. In problem 9–6 probabilities are surely implied, and their values are suggested by the problem.

Now consider n objects taken n at a time, with i of the objects alike, j of them alike but different from those of the first group, k of them alike but different from those of either of the first two groups, with $i + j + k = n$. Were all objects different, the number of permutations would be $P_n^n = n!$ But the i alike objects reduce this number to $n!/i!$, the j alike objects *further* reduce this number to $n!/i!j!$, and so on. We have (Montmort, 1708)

$$P_n^n(i, j, k) = \frac{n!}{i!j!k!}, \tag{9–3}$$

and the extension to more than three groups should be evident.

Example. Four objects a, b, c, d, but with b and c alike, are to be taken 4 at a time:

$$P_n^n(1, 2, 1) = \frac{4!}{1!2!1!} = 12.$$

The twelve permutations are

abcd	*bacd*	*cbda*	*dabc*
abdc	*badc*	*cdab*	*dbac*
adbc	*bcad*	*cdba*	*dbca.*

A natural concept here is $P_m^n(i, j, k)$ with $m < n$, but this appears to be of little utility in probability and statistics. In the preceding example, for $m = 2$,

$$P_2^4(1, 2, 1) = 7;$$

the permutations are

$$ab \quad ba \quad ad \quad da \quad bd \quad db \quad bc.$$

A link between (9–2) and (9–3) can be seen by reinterpreting them in terms of the following problem. In how many ways can n objects be divided into two groups, one containing r objects and the other $n - r$ objects? There are C_r^n ways of obtaining r (unordered) objects from n objects; once r objects are obtained, the remaining group *must* consist of $n - r$ objects. Therefore the answer is C_r^n. For example, consider the number of ways of dividing 7 objects into groups of 3 and 4. The answer is $C_3^7 = 35$, or $C_4^7 = 35$.

Extending the combinatorial argument to three groups, in how many ways can n objects be divided into 3 groups, one containing r objects, one containing s objects, and one containing $n - r - s$ objects? There are C_r^n ways of obtaining r (unordered) objects from n objects. With the remaining $n - r$ objects, there are C_s^{n-r} ways of obtaining s unordered objects from them; once r and s objects have been obtained, the third group *must* contain $n - r - s$ objects. As each of the C_r^n ways can be associated with each of the C_s^{n-r} ways, the final answer is $C_r^n C_s^{n-r}$. For example consider the number of ways of dividing 7 objects into groups of 3, 2, and 2. The answer is $C_3^7 C_2^4 = 210$ or $C_2^7 C_3^5 = 210$ or $C_2^7 C_2^5 = 210$.

Now note that C_r^n, the answer to the first problem, is

$$C_r^n = P_n^n(r, n - r),$$

and note that $C_r^n C_s^{n-r}$, the answer to the second problem, is

$$C_r^n C_s^{n-r} = P_n^n(r, s, n - r - s),$$

the right-hand answers already reached by different arguments.

9–2 A note on 0!. If in (9–1) or (9–2) we let $m = n$, the quantity 0! appears. Since P_n^n is sensibly $n!$ and C_n^n is sensibly 1, it is evidently necessary for 0! to be 1. But by the argument by which we reached (9–1), any factorial quantity, say $k!$, is defined by

$$k! = k(k - 1)(k - 2) \ldots 3 \cdot 2 \cdot 1 \tag{9–4}$$

only for positive integral k, and such expressions as 0! must be settled otherwise. Most simply, we may settle 0! to be 1 by *definition*. A more general view, unnecessary here but useful in other areas of statistics and in other branches of applied mathematics, is possible. For positive integral k, we can show, by repeated integration by parts, that

$$\int_0^\infty x^k e^{-x}\,dx = k(k-1)(k-2)\ldots 3\cdot 2\cdot 1; \qquad (9\text{–}5)$$

the integral is sometimes called the gamma function of $k+1$ and is often written $\Gamma(k+1)$. If $k!$ is defined by (9–5) rather than by (9–4), *this* integral definition of $k!$, already compatible with (9–4) for positive integral k, may be extended to fractional (and negative) values of k, as well as to $k = 0$. The desired result, $0! = 1$, as well as such curious and elsewhere valuable results as the following can then be *derived:*

$$(\tfrac{1}{2})! = \sqrt{\pi/2}, \qquad (-\tfrac{1}{2})! = \sqrt{\pi}, \qquad (-1)! = \infty.$$

9–3 Motivation. Section 9–1 is motivated by our need to find expressions for the *number* of outcomes favorable to a molecular event of interest, in experiments in which the total number of outcomes is large. If a trial consists of one toss of a coin, a typical outcome is T, and there are only 2 possible outcomes, H and T. If a trial consists of one throw of a die, a typical outcome is 3, and there are only 6 possible outcomes, 1, 2, . . . , 6. Event spaces of 2 and 6 points are hardly large, and the number of outcomes favoring any molecular event of interest within them is readily calculated. But we are often concerned with the outcomes of experiments which consist of m *repeated* trials, with m large. If a coin is tossed $m = 10$ times, a typical outcome is the 10-tuple

$$H\ T\ H\ T\ T\ T\ T\ H\ H\ T,$$

and there are $2^{10} = 1024$ possible outcomes. If a die is thrown 10 times, a typical outcome is the 10-tuple

$$1\ 3\ 6\ 2\ 5\ 2\ 3\ 1\ 2\ 6,$$

and there are $6^{10} = 60{,}466{,}176$ possible outcomes. It would be lengthy to determine, by enumeration, the number of these outcomes favorable to a particular molecular event, say five 3's.

The character and number of outcomes depend on whether the repeated trials are made *with or without replacement.* In our two examples, the occurrence of a letter or number in an early trial does not preclude its appearance later. The trials are made "with replacement"; the letter or number is, so to speak, restored after appearing and is again available, and the number of possible outcomes are special cases of n^m reached in Section 9–1 in the discussion of replacement. By the nature of a trial involving a coin or die, this is natural and almost inevitable; but it need not be natural or inevitable in repeated trials having to do with (a) drawing cards, (b) testing objects, (c) interviewing people. The appearance of a particular card or object or person on a trial may preclude the appearance of that card or object or person on a later trial; in fact, it may be, as we shall see, statistically advantageous to enforce such a restriction. In such cases, the principal formulae of Section 9–1 become relevant. If n is the number of possible outcomes of the first trial, $n - 1$ the number of possible outcomes of the second trial, . . . , $n - (m - 1)$ the number of possible outcomes of the mth trial, the number of ordered outcomes of the experiment of m trials is given by (9–1) and the number of unordered outcomes by (9–2). For example, consider a lot of 10 objects numbered 1, 2, . . . , 10; if by 4 trials we mean 4 objects selected without replacement, a typical outcome is the 4-tuple

$$3\ 4\ 1\ 7,$$

and there are $P_4^{10} = 5040$ possible ordered and $C_4^{10} = 210$ possible unordered outcomes. If the four trials are with replacement, the number of ordered outcomes rises to $10^4 = 10{,}000$.

In probability, it is often merely a matter of taste as to whether we examine ordered or unordered outcomes; they always differ in number by the constant factor $m!$.

Looking ahead, in Chapter 10 where we will be concerned with the number of sets of n objects of which i are alike and $j(=n - i)$ are alike (for example, the number of different sets each containing n units of product of which i are defective and j are good),

$$P_n^n(i, j)[=C_i^n]$$

is the number of such sets, and this number will appear as the coefficient of the term $q^i p^j$ in the algebraic expansion of the *binomial* $(q + p)^n$, with the quantities q and p having probability meanings to be described there. In Chapter 14, where we will be concerned with the number of sets of n objects of which i are alike, j are alike, . . . , m are alike (for example, the number of different groups each containing n voters of whom i favor candidate A, j favor candidate B, . . . , m favor candidate R),

$$P_n^n(i, j, \ldots, m)$$

is the number of such sets, and will appear as the coefficient of the term $p_A^i p_B^j \ldots p_R^m$ in the algebraic expansion of the *multinomial*

$$(p_A + p_B + \cdots + p_R)^n,$$

with the quantities $p_A, p_B, \ldots, p_R$ having probability meanings to be described there.

Tables

K. Hayashi, *Fünfstellige Funktionentafeln.* Berlin, Springer-Verlag, 1930.
Royal Society Mathematical Tables, Vol. 3, "Table of binomial coefficients," Cambridge (Eng.), Cambridge University Press, 1954.

Collateral Reading

W. Feller, *An Introduction to Probability Theory and Its Applications.* 2nd ed., New York, John Wiley and Sons, 1957. A sparkling discussion with a variety of interesting examples.

D. A. S. Fraser, *Statistics: An Introduction.* New York, John Wiley and Sons, 1957.

S. Goldberg, *Probability, an Introduction.* Englewood Cliffs (N.J.), Prentice-Hall, 1960.

A. M. Mood, *Introduction to the Theory of Statistics.* New York, McGraw-Hill Book Co., 1950.

E. Parzen, *Modern Probability Theory and Its Applications.* New York, John Wiley and Sons, 1960.

J. Riordan, *An Introduction to Combinatorial Analysis.* New York, John Wiley and Sons, 1958. The leading work.

W. A. Whitworth, *Choice and Chance.* 5th ed., New York, Hafner Publishing Co., 1951; same author, *DCC Exercises.* New York, G. E. Stechert and Co. (Hafner), 1945. A reprinted classic; a remarkable variety of simple and difficult combinatorial problems, with solutions.

Problems

Purely combinatorial problems.

9–1. Explain, as well as show, that $C^n_m = C^n_{n-m}$.

9–2. Prove the recursion relationship $C^{n-1}_m + C^{n-1}_{m-1} = C^n_m$, which provides a method of building the Pascal triangle.

9–3. A more difficult combinatorial problem which we did not need to solve but which the student may want to ponder over is an expression for $P^n_m(i, j, k)$. For discussion and answer, see Riordan [170].

9–4. Mood [129]. Find the number of ways in which n different objects can be arranged in k groups containing $n_1, n_2, \ldots, n_k$ objects, respectively, where $n_1 + n_2 + \cdots + n_k = n - m$.

9–5. Show that the total number of different subsets (events) including the null subset (impossible event) that can be formed from n different objects (an elementary event space of n points) is 2^n.

Combinatorial-probability problem.

9–6. From a pile of n counters, a handful is taken. Assuming that the 2^n subsets (note problem 9–5) are equally likely, study the proposition that the probability is greater than one-half that the number taken is odd!

CHAPTER 10

BINOMIAL DISTRIBUTION

10–1 The distribution. Let the outcome of a single trial be one of the two mutually exclusive and exhaustive events E or $\bar{E}$; each trial has only two possible outcomes and, on any single trial, both cannot, and one must, occur. We will call E a success, $\bar{E}$ a failure.

The event space S consisting of E and $\bar{E}$ will now be replaced by a new event space on the real line. Let the random function or random variable **y** transform the events E and $\bar{E}$ into the numbers 1 and 0:

$$\mathbf{y}(E) = 1, \qquad \mathbf{y}(\bar{E}) = 0.$$

The random variable **y** can be interpreted as the *number* of successes in a single trial.

For the probabilities of these events, we write

$$g(y) = p \quad \text{for } y = 1, \qquad g(y) = q \quad \text{for } y = 0, \qquad p + q = 1,$$

from which

$$E(\mathbf{y}) = 1 \cdot p + 0 \cdot q = p, \qquad \sigma^2(\mathbf{y}) = (1 - p)^2 p + (0 - p)^2 q = pq. \tag{10–1}$$

Now consider an experiment in which n (a positive integer) such trials are made. Let the probability of success on any trial be constant and equal to p; the (constant) probability of failure will be $q = 1 - p$. The n trials will be statistically independent, for the probability of success (or failure) on any trial is not affected by the outcome of other trials.

What is the probability $f(x)$ of x successes in n trials? The outcome of an experiment involving n trials each of which can result only in 1 or 0 will be an n-tuple of 1's and 0's and, as we have already seen, the representation of n-tuples generally requires a space of n dimensions. For example, for $n = 3$, a three-dimensional space (of 8 points) would be required, as shown in Fig. 10–1, where x_1 is the outcome of the first trial, x_2 is the outcome of the second trial, and x_3 is the outcome of the third trial. But as we are here interested only in the *number* of successes in n trials, and not in their order, we can, without concern for the loss of information on order, map all n-tuples by a *single* random variable **x** onto the real line, x being simply the number of 1's (successes) in an n-tuple. For example, all n-tuples with two successes (of which there are three in Fig. 10–1)

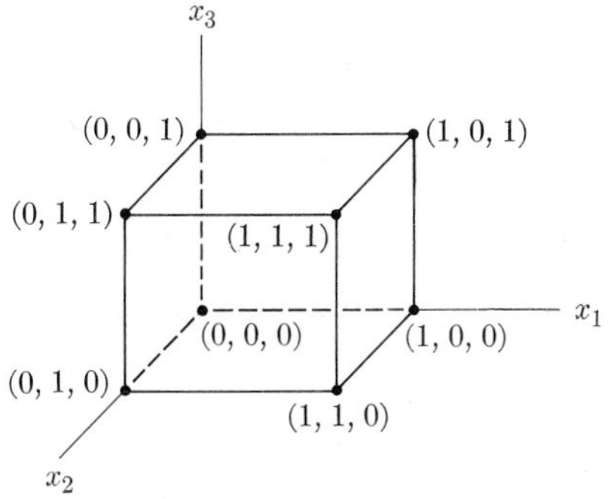

FIG. 10–1. A three-dimensional event space showing eight events.

will map into the same real number 2; the loss of information on the order by which two successes occur (whether 110 or 101 or 011) is not of concern.

One such ordered n-tuple in which x successes (and $n - x$ failures) occur (the order of trials is from left to right) is

$$\overbrace{1\ 1\ \ldots\ 1}^{\substack{x \\ \text{successes}}}\ \overbrace{0\ 0\ \ldots\ 0}^{\substack{n-x \\ \text{failures}}}. \tag{10–2}$$

Using the fact that the n trials are independent, the probability of occurrence of (10–2) is

$$p^x q^{n-x}. \tag{10–3}$$

But (10–2) is only one (though one of the most dramatic, all x successes occurring during the first x trials, all $n - x$ failures during the last $n - x$ trials) of a total of

$$P_n^n(x, n - x) = C_x^n$$

ordered n-tuples, each containing x successes and $n - x$ failures. All these n-tuples, dramatic or not, have the same probability (10–3) of occurrence. Moreover, they are mutually exclusive events, for in any one experiment of n trials, no more than one n-tuple can occur. We therefore *add* their individual probabilities to reach $f(x)$, the total probability of x successes and $n - x$ failures in n trials, and we have (James Bernoulli, 1713)

Formula for Binomial

$$f(x) = C_x^n p^x q^{n-x}, \quad x = 0, 1, \ldots, n, \tag{10–4}$$

a two-parameter (p and n are parameters, the former continuous, the latter discrete) family of probability distributions called the binomial distribution. We call p and n parameters; they are likely to differ from one binomial distribution to another, but are constant within one binomial distribution.

To summarize, we have n independent trials, two possible outcomes per trial, probabilities of success constant from trial to trial, a space S of 2^n possible compound atomic events which we map into the nonnegative integers on the real line. Such trials are often spoken of as Bernoulli trials. A set of n such trials can be viewed, as we shall see, as a reasonable description of the outcome of drawing a sample of n objects from a population of dichotomous (good and defective) objects; (10–4) can be viewed, as we shall see, as the probability that the sample of n objects will contain x good objects and $n - x$ defective objects. More exactly, the mathematical model leading to (10–4) is closely realized in the real world when (what we shall later call) a *random* sample of n objects is drawn without replacement from a population containing an indefinitely large number of good and bad objects in proportion p to q, or when random samples are drawn with replacement from any such population. In both cases, constant probability p from trial to trial (from drawing one object to drawing another) is realized. The model leading to (10–4) is *approximately* realized when sampling without replacement from a large, finite population; changes in p from trial to trial will be small.

We also have the binomial distribution function

$$p(\mathbf{x} \leq b) = F(b) = \sum_{x=0}^{x=b} C_x^n p^x q^{n-x}, \tag{10–5}$$

which along with (10–4), has now been extensively tabled by the Harvard Computation Laboratory [86] and by the National Bureau of Standards [131]. A small illustrative table of (10–4) and (10–5), formed from [131], is shown on pages 404–407.

Note that (10–4) is the general term in the expansion of the binomial

$$(q + p)^n$$

from which (10–4) gets its name. Therefore

$$p(\mathbf{x} \leq n) = \sum_{x=0}^{n} C_x^n p^x q^{n-x} = (q + p)^n \tag{10–6}$$

which is 1, as would be expected.

10–2 Applications. Areas to which the binomial distribution has been applied are many. Among the more uncommon are the power-supply and vaccine examples in Feller [57], the Mendelian theory example in Neyman [140], the random-walk example in Munroe [130], and the effect-of-stress example in Siegel [178]. We give a single example from industrial quality control.

Let an industrial process producing a large number of pieces be regarded as satisfactory if the proportion of defective output is less than or equal to p_1, and unsatisfactory if the proportion of defective output is equal to or greater than p_2. Proportions between p_1 and p_2 are regarded as describing product of intermediate quality. Let the sampling plan by which decisions are to be made to accept or reject (to be satisfied with or to improve) the process be defined by two quantities: n, the number of pieces taken from the process and inspected, and b, the maximum allowable number of defective pieces in the sample of n for the process to be called satisfactory. Let the probability that the sampling plan and the decision rule (n, b) will call unsatisfactory (reject) the worst of the satisfactory processes (one with proportion defective p_1) be a small number α; let the probability that the same sampling plan and decision rule will call satisfactory (accept) the best of the unsatisfactory processes (one with proportion defective p_2) be a small number β. We would also want the sampling plan and decision rule to be such that if even better processes were operating ($p \leq p_1$), the probability of calling them unsatisfactory would be even less than α, and if even poorer processes were operating ($p \geq p_2$), the probability of calling them satisfactory would be even less than β; for the class of binomial sampling plans considered here, these requirements are automatically satisfied. The problem: for given p_1, p_2, α, and β, determine the sampling plan and decision rule (n, b).

The statement of the problem is lengthy but the solution is brief. If the sample of size n constitutes n Bernoulli trials, we have

$$\alpha = \sum_{x=b+1}^{n} C^n_x p_1^x (1 - p_1)^{n-x}, \qquad \beta = \sum_{x=0}^{x=b} C^n_x p_2^x (1 - p_2)^{n-x},$$

which, for given p_1, p_2, α, and β, can be approximately (exactly only for certain α, β combinations) solved for n and b by maneuvering in the binomial tables (an analytical solution for n and b is impossible). For example, for

$$p_1 = 0.02, \qquad p_2 = 0.07, \qquad \alpha = 0.05, \qquad \beta = 0.10,$$

we find $n = 130$, $b = 5$. A sample of 130 pieces is drawn from the output; the process is judged satisfactory if and only if 5 or fewer defective pieces are found. With this sampling plan and decision rule, the risk of indicting a satisfactory process ($p = 0.02$) is 0.05 (the risk of indicting an even better one is *less* than 0.05), and the risk of approving an unsatisfactory process ($p = 0.07$) is 0.10 (the risk of approving an even poorer process is less than 0.10).

The logic of this example will, in more general form, play a central role in Chapter 28 on testing hypotheses, where p_1 and p_2 will be described as hypotheses on the value of the parameter p, α will be described as the probability of rejecting the hypothesis $p = p_1$ when it is true, and β as the probability of accepting the hypothesis $p = p_1$ when the hypothesis $p = p_2$ is true.

In most applications of the binomial (and of any other probability distribution), the practical problem is seldom merely one of determining the probability of a particular value of the random variable (here the number of successes x) given knowledge of the form of the distribution (here the binomial) and values of the parameters (here p and n). Generally, the problem is one in which values of the parameters are historical, hypothetical, or conjectural, and the practical problem is to determine whether or not, in the light of observed data (a particular known value of $\mathbf{x}$) these historical, hypothetical, or conjectural values are sustained.

10–3 Moments. The moment generating function

$$M_{\mathbf{x}}(\theta) = \sum_{x=0}^{n} e^{\theta x} C_x^n p^x q^{n-x} = \sum_{x=0}^{n} C_x^n (pe^{\theta})^x q^{n-x} = (q + pe^{\theta})^n \qquad (10\text{–}7)$$

yields the moments; the expression on the far right follows from that to its immediate left by virtue of the second equality of (10–6), which is an algebraic result *not* requiring $p + q = 1$ for its validity. By expansion or differentiation of (10–7) we obtain

$$E(\mathbf{x}) = np, \qquad \sigma^2(\mathbf{x}) = npq. \qquad (10\text{–}8)$$

From the fact that the binomial random variable $\mathbf{x}$ is the sum of n random variables $\mathbf{y}$ each of which has expectation p, the first result of (10–8) also follows from (7–15). From the fact that $\mathbf{x}$ is the sum of n independent $\mathbf{y}$ each of which has variance pq, the second result of (10–8) follows by applying (7–16) with $\rho = 0$.

Instead of the random variable $\mathbf{x}$, the *number* of successes in n trials, we may prefer to consider the random variable $\mathbf{x}/n$, the *proportion* of successes in n trials. The mean and variance of $\mathbf{x}/n$ which follows the binomial distribution with arguments $0/n, 1/n, 2/n, \ldots, 1$, are

$$E\left(\frac{\mathbf{x}}{n}\right) = \frac{1}{n} E(\mathbf{x}) = \frac{np}{n} = p,$$

$$\sigma^2\left(\frac{\mathbf{x}}{n}\right) = E\left(\frac{\mathbf{x}}{n} - p\right)^2 = \frac{1}{n^2} E(\mathbf{x} - np)^2 = \frac{npq}{n^2} = \frac{pq}{n}. \qquad (10\text{–}9)$$

10–4 Law of large numbers. The combination of (10–9) with the Bienaymé-Chebychev inequality (3–12) yields the classical "weak law of large numbers" for dichotomous populations. Let

$$k = \epsilon\sqrt{\frac{n}{pq}},$$

variance of proportion of successes Max

and let ϵ be a *given* small positive constant. Noting that the maximum possible value of pq is $\frac{1}{4}$, (3–12) here reduces to

$$\left(\frac{x}{n} - E\,\frac{x}{n}\right)$$

$$p\left\{\left|\frac{\mathbf{x}}{n} - p\right| \geq \epsilon\right\} \leq \frac{1}{4n\epsilon^2}, \tag{10–10}$$

due to James Bernoulli, 1713, by, of course, a different method. For any *given* positive ϵ however small, the probability that the absolute difference between the observed proportion of successes x/n and the expected proportion of successes p will exceed ϵ approaches 0 and n approaches ∞. This is a simple but profound result, connecting the performance of the *random variable* $\mathbf{x}/n$ with the *parameter p*, for n large.

10–5 Relationship to the incomplete beta-function. The binomial distribution is closely related to the incomplete beta-function, often met in statistics and applied mathematics, and itself closely related to the **F** distribution of Chapter 23; the latter distribution plays a central role in the statistical theory of modern experimental design. As direct tabling of the two-parameter binomial distribution was laborious, early tables made use of this relationship, for the incomplete beta-function had been tabled in 1934 by Karl Pearson [157]. While the advent of excellent tables of the binomial has reduced the utility of this relationship, we give it here for its mathematical interest.

Consider a continuous function $\phi(y)$ which may be expanded near $y + h$ in a Taylor series of $k + 1$ terms. Writing the remainder of the series in the somewhat uncommon integral form, we have

$$\phi(y + h) = \sum_{j=0}^{k} \frac{h^j}{j!} \frac{d^j\phi(y)}{dy^j} + \frac{1}{k!}\int_0^h (h - \theta)^k \frac{d^{k+1}\phi(y + \theta)}{dy^{k+1}}\, d\theta.$$

Replace y by the binomial q, h by the binomial p (note that q and p are continuous and bounded), $\phi(z)$ by z^n. We obtain

$$\frac{d^j\phi(y)}{dy^j} = \frac{n!}{(n-j)!}q^{n-j}, \qquad \frac{d^{k+1}\phi(y + \theta)}{dy^{k+1}} = \frac{n!}{(n-k-1)!}(q + \theta)^{n-k-1},$$

and finally,

$$(q + p)^n = 1 = \sum_{j=0}^{k} C_j^n p^j q^{n-j} + \frac{n!}{k!(n-k-1)!} \int_0^p (p - \theta)^k (q + \theta)^{n-k-1} \, d\theta. \qquad (10\text{–}11)$$

Given p, n, and k, the value of the integral (known as the incomplete beta-function) can be found in Pearson's tables, and the value of the binomial distribution function $F(k)$ is thereby determined; the Taylor series expansion merely provides a plausible avenue to the integral.

Before 1763, Thomas Bayes used (10–11) in the reverse direction; he evaluated (Deming; see [11]) the incomplete beta-function by use of the binomial distribution function.

10–6 Applicability and variants of the binomial model. In the binomial model, p is constant from trial to trial. It is not always easy to determine if this (or any other) model is the proper one for a particular practical problem. For example, in industrial sampling inspection of product each unit of which is classified as good or defective (or of naturally continuous industrial variables, reduced for practical reasons to the "good or defective" categories of the binomial) it can often be estimated or argued that the proportion p of good product coming from a process or plant is constant through the sampling. Successive trials, that is, successive drawings of product, satisfy the condition of independence, and the binomial model is appropriate. But p sometimes varies from trial to trial (the industrial process may produce runs of good or defective material as a consequence of dependence of trials) and the binomial model is invalid. Only repeated support by data can validate any model, and this is true even with coins, dice, and cards where correct models may often seem *a priori* obvious.

Variations of the binomial model include (a) n trials with p changing from trial to trial (Poisson, 1837) and considered in problem 10–4; (b) N groups each of n trials, p constant within the n trials but varying from group to group (Lexis, 1877 see A. Fisher [58]); (c) the combination of these two variants (Coolidge [31], 1925).

Tables

Harvard Computation Laboratory, *Tables of the Cumulative Binomial Probability Distribution.* Cambridge (Mass.), Harvard University Press, 1955.

National Bureau of Standards, *Tables of the Binomial Probability Distribution.* Applied Mathematics Series 6. Washington, D.C., 1950.

U.S. Dept. of Commerce, *Tables of the Cumulative Binomial Probabilities.* Washington, Ordnance Corps Pamphlet PB 111389, 1952.

Collateral Reading

H. D. Brunk, *An Introduction to Mathematical Statistics.* Boston, Ginn and Co., 1960.

W. Feller, *An Introduction to Probability Theory and Its Applications.* 2nd ed., New York, John Wiley and Sons, 1957.

A. Hald, *Statistical Theory with Engineering Applications.* New York, John Wiley and Sons, 1952.

M. G. Kendall and A. Stuart, *The Advanced Theory of Statistics,* Volume 1. London, Charles Griffin and Company, 1958.

R. von Mises, *Notes on Mathematical Theory of Probability and Statistics.* Cambridge (Mass.), Harvard University, Graduate School of Engineering, Special Publication No. 1, 1946.

E. Parzen, *Modern Probability Theory and Its Applications.* New York, John Wiley and Sons, 1960. A particularly valuable reference.

Binomial $f(x) = \binom{n}{x} p^x q^{n-x}$ $p(x \le b) = F(b) = \sum_{x=0}^{x=b} \binom{n}{x} p^x q^{n-x}$

A Characteristics:
1. Only 2 outcomes — mutually exclusive and exhaustive.
2. p(success) is constant, n is constant within one binomial distribution
3. p, n are parameters
4 Random sample n is drawn without replacement or $\bar{c}$ from population contain p good and q bad objects or visa versa.

B. Applications
1. Drawing n objects from a population of dichotomus (good + bad) eg. process is satisfactory iff 5 or fewer defective pieces from 130
   Given α (reject worse) $x > b$
   β (accept Best) $x \le b$
   p_1
   p_2
   } may also solve for α, β
2. May evaluate Beta function by use of binomial

C. Moment Generating function

$$M_x(\theta) = \sum_{x=0}^{n} e^{\theta x} \binom{n}{x} p^x q^{n-x} = (q + pe^{\theta})^n$$

yields moments by expansion or differentiation

$E(x) = np$

$\sigma^2(x) = npq$ $\sigma = \sqrt{npq}$ = S.D.

$M_{x-M}(\theta) = e^{-np\theta}(pe^{\theta} + q)^n$

diff to find variance.

PROBLEMS

10–1. For a binomial random variable **x** prove the useful recursive relationship

$$f(x+1) = \frac{n-x}{x+1} \cdot \frac{p}{1-p} f(x).$$

10–2. Show that the sum of two independent binomial variables with parameters p_1 and p_2 follows the binomial distribution if $p_1 = p_2$.

10–3. As binomial **x** moves from 0 to n, the probability of **x** increases so long as $x < np - q$.

10–4. The Poisson variant of the binomial. Let the probability of success change from trial to trial; let p_i be the probability of success on the ith trial. As in the text, let x be the number of successes in n independent trials. Show that

$$\mathbf{E}(\mathbf{x}) = np, \qquad \text{and} \qquad \sigma^2(\mathbf{x}) = npq - \sum_{i=1}^{n} (p_i - p)^2,$$

where

$$p = \frac{1}{n} \sum_{i=1}^{n} p_i.$$

Note that the variance is *smaller* than (10–8), except when $p_1 = p_2 = \cdots = p_n$.

10–5. (Kendall [103]). The variable **x** takes on only two values, 0 and 1; the respective probabilities are q and p. Let n independent trials $\mathbf{x}_1, \mathbf{x}_2, \ldots, \mathbf{x}_n$ be made on **x**. Let **r** be the number of trials which yield values as large as x_1, the outcome of the first trial. Determine $p(\mathbf{r} = n)$ and the density function of **r**.

10–6. (Prendergast). F, the rocket designer, has come to B, the reliability expert, with a problem:

"The vehicle is designed. We can use two large motors or four small motors and get the same thrust and the same weight. However, we know that the motors are subject to catastrophic failure and we have designed so that we will still get into orbit if half the motors fail. Now, if you will tell me the probability of a motor failing in the time required to get into orbit, I can decide whether to use two or four."

B replied, "We have analyzed the test data on the motors, and have found that the large and small motors have the same probability of failing in a given time. I can assure you that it makes no difference whether you use two or four motors. However, this failure probability is classified top secret and I cannot give it to you."

F said, "Never mind. From what you have just told me, I can calculate the failure probability for myself, for a motor and for the rocket."

What is the failure probability for a motor and for the rocket?

CHAPTER 11

POISSON DISTRIBUTION

11–1 The Poisson distribution as an approximation to the binomial. In the binomial density function,

$$f(x) = C_x^n p^x q^{n-x}, \quad x = 0, 1, \ldots, n, \tag{11–1}$$

let $n \to \infty$, and as $n \to \infty$ let $p \to 0$ in such a way that np remains finite and approaches a positive constant A. Conservatively, this may mean that our results are applicable if $p < 0.05$, $n > 100$; practically $p < 0.10$, $n > 50$ may be satisfactory.

We consider $f(x)$ for every *fixed* finite x; we replace p by A/n and we arrange (11–1) so that it can be evaluated at the limiting value, $n \to \infty$:

$$\begin{aligned} &\frac{n!}{x!(n-x)!}\left(\frac{A}{n}\right)^x \frac{[1-(A/n)]^n}{[1-(A/n)]^x} \\ &\quad = \frac{A^x}{x!}\,\frac{n(n-1)\cdots(n-x+1)}{n^x}\,\frac{[1-(A/n)]^n}{[1-(A/n)]^x} \\ &\quad = \frac{A^x}{x!}\left(1-\frac{A}{n}\right)^n \cdot \left\{\frac{[1-(1/n)][1-(2/n)]\cdots[1-(x-1)/n]}{[1-(A/n)]^x}\right\}. \end{aligned}$$

As

$$\lim_{n\to\infty}\left(1-\frac{A}{n}\right)^n = e^{-A}$$

and as the numerator and denominator of the expression in brackets consists of a finite number (x) of terms each of which approaches 1, (11–1) reduces to

$$f(x) = \frac{e^{-A}A^x}{x!}, \quad x = 0, 1, 2, \ldots, \tag{11–2}$$

a *single-parameter* density function of discrete **x** due to Poisson, 1837. As here obtained, as an approximation to the binomial, it is particularly useful in a class of binomial problems in which neither p nor n are known, but their product A is known or can be estimated.

Tables of the density function $f(x)$ and the distribution function

$$F(x) = \sum_{r=0}^{r=x} \frac{e^{-A}A^r}{r!} \tag{11–3}$$

have been prepared by Kitigawa [106], Whittaker [201], and Molina [128]. A small table, of $f(x)$ and $1 - F(x - 1)$ abstracted from [128], is shown on pages 408–411.

Note that

$$\sum_{x=0}^{\infty} \frac{e^{-A}A^x}{x!} = e^{-A}e^{A} = 1,$$

the sum being taken to ∞, the limiting value of n, and not to n.

The moment generating function is

$$M_{\mathbf{x}}(\theta) = \sum_{x=0}^{\infty} e^{\theta x}\frac{e^{-A}A^x}{x!} = e^{-A}\sum_{x=0}^{\infty}\frac{(e^{\theta}A)^x}{x!} = e^{-A}e^{e^{\theta}A} = e^{A(e^{\theta}-1)}, \qquad (11\text{–}4)$$

from which $E(\mathbf{x}) = A$ and $\sigma^2(\mathbf{x}) = A$; both the mean and the variance of the Poisson variable are equal to the parameter A.

Even as developed here, simply as an approximation to the binomial, applications of this distribution are numerous; they range from the number of articles lost in subways to the frequency of comets. Before the appearance of elaborate tables of the binomial distribution, one of the most successful applications of the Poisson approximation to the binomial was to sampling inspection of industrial product. The probability p of a defective unit of product is typically small, and the number of units inspected n (the number of trials) is often fairly large. A lot may be judged satisfactory only if the number of defective units r in an inspected sample of n units is less than or equal to a fixed number b; as in Chapter 10, (n, b) constitute a sampling plan and decision rule. If the conditions for Bernoulli trials of Chapter 10 are met, the Poisson approximation may apply to the calculation of probabilities.

Examples. Given the sampling plan and decision rule (50, 3). Units are drawn without replacement from a *large* lot; the conditions for Bernoulli trials approximately hold. Consider $p = 0.02$ (a good lot) and $p = 0.10$ (a bad lot). From Table A-7 on pages 410–411,

$$\sum_{r=0}^{3}\frac{e^{-1.0}(1.0)^r}{r!} = 0.981012, \qquad \sum_{r=4}^{\infty}\frac{e^{-5.0}(5.0)^r}{r!} = 0.734974$$

are the Poisson approximations to the probabilities of accepting a good lot and rejecting a bad lot, respectively, with the given sampling plan and decision rule.

Our second example involves the classical experimental data of Rutherford and Geiger showing the number of alpha particles emitted from a radioactive specimen in 2608 periods of time each of $7\frac{1}{2}$ seconds; the data are shown in Table 11–1. The parameter A of the Poisson approximation

TABLE 11-1

THE RUTHERFORD-GEIGER DATA, AND CORRESPONDING POISSON FREQUENCIES

Number of emissions, x_i	Observed frequency, f_i	Poisson frequency
0	57	54
1	203	210
2	383	407
3	525	525
4	532	508
5	408	394
6	273	254
7	139	140
8	45	68
9	27	29
10	10	11
11	4	4
12	0	1
13	1	1
14	1	1
	2608	2607

is, as we have seen, the arithmetic mean of the Poisson variable. As A is here (and generally) unknown, we estimate it reasonably by the arithmetic mean of the *data*:

$$\bar{x} = \frac{x_1 f_1 + \cdots + x_{15} f_{15}}{f_1 + f_2 + \cdots + f_{15}} = \frac{10{,}097}{2608} = 3.87. \quad = A$$

The Poisson frequencies, calculated from (11–2) using 3.87 as an estimate of A, are also shown in Table 11–1. Note that while $A = np$ is estimated by 3.87, we *cannot* (and need not) determine p and n here; we are approximating a binomial distribution which we cannot and need not fully identify.

The Poisson frequencies agree quite well with the observed frequencies, and this suggests that the conditions underlying the Poisson approximation may be satisfied here. For example, we may possibly argue that the probability p of an atom emitting an alpha particle in $7\frac{1}{2}$ seconds is small, that the number of atoms in the specimen available to emit (the number of independent trials) is very large, of the order of 10^{20} (and not 2608), that p is constant from trial to trial (that the various atoms have the same chance of emitting particles), that no chain effects are present; in short, that we have here Bernoulli trials with small p and large n. The data of Table 11–1 are, however, better interpreted in terms of a *Poisson process* and we now turn to consideration of such processes.

11–2 The Poisson distribution as a stochastic process. The Poisson distribution can be produced, not merely as an approximation to the binomial distribution, but directly from a simple and popular model of stochastic behavior dealing with events distributed randomly over time or space. This is an area of great and expanding interest in modern probability; the Poisson has, therefore, importance far beyond that of an approximation.

Beginning with some fixed but *arbitrary* time, say $t = 0$, we seek the probability $f(x, t)$ that x events occur up to time t. The random variable is the frequency $\mathbf{x}$, and the parameter of the distribution is (some fixed but arbitrary value of) t. With respect to the occurrence of events, the postulates underlying our model are:

(a) The occurrence of events in the current interval of time is independent of their occurrence or nonoccurrence during past intervals.

(b) The probability of exactly one event in the short interval $(t, t + \Delta t)$ is $B\,\Delta t$ $(+0\,\Delta t$ which we neglect) where B is a constant. If Δt is doubled, the probablity of one event is doubled. By assumption, this probability evidently depends only on the length of the interval and not on its starting point; the probability is the same for all intervals of length Δt.

(c) The probability of more than one event in the short interval Δt is assumed to be of smaller order of magnitude than Δt and is neglected; that is, the probability of more than one event in the interval Δt approaches 0 faster than does Δt.

Regarding $(0, t)$ and $(t, t + \Delta t)$ as nonoverlapping time intervals, it is clear that, given our postulates, x events up to time $t + \Delta t$ can occur in only two ways: x events in the interval $(0, t)$ and 0 events in the interval $(t, t + \Delta t)$, or $x - 1$ events in the interval $(0, t)$ and 1 event in the interval $(t, t, \Delta t)$. We have

$$f(x, t + \Delta t) = f(x, t) \cdot (1 - B\,\Delta t) + f(x - 1, t) \cdot B\,\Delta t,$$

which, as $\Delta t \to 0$, defines the differential equation,

$$\lim_{\Delta t \to 0} \frac{f(x, t + \Delta t) - f(x, t)}{\Delta t} = \frac{df(x, t)}{dt} = -Bf(x, t) + Bf(x - 1, t). \quad (11\text{–}5)$$

We want to determine $f(x, t)$. Begin with $x = 0$; since $f(-1, 0) = 0$, (11–5) reduces to

$$\frac{df(0, t)}{dt} = -Bf(0, t),$$

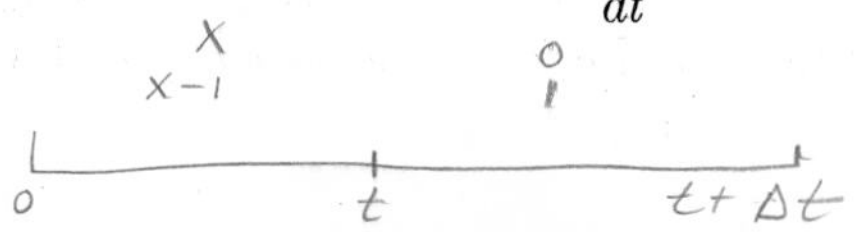

a solution of which is

$$f(0, t) = e^{-Bt}, \tag{11–6}$$

the constant of integration being determined by the reasonable assumption $f(0, 0) = 1$.

Now consider $x = 1$; inserting (11–6) in (11–5), we have

$$\frac{df(1, t)}{dt} = -Bf(1, t) + Be^{-Bt},$$

a solution of which is

$$f(1, t) = Bte^{-Bt}, \tag{11–7}$$

the constant of integration being determined by the reasonable assumption $f(1, 0) = 0$. Finally, x repetitions of the argument (and postulates) produce

$$f(x, t) = \frac{e^{-Bt}(Bt)^x}{x!}, \quad x = 0, 1, \ldots, \tag{11–8}$$

the Poisson density function of the discrete random variable $\mathbf{x}$, with continuous time parameter t; Bt is the mean value of $\mathbf{x}$.

Finally, a heuristic reinterpretation of one of the values of (11–8) and a simple but significant physical model underlying the reinterpretation is considered. The probability of 0 events up to time t is e^{-Bt}, and as "at least one" event is complementary to 0 events, the probability of at least one event is $1 - e^{-Bt}$, a function of the continuous time parameter t. But the occurrence of at least one event is surely equivalent to the occurrence of the first event by time t ("at least one" may, of course, include the second, third, . . . , events as well).

Thus, given t (say $\frac{1}{2}$ second), we have the probability (say $\frac{1}{20}$) of the first emission of an alpha particle, a probability evidently increasing with t. But this probability may be read in terms of the time to the first emission. The probability is $\frac{1}{20}$ that the time to the first emission (event) is equal to or less than $\frac{1}{2}$ second; $1 - e^{-Bt}$ is the exponential *distribution* function $F(t)$ of the continuous parameter t, with t now viewed (and this can be justified) as a continuous random variable (Fig. 11–1). The quantity t

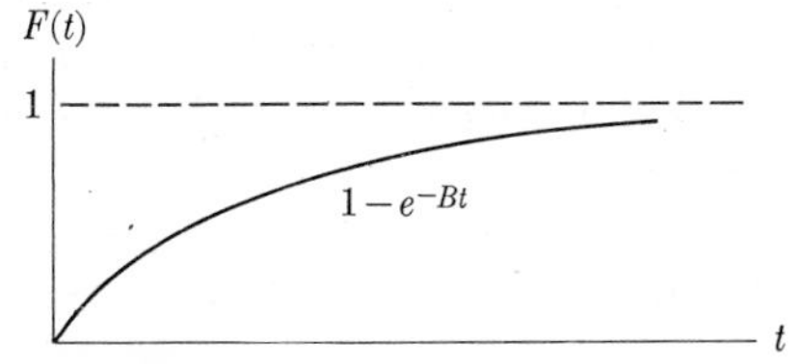

FIG. 11–1. The distribution function $F(t)$ of the time to the first event in a Poisson process.

is the time from 0 to the first event, emission or disintegration, and in this model is sometimes known as the duration time. Differentiating $F(t)$, we get the density function of **t**

$$f(t) = Be^{-Bt},$$

with mean duration time

$$E(\mathbf{t}) = \int_0^\infty tBe^{-Bt}\,dt = \frac{1}{B}.$$

These results also follow from a simple but important physical model in which the time-rate of decay is proportional to the mass M of the specimen

$$\frac{dM}{dt} = -BM, \qquad M = M_0e^{-Bt},$$

where M_0 is the original mass. The proportion (ranging from 0 to 1) of mass disappearing in time t is

$$\frac{M_0 - M}{M} = 1 - e^{-Bt}$$

and if (hopefully) we apply to a single atom this model of the behavior of a specimen of billions of atoms, we have the distribution function

$$p\text{ (an atomic disintegration before time } t) = 1 - e^{-Bt}$$

and, as already discussed, this can be replaced by

$$p\text{ (time to first disintegration } \leq t) = 1 - e^{-Bt},$$

as already given.

The stochastic process argument here, in somewhat more general form, leads to the Erlang distribution (Cramér [37]), which plays a central role in modern waiting-line theory.

11–3 Concluding remarks. The Poisson distribution is one (and the only one we consider) of a broad class of random or stochastic processes (processes which have random elements in their behavior and which exist over time or space) which are of great interest in modern probability. Examples of Poisson processes are many: from the distribution of raisins in fruit cake to the distribution of stars in the sky, from the frequency of death by horsekick in the Prussian Army to the distribution of yeast cells over the squares of a hemacytometer, the frequency of Supreme Court vacancies, electron emission times, decomposition of radioactive atomic nuclei, frequency of dust particles in the air, calls on a telephone line, daily insurance claims; all have been profitably studied as Poisson phenomena. Some of these have been viewed as examples of the Poisson as an approximation *and* as examples of the Poisson as a stochastic process.

The example of radioactive disintegration in Section 11–1, described there in terms of the Poisson approximation to the binomial distribution, may perhaps be better viewed as a Poisson stochastic process. If the time interval is small enough so that the number of atoms available for disintegration does not markedly decrease during the interval, the postulates underlying the Poisson process are probably met. If so, we may face the dilemma of observed data which are compatible with the same distribution as following from two different (but compatible) underlying arguments, the Poisson as approximation to a binomial and the Poisson as a stochastic process. Which argument to prefer is hard to say; the current trend is surely toward the latter.

Finally, note that more subtle models (for example, allowing for the declining number of atoms, for chain effects, etc.) may lead to distributions other than the Poisson; such models will (almost) always better fit the observed data.

Tables

T. Kitigawa, *Tables of the Poisson Distribution.* Tokyo, Baifukon Book Publishing Co., 1952.

E. C. Molina, *Poisson's Exponential Binomial Limit.* New York, D. Van Nostrand Co., 1942.

L. Whittaker in K. Pearson, *Tables for Statisticians and Biometricians, Part I.* 3rd ed., Cambridge (Eng.), Cambridge University Press, 1930.

Collateral Reading

A. T. Bharucha-Reid, *Elements of the Theory of Markov Processes and Their Applications.* New York, McGraw-Hill Book Co., 1960. Sophisticated applications.

K. A. Brownlee, *Statistical Theory and Methodology in Science and Engineering.* New York, John Wiley and Sons, 1960.

H. Cramér, *The Elements of Probability Theory and Some of Its Applications.* New York, John Wiley and Sons, 1955.

W. Feller, *An Introduction to Probability Theory and Its Applications.* 2nd ed., New York, John Wiley and Sons, 1957. The definitive reference as usual. Particularly impressive on this subject.

T. C. Fry, *Probability and Its Engineering Uses.* New York, D. Van Nostrand Co., 1928. Good applications.

M. E. Munroe, *Theory of Probability.* New York, McGraw-Hill Book Co., 1951. An unusual and provocative discussion.

E. Parzen, *Modern Probability Theory and Its Applications*. New York, John Wiley and Sons, 1960. Including an extensive set of serious applications.

G. P. Wadsworth and J. G. Bryan, *Introduction to Probability and Random Variables*. New York, McGraw-Hill Book Co., 1960. The Poisson process in more general form, and good examples.

Problems

11–1. Given that **x** and **y** independently follow Poisson distributions with parameters A and B. Show that $\mathbf{x} + \mathbf{y}$ follows a Poisson distribution with parameter $A + B$. Show that $\mathbf{x} - \mathbf{y}$ does not follow a Poisson distribution.

11–2. The random variable **x** follows the Poisson distribution with parameter y, and **y** is a continuous random variable with density function

$$h(y) = e^{-y}, \quad y \geq 0,$$

the latter a distribution of much interest (for example, in the description of equipment failures). Find the unconditional distribution of **x**.

Characteristics $f(x) = \frac{e^{-A} A^x}{x!}$ density $F(x) = \sum_{r=0}^{x} \frac{e^{-A} A^r}{r!}$ distribution

1. Approx. Binomial
   (a) single parameter A – discrete X – p is constant from trial to trial independent trials – without replacement
   (b) $E(x) = A$ $\sigma^2(x) = A$
   (c) $F(x) = \sum_{x=0}^{\infty} = 1$ (d) May estimate A by taking arith mean.

2. Stochastic (events distributed randomly over time or space
   $f(x,t)$ = prob. that x events occur up to time t.
   $= \frac{e^{-Bt}(Bt)^x}{x!}$ x is discrete
   t is continuous parameter
   B is constant — p same for all intervals

   p(at least one event in time t) = 1 – prob. no events = $1 - e^{-Bt}$
   $F(t) = 1 - e^{-Bt}$ expon. dist function of t
   $f(t) = Be^{-Bt}$ (differe Ft) $E(t) = \frac{1}{B}$

3. Mgf $M_x(\theta) = \sum_{x=0}^{\infty} e^{\theta x} \frac{e^{-A} A^x}{x!} = e^{A(e^{\theta}-1)}$

4. Hypothesis
   p(accepting good lot) = $p(r \leq b) = \sum_{r=0}^{x} \frac{e^{-\bar{x}}(\bar{x})^r}{r!}$
   p(rejecting bad lot) = $p(r > b) = \sum_{r=x}^{\infty} \frac{e^{-\bar{x}}(\bar{x})^r}{r!}$

CHAPTER 12

HYPERGEOMETRIC DISTRIBUTION

12–1 The distribution itself. As in Chapter 10 on the binomial distribution, we begin with a single trial which must have one of two possible outcomes: success or failure. As earlier, by a random variable **y**, we map these outcomes into the real numbers 1 and 0. On any single trial, let $p(1)$ be the probability of a success, $p(0)$ the probability of a failure; $p(1) + p(0) = 1$. Now consider n trials, but, unlike the binomial model, let the probabilities of success and failure change from trial to trial. The whole pattern of the changes we have in mind, as well as the motivation underlying them, is seen in the following model. Let the n trials consist of drawing, *without replacement*, n units (numbers: 1's or 0's) from a *finite* population containing A successes (1's) and B failures (0's), $A + B = N$. Let **x** be the number of successes (1's) in n such trials. We seek the probability $f(x)$ of x successes (and $n - x$ failures) in n trials.

The n trials are no longer mutually independent; as we are sampling without replacement, the appearance of a success or failure in a trial depends on the outcome of earlier trials. The probability that the *first* x trials yield successes is

$$\frac{A}{n} \cdot \frac{A-1}{N-1} \cdot \cdots \cdot \frac{A-(x-1)}{N-(x-1)}, \tag{12–1}$$

and the probability that the *last* $n - x$ trials yield failures is

$$\frac{B}{N-x} \cdot \frac{B-1}{N-(x+1)} \cdot \cdots \cdot \frac{B-(n-x-1)}{N-(n-1)}. \tag{12–2}$$

But x successes (all alike, all 1's) and $n - x$ failures (all alike, all 0's) can be obtained in $P_n^n(x, n-x)$ $(=C_x^n)$ mutually exclusive ways. Moreover, these are *equally likely* ways, as in the case of the binomial distribution, for though many n-tuples containing x successes and $n - x$ failures will be less striking than the ordered n-tuple described above, the probabilities associated with *all* n-tuples will turn out to be the product of (12–1) and (12–2), as the student can readily verify by trial. Finally,

$$f(x) = C_x^n$$
$$\times \frac{A(A-1)\cdots[A-(x-1)]\cdot B(B-1)\cdots[B-(n-x-1)]}{N\cdot(N-1)\cdots[N-(x-1)](N-x)[N-(x+1)]\cdots[N-(n-1)]}, \tag{12–3}$$

which (for later purposes) may be written in the form

$$f(x) = \frac{n^{[x]}}{x!} \cdot \frac{A^{[x]}B^{[n-x]}}{N^{[n]}}, \tag{12-4}$$

where

$$n^{[x]} = \frac{n!}{(n-x)!} = n(n-1)\cdots[n-(x-1)],$$

$$A^{[x]} = \frac{A!}{(A-x)!} = A(A-1)\cdots[A-(x-1)],$$

but which now is reduced to and written in the simpler form,

$$f(x) = \frac{C_x^A C_{n-x}^B}{C_n^N}, \quad x = 0, 1, 2, \ldots, n. \tag{12-5}$$

If the number of failures in the lot B is less than the number of trials n, the upper limit of x is B instead of n.

Expression (12–5) is the density function of the *hypergeometric* discrete random variable $\mathbf{x}$, with parameters N, A, and n. The historical origin of (12–5) is unclear.

The result, written in form (12–5), may seem self-evident, for the probability of x successes and $n - x$ failures in n trials should be the product of the number of ways C_x^A of obtaining x successes from the available A successes and the *independent* number of ways C_{n-x}^B of obtaining $n - x$ failures from the available B failures, divided by C_n^N, the total number of different samples of size n that can be drawn from a population of size N.

Tables of (12–5) and of the distribution function of $\mathbf{x}$ have been constructed by Neild [134] in nine volumes and by Lieberman and Owen [116]; a small illustrative table from [116] is given on pages 412–415. Neild's tables reveal that the approximation of the hypergeometric by the binomial distribution (see problem 12–4) is often not as good as has sometimes been supposed.

12–2 Examples. As most sampling is without replacement (from human and animal populations for reasons of convenience as well as, as we shall see, lower errors of inference, and from industrial product for the additional reason that even good product is often marred, or worse, in inspection and cannot be returned to the finite lot), the hypergeometric distribution has important applications. The principal application has been to sampling without replacement of finite lots of industrial product. A more recent and somewhat unusual application is to the estimation of the size of zoological populations from recapture data; we give such an example here.

In one of the simpler plans, k fish are netted from a pond, marked, and returned to the pond. Somewhat later (the experimenter having waited for mixing but not long enough for another generation to be born), k fish are netted; it is noted that m of these are marked. The probability associated with this latter event is

$$\frac{C_m^k C_{k-m}^{N-k}}{C_k^N}, \tag{12-6}$$

where N is the (unknown) number of the fish in the pond. We wish to determine N.

It is evident that the pond must contain at least $2k - m$ fish and if, for a moment, we postulate that it contains precisely this minimum number, the probability associated with the data actually obtained in the second netting is

$$\frac{C_m^k C_{k-m}^{k-m}}{C_k^{2k-m}}.$$

If this is very small (for $k = 1000$ and $m = 100$, Feller [57] calculates this probability to be of the order of 10^{-430}(!)), we may have evidence enough that a better postulate for the number of fish in the pond is one above the minimum, one which associates a higher probability with the data actually observed. In Chapter 26 we will complete the process of estimation, increasing our estimate of N until (12–6) is maximized (at $N = 10{,}000$); the only point here is that the relevant probability distribution is the hypergeometric.

12–3 Moments. To find the lower moments $E(\mathbf{x})$, $\sigma^2(\mathbf{x})$ of the hypergeometric random variable $\mathbf{x}$, it is best to work directly from the definitions of those moments, but for reasons of mathematical interest and the relevance of the argument to the statistical theory of experimental design, the moment generating function of $\mathbf{x}$ will be used; it leads into an unusual development. A second and simpler derivation follows.

We replace A by pN and B by qN; p can be regarded as the *initial* probability of success ($p = A/N$) and q the *initial* probability of failure ($q = B/N$):

$$M_{\mathbf{x}}(\theta) = \sum_{x=0}^{n} \frac{(n)^{[x]}(pN)^{[x]}(qN)^{[n-x]}}{x!N^{[n]}} e^{\theta x}. \tag{12-7}$$

Since

$$(qN)^{[n-x]} = \frac{(qN)^{[n]}}{[qN - (n - x)]^{[x]}},$$

a fact which the student can verify, (12–7) may be written

$$M_{\mathbf{x}}(\theta) = \frac{(qN)^{[n]}}{N^{[n]}} \sum_{x=0}^{n} \frac{n^{[x]}(pN)^{[x]}}{x![qN - (n - x)]^{[x]}} e^{\theta x}, \tag{12–8}$$

a form in which all factorials under the summation sign have x terms.

The sum in (12–8) is now evaluated by an indirect method. Consider the infinite hypergeometric series, from which (12–5) gets its name,

$$G(\alpha, \beta, \gamma, u) = 1 + \frac{\alpha\beta}{\gamma}\frac{u}{1!} + \frac{\alpha(\alpha+1)\beta(\beta+1)}{\gamma(\gamma+1)}\frac{u^2}{2!} + \cdots, \quad \gamma \geq 0; \tag{12–9}$$

this series was extensively studied by Gauss and Riemann (see Forsythe [68]) and was applied to our problem by Karl Pearson [155]. Obviously α and β are interchangeable, and if either α or β is negative, (12–9) has a finite number of terms [(12–9) converges for $u < 1$, diverges for $u > 1$]. We replace the three constants α, β, γ of (12–9) by our three statistical parameters p, n, N, as follows:

$$\alpha = -n, \qquad \beta = -pN, \qquad \gamma = qN - n + 1,$$

(12–9) now terminating since α is negative. We find

$$G(-n, -pN - n + 1, u) = \sum_{x=0}^{n} \frac{n^{[x]}(pN)^{[x]}u^x}{x![qN - (n - x)]^{[x]}}, \tag{12–10}$$

and for the moment generating function (12–8),

$$M_{\mathbf{x}}(\theta) = \frac{(qN)^{[n]}}{N^{[n]}} G(-n, -pN, qN - n + 1, e^{\theta}), \tag{12–11}$$

a hypergeometric function multiplied by a constant. But the infinite series (12–9) is known to be the solution of the differential equation

$$u(1-u)\frac{d^2G(u)}{du^2} + [\gamma - (\alpha + \beta + 1)u]\frac{dG(u)}{du} - \alpha\beta G(u) = 0. \tag{12–12}$$

Of what differential equation is $G(\theta)$ of (12–11) the solution? Systematically replacing terms in u by terms in θ in (12–12), we have

$$u = e^{\theta}, \qquad \frac{dG}{du} = \frac{dG}{d\theta}e^{-\theta},$$

$$\frac{d^2G}{du^2} = \frac{dG}{d\theta}(-e^{-2\theta}) + \frac{d^2G}{d\theta^2}e^{-2\theta},$$

and since, from (12–11), $M_{\mathbf{x}}(\theta)$ is simply $G(\theta)$ multiplied by a constant, (12–12) becomes

$$e^{-\theta}(1 - e^{\theta})\left[\frac{d^2M}{d\theta^2} + \frac{dM}{d\theta}\right]$$

$$+ [qN - n + 1 - (-n - pN + 1)e^{\theta}]e^{-\theta}\frac{dM}{d\theta} + nNpM = 0, \qquad (12\text{–}13)$$

the constant of (12–11) vanishing.

We know from (3–14) that the moment generating function of any random variable may be expressed as

$$M_{\mathbf{x}}(\theta) = \frac{\theta^0}{0!}\lambda_0 + \frac{\theta^1}{1!}\lambda_1 + \frac{\theta^2}{2!}\lambda_2 + \cdots, \qquad (12\text{–}14)$$

the term multiplying $\theta^0/0!$ being λ_0, the 0th moment about the origin (=1), the term multiplying $\theta^1/1!$ being λ_1 or $E(\mathbf{x})$, the first moment about the origin, the term multiplying $\theta^2/2!$ being λ_2 or $E(\mathbf{x}^2)$, the second moment about the origin, etc. But the moment generating function of a hypergeometric variable must satisfy (12–14) *and* (12–13). Therefore, if in (12–13) we replace e^{θ} by (3–13), M, $dM/d\theta$, and $d^2M/d\theta^2$ by their equivalents from (12–14), and then separately collect terms in $\theta^0/0!$, $\theta^1/1!$, $\theta^2/2!$, we get equations which yield hypergeometric moments. For example, collecting terms in θ^0, we find

$$[qN - n + 1 - (-n - pN + 1)]E(\mathbf{x}) - nNp = 0,$$

$$E(\mathbf{x}) = np, \qquad (12\text{–}15)$$

independent of N and identical with the binomial mean, as it should be, since the marginal probability of success on the kth trial is p, for any k. Collecting terms in θ^1, we find, after some reduction and using (3–9) and (12–15),

$$\sigma^2(\mathbf{x}) = npq\frac{N - n}{N - 1}, \qquad (12\text{–}16)$$

not independent of N and *less* than the binomial variance.

An entirely different (and instructive) method is used by Feller [57]. Let

$$x_k = 1 \quad \text{if the } k\text{th element of the sample is a success,}$$
$$= 0 \quad \text{if the } k\text{th element of the sample is a failure.}$$

By symmetry (though in Chapter 21 this will be formally discussed),

$$p(\mathbf{x}_k = 1) = A/N = \text{say, } p,$$
$$p(\mathbf{x}_k = 0) = 1 - p.$$

We find

$$E(\mathbf{x}_k) = 1 \cdot p + 0 \cdot (1 - p) = p$$
$$\sigma^2(\mathbf{x}_k) = (1 - p)^2 p + (0 - p)^2 (1 - p) = p(1 - p).$$

Now consider $\mathbf{x}_j\mathbf{x}_k$, $j \neq k$. Clearly

$$\begin{aligned} x_j x_k &= 1 \quad \text{only if the } j\text{th and } k\text{th elements of the sample are successes,} \\ &= 0 \quad \text{otherwise.} \end{aligned}$$

Again by symmetry, it is fairly evident that

$$p(\mathbf{x}_k = 1) = (A/N) \cdot (A - 1)/(N - 1) = \text{say, } r.$$

We find

$$E(\mathbf{x}_j\mathbf{x}_k) = 1 \cdot r + 0 \cdot (1 - r) = r$$
$$\text{cov}\,(\mathbf{x}_j\mathbf{x}_k) = E(\mathbf{x}_j - p)(\mathbf{x}_k - p) = E(\mathbf{x}_j\mathbf{x}_k) - p^2 = r - p^2.$$

Let x be the number of successes in a sample of size n. We find

$$E(\mathbf{x}) = nE(\mathbf{x}_k) = np$$
$$\sigma^2(\mathbf{x}) = \sum_{i=1}^{n} \sigma_i^2 + 2 \sum \text{cov}\ \mathbf{x}_j\mathbf{x}_k,$$

the covariance sum taken over $n(n - 1)/2$ pairs for which $j < k$. We find

$$\sigma^2(\mathbf{x}) = np(1 - p) + n(n - 1)(r - p^2)$$
$$= np(1 - p)\frac{N - n}{N - 1}.$$

12–4 Generalization. The generalization to a random variable which on each of N trials may fall in one of three or more categories, as against the two categories considered here, is simple. For three categories we would have, with obvious symbolism,

$$f(x_1, x_2, x_3) = \frac{C_{x_1}^{N_1} C_{x_2}^{N_2} C_{x_3}^{N_3}}{C_n^N}, \quad x_1, x_2, x_3 = 0, 1, 2, \ldots, n, \tag{12–17}$$

with

$$x_1 \leq N_1, x_2 \leq N_2, x_3 \leq N_3, \ N = N_1 + N_2 + N_3, n = x_1 + x_2 + x_3.$$

Tables

G. J. Lieberman and D. B. Owen, *Tables of the Hypergeometric Distribution.* Stanford, Stanford University Press, 1961.

E. F. Neild, III, *Hypergeometric Distribution, Volumes I–IX.* Cambridge (Mass.), M.I.T. thesis (S.B.), 1960.

Collateral Reading

K. A. Brownlee, *Statistical Theory and Methodology in Science and Engineering.* New York, John Wiley and Sons, 1960. A good discussion.

D. G. Chapman, "Some properties of the hypergeometric distribution with application to zoological sample censuses," *Univ. of California Publications in Statistics,* Vol. 1 (1951). For more complex models of capture-recapture censuses, see N. J. Bailey, "On estimating the size of mobile populations from recapture data," *Biometrika,* Vol. 38 (1951), pp. 293–306.

W. Feller, *An Introduction to Probability Theory and Its Applications.* 2nd ed., New York, John Wiley and Sons, 1957.

A. R. Forsythe, *A Treatise on Differential Equations,* 6th ed., London, Macmillan Co., 1933.

A. Hald, "The compound hypergeometric distribution and a system of single sampling inspection plans based on prior distributions and costs," *Technometrics,* Vol. 2 (1960), pp. 275–340.

M. G. Kendall and A. Stuart, *The Advanced Theory of Statistics, Volume 1.* 2nd ed., Charles Griffin and Co., London, 1958.

J. Neyman, *First Course in Probability and Statistics.* New York, Henry Holt and Co., 1950.

L. Schmetterer, *Einführung in die Mathematische Statistik.* Vienna, Springer Verlag, 1956. A very good reference.

G. P. Wadsworth and J. G. Bryan, *Introduction to Probability and Random Variables.* New York, McGraw-Hill Book Co., 1960. With good applications.

1. Sampling without replacement (destroy sample) $f(x) = \frac{C^A_x \, C^B_{n-x}}{C^N_n} = \frac{\binom{A}{x}\binom{B}{n-x}}{\binom{A+B}{n}}$
2. n is sample drawn from N = A + B
   p changes from trial to trial (w/o replacement cause dependent)
   f(x) = p(x successes in n trials) $E(x) = np$
3. This gives estimate of N $\sigma^2(x) = npq\,\frac{N-n}{N-1}$
4. May generalize for variables that may fall into 1 of 3 or more categories
5. Don't use Mgf use:
   $P(X_k = 1) = \frac{A}{N} = p$, $P(X_k = 0) = (1-p) = \frac{B}{N}$ $E(X_k) = 1 \cdot p + 0(1-p) = p$ (for 1 variable) etc. $V(X_k) = \sum X_k^2 P(X_k) - E^2(X_k) = pq$
6. Lower Limit of N = 2K − m for fish problem

Generalize $\frac{\binom{N_1}{n_1}\binom{N_2}{n_2}\cdots\binom{N_K}{n_K}}{\binom{N}{n}}$ $V(X_1 + X_2) = V(X_1) + V(X_2) - \text{cov.}(X_1, X_2)$

PROBLEMS

12–1. Show that

$$\sum_{x=0}^{x=n} \frac{C_x^{pN} C_{n-x}^{qN}}{C_n^N} = 1.$$

12–2. Given A, n, and x, for what value of N is the hypergeometric probability (12–5) maximum?

12–3. Solve problem 4–2, using the methods of this chapter.

12–4. It is intuitive that for N and A large and n small relative to N, the hypergeometric model approaches the binomial. For in this case, sampling without replacement has negligible effect on the original proportion of successes A/N in the population. Show formally the equality of hypergeometric and binomial probabilities under the stated conditions.

To show V(sampling w/out repl.) < V(sampl. w/ repl.)

$$V(X) = npq + n(n-1)(r - p^2)$$

where $r = p \cdot \frac{A-1}{N-1}$

As N gets large $npq(X) \to npq$

$$\text{Mgf} = \sum e^{\theta x} f(x)$$

replacing:

$$= \sum_{x=0}^{n} \frac{n^{[x]} (pN)^{[x]} (qN)^{[N-x]}}{x!\, N^{[n]}} e^{\theta x}$$

ex. $N = 50 \begin{cases} 20 \text{ w} \\ 18 \text{ y} \\ 12 \text{ g} \end{cases}$ $n = 10 \left\{ \begin{matrix} 2 \text{ w} \\ 3 \text{ y} \\ 5 \text{ G} \end{matrix} \right\} = E$

$$P(E) = \frac{\binom{20}{2}\binom{18}{3}\binom{12}{5}}{\binom{50}{10}}$$

CHAPTER 13

NEGATIVE BINOMIAL DISTRIBUTION

13–1 The distribution and one of its origins. The negative binomial distribution was discovered in the course of study of the frequency of death following repeated exposure to a disease or to a toxicant. In the earlier part of this chapter, we retain the language of this area of application.

We shall consider the event death during the nth exposure (to a disease or toxicant), and the probability of death during that exposure. Let p be the probability of an individual being attacked during any single exposure to a disease. In the simple model of this chapter, this probability is *constant* from exposure to exposure; exposure neither conveys immunity nor increases susceptibility. Let q be the (constant) probability of not being attacked; $p + q = 1$. Let n (which may be large) be the number of exposures to the disease; in the language of the binomial distribution, n is the number of independent trials. The successive terms of $(q + p)^n$ are the probabilities that an individual will be attacked $0, 1, \ldots, n$ times in n exposures.

Assume that death ensues after r attacks; when an individual has been repeatedly exposed to a poison, r has sometimes been described as the lethal dose. The probability that an individual will survive n exposures is identical with the probability that he will be attacked 0 or 1 or . . . or $r - 1$ times during n exposures, that is,

$$\sum_{j=0}^{r-1} C_j^n q^{n-j} p^j. \tag{13–1}$$

Now consider $n - 1$ exposures, death, as before, ensuing after r attacks. The probability of surviving $n - 1$ exposures is

$$\sum_{j=0}^{r-1} C_j^{n-1} q^{n-1-j} p^j. \tag{13–2}$$

The probability of dying during the nth exposure [a function of n and therefore written $f(n)$] is the probability of surviving $n - 1$ exposures minus the probability of surviving n exposures, that is, (13–2) — (13–1). To reduce this difference to its simplest form, we recall from problem 9–2

$$C_j^n = C_j^{n-1} + C_{j-1}^{n-1},$$

and we obtain

$$f(n) = \sum_{j=0}^{r-1} (C_j^{n-1} p^{j+1} q^{n-j-1} - C_{j-1}^{n-1} p^j q^{n-j}). \qquad (13\text{–}3)$$

Putting, successively, $j = 0, 1, \ldots, r - 1$ in (13–3), the positive term at each value of j is cancelled by the negative term at the next higher value of j. Only the positive term at $j = r - 1$ remains; the probability of death during the nth exposure reduces to the simple density function of the discrete random variable **n**,

$$f(n) = C_{r-1}^{n-1} p^r q^{n-r}, \quad n \geq r, \qquad (13\text{–}4)$$

and 0 for smaller values of n (Yule, 1910). This is the negative binomial distribution, known also as the Pascal or the Pólya distribution. The probability $f(n)$ of death during the nth exposure depends on two parameters: r the lethal number of attacks and p the constant probability of attack. Extensive tables, a small extract of which is shown on page 416, have been constructed by Beale [12]; tables of the positive binomial readily yield negative binomial probabilities, as we shall see in Section 13–3.

We have

$$n = r, \qquad f(n) = p^r,$$

(which is natural, since at $n = r$, the individual to die must be attacked at each exposure);

$$n = r + 1, \qquad f(n) = p^r rq,$$

and

$$n = r + 2, \qquad f(n) = p^r \cdot \frac{r(r+1)}{2} q^2,$$

and so on, the probabilities being successive terms in the expansion of

$$p^r(1 - q)^{-r} \qquad (13\text{–}5)$$

(equal to 1) from which the name "negative binomial" comes. Note that while the probability of death during the $(r + 1)$th exposure may be smaller than during the rth, the $(r + 2)$th smaller than the $(r + 1)$th, and so on, the *cumulative* probability [the distribution function $F(n)$] of death increases with additional exposure. Note also the similarity to

TABLE 13–1

COMPARISON OF THE NEGATIVE AND THE POSITIVE BINOMIAL

Binomial	Random variable	Parameters	Range of random variable
Negative	**n**	p, r	$n = r, r + 1, \ldots$
Positive	**x**	p, n	$x = 0, 1, \ldots, n.$

(and differences from) the notation of the positive binomial (10–4) shown in Table 13–1.

The following small numerical example may throw light on this curious distribution. While the theory was developed in terms of the probability of death to a single individual, this example is formulated in terms of a community of 900 persons. We have a population of 900 persons, $p = \frac{2}{3}$, $r = 2$. The example is tabulated for two to six exposures in Table 13–2A and in greater detail for one to three exposures in Table 13–2B.

TABLE 13–2A

NEGATIVE BINOMIAL PROBABILITIES, WITH $p = \frac{2}{3}$, $r = 2$, AND CORRESPONDING FREQUENCIES IN A POPULATION OF SIZE $N = 900$

Number of exposures, n	$C_{r-1}^{n-1}p^rq^{n-r}$	Number who die
2	4/9 = 324/729	400
3	8/27 = 216/729	267
4	12/81 = 108/729	133
5	16/243 = 48/729	59
6	20/729 = 20/729	25

TABLE 13–2B

DETAILED VIEW OF THE EFFECT OF ONE TO THREE EXPOSURES, WITH $p = \frac{2}{3}$, $r = 2$, $N = 900$.

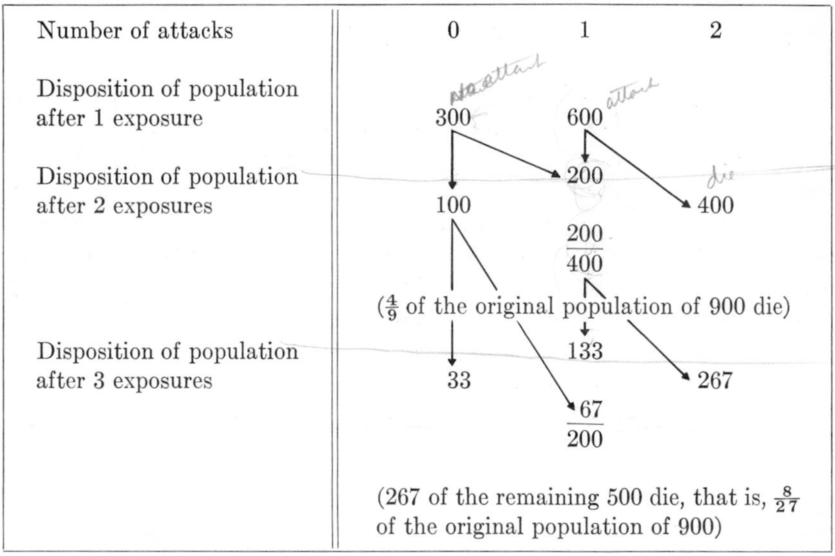

13–2 Moments. The moments of the negative binomial random variable **n** are readily obtained from its moment generating function. We have

$$
\begin{aligned}
M_{\mathbf{n}}(\theta) &= \sum_{n=r}^{\infty} e^{\theta n} C_{r-1}^{n-1} p^r q^{n-r} \\
&= e^{\theta r} \sum_{n=r}^{\infty} C_{r-1}^{n-1} p^r (qe^{\theta})^{n-r} \\
&= e^{\theta r} p^r (1 - qe^{\theta})^{-r}. \qquad (13\text{–}6)
\end{aligned}
$$

Expansion or differentiation of (13–6) yields

$$
\begin{aligned}
E(\mathbf{n}) &= \frac{r}{p}, \\
\sigma^2(\mathbf{n}) &= \frac{rq}{p^2}.
\end{aligned} \qquad (13\text{–}7)
$$

13–3 Other approaches to the negative binomial. The distribution (13–4) may be reached by a different and simpler path. Again, let p be the constant probability of being attacked and q the (constant) probability of not being attacked; what is the probability that exactly n exposures are required for an individual to be attacked exactly r times (and therefore die during the nth exposure)? For this to happen, the individual must be attacked $r - 1$ times during the first $n - 1$ exposures, the probability of which is, by the positive binomial,

$$C_{r-1}^{n-1} p^{r-1} q^{n-r},$$

and he must be attacked during the nth exposure, the probability of which is p. The product of these two probabilities is identical with (13–4).

The problem as phrased here, of determining the probability that the rth "success" occurs on the nth independent trial, is one of a general class of "waiting-time" problems, including problems found in the literature under such titles as pure birth processes, branching processes, and (limiting forms of) occupancy problems, which more and more are seen to unify much of the area of modern applied probability.

Finally, we present a somewhat more specialized view of the negative binomial distribution, though one which in fact generalizes the Poisson distribution. Consider a random variable **x** which follows the Poisson distribution (11–2), but with its (single) parameter here regarded as a *random variable* independently following a particular continuous probability distribution. Such a model was proposed by Greenwood and Yule [78] in 1920. They argued that while for one person the probability of 0, 1, 2, . . .

illnesses or accidents may follow a Poisson distribution with parameter λ,

$$f(x) = \frac{e^{-\lambda}\lambda^x}{x!}, \quad x = 0, 1, 2, \ldots,$$

the mean number (also λ) of illnesses or accidents (which has been interpreted as proneness to illness or accident) may vary from person to person.

The particular probability distribution of λ proposed (merely for mathematical convenience) by Greenwood and Yule was the *gamma* or *Pearson Type III* distribution whose continuous density function is

$$g(\lambda) = \frac{1}{\alpha!\beta^{\alpha+1}} \lambda^\alpha e^{-\lambda/\beta}, \quad \beta > 0, \quad \alpha \geq 0 \text{ integral}, \quad \lambda \geq 0. \tag{13-8}$$

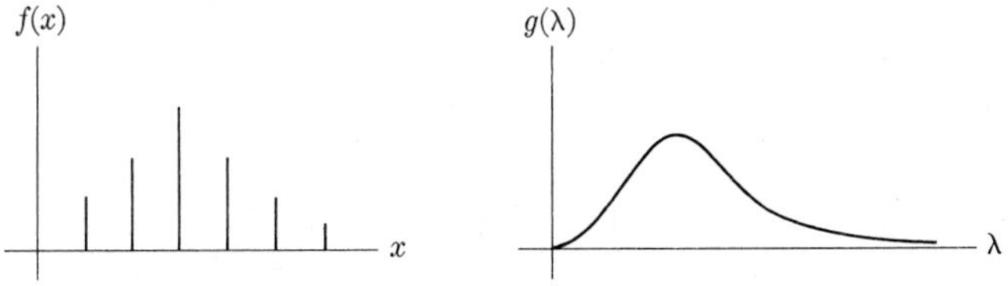

FIG. 13-1. Density functions of Poisson and gamma random variables.

Density functions of (discrete) Poisson and (continuous) gamma random variables are shown in Fig. 13-1. The joint density function of **x** and λ is the product of (11-2) and (13-8),

$$h(x, \lambda) = \frac{1}{x!\alpha!\beta^{\alpha+1}} \lambda^{x+\alpha} e^{-\lambda[1+(1/\beta)]}; \tag{13-9}$$

integrating (13-9) on λ over its range 0 to ∞, we get the density function of **x**. The result is

$$\frac{1}{x!\alpha!\beta^{\alpha+1}} \cdot \frac{(x+\alpha)!}{[1+(1/\beta)]^{x+\alpha+1}}, \tag{13-10}$$

which, on putting $x + \alpha = n - 1$, $\beta/(1 + \beta) = q$, then $x = n - r$, turns out to be exactly (13-4), though now with *nonintegral* values of the parameter r as well as p.

This is a simple model; any cumulative impact of exposure to illness or accident, either toward lowering resistance or conveying immunity, is not considered. No concept of degree of exposure is present. But it is path-breaking; it is an early (1920) example of a stochastic process such as is discussed in Chapter 11 (the number of small time intervals of exposure to illness or accident is large, two accidents cannot occur in one time-interval, etc.).

TABLE 13–3

FREQUENCY OF ACCIDENTS IN FIVE WEEKS TO 648 WOMEN (WITH FITTED POISSON AND NEGATIVE BINOMIAL DISTRIBUTIONS)

Number of accidents	Number of women	Frequencies calculated from	
		Poisson distribution	Negative binomial distribution
0	447	406	442
1	132	189	140
2	42	45	45
3	21	7	14
4	3	1	5
5	2	0	2
Total	647	648	648

Table 13–3 was given by Greenwood and Yule [78]; it relates to the frequency of accidents in five weeks to 648 (their table totals 647) women working on artillery shells.

TABLES

R. BEALE, *Generation of the Pascal Distribution Tables on the 704 Electronic Computer*. M.I.T. term paper, 1959.

G. TAGUCHI, "Tables of 5 per cent and 1 per cent points for the Pólya-Eggenberger distribution function," *Reports of Statistical Application Research, Union of Japanese Scientists and Engineers*, Vol. 2, 1952–1953, pp. 27–32.

COLLATERAL READING

F. J. ANSCOMBE, "Sampling theory of the negative binomial and logarithmic series distributions," *Biometrika*, Vol. 37 (1950), pp. 358–382.

N. ARLEY, *Stochastic Processes and Cosmic Radiation*. Copenhagen, G.E.C. Gads Forlag, 1943. An important reference. Including the simplified Bhabha-Heitler model of cosmic-ray showers, leading to a Poisson process; see "On the theory of stochastic processes and their application to the theory of cosmic radiation," (pp. 88–90). There is much additional information on related negative binomial models (pp. 92–97).

GRACE E. BATES and J. NEYMAN, "Contributions to the theory of accident proneness, I, II," *Univ. of California Publications in Statistics*, Vol. 1 (1952), Nos. 9, 10, pp. 215–234, 255–276.

R. R. BUSH and F. MOSTELLER, *Stochastic Models for Learning*. New York, John Wiley and Sons, 1955. Application in learning-process models.

F. EGGENBERGER and G. PÓLYA, "Über die Statistik verketterer Vorgänge," *Zeitschrift für Angewandte Mathematik und Mechanik*, Vol. 1 (1923), pp. 279–289.

W. FELLER, *An Introduction to Probability Theory and Its Applications.* 2nd ed., New York, John Wiley and Sons, 1957. The best reference. Includes, among examples, the already celebrated Banach match-box problem.

M. GREENWOOD, "Accident proneness," *Biometrika*, Vol. 37 (1950), pp. 24–29.

M. GREENWOOD and G. U. YULE, "An enquiry into the nature of frequency distributions representative of multiple happenings . . . ," *Jour. Royal Stat. Soc.*, Vol. 83 (1920), pp. 255–279.

H. JEFFREYS, *Theory of Probability.* 3rd ed., Oxford, Clarendon Press, 1961.

N. L. JOHNSON, "Uniqueness of a result in the theory of accident proneness," *Biometrika*, Vol. 44 (1957), pp. 530–531. Includes a generalization of the Greenwood-Yule model.

O. LUNDBERG, "On random processes and their application to sickness and accident statistics." *Uppsala*, Almqvist and Wiksells, 1940. Extension of the negative binomial to complex models and a unification of the general theory.

"Student's" Collected Papers. Edited by E. S. PEARSON and J. WISHART, Cambridge (Eng.), Cambridge University Press, 1942. Paper No. 9.

G. U. YULE, "On the distribution of deaths with age when causes of death act cumulatively, and similar frequency distributions," *Jour. Royal Stat. Soc.*, Vol. 73 (1910), pp. 26–38.

f(n) = p(death on n^{th} exposure)

Negative binomial

$f(n) = P(n-1) - P(n)$

$P(n)$ = prob(surviving n exp) = $\sum_{j=0}^{r-1} C_j^n q^{n-j} p^j$

$P(n-1)$ = p(surv, n-1 exposures) = $\sum_{j=0}^{r-1} C_j^{n-1} q^{n-1-j} p^j$

$f(n) = C_{r-1}^{n-1} p^r q^{n-r}, \quad n \geq r$

p = constant prob. of attack
r = lethal # of attacks } 2 parameters { n = variable

$M_n(\theta) = \sum e^{\theta n} C_{r-1}^{n-1} p^r q^{n-r} = e^{\theta r} p^r (1-qe^{\theta})^{-r}$

$E(n) = \frac{r}{p}$ $\sigma^2(n) = \frac{rq}{p^2}$ check $\sum_{r}^{\infty} f(n) = 1$

Ways to derive neg. Binomial

(1) binomial $\binom{n-1}{r-1} p^{r-1} q^{(n-1)-(r-1)}$

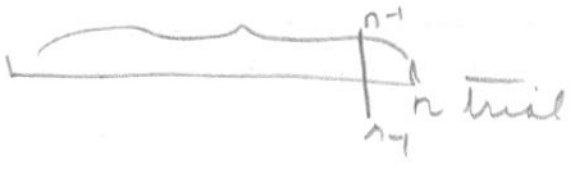

f(0) ... f(r-1) } = 0
f(r) ... f(n) } = 1

expansion $f(r) = p^r$ $f(r+1) = r p^r q + \dots + \dots = 1$

$f(n) = \binom{-r}{\ } p^r (-q)^{n-r}$

1st success when r = n

(2) TREE

\# of exposures
Attack or success = p
non attack failure q

$f(n) = \sum_{x=0}^{r-1} \binom{n-1}{x} p^x q^{n-x-1} - \sum_{x=0}^{r-1} \binom{n}{x} p^x q^{n-x}$

CHAPTER 14

MULTINOMIAL DISTRIBUTION

14–1 The distribution and its moments. We now consider the multinomial distribution, a straightforward generalization of the binomial distribution. In the binomial, the mutually exclusive and exhaustive outcomes of a single trial are E and $\bar{E}$; in the multinomial, they are E_1, $E_2, \ldots, E_h$. In the binomial, the probabilities are constant from trial to trial, with $p(E) + p(\bar{E}) = 1$; in the multinomial, the probabilities are constant from trial to trial, with $p(E_1) + p(E_2) + \cdots + p(E_h) = 1$.

As in the binomial, let n independent trials be made. The random variables will be $\mathbf{x}_1, \mathbf{x}_2, \ldots, \mathbf{x}_h$ where x_1 is the number of occurrences of E_1, x_2 the number of occurrences of $E_2, \ldots, x_h$ the number of occurrences of E_h, where $x_1, x_2, \ldots, x_h$ are nonnegative integers subject to the condition $x_1 + x_2 + \cdots + x_h = n$. We seek to determine the joint density function $f(x_1, x_2, \ldots, x_h)$.

The first x_1 trials could yield E_1, the second x_2 trials $E_2, \ldots,$ the last x_h trials E_h. The probability of this *ordered* result is

$$p_1^{x_1} p_2^{x_2} \ldots p_h^{x_h}.$$

But x_1 alike values E_1, x_2 alike values $E_2, \ldots, x_h$ alike values E_h can be obtained in

$$P_n^n(x_1, x_2, \ldots, x_h) = \frac{n!}{x_1! x_2! \ldots x_h!}$$

mutually exclusive ways. Moreover, these are readily seen to be *equally likely* ways. Therefore (J. Bernoulli, 1713),

$$f(x_1, x_2, \ldots, x_h) = \frac{n!}{x_1! x_2! \ldots x_h!} p_1^{x_1} p_2^{x_2} \ldots p_h^{x_h}, \tag{14–1}$$

for $x_1, x_2, \ldots, x_h$ nonnegative integers whose sum is n. This is the *multinomial* density function, depending on h continuous parameters $p_1, p_2, \ldots, p_h$ subject to one constraint, one discrete parameter n, and h discrete random variables $\mathbf{x}_1, \mathbf{x}_2, \ldots, \mathbf{x}_h$. Tables have been constructed by Pierce [158].

Expression (14–1) is the general term in the expansion of the multinomial

$$(p_1 + p_2 + \cdots + p_h)^n,$$

that is,

$$\sum \frac{n!}{x_1!x_2!\ldots x_h!} p_1^{x_1} p_2^{x_2} \ldots p_h^{x_h} = (p_1 + p_2 + \cdots + p_h)^n = 1, \quad (14\text{–}2)$$

where the sum in (14–2) [as well as the sum in (14–3) below] is over all possible values of $x_1, x_2, \ldots, x_h$ which satisfy $x_1 + x_2 + \cdots + x_h = n$. The argument is seen to be entirely analogous to that for the binomial. Also analogous is the moment generating function

$$\begin{aligned} M_{\mathbf{x}_1,\mathbf{x}_2,\ldots,\mathbf{x}_h}(\theta_1, \theta_2, \ldots, \theta_h) &= \sum e^{(\theta_1 x_1 + \theta_2 x_2 + \cdots + \theta_h x_h)} \frac{n!}{x_1!x_2!\ldots x_h!} p_1^{x_1} p_2^{x_2} \ldots p_h^{x_h} \\ &= \sum \frac{n!}{x_1!x_2!\ldots x_h!} (p_1 e^{\theta_1})^{x_1} (p_2 e^{\theta_2})^{x_2} \ldots (p_h e^{\theta_h})^{x_h}; \quad (14\text{–}3) \end{aligned}$$

as with the similar expression which appeared in the discussion of the binomial distribution, the first equality of (14–2) is an *algebraic* identity not requiring for its validity the probability condition $p_1 + p_2 + \cdots + p_h = 1$. Applying this fact to (14–3), the moment generating function reduces to

$$M_{\mathbf{x}_1,\mathbf{x}_2,\ldots,\mathbf{x}_h}(\theta_1, \theta_2, \ldots, \theta_h) = (p_1 e^{\theta_1} + p_2 e^{\theta_2} + \cdots + p_h e^{\theta_h})^n. \quad (14\text{–}4)$$

Expansion of (14–4) or differentiation of (14–4) with respect to θ_i yields

$$E(\mathbf{x}_i) = np_i, \qquad \sigma^2(\mathbf{x}_i) = np_i(1 - p_i), \quad (14\text{–}5)$$

while the cross-derivative of (14–4) with respect to θ_i and θ_j yields the covariance $-np_ip_j$, the negative sign indicating *negative* correlation between $\mathbf{x}_i$ and $\mathbf{x}_j$. Again note that all arguments here are generalizations of those used for the binomial distribution; the latter is the special case of the multinomial $h = 2$.

14–2 Examples. The multinomial model is clearly appropriate to several dice as the binomial is to two; it is also appropriate to polling inquiries, market surveys, and industrial inspection in which multiple outcomes of a single trial as well as sampling with replacement, or without replacement from *large* populations are met (the two sampling models implying constant p from trial to trial). The sampling of a consumer population faced by several brands, the sampling of industrial output, each unit of which can fall into one of several quality categories, often suggest the multinomial model.

We give an industrial example from Hald [81]. Upper and lower engineering tolerances on a unit of product are such that, given normal

TABLE 14–1

THE TRINOMIAL $f(x_1, x_2, x_3)$ FOR $p_1 = 0.06,\ p_2 = 0.02,\ p_3 = 0.92,$
$n = 100$

x_1 \ x_2	0	1	2	3	4	5	6	7	8	Marginal distribution
0	0.0002	0.0005	0.0006	0.0004	0.0002	0.0001	0.0000	0.0000	0.0000	0.0020
1	0.0016	0.0034	0.0036	0.0025	0.0013	0.0005	0.0002	0.0001	0.0000	0.0132
2	0.0050	0.0107	0.0113	0.0079	0.0041	0.0017	0.0006	0.0002	0.0000	0.0415
3	0.0107	0.0226	0.0236	0.0163	0.0083	0.0034	0.0011	0.0003	0.0001	0.0864
4	0.0170	0.0354	0.0366	0.0249	0.0126	0.0050	0.0017	0.0005	0.0001	0.1338
5	0.0213	0.0439	0.0448	0.0302	0.0151	0.0060	0.0020	0.0005	0.0001	0.1639
6	0.0219	0.0448	0.0453	0.0302	0.0149	0.0059	0.0019	0.0005	0.0001	0.1655
7	0.0192	0.0389	0.0389	0.0256	0.0125	0.0048	0.0016	0.0004	0.0001	0.1420
8	0.0146	0.0291	0.0288	0.0188	0.0091	0.0035	0.0011	0.0003	0.0001	0.1054
9	0.0097	0.0192	0.0188	0.0121	0.0058	0.0022	0.0007	0.0002	0.0000	0.0687
10	0.0058	0.0113	0.0109	0.0069	0.0033	0.0012	0.0004	0.0001	0.0000	0.0399
11	0.0031	0.0060	0.0057	0.0036	0.0017	0.0006	0.0002	0.0001	0.0000	0.0210
12	0.0015	0.0029	0.0027	0.0017	0.0008	0.0003	0.0001	0.0000	0.0000	0.0100
13	0.0007	0.0012	0.0012	0.0007	0.0003	0.0001	0.0000	0.0000	0.0000	0.0042
14	0.0003	0.0005	0.0005	0.0003	0.0001	0.0001	0.0000	0.0000	0.0000	0.0018
15	0.0001	0.0002	0.0002	0.0001	0.0001	0.0000	0.0000	0.0000	0.0000	0.0007
Marginal distribution	0.1327	0.2706	0.2735	0.1822	0.0902	0.0354	0.0116	0.0032	0.0006	1.0000

production, 6% of the output will lie above the upper tolerance, 2% below the lower tolerance. In 100 independent trials with replacement, that is, in a sample of 100 units of product, the trinomial density function

$$\frac{100!}{x_1!x_2!(100 - x_1 - x_2)!}(0.06)^{x_1}(0.02)^{x_2}(0.92)^{100-x_1-x_2}$$

gives the probabilities of various frequencies x_1 and x_2 of the two kinds of defective product and (residually, since $x_3 = n - x_1 - x_2$) of good product. These probabilities are tabulated in Table 14–1. Note that the probability of the most likely outcome ($x_1 = 6$, $x_2 = 2$, $x_3 = 92$) is 0.0453, hardly large.

As a second example let us hypothesize that the voting public is equally divided among three candidates A, B, C. Then the probability that 10 independent trials with replacement will yield $8A$, $0B$, $2C$, is

$$\frac{10!}{8!0!2!}\left(\frac{1}{3}\right)^8\left(\frac{1}{3}\right)^0\left(\frac{1}{3}\right)^2 = \frac{45}{59{,}049}. \tag{14–6}$$

Note that, given the hypothesis, the probability of the *most* likely outcome $3A$, $3B$, $4C$ (or $3A$, $4B$, $3C$ or $4A$, $3B$, $3C$) is hardly large:

$$\frac{10!}{3!3!4!}\left(\frac{1}{3}\right)^3\left(\frac{1}{3}\right)^3\left(\frac{1}{3}\right)^4 = \frac{4200}{59{,}049}.$$

As we have noted earlier, in real problems the values of the parameters of a probability distribution are often given not by fact but by hypothesis, and the purpose of the n independent trials is to provide information to serve as the basis for a decision as to the reasonableness of the hypothesis. So it is here. In the example with three candidates, the hypothesis is that the voting public is equally divided; the data from the 10 independent trials will later lead, on analysis of calculations such as (14–6), to acceptance or rejection of this hypothesis.

Similarly, in the industrial example, it is hardly the ultimate purpose of statistical analysis to study the behavior of repeated trials (here 100), given *full* knowledge of the distribution function (here the multinomial) and *full* knowledge of the values of the parameters (here $p_1 = 0.06$, $p_2 = 0.02$, $p_3 = 0.92$). More commonly, we *hypothesize* that the industrial process is described by $p_1 = 0.06$, $p_2 = 0.02$, $p_3 = 0.92$, and the data yielded by $n = 100$ trials will later lead, on analysis, to acceptance or rejection of this hypothesis.

Tables

J. Pierce, *Tables of the Multinomial Distribution.* Cambridge (Mass.), M.I.T. thesis (M.S.), 1957.

Collateral Reading

H. D. Brunk, *An Introduction to Mathematical Statistics.* Boston, Ginn and Co., 1960.

F. N. David, *Probability Theory for Statistical Methods.* Cambridge (Eng.), Cambridge University Press, 1949.

J. G. Smith and A. J. Duncan, *Sampling Statistics and Applications.* New York, McGraw-Hill Book Co., 1945. Includes a detailed example.

Problem

14–1. Discuss the marginal and conditional distributions associated with the multinomial distribution.

$$f(x_1, \dots x_h) = \frac{n!}{x_1!\, x_2! \cdots x_n!} p_1^{x_1} \cdots p_h^{x_h}$$

$$E(x_i) = np_i \qquad \sigma^2(x_i) = np_i(1-p_i)$$

Outcomes $E_1, E_2, \dots, E_h$

any p_i is constant from trial to trial

$p_1, p_2 \dots p_h$ are contin parameters

$x_1 \dots x_h$ " discrete random variables

n — discrete parameter — indep. trials.

$p(E_1) + \dots + p(E_h) = 1$

CHAPTER 15

UNIFORM, LAPLACE, CAUCHY DISTRIBUTIONS

In this chapter we consider briefly three probability distributions which appear in theoretical and applied statistical literature and which will reappear later in this text.

15–1 Uniform distribution. The uniform or rectangular random variable $\mathbf{x}$ is continuous and has constant probability over the range $a < \mathbf{x} < b$, with discontinuities at the endpoints $\mathbf{x} = a$ and $\mathbf{x} = b$, where a and b are finite and real. The density function of $\mathbf{x}$ is

$$f(x) = \frac{1}{b - a}, \quad a < x < b, \quad f(x) = 0 \text{ elsewhere.} \tag{15–1}$$

The density is constant over the range $a < x < b$ (Fig. 15–1). Equivalently, the distribution function of the uniform random variable is

$$F(x) = \frac{x - a}{b - a}, \quad a < x < b, \tag{15–2}$$

with $F(x) = 0$ for $x \leq a$ and $F(x) = 1$ for $x \geq b$; differentiation of (15–2) produces (15–1). The first two moments are

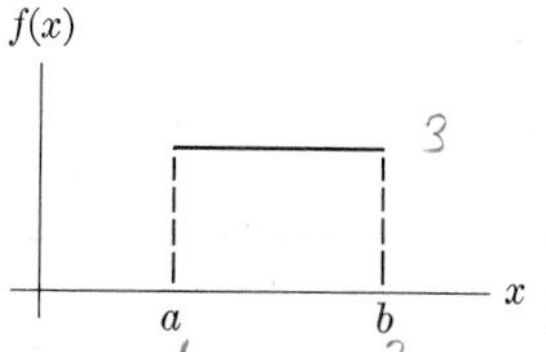

FIG. 15–1. Uniform density function.

$$E(\mathbf{x}) = \int_a^b xf(x)\,dx = \frac{a + b}{2}, \tag{15–3}$$

and

$$\sigma^2(\mathbf{x}) = \int_a^b [x - E(\mathbf{x})]^2 f(x)\,dx = \frac{(b - a)^2}{12}. \tag{15–4}$$

The uniform distribution is hardly a likely description of the probability behavior of many physical, biological, and social phenomena; its importance lies in its utility in statistical theory. As we have seen, any continuous probability distribution can be readily transformed first into the simple uniform distribution, then into a *given* continuous distribution. This often facilitates the study of properties of distributions which themselves are somewhat intractable. As we have seen by the argument of page

73, given the density function $g(y)$ of any continuous random variable **y**, the change of variable

$$x = G(y) = \int_{-\infty}^{y} g(y)\, dy, \qquad \text{that is,} \qquad \frac{dx}{dy} = g(y), \tag{15-5}$$

where x is evidently the *distribution function* of **y**, converts the probability element $g(y)\,dy$ of the random variable **y** into the *uniform* probability element dx of the random variable **x**. From (15–5), the range of x is seen to be $0 < x < 1$.

15–2 Laplace distribution. The Laplace or double exponential random variable **x** is continuous over the entire real line with density function (Fig. 15–2)

$$f(x) = \tfrac{1}{2}e^{-|x-\theta|}, \qquad -\infty < x < \infty. \tag{15-6}$$

We find

$$E(\mathbf{x}) = \int_{-\infty}^{\infty} xf(x)\, dx = \int_{-\infty}^{\theta} x\tfrac{1}{2}e^{(x-\theta)}\, dx + \int_{\theta}^{\infty} x\tfrac{1}{2}e^{(\theta-x)}\, dx = \theta, \tag{15-7}$$

as is obvious by symmetry (assuming the mean exists), and

$$\sigma^2(\mathbf{x}) = \int_{-\infty}^{\infty} [x - E(\mathbf{x})]^2 f(x)\, dx = 2. \tag{15-8}$$

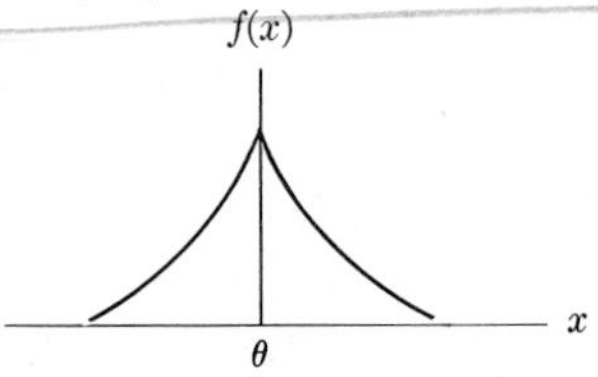

FIG. 15–2. Laplace density function.

The Laplace distribution appears not to follow from any model of general interest, and therefore is itself not of major interest; we introduce it here because of interesting problems which arise later when we consider the problem of estimating from observed data the value of the parameter θ.

15–3 Cauchy distribution. Consider the physical model illustrated in Fig. 15–3. The practical problem, which we will consider in Chapter 26, is to estimate from the distribution of particles on the screen the location of the source of radioactive energy. This is equivalent to estimating θ, since the source is a known perpendicular distance (say unity) from the screen. But here we are concerned only with the form of the distribution of particles and its moments.

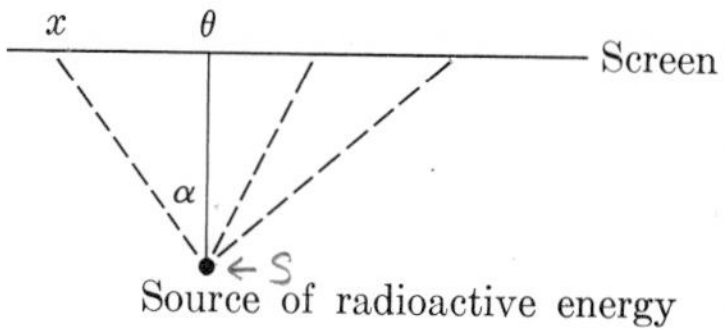

FIG. 15–3. A physical model leading to the Cauchy distribution.

In one common physical model, all angles of emission α are postulated to be equally likely; the density function of $\boldsymbol{\alpha}$ is uniform over the interval $-\pi/2$ to $\pi/2$,

$$f(\alpha)\,d\alpha = \frac{1}{\pi}\,d\alpha. \tag{15–9}$$

What does this imply relative to x, the distance along the screen from θ? Note that doubling α *more* than doubles x; the frequency of particles along the screen must thin out as we move along the screen away from θ. Let $\alpha = \phi(x)$. From Chapter 8 on change of variable, we have

$$f(\alpha)\,d\alpha = f[\phi(x)]\,\frac{d\phi(x)}{dx}\,dx. \tag{15–10}$$

We have

$$f(\alpha) = f[\varphi(x)] = \frac{1}{\pi},$$

$$\tan\alpha = x - \theta, \qquad \sec^2\alpha\,d\alpha = dx,$$

$$d\alpha = \frac{dx}{\sec^2\alpha} = \frac{dx}{1+\tan^2\alpha} = \frac{dx}{1+(x-\theta)^2}.$$

So $f(\alpha)\,d\alpha$, $-\pi/2 < \alpha < \pi/2$ becomes

$$\frac{1}{\pi}\,\frac{1}{1+(x-\theta)^2}\,dx, \qquad -\infty < x < \infty, \tag{15–11}$$

the probability element of the Cauchy random variable $\mathbf{x}$. The density function is illustrated in Fig. 15–4.

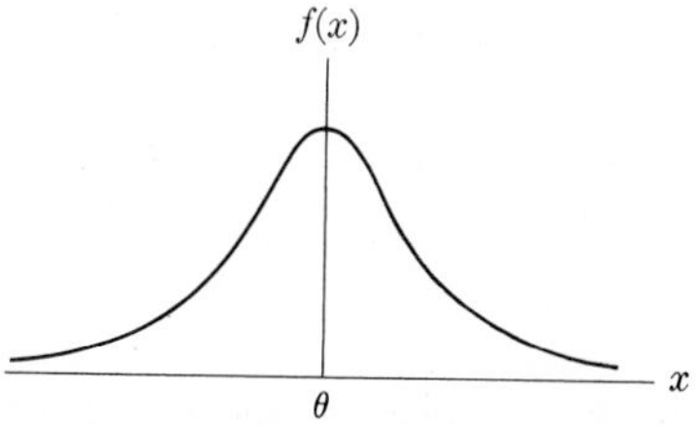

FIG. 15–4. Cauchy density function.

The distribution function $F(x)$ is

$$F(x) = \int_{-\infty}^{x} \frac{1}{\pi} \frac{1}{1 + (x - \theta)^2}\, dx = \frac{1}{2} + \frac{1}{\pi} \tan^{-1}(x - \theta),$$

which follows even more immediately from (15–9) and $\tan \alpha = x - \theta$.

The Cauchy distribution arises in a variety of problems centering on disintegration of atomic nuclei; scattering problems, capture processes, intensity of spectral lines. It also arises in the study of the probability behavior of the *ratios* of certain random variables; this will be explored in Chapter 17.

Although the Cauchy distribution has an evident center (θ) and evident variability about this center, these characteristics cannot be described by the customary parameters, $E(\mathbf{x})$ and $\sigma^2(\mathbf{x})$. For the Cauchy distribution, $E(\mathbf{x})$, $\sigma^2(\mathbf{x})$, [and for real θ, $M_{\mathbf{x}}(\theta)$] do not exist. The integrals

E(X) and σ²(X) do not exist cause do not converge.

$$\int_{-\infty}^{\infty} x \frac{1}{\pi} \frac{1}{1 + (x - \theta)^2}\, dx,$$

$$\int_{-\infty}^{\infty} x^2 \frac{1}{\pi} \frac{1}{1 + (x - \theta)^2}\, dx, \tag{15–12}$$

$$\int_{-\infty}^{\infty} e^{\theta x} \frac{1}{\pi} \frac{1}{1 + (x - \theta)^2}\, dx,$$

defining these quantities are not absolutely convergent. Among distributions encountered in statistical inference, this is unusual.

Collateral Reading

Uniform distribution

D. A. S. Fraser, *Statistics: an Introduction.* New York, John Wiley and Sons, 1958.

A. M. Mood, *Introduction to the Theory of Statistics.* New York, McGraw-Hill Book Co., 1950.

E. Parzen, *Modern Probability Theory and Its Applications.* New York, John Wiley and Sons, 1960.

G. P. Wadsworth and J. G. Bryan, *Introduction to Probability and Random Variables.* New York, McGraw-Hill Book Co., 1960.

Laplace distribution

N. Arley and K. R. Buch, *Introduction to the Theory of Probability and Statistics.* New York, John Wiley and Sons, 1950.

H. Cramér, *Mathematical Methods of Statistics.* Princeton, Princeton University Press, 1946.

Cauchy distribution

H. Cramér, *Mathematical Methods of Statistics.* Princeton, Princeton University Press, 1946.
R. A. Fisher, *Statistical Theory of Estimation.* Calcutta, University of Calcutta, 1938.

Rectangular dist

X continuous c̄ constant prob. $a < x < b$ in range

$f(x) = \frac{1}{b-a}$, $a < x < b$ $f(x) = 0$ elsewhere

$F(x) = \frac{x-a}{b-a}$ $a < x < b$ $= 0$ $x \le a$; $= 1$ $x \geq b$

CHAPTER 16

NORMAL DISTRIBUTION IN ONE VARIABLE

We consider a random variable which is *normally* distributed. In the first two sections of this chapter, the form, properties, and tables of the normal distribution are discussed; the final four sections consist of remarks on its importance and a discussion of its origins.

16–1 Form and properties. Consider the symmetrical function of the continuous variable x,

$$f(x) = ae^{-b(x-c)^2}, \quad -\infty < x < \infty, \tag{16–1}$$

with a, b, c constants. For $f(x)$ to be a density function, $\int_{\infty}^{\infty} f(x)\,dx = 1$ is required; this places a condition on the constants. To find the condition, let $b(x-c)^2 = u^2$. We have

$$\int_{-\infty}^{\infty} f(x)\,dx = 1 = \frac{a}{\sqrt{b}}\int_{-\infty}^{\infty} e^{-u^2}\,du = \frac{2a}{\sqrt{b}}\int_0^{\infty} e^{-u^2}\,du.$$

The last integral, which we call I, is easily evaluated by the curious device* of forming I^2 and shifting to polar coordinates. Thus,

$$I = \int_0^{\infty} e^{-u^2}\,du = \int_0^{\infty} e^{-v^2}\,dv, \qquad I^2 = \int_0^{\infty}\int_0^{\infty} e^{-(u^2+v^2)}\,du\,dv.$$

Integrating over the infinite square,

$$u = r\sin\theta, \qquad v = r\cos\theta, \qquad u^2+v^2 = r^2, \qquad du\,dv = r\,dr\,d\theta,$$

$$I^2 = 4\int_0^{\infty}\int_0^{\pi/2} e^{-r^2} r\,d\theta\,dr.$$

Integrating over the infinite circle,

$$I = \frac{\sqrt{\pi}}{2},$$

---

* More naturally, via change of variable and gamma functions

$$\int_0^{\infty} e^{-u^2}\,du = \frac{1}{2}\int_0^{\infty} e^{-v}v^{1/2}\,dv = \frac{1}{2}\Gamma\left(\frac{1}{2}\right) = \frac{\sqrt{\pi}}{2}.$$

and the condition is given by

$$a\sqrt{\frac{\pi}{b}} = 1. \tag{16–2}$$

As the density function $f(x)$ in (16–1) already involves $x - c$, and as it is evident that if the mean of this (symmetrical) distribution of $\mathbf{x}$ exists (that it does, see problem 16–4), c will be that mean, we work with $M_{\mathbf{x}-c}(\theta)$ rather than with $M_{\mathbf{x}}(\theta)$, and directly obtain moments of $\mathbf{x}$ about c:

$$M_{\mathbf{x}-c}(\theta) = \int_{-\infty}^{\infty} e^{\theta(x-c)} a e^{-b(x-c)^2}\, dx.$$

Let $u = x - c$; complete the square in the exponent in order to bring to bear the method used for determining I:

$$M_{\mathbf{x}-c}(\theta) = a \int_{-\infty}^{\infty} \exp\left\{-\left[bu^2 - \theta u + \frac{\theta^2}{4b} - \frac{\theta^2}{4b}\right]\right\} du$$

$$= a \exp\left[\frac{\theta^2}{4b}\right] \int_{-\infty}^{\infty} \exp\left\{-\left[bu - \frac{\theta}{2\sqrt{b}}\right]^2\right\} du = \exp\left[\frac{\theta^2}{4b}\right]. \tag{16–3}$$

Either by expansion or differentiation of $e^{\theta^2/4b}$, we find the mean and the variance of the *normal* random variable of (16–1) to be

$$E(\mathbf{x}) = c, \qquad \sigma^2(\mathbf{x}) = \frac{1}{2b}, \tag{16–4}$$

thus, with (16–2), expressing all constants a, b, c in terms of moments. We also find

$$\mu_{2k+1} = 0;$$

the odd moments of a normal random variable about its mean (in fact, if they exist, of *any* random variable with symmetrical distribution) are zero. Also, a unique result which we shall use in Chapter 18 and which the student is asked to show in problem 16–5,

$$\mu_{2k} = \frac{(2k)!}{2^k k!} \mu_2^k; \tag{16–5}$$

all higher even moments of a normal random variable about its mean are functions of the second moment about the mean.

By use of (16–4), the normal density and distribution functions can be written in terms of their two parameters, the mean λ (or $\lambda_{\mathbf{x}}$) and the variance σ^2 (or $\sigma^2_{\mathbf{x}}$), as follows:

$$f(x) = \frac{1}{\sigma\sqrt{2\pi}} \exp\left[-\frac{1}{2}\left(\frac{x-\lambda}{\sigma}\right)^2\right], \quad -\infty < x < \infty, \tag{16–6}$$

$$F(x) = \int_{-\infty}^{x} \frac{1}{\sigma\sqrt{2\pi}} \exp\left[-\frac{1}{2}\left(\frac{t-\lambda}{\sigma}\right)^2\right] dt, \quad -\infty < x < \infty, \tag{16–7}$$

due to de Moivre, 1733, explored by Laplace, 1783, and later by Gauss.

16–2 Tables. For the purpose of tabling it is (always) desirable to remove from the integrand parameters (here the mean and the variance) which change in value from problem to problem. Therefore we introduce (without interfering with the possibility of computing what we really want) the standardized normal variable

$$\mathbf{u} = \frac{\mathbf{x} - \lambda_{\mathbf{x}}}{\sigma_{\mathbf{x}}}, \tag{16–8}$$

whose mean and variance are, as noted in problem 3–3, *fixed*; $E(\mathbf{u}) = 0$ and $\sigma^2(\mathbf{u}) = 1$. Moreover, by problem 16–3, any linear function [of which (16–8) is one] of a normal variable is itself normally distributed. Therefore,

$$p(x_1 \le \mathbf{x} \le x_2) = \int_{x_1}^{x_2} \frac{1}{\sigma_{\mathbf{x}}\sqrt{2\pi}} \exp\left[-\frac{1}{2}\left(\frac{x-\lambda_x}{\sigma_{\mathbf{x}}}\right)^2\right] dx$$

$$= p(u_1 \le \mathbf{u} \le u_2) = \int_{u_1}^{u_2} \frac{1}{\sqrt{2\pi}} e^{-(1/2)u^2}\, du, \tag{16–9}$$

where

$$u_1 = \frac{x_1 - \lambda_{\mathbf{x}}}{\sigma_{\mathbf{x}}}, \qquad u_2 = \frac{x_2 - \lambda_{\mathbf{x}}}{\sigma_{\mathbf{x}}}.$$

Tables of the right side of (16–9), or more commonly, of the standardized *distribution* function

$$F(x) = \int_{-\infty}^{x} \frac{1}{\sqrt{2\pi}} e^{-(1/2)u^2}\, du, \tag{16–10}$$

may be constructed by expanding the integrand in a series, for example,

$$e^{-(1/2)u^2} = 1 - \frac{u^2}{2} + \frac{1}{2!}\left(\frac{u^2}{2}\right)^2 - \cdots$$

(this series unfortunately converges slowly and is suitable only for small u;

other series are used for large u), and integrating term by term from 0 to given x. [The integral (16–10) to $x = 0$ is $\frac{1}{2}$.] Many tables have been constructed, the most notable to 24 decimals by Sheppard [177] in 1939, and to 15 decimals by the National Bureau of Standards [132] in 1953. A brief table from [132] is shown on page 417.

16–3 Importance of the normal distribution. The normal distribution appears to be a reasonable description of the behavior of certain observable phenomena. Such phenomena as repeated measurements in biology, physics, and astronomy, ballistic measurements, insurance payments, cranial lengths, molecular velocities, and the quality of mass-produced product are among them. Despite such nonsense as negative velocities and negative lengths allowed by the normal, these and many more could be considered. Apart from such extensive, if often controversial (and since 1901 and the work of Karl Pearson, perhaps declining), application of the normal distribution as a generally valid description of observable data, its importance in statistical theory is deep and unquestionable. It is a limiting form of many other probability distributions, some of which we have met (the binomial, Poisson, and the hypergeometric) and some of which will appear in Part III (χ^2 and **t**). It has a number of properties in decomposition which are unique to it; these will be noted later.

We now show three results. First, we show that the limiting distribution, as the number of trials becomes large, of a binomial variable is normal. Second, using this result, we will describe a simple model which may often underlie observable phenomena and which leads to a normal distribution. Third, we describe a second and different model of somewhat lesser generality which also leads to a normal distribution.

16–4 The normal as a limit of the binomial. For a, b (and p) fixed, we wish to show that as $n \to \infty$ the binomial

$$p(a \leq \mathbf{x} \leq b) = \sum_{x=a}^{x=b} C_x^n p^x q^{n-x}$$

converges to

$$\int_{(a-\lambda)/\sigma}^{(b-\lambda)/\sigma} \frac{1}{\sqrt{2\pi}} e^{-(1/2)u^2}\, du, \tag{16–11}$$

with $\lambda = np$ and $\sigma = \sqrt{pqn}$. One method is to work directly with the binomial sum, replacing the three factorials by their Stirling approximations for large n. A second method is to examine the moment generating function of the binomial variable for $n \to \infty$. We follow the latter. The argument will be heuristic; justification of the limiting processes employed here lies beyond elementary mathematics.

Consider the standardized binomial variable

$$\mathbf{t} = \frac{\mathbf{x} - np}{\sqrt{npq}} \tag{16–12}$$

which, recalling (3–3), has mean 0 and variance 1. The moment generating function of $\mathbf{t}$ is

$$M_{\mathbf{t}}(\theta) = M_{(\mathbf{x}-\lambda)/\sigma}(\theta) = M_{\mathbf{x}-\lambda}\left(\frac{\theta}{\sigma}\right) = e^{-\theta\lambda/\sigma}M_{\mathbf{x}}(\theta/\sigma) = e^{-\theta\lambda/\sigma}(q + pe^{\theta/\sigma})^n,$$

$$\log_e M_{\mathbf{t}}(\theta) = \frac{-\theta\lambda}{\sigma} + n \log_e (q + pe^{\theta/\sigma});$$

expanding $e^{\theta/\sigma}$ and using the fact that $p + q = 1$,

$$\log_e M_{\mathbf{t}}(\theta) = -\frac{\theta\lambda}{\sigma} + n \log_e \left\{1 + p\left[\left(\frac{\theta}{\sigma}\right) + \frac{1}{2!}\left(\frac{\theta}{\sigma}\right)^2 + \frac{1}{3!}\left(\frac{\theta}{\sigma}\right)^3 + \cdots\right]\right\}.$$

Write z for p multiplied by the expression in square brackets. For $|z| < 1$, the Taylor expansion of $\log_e(1 + z)$ is

$$\log_e (1 + z) = z - \frac{z^2}{2} + \frac{z^3}{3} - \cdots.$$

So, for (arbitrary) sufficiently small z, we have

$$\log_e M_{\mathbf{t}}(\theta) = -\frac{\theta\lambda}{\sigma} + n\left(z - \frac{z^2}{2} + \frac{z^3}{3} - \cdots\right).$$

Substituting for z, collecting terms separately in θ^1, in θ^2, . . . , and using $\lambda = np$ and $\sigma^2 = npq$, we find

$$\log_e M_{\mathbf{t}}(\theta) = \frac{\theta^2}{2} + \text{terms in } \theta^k, \quad k \geq 3.$$

But all terms in θ^k involve n only by terms of the form

$$\frac{n}{\sigma^k} = \frac{n}{(npq)^{k/2}}, \quad k \geq 3,$$

which vanish when $n \to \infty$. Therefore

$$\lim_{n\to\infty} \log_e M_{\mathbf{t}}(\theta) = \frac{\theta^2}{2} \quad \text{and} \quad \lim_{n\to\infty} M_{\mathbf{t}}(\theta) = e^{\theta^2/2}. \tag{16–13}$$

From (16–3), this is precisely the moment generating function of a normal variable of mean 0 and variance 1. By the heuristic arguments of page 33 which relate moment generating functions and probability distributions, we conclude that for n large, $\mathbf{t}$ is normally distributed with mean 0 and variance 1; for n large, (16–11) is true. A more general asymptotic theorem, of which the result of this section is a special case, will be given in Chapter 21.

16–5 The Laplace model of cumulated errors. We now consider a simple model which, utilizing the result (16–13), leads to a normal distribution. It centers on the cumulative effect of small independent errors of observation on a measurement or, in language of greater contemporary interest, on the cumulative effect on a parameter of a large number of independently acting causes, each of which has small effect. Without justifying the limiting arguments in detail, we shall give Jeffreys' [93] formulation of this classical result probably best credited to Laplace, 1783.

Let λ be the true value of some constant and let n possible disturbances affect our determination x of λ. We consider the simple model in which the magnitude of each disturbance is $\pm\epsilon$ where the signs from one disturbance to another are statistically independent and where the probabilities of $+\epsilon$ and of $-\epsilon$ are equal. The probability $f(r)$ of r positive disturbances (and $n - r$ negative disturbances) is given by the binomial

$$f(r) = C_r^n(\tfrac{1}{2})^n$$

with $E(\mathbf{r}) = n/2$ and $\sigma^2(\mathbf{r}) = n/4$. From (16–12), for large n, $\mathbf{r}$ is normally distributed,

$$f(r) = \sqrt{\frac{2}{\pi n}}\exp\left[-\frac{1}{2}\left(\frac{r - n/2}{\sqrt{n}/2}\right)^2\right].$$

But after r positive disturbances and $n - r$ negative disturbances, the relationship between the estimate x and the true value λ is

$$x = \lambda + r\epsilon + (n - r)(-\epsilon) = \lambda + \epsilon(2r - n);$$

$\mathbf{x}$ is a *linear* function of the (asymptotically) normally distributed random variable $\mathbf{r}$, and therefore is itself *normally* distributed, with mean and variance

$$E(\mathbf{x}) = \lambda, \qquad \sigma^2(\mathbf{x}) = 4\epsilon^2\sigma^2(\mathbf{r}) = \epsilon^2 n,$$

where, since n is very large, we must have ϵ very small for $\epsilon^2 n$ to be finite.

16–6 Herschel's model. We now give a derivation due to the English astronomer (John) Herschel and employed (in three dimensions) by Maxwell in his *early* work on the distribution of particle velocities within a perfect gas. On a board ruled with rectangular axes OX and OY, shot is dropped with the object of hitting O. Assume first, and this is not a severe condition, symmetry about O for the resulting distribution of shot; any *systematic* departures from the target have been removed. Then the probability of shot falling between radii r and $r + dr$ is $h(r)\,dr$. Assume second, and this is somewhat harder to defend as always reasonable (in fact, in certain problems, it is precisely what needs to be proved), that the probability of shot falling between x and $x + dx$ is $f(x)\,dx$ independent of any knowledge we may have of **y**, and of shot falling between y and $y + dy$ is $g(y)\,dy$ independent of any knowledge we may have of **x**. That is, the x and y coordinates are stochastically independent.

We seek $f(x)$ and $g(y)$. We have

$$f(x)\,dx\,g(y)\,dy = h(r)\,dr\,d\theta,$$

$$\log f(x) + \log g(y) = \log h(r),$$

$$\log f(r\cos\theta) + \log g(r\sin\theta) = \log h(r).$$

Differentiating with respect to θ, we have

$$-r\sin\theta\,\frac{f'(r\cos\theta)}{f(r\cos\theta)} + r\cos\theta\,\frac{g'(r\sin\theta)}{g(r\sin\theta)} = 0,$$

or

$$\frac{f'(x)}{xf(x)} = \frac{g'(y)}{yg(y)}\,. \tag{16–14}$$

But **x** and **y** are unconstrained; for the equality to hold for all values of **x** and **y**, both ratios of (16–14) must be equal to a constant. Since $f(x)$ cannot indefinitely increase with x, this constant must be negative, say $-h^2$. Integrating, we obtain

$$\log f(x) = -\frac{h^2x^2}{2} + \log C, \qquad f(x) = Ce^{-h^2x^2/2}.$$

As for the constant C, from

$$\int_{-\infty}^{\infty}\int_{-\infty}^{\infty} f(x)g(y)\,dx\,dy = \int_0^{2\pi}\int_0^{\infty} h(r)r\,dr\,d\theta$$

we get $C = h/\sqrt{2\pi}$; finally,

$$f(x) = \frac{h}{\sqrt{2\pi}}\,e^{-h^2x^2/2},$$

with h evidently equal to $1/\sigma$. The variable **x** is normally distributed; similarly **y** is normally distributed.

16–7 Final remarks. So it appears that under several reasonable hypotheses or models, including others than those discussed here (see, for example, Levy-Roth [115]), the normal distribution will arise. But caution is well advised. Gauss remarked that no general law of error of observation exists. A large number of independent disturbances, each with small effect, may not lead to a normal distribution (see, for example, Whittaker-Robinson [200]). Moreover, models differing but slightly from those employed here (for example, models in which one cause of variation is dominant or models in which transformations are made after all small-effect causes have been cumulated) lead to nonnormal distributions.

Tables

National Bureau of Standards, *Tables of Probability Functions, Volume 2.* New York, 1942.

National Bureau of Standards, *Tables of Normal Probability Functions.* Applied Mathematics Series 23, Washington, 1953.

National Bureau of Standards, *Tables of the Error Function and Its Derivatives.* Applied Mathematics Series 41, Washington, 1954.

W. F. Sheppard, "The probability integral," *British Assoc. Math. Tables, Volume 7.* Cambridge (Eng.), University Press, 1939.

Collateral Reading

E. Borel, *Mécanique Statistique Classique.* Paris, Gauthier-Villars et Cie, 1925. On the earlier (given on page 144) and the later (much superior) derivations of the gas laws by Maxwell.

H. Cramér, *The Elements of Probability Theory and Some of Its Applications.* New York, John Wiley and Sons, 1955.

T. C. Fry, *Probability and Its Engineering Uses.* New York, D. van Nostrand Co., 1928. Includes a valuable discussion of Maxwell's work.

H. Jeffreys, *Theory of Probability.* 3rd ed., Oxford, Clarendon Press, 1961.

H. Levy and L. Roth, *Elements of Probability.* Oxford, Clarendon Press, 1936. Particularly on the origins of the normal distribution.

H. C. Plummer, *Probability and Frequency.* London, Macmillan Co., 1940.

J. Topping, *Errors of Observation and Their Treatment.* London, Chapman and Hall, Revised ed., 1957.

E. T. Whittaker and G. Robinson, *The Calculus of Observations.* 3rd ed., London, Blackie and Sons, 1940. In particular, that a large number of independent causes, each with small effect, may not lead to a normal distribution.

Problems

16–1. For a normal random variable, show that the density has two points of inflection, located one standard deviation on either side of the axis of symmetry.

16–2. Compare the probabilities of various deviations of a normal random variable from its mean with those guaranteed for any random variable by the Bienaymé-Chebychev inequality.

16–3. Using moment generating functions, show that for $\mathbf{x}$ normal, $\mathbf{y} = a + \beta\mathbf{x}$ is normal [with mean $a + \beta E(\mathbf{x})$ and variance $\beta^2\sigma^2(\mathbf{x})$]. More generally, if $\mathbf{x}_1, \ldots, \mathbf{x}_n$ are independently normal with means $\lambda_1, \ldots, \lambda_n$ and variances $\sigma_1^2, \ldots, \sigma_n^2$, find the probability behavior of the linear sum $a_1\mathbf{x}_1 + \cdots + a_n\mathbf{x}_n$.

16–4. The random variable $\mathbf{x}$ is normal with mean 0 and variance σ^2. Find $E|\mathbf{x}|$.

16–5. For a normal random variable, show that

$$\mu_{2k} = \frac{(2k)!}{2^k k!}\,\mu_2^k.$$

Any linear function of a normal variable is itself normally dist.

repeated measurements

two param. λ σ^2

CHAPTER 17

NORMAL DISTRIBUTION IN TWO VARIABLES

Of all probability distributions in two or more random variables, the normal distribution is the most important. Bivariate and multivariate binomial, Poisson and exponential (Gumbel [80]) distributions are met in the literature, but the bivariate and multivariate normal are, for reasons identical with those given in behalf of the univariate normal, the critical distributions. In two and more variables, the role of the normal distribution in sampling theory relative to its role as a description of observed phenomena must be even further emphasized.

17–1 The distribution and its moments. We restrict the discussion to the form and properties of the bivariate normal. We begin with

$$f(x, y) \, dx \, dy$$

$$= K \exp \{-\tfrac{1}{2}[a(x - A)^2 + 2b(x - A)(y - B) + c(y - B)^2]\} \, dx \, dy,$$

$$-\infty < x, y < \infty, \tag{17-1}$$

with K, A, B, a, b, c constants. Let $u = x - A$ and $v = y - B$; (17–1) becomes

$$f(u, v) \, du \, dv = K \exp \{-\tfrac{1}{2}[au^2 + 2buv + cv^2]\} \, du \, dv, \quad -\infty < u, v < \infty. \tag{17-2}$$

The expression within square brackets is a second-degree function in the variables u and v. We label it Q; it is called a quadratic form. Unless $u = v = 0$, Q must be positive. Even further, Q must be nonnegative separately in **u** and **v**; otherwise the integral of (17–2) over the whole range of **u** and **v** (noting the sign in front of the brackets) could exceed 1 (in fact, would be infinite).

A quadratic form Q which, except when the variables u and v are 0, always has the same sign is *definite.* If the sign of Q is positive, the form is *positive definite.* For Q to be positive definite, three conditions on $a, b,$ and c are needed:

① when $u = 0$, $Q = cv^2$; therefore, $c > 0,$

② when $v = 0$, $Q = au^2$; therefore, $a > 0,$

③ $(ac - b^2) > 0$

147

3 conditions that a be positive

For Q to be positive definite

and since

$$Q = \frac{1}{a}\,[(au + bv)^2 + (ac - b^2)v^2], \tag{17-3}$$

it is evident that, should $au + bv = 0$, a third condition must be imposed on a, b, and c, namely $(ac - b^2) > 0$. These three necessary conditions are also, in fact, *sufficient* for Q to be positive definite.

Our axioms require

$$\int_{-\infty}^{\infty}\int_{-\infty}^{\infty} K \exp\{-\tfrac{1}{2}[au^2 + 2buv + cv^2]\}\, du\, dv = 1, \tag{17-4}$$

which places a condition on the constant K. To determine this condition, we perform the integration in (17–4) by completing the *square* of the first two terms in brackets, thereby reducing part of the exponent to the square form already encountered in dealing with the normal distribution in one variable. Completion of the square of the first two terms in brackets requires the addition (and subtraction) of a term in v^2. The term subtracted is then combined with cv^2 and a second additive square term is created. The student should do this in detail; he will find that the bracket reduces, exactly, to Q. Writing g^2 for the first square term and h^2 for the second square term, we find that the inverse of the Jacobian of the transformation from **u**, **v** to **g**, **h** is

$$J^{-1} = \begin{vmatrix} \dfrac{\partial g}{\partial u} & \dfrac{\partial g}{\partial v} \\[2ex] \dfrac{\partial h}{\partial u} & \dfrac{\partial h}{\partial v} \end{vmatrix} = \sqrt{ac - b^2},$$

and

$$K \cdot \frac{1}{\sqrt{ac - b^2}} \int_{-\infty}^{\infty}\int_{-\infty}^{\infty} \exp[-\tfrac{1}{2}(g^2 + h^2)]\, dg\, dh = 1. \tag{17-5}$$

As already seen, a shift to polar coordinates permits the integral of (17–5) to be readily determined; it reduces to 2π and we have, finally,

$$K = \frac{\sqrt{ac - b^2}}{2\pi}. \tag{17-6}$$

Thus (17–1) is now a *bonafide* probability element of the normal random variable pair **x**, **y**.

To find the moments of **u**, **v** (rather than of **x**, **y**), we write the moment generating function of **u**, **v**

$$M_{\mathbf{u},\mathbf{v}}(\alpha, \beta) = K \int_{-\infty}^{\infty}\int_{-\infty}^{\infty} \exp[\alpha u + \beta v] \exp\{-\tfrac{1}{2}[au^2 + 2buv + cv^2]\}\, du\, dv. \tag{17-7}$$

Let H be the exponent of e:

$$H = -\tfrac{1}{2}au^2 - buv - \tfrac{1}{2}cv^2 + \alpha u + \beta v. \tag{17–8}$$

The integration in (17–7) is routine, if somewhat painstaking, the idea once again being to replace H by one or more additive *square* terms since, as we have repeatedly seen, only integrands of the form $e^{-\text{square}}$ are simply integrable over the whole range $-\infty, \infty$. We encompass the terms in u^2, uv, and u by a square term

$$-\gamma^2 = -\frac{a}{2}\left(u + \frac{b}{a}v - \frac{\alpha}{a}\right)^2,$$

which leaves only terms in v^2, v, and constants. Thus (17–8) becomes

$$H = \gamma^2 + \frac{1}{2a}[v^2(b^2 - ac) + 2v(a\beta - b\alpha) + \alpha^2]. \tag{17–9}$$

Now completing the square of the expression in brackets in (17–9) and calling this completed square δ^2, we find

$$H = -\gamma^2 - \frac{1}{2a}\delta^2 + L,$$

where

$$L = \frac{1}{2(ac - b^2)}[c\alpha^2 - 2\alpha b\beta + a\beta^2],$$

a term not involving u or v. The inverse of the Jacobian of the transformation from $\mathbf{u}$, $\mathbf{v}$ to $\boldsymbol{\gamma}$, $\boldsymbol{\delta}$ is

$$J^{-1} = \begin{vmatrix} \dfrac{\partial\gamma}{\partial u} & \dfrac{\partial\gamma}{\partial v} \\ \dfrac{\partial\delta}{\partial u} & \dfrac{\partial\delta}{\partial v} \end{vmatrix} = \left[\frac{a}{2}(ac - b^2)\right]^{1/2}$$

and

$$M_{\mathbf{u},\mathbf{v}}(\alpha, \beta) = K\sqrt{\frac{2}{a}}\sqrt{\frac{1}{ac - b^2}}\int_{-\infty}^{\infty}\int_{-\infty}^{\infty}\exp\left[-\gamma^2 - \frac{1}{2a}\delta^2 + L\right]d\gamma\, d\delta,$$

which, on integrating out and replacing K by (17–6), reduces to

$$M_{\mathbf{u},\mathbf{v}}(\alpha, \beta) = \mathrm{e}^L. \tag{17–10}$$

Differentiating (17–10) with respect to α and with respect to β, we have

$E(\mathbf{u}) = 0$, $\quad E(\mathbf{x}) = A$, $\quad$ which we write $\lambda(\mathbf{x})$;

$E(\mathbf{v}) = 0$, $\quad E(\mathbf{y}) = B$, $\quad$ which we write $\lambda(\mathbf{y})$.

Differentiating (17–10) twice with respect to α and twice with respect to β, we have

$$E(\mathbf{u}^2) = E[\mathbf{x} - \lambda(\mathbf{x})]^2 = \sigma^2(\mathbf{x}) = \frac{c}{ac - b^2},$$
$$E(\mathbf{v}^2) = E[\mathbf{y} - \lambda(\mathbf{y})]^2 = \sigma^2(\mathbf{y}) = \frac{a}{ac - b^2}, \tag{17–11}$$

and from the cross-derivative of (17–10) with respect to α and β, the covariance term, not of course found in the univariate normal discussion,

$$E(\mathbf{uv}) = \left.\frac{\partial^2 M}{\partial\alpha\,\partial\beta}\right|_{\alpha=\beta=0}$$
$$= E[\mathbf{x} - \lambda(\mathbf{x})][\mathbf{y} - \lambda(\mathbf{y})] = \rho\sigma(\mathbf{x})\sigma(\mathbf{y}) = -\frac{b}{ac - b^2}. \tag{17–12}$$

From (17–11) and (17–12), we have

$$a = \frac{1}{(1 - \rho^2)\sigma^2(\mathbf{x})}, \qquad b = \frac{-\rho}{(1 - \rho^2)\sigma(\mathbf{x})\sigma(\mathbf{y})}, \qquad c = \frac{1}{(1 - \rho^2)\sigma^2(\mathbf{y})},$$

and the joint density function (17–1) may now be written in terms of the first and second moments,

$$f(x, y) = \frac{1}{2\pi\sigma(\mathbf{x})\sigma(\mathbf{y})\sqrt{1 - \rho^2}}$$
$$\times \exp\left[-\frac{1}{2(1 - \rho^2)}\left\{\left[\frac{x - \lambda(\mathbf{x})}{\sigma(\mathbf{x})}\right]^2 - 2\rho\frac{[x - \lambda(\mathbf{y})][y - \lambda(\mathbf{y})]}{\sigma(\mathbf{x})\sigma(\mathbf{y})} + \left[\frac{y - \lambda(\mathbf{y})}{\sigma(\mathbf{y})}\right]^2\right\}\right], \tag{17–13}$$

the joint density function of the normal random variables $\mathbf{x}$, $\mathbf{y}$, with five parameters, the means $\lambda(\mathbf{x})$ and $\lambda(\mathbf{y})$, the standard deviations $\sigma(\mathbf{x})$ and $\sigma(\mathbf{y})$, and the correlation coefficient ρ. The expression (17–13) is due to Laplace, 1810, and Plana, 1812, and was explored by Gauss, 1823.

The standardizing change of variables

$$\mathbf{u} = \frac{\mathbf{x} - \lambda(\mathbf{x})}{\sigma(\mathbf{x})}, \qquad \mathbf{v} = \frac{\mathbf{y} - \lambda(\mathbf{y})}{\sigma(\mathbf{y})} \tag{17–14}$$

reduces the five parameters of (17–13) to one parameter ρ. The bivariate normal distribution function of $\mathbf{u}$, $\mathbf{v}$ was first tabled by K. Pearson [156].

17–2 Marginal distributions. We now show that *each* of the random variables in (17–13) is normally distributed. We have, generally, for the marginal distribution $g(x)$ of any continuous random variable $\mathbf{x}$,

$$g(x) = \int_{-\infty}^{\infty} f(x, y)\,dy. \tag{17–15}$$

Specializing (17–15) to the bivariate normal (17–13) and changing to the standardized variables **u** and **v** of (17–14), we find

$$g(x) = \frac{\exp[-u^2/2]}{2\pi\sigma(\mathbf{x})\sqrt{1-\rho^2}} \int_{-\infty}^{\infty} \exp\left[\frac{1}{2(1-\rho^2)}(v-\rho u)^2\right] dv.$$

Letting

$$z = \frac{v - \rho u}{\sqrt{1-\rho^2}},$$

we find

$$g(x) = \frac{\exp[-u^2/2]}{2\pi\sigma(\mathbf{x})} \int_{-\infty}^{\infty} \exp[-z^2/2]\, dz$$

$$= \frac{1}{\sigma(\mathbf{x})\sqrt{2\pi}} \exp\left\{-\frac{1}{2}\left[\frac{\mathbf{x} - \lambda(\mathbf{x})}{\sigma(\mathbf{x})}\right]^2\right\}, \tag{17–16}$$

the normal density function of **x**; similarly for the marginal distribution of **y**. The converse is *not* true; if the marginal distributions $g(x)$ and $h(y)$ are normal, the joint density function $f(x, y)$ is not necessarily bivariate normal.

Note in (17–13) that a sufficient condition that two *normally* distributed variables **x** and **y** be statistically independent, that is, that the joint density function be factorable,

$$f(x, y) = g(x) \cdot h(y),$$

is simply $\rho = 0$. As we saw in Chapter 6, this is not *generally* true. Note also that the Herschel model of Chapter 16 can be stated as follows: the only circularly symmetric distribution for which two random variables **x** and **y** are independent is the bivariate normal with $\sigma(\mathbf{x}) = \sigma(\mathbf{y})$, and $\rho(\mathbf{x}, \mathbf{y}) = 0$. For a sharp discussion, see Lancaster [109].

17–3 Normal regression functions. We now show the remarkable fact that the regression functions of bivariate normal variables are *linear*. We have, generally,

$$h(y|x) = \frac{f(x, y)}{g(x)}. \tag{17–17}$$

For normal **x**, **y** both the numerator and the denominator of the right-hand side of (17–17) have already been given in (17–13) and (17–16). Inserting these results in (17–17) we get, after some reduction,

$$h(y|x) = \frac{1}{\sigma(\mathbf{y})\sqrt{2\pi}\sqrt{1-\rho^2}}$$

$$\times \exp -\frac{1}{2}\left[\frac{y - \{\lambda(\mathbf{y}) + [\rho\sigma(\mathbf{y})/\sigma(\mathbf{x})][x - \lambda(\mathbf{x})]\}}{\sigma(\mathbf{y})\sqrt{1-\rho^2}}\right]^2; \tag{17–18}$$

the *conditional* density function of $\mathbf{y}$ is normal with conditional mean

$$\lambda(\mathbf{y}) + \rho \frac{\sigma(\mathbf{y})}{\sigma(\mathbf{x})} [x - \lambda(\mathbf{x})]$$

and conditional standard deviation $\sigma(\mathbf{y})\sqrt{1 - \rho^2}$. Thus, the conditional, as well as the marginal, distributions of the bivariate normal distribution are normal.

But the regression function of $\mathbf{y}$ on x *is* the conditional mean of $\mathbf{y}$ given x:

$$E(\mathbf{y}|x) = \lambda(\mathbf{y}) + \rho \frac{\sigma(\mathbf{y})}{\sigma(\mathbf{x})} [x - \lambda(\mathbf{x})], \tag{17–19}$$

a *linear* function of x. Analogous results for $E(\mathbf{x}|y)$ and $\sigma(\mathbf{x}|y)$ may be obtained. A drawing of the bivariate normal, with its marginal distributions and regression functions, is reproduced in Fig. 17–1 from Hoel [89].

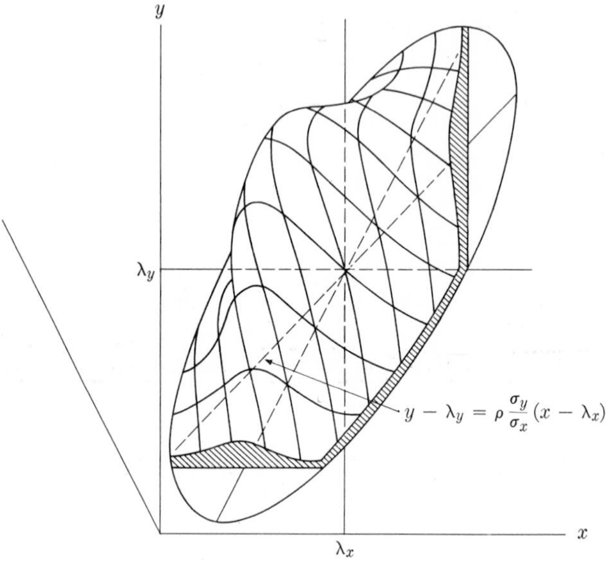

FIG. 17–1. Normal correlation surface.

17–4 Distribution of x/y. The distribution of $\mathbf{x}/\mathbf{y}$, where $\mathbf{x}$ and $\mathbf{y}$ are normal random variables, is of interest; scientists are frequently concerned with the probability behavior of the ratio of random variables.

First we consider the general bivariate probability element

$$f(x, y)\, dx\, dy, \tag{17–20}$$

and let $\mathbf{u} = \mathbf{x}/\mathbf{y}$ (with $y \neq 0$, an often overlooked condition in application), and $\mathbf{y} = \mathbf{y}$. The Jacobian of the transformation from $\mathbf{x}$, $\mathbf{y}$ to $\mathbf{u}$, $\mathbf{v}$ is $|v|$. We have

transformation

$$f(x, y)\,dx\,dy = |v| f(uv, v)\,du\,dv. \tag{17–21}$$

Integrating out v, we have the desired marginal distribution of $\mathbf{u}$,

marginal of u

$$g(u) = \int_{-\infty}^{\infty} |v| f(uv, v)\,dv. \tag{17–22}$$

For $\mathbf{u}$ and $\mathbf{v}$ independent, the integration in (17–22) is straightforward; several examples will be met in Chapter 23. But for $\mathbf{u}$ and $\mathbf{v}$ not independent, the integral is tractable only for certain joint density functions. Among these is the bivariate normal with correlation ρ and means *zero* (if the means are not zero, the problem is more difficult).

The probability element of such a pair of normal random variables is

$$f(x, y)\,dx\,dy = \frac{1}{2\pi\sigma(\mathbf{x})\sigma(\mathbf{y})\sqrt{1 - \rho^2}}$$

$$\times \exp\left[-\frac{1}{2(1-\rho^2)}\left\{\left[\frac{x}{\sigma(\mathbf{x})}\right]^2 - 2\rho\frac{xy}{\sigma(\mathbf{x})\sigma(\mathbf{y})} + \left[\frac{y}{\sigma(\mathbf{y})}\right]^2\right\}\right] dx\,dy.$$

Replacing $\mathbf{v}$ by $\mathbf{y}$, (17–22) becomes

$$g(u) = \int_{-\infty}^{\infty} |y| \frac{1}{2\pi\sigma(\mathbf{x})\sigma(\mathbf{y})\sqrt{1 - \rho^2}}$$

$$\times \exp\left\{-\frac{1}{2(1-\rho^2)}\left[\left(\frac{uy}{\sigma(\mathbf{x})}\right)^2 - 2\rho\frac{uy^2}{\sigma(\mathbf{x})\sigma(\mathbf{y})} + \left(\frac{y}{\sigma(\mathbf{y})}\right)^2\right]\right\} dy.$$

Letting

$$\frac{1}{2\pi\sigma(\mathbf{x})\sigma(\mathbf{y})\sqrt{1-\rho^2}} = A, \qquad \frac{u^2}{\sigma^2(\mathbf{x})} - 2\rho\frac{u}{\sigma(\mathbf{x})\sigma(\mathbf{y})} + \frac{1}{\sigma^2(\mathbf{y})} = B,$$

$$\frac{B}{2(1-\rho^2)} = C,$$

we find

$$g(u) = A\int_{-\infty}^{\infty} |y| e^{-Cy^2}\,dy,$$

which finally integrates out to

$$\frac{1}{\dfrac{\pi\sigma(\mathbf{x})\sigma(\mathbf{y})}{\sqrt{1-\rho^2}}\left[\left(\dfrac{u}{\sigma(\mathbf{x})} - \dfrac{\rho}{\sigma(\mathbf{y})}\right)^2 + \dfrac{1}{\sigma^2(\mathbf{y})}\sqrt{1-\rho^2}^{\,2}\right]}. \tag{17–23}$$

Noting that all quantities in (17–23) other than u are constants, (17–23) is recognized, by comparison with (15–11), as the Cauchy density function, with maximum ordinate at

$$u = \rho\,\frac{\sigma(\mathbf{x})}{\sigma(\mathbf{y})}\,;$$

the latter is the *regression coefficient* of **x** on **y**. For $\rho = 0$, (17–23) reduces to a Cauchy density function of simpler form.

For **x** and **y** normal, a discussion of the probability behavior of $1/\mathbf{x}$ and $\mathbf{x}/\mathbf{y}$ is found in Curtiss [38] and Purohit and Phatarford [161].

Tables

KARL PEARSON's original tables of volumes under the normal bivariate surface are found in *Tables for Statisticians and Biometricians, Part II.* Cambridge (Eng.), Cambridge University Press, 1931. Pearson tabled

$$\frac{1}{2\pi\sqrt{1-\rho^2}}\int_h^\infty\int_k^\infty \exp\left[-\frac{1}{2}\left(\frac{x^2-2\rho xy+y^2}{1-\rho^2}\right)\right]dx\,dy = M(h, k, \rho).$$

His tables have been reworked by E. S. PEARSON and H. O. HARTLEY, *Biometrika Tables for Statisticians.* Cambridge (Eng.), Cambridge University Press, 1954.

KARL PEARSON's integral was later reduced to two parameters and studied by C. NICHOLSON in *Biometrika,* Vol. 33 (1943), pp. 59–72. History, properties, analysis, fresh tables (of the two-parameter formulation), and examples are given by D. B. OWEN, *The Bivariate Normal Probability Distribution.* Washington, U.S. Department of Commerce, Office of Technical Services, 1957.

The definitive tables (also including properties and examples) are *Tables of the Bivariate Normal Distribution Function and Related Functions.* Preparation under the direction of GERTRUDE BLANCH, Washington, Department of Commerce, National Bureau of Standards, Applied Mathematics Series 50, 1959.

Collateral Reading

C. A. BENNETT and N. L. FRANKLIN, *Statistical Analysis in Chemistry and the Chemical Industry.* New York, John Wiley and Sons, 1954.

K. A. BROWNLEE, *Statistical Theory and Methodology in Science and Engineering.* New York, John Wiley and Sons, 1960.

C. C. CRAIG, "On frequency distributions of the quotient and of the product of two statistical variables, *Amer. Math. Monthly,* Vol. 49 (1942), pp. 24–32.

J. H. CURTISS, "On the distribution of the quotient of two chance variables," *Ann. Math. Stat.,* Vol. 12 (1941), pp. 409–421.

E. J. GUMBEL, "Bivariate exponential distributions," *Jour. Amer. Stat. Assoc.,* Vol. 55 (1960), pp. 698–707.

P. G. HOEL, *Introduction to Mathematical Statistics.* 3rd ed., New York, John Wiley and Sons, 1962.

A. M. MOOD, *Introduction to the Theory of Statistics.* New York, McGraw-Hill Book Co., 1950. A particularly good discussion.

E. PARZEN, *Modern Probability Theory and Its Applications.* New York, John Wiley and Sons, 1960.

J. N. PUROHIT and R. M. PHATARFORD, "Use of the reciprocal of a normal variate in textile statistics," *Calcutta Stat. Assoc. Bull.*, Vol. 6 (1955), pp. 99–101.

S. S. WILKS, *Mathematical Statistics.* Princeton, Princeton University Press, 1943. Very good discussion. Also the 1962 Wiley volume.

PROBLEMS

17–1. The random variables $\mathbf{x}$ and $\mathbf{y}$ are normally and independently distributed with means 0 and variances 1. Determine the radius of the circle with center at the origin such that the probability is 0.95 that x, y falls inside the circle.

17–2. If $\mathbf{x}$ and $\mathbf{y}$ are normally and independently distributed with zero means and variances σ^2, show that the radius $\mathbf{r} = \sqrt{\mathbf{x}^2 + \mathbf{y}^2}$ has density function

$$f(r) = \frac{r}{\sigma^2} \exp\left[-\frac{r^2}{2\sigma^2}\right].$$

This is the Rayleigh distribution, important in noise theory and weapons testing. See, for example, Woodward [210], Kamat [97].

17–3. The change of variable

$$\mathbf{u} = \frac{\mathbf{x}}{\sigma_{\mathbf{x}}} - \frac{\rho_{\mathbf{xy}}\mathbf{y}}{\sigma_{\mathbf{y}}}, \qquad \mathbf{v} = \sqrt{(1 - \rho_{\mathbf{xy}}^2)} \cdot \frac{\mathbf{y}}{\sigma_{\mathbf{y}}}$$

of normal $\mathbf{x}$, $\mathbf{y}$ with correlation $\rho_{\mathbf{xy}}$, leaves $\mathbf{u}$, $\mathbf{v}$ normal and independent.

17–4. An object tends to fall on a point target, without systematic error. The distribution about the target is bivariate normal with zero correlation and equal variances. Consider a circle and a square, of equal area, centering on the target. Do they have equal probability?

CHAPTER 18*

GRAM-CHARLIER SERIES AND HERMITE POLYNOMIALS

18–1 Gram-Charlier series. Several attempts have been made to describe at once a large class of density functions. One such attempt, which arose out of interest in graduating actuarial data, is the Gram-Charlier system, in which a density function is represented by an infinite series consisting of a linear function of a *normal* density function and its successive derivatives. Density functions other than the normal have also been used to generate other infinite series.

We regard Gram-Charlier series here as an algebraically simple and fairly practical system of describing a class of density functions; but they have also been proposed, by Charlier in 1905, as following from certain theoretical models of error distribution.

Though the issue of describing entire density functions neither is altogether clear nor currently has the appeal of its early years, there remain situations, among them actuarial problems, in which knowledge of the entire probability distribution, rather than of several of its parameters, is required. To the extent that a strong case can be made for the normal distribution supported on the one hand by a reasonable hypothesis of error distribution and on the other by many sets of data themselves, a case can be made out for the Gram-Charlier system which has both these supports, the former in lesser and the latter in greater degree than the normal distribution alone.

Consider the standardized random variable $\mathbf{u} = (\mathbf{x} - \lambda)/\sigma$, where $\mathbf{x}$ is a *normal* variable with mean λ and standard deviation σ. Recall from problems 16–3 and 3–3 that $\mathbf{u}$ is normally distributed with mean 0 and variance 1,

$$f(u) = \frac{1}{\sqrt{2\pi}} e^{-(1/2)u^2}, \quad -\infty < u < \infty. \tag{18–1}$$

Letting $f^{(k)}$ be the kth derivative of $f(u)$, we find

$$\begin{aligned} f^{(0)}(u) &= (1)f(u), \qquad f^{(1)}(u) = -uf(u), \\ f^{(2)}(u) &= (u^2 - 1)f(u), \qquad f^{(3)}(u) = -(u^3 - 3u)f(u), \end{aligned} \tag{18–2}$$

* This chapter may be omitted.

the general polynomial in u being

$$u^k - \frac{k(k-1)}{2}u^{k-2} + \frac{k(k-1)(k-2)(k-3)}{2 \cdot 4}u^{k-4} - \cdots, \qquad (18\text{–}3)$$

tabled by Jørgensen in 1916 for $u = 1(0.01)6$. Kendall and Stuart [104] list the polynomials to $f^{(11)}$, Fry [72] to $f^{(12)}$. The Harvard tables [85] are probably best.

Consider $\phi(u)$, an arbitrary function of **u**, continuous, with continuous derivatives all vanishing at $\pm\infty$, all moments finite. We express $\phi(u)$ by the following infinite series:

$$\phi(u) = b_0 f(u) + b_1 f^{(1)}(u) + b_2 f^{(2)}(u) + \cdots, \qquad (18\text{–}4)$$

where $f(u)$ is the standardized normal density function (18–1) and $f^{(k)}(u)$ is the kth derivative of $f(u)$. In the Gram-Charlier system, the constants b_k are determined by equating each moment of the function to be fitted [$\phi(u)$, the left-hand side of (18–4)] and the corresponding moment of the fitting function [the right-hand side of (18–4)], with the natural additional proviso that $\phi(u)$ first be expressed as a probability distribution in standardized form, that is, the mean of **u** is 0 and the variance of **u** is 1.

Equate the zeroth moments of both sides of (18–4):

$$\int_{-\infty}^{\infty} \phi(u)\,du = b_0 \int_{-\infty}^{\infty} f(u)\,du + b_1 \int_{-\infty}^{\infty} f^{(1)}(u)\,du + b_2 \int_{-\infty}^{\infty} f^{(2)}(u)\,du + \cdots.$$

With the exception of

$$\int_{-\infty}^{\infty} f(u)\,du,$$

which is 1, each right-hand integral vanishes. All integrals involving odd moments vanish because the normal distribution is symmetrical about its mean. All integrals involving even moments also vanish; for example,

$$\int_{-\infty}^{\infty} f^{(6)}(u)\,du = \int_{-\infty}^{\infty} (u^6 - 15u^4 + 45u^2 - 15)f(u)\,du$$
$$= \mu_6 - 15\mu_4 + 45\mu_2 - 15,$$

which, recalling (16–5), vanishes. Equating the zeroth moments of $\phi(u)$ and the fitting function thus leads to

$$\int_{-\infty}^{\infty} \phi(u)\,du = b_0 \int_{-\infty}^{\infty} f(u)\,du$$

from which $b_0 = 1$.

Equate the first moments:

$$\int_{-\infty}^{\infty} u\phi(u)\,du = \int_{-\infty}^{\infty} uf(u)\,du + b_1 \int_{-\infty}^{\infty} uf^{(1)}(u)\,du + b_2 \int_{-\infty}^{\infty} uf^{(2)}(u)\,du + \cdots.$$

The left-hand integral is 0; integrating on the right-hand side, we finally get $b_1 = 0$.

Equating second moments, the student should verify that a similar argument leads to $b_2 = 0$.

Equate the third moments:

$$\int_{-\infty}^{\infty} u^3\phi(u)\,du = \int_{-\infty}^{\infty} u^3f(u)\,du + b_3 \int_{-\infty}^{\infty} u^3f^{(3)}(u)\,du + \cdots.$$

Except for the integral multiplied by b_3, that is,

$$\int_{-\infty}^{\infty} u^3f^{(3)}(u)\,du = \int_{-\infty}^{\infty} (-u^6 + 3u^4)f(u)\,du$$
$$= -\mu_6 + 3\mu_4 = -6\mu_2 = -6,$$

all right-hand integrals vanish, and we have

$$b_3 = \frac{-1}{3!}\,\mu_3[\phi(u)].$$

Entirely similarly,

$$b_4 = \frac{1}{4!}\,(\mu_4[\phi(u)] - 3),$$

and finally,

$$\phi(u) = f(u) - \frac{\mu_3[\phi(u)]}{3!}f^{(3)}(u) + \frac{\mu_4[\phi(u)] - 3}{4!}f^{(4)}u + \cdots, \tag{18–5}$$

the Gram-Charlier series, due to Gram (his doctoral thesis) in 1879, and Charlier in 1905.

The problem of fitting $\phi(u)$ is reduced to the problem of determining (several of) the moments of $\phi(u)$,

$$\mu_3[\phi(u)] = \int_{-\infty}^{\infty} u^3\phi(u)\,du, \qquad \mu_4[\phi(u)] = \int_{-\infty}^{\infty} u^4\phi(u)\,du, \ldots,$$

or, in practical applications, of closely *estimating* (several of) the moments, generally by calculating the corresponding moments of a sample of n

observations on u,

$$\sum_{i=1}^{n} \frac{u_i^3}{n}, \qquad \sum_{i=1}^{n} \frac{u_i^4}{n}, \ldots .$$

Though we have represented $\phi(u)$ by the series (18–4), the convergence of the right-hand side of (18–4) to $\phi(u)$ is certain only if $\phi(u)$ and its first two derivatives are continuous and vanish at infinity. That is, while we can formally associate a Gram-Charlier series with *every* $\phi(u)$, the series may not converge to represent $\phi(u)$. But convergence or no, the important question is whether $\phi(u)$ can be approximated by a finite, or more usefully, a few terms of (18–4). The answer is not altogether satisfactory. The order of magnitude of the successive terms of (18–4), not discussed here, does not steadily decrease (Cramér [36]); more terms may fit less closely than fewer terms. Still, two inferences seem reasonable; terms up to and including b_6, and large sample sizes for the estimates are needed for (and will almost always provide) a good approximation to $\phi(u)$. Fewer terms and small samples can lead to a poor approximation, including negative probabilities.

18–2 Hermite polynomials. The expressions for $b_0, b_1, b_2, \ldots,$ in the Gram-Charlier series are properly obtained in the following way. Write the ith derivative of the normal density function in the form

$$f^{(i)}(u) = (-1)^i H_i(u) f(u).$$

The polynomials $H_i(u)$, shown in (18–2), were known to Laplace and were explicitly introduced by Chebychev in 1860, but are named for Hermite, who explored them in 1864. They play important roles in many areas of mathematics and statistics, including stochastic processes in which normal distributions are present. We have

$$\phi(u) = \sum_i b_i f^{(i)}(u)$$

which, in terms of the Hermite polynomials, is equal to

$$\sum_i b_i (-1)^i H_i(u) f(u).$$

Form the integral

$$\int_{-\infty}^{\infty} \phi(u) H_j(u)\, du = \int_{-\infty}^{\infty} \Big[H_j(u) \sum_i b_i (-1)^i H_i(u) f(u)\Big]\, du. \tag{18–6}$$

The left-hand side of (18–6) is composed of moments of $\phi(u)$. The right-hand side reduces to a simple form, for the integral

$$\int_{-\infty}^{\infty} H_j(u)H_i(u)f(u)\,du$$

has a remarkable orthogonal property; it is 1 for $i = j$ and is 0 for $i \neq j$. Thus, we can solve (18–6) at once for b_i in terms of the moments of $\phi(u)$.

The orthogonal property of Hermite polynomials is shown as follows: write

$$\int_{-\infty}^{\infty} H_j(u)H_i(u)f(u)\,du = (-1)^j \int_{-\infty}^{\infty} H_i(u)f^{(j)}(u)\,du$$

and integrate by parts. Let

$$v = H_i(u), \qquad dw = f^{(j)}(u)\,du.$$

Applying $\int v\,dw = vw - \int w\,dv$, we have

$$\int_{-\infty}^{\infty} H_i(u)f^{(j)}(u)\,du = H_i(u)f^{(j-1)}(u)\Big]_{-\infty}^{\infty} - \int_{-\infty}^{\infty} f^{(j-1)}(u)H_i'(u)\,du.$$

The bracket is zero, for $f^{(j-1)}(u)$ is a nonvanishing function of u multiplied by normal u, and normal u is zero at $-\infty$ and $+\infty$. The right-hand integral is

$$\int_{-\infty}^{\infty} f^{(j-1)}(u)H_i'(u)\,du = i\int_{-\infty}^{\infty} f^{(j-1)}(u)H_{i-1}(u)\,du.$$

Note carefully the i in front of the right-hand integral. If we continue to integrate by parts, it is evident that for $i = j$, we end with $i!$, for the final integral

$$\int_{-\infty}^{\infty} f^{(0)}(u)H_0(u)\,du$$

is 1; while for $i \neq j$, we end with 0, for either

$$\int_{-\infty}^{\infty} f^{(0)}(u)H_i(u)\,du$$

or

$$\int_{-\infty}^{\infty} f^{(i)}(u)H_0(u)\,du$$

(we may reach one or the other first) vanishes for $i \neq 0$. Summarizing,

$$\int_{-\infty}^{\infty} H_j(u)H_i(u)f(u)\,du = \begin{cases} 1, & i = j, \\ 0, & \text{otherwise.} \end{cases}$$

From (18–6),

$$b_i = \frac{(-1)^i}{i!} \int_{-\infty}^{\infty} \phi(u) H_i(u)\, du,$$

identical with the values of b_i of Section 18–1, and often a more convenient formula for obtaining them.

Tables

Harvard University, Computation Laboratory, *Tables of the Error Function and Its First Twenty Derivatives.* Volume 23. Cambridge (Mass.), Harvard University Press, 1952.

N. R. Jørgensen, *Undersøgelser over Frequensflader og Korrelation.* Copenhagen, Busck, 1916.

Collateral Reading

C. V. L. Charlier, *Application de la théorie des probabilités à l'astronomie.* Paris, Gauthier-Villars et Cie, 1931.

H. Cramér, *Mathematical Methods of Statistics.* Princeton, Princeton University Press, 1946. Includes an analysis of the convergence problem with some of the author's important results.

F. Y. Edgeworth, "The law of error," *Cambridge Philosophical Trans.*, Vol. 20, 1905, p. 36, p. 113 and an appendix bound (only) with the reprints. A modified form of Gram-Charlier series in which the order of magnitude of successive terms steadily decreases.

A. Fisher, *The Mathematical Theory of Probabilities and Its Application to Frequency Curves and Statistical Methods.* 2nd ed., New York, Macmillan Co., 1922.

T. C. Fry, *Probability and Its Engineering Uses.* New York, D. Van Nostrand Co., 1928.

M. G. Kendall and A. Stuart, *The Advanced Theory of Statistics, Vol.* 1. London, Charles Griffin and Co., 1958. Excellent discussion with numerical examples.

H. L. Rietz, *Mathematical Statistics.* Carus Mathematical Monograph No. 3, Chicago, Math. Assoc. of America, Open Court Publishing Company, 1927. Includes a discussion of Charlier's "Type B" series which employs, in place of the normal distribution and its derivatives, the Poisson distribution and its differences.

J. G. Smith and A. J. Duncan, *Sampling Statistics and Applications, Volume* 2. New York, McGraw-Hill Book Co., 1945.

E. T. Whittaker and G. Robinson, *The Calculus of Observations.* 3rd ed., London, Blackie and Sons, 1940. Includes a linear model of error accumulation which produces the Gram-Charlier series.

Part III

Sampling Theory

In Part III we concentrate on the behavior of characteristics of samples drawn from known but quite generally described populations, ending, however, with a chapter on the characteristics of samples drawn from a particular (normal) population. The motivation is both interest in the behavior of samples as well as preparation for Part IV where the arguments of Part III are essentially reversed (the population there, in certain respects unknown, will be characterized from information contained in samples drawn from it).

CHAPTER 19

RANDOM SAMPLING

19–1 The purpose and nature of sampling. In Part I we dealt with general probability distributions. In Part II, discussion centered on ten specific probability distributions. In both parts, all probability distributions, general or specific, were considered only as mathematical functions.

In Part III the discussion has its motivation in an argument which implies a physical operation. We wish to draw a sample of data from a population and study the behavior of the sample data in the light of known or hypothesized values of characteristics of the population. In Parts IV and V, we shall reverse the analysis (and in so doing further motivate the whole of Part III) by using the sample data to reach conclusions as to the *unknown* values of the characteristics of the population from which the sample was drawn. This is the real-world problem and is the central problem of statistical inference.

In Part III the probability distributions of Parts I and II become, so to speak, living things. Probability distributions become descriptions of a population of weights of all persons in a community, a (bivariate) population of the compression and tensile strengths of all units of a certain type of spring in inventory, a population of all measurements of the speed of light. A sample of size n from the population means the weights of n persons, the compression and tensile strengths of n springs, n measurements of speed from the respective populations. The ultimate motivation for sampling is simple. We generally have incomplete knowledge of the distribution function, here $F(x)$ or $F(x, y)$, of the random variables (weight, compression and tensile strength, speed) and we imagine that we can improve our knowledge of $F(x)$ or $F(x, y)$ by drawing and studying a sample from the population described by $F(x)$ or $F(x, y)$.

Let us focus on one random variable $\mathbf{x}$. A sample of size n will yield n numbers (weights or speeds)

$$x_1, x_2, \ldots, x_n.$$

We require that the sampling process by which these numbers are obtained be such that the numbers will yield estimates of characteristics of $F(x)$ with *determinable errors of estimates.*

It is the latter requirement which severely limits the ways and means by which $x_1, x_2, \ldots, x_n$ can be obtained. For example, selecting the apparent lightest and apparent heaviest persons in a population, weighing them and averaging their weights as an estimate of the mean weight $E(\mathbf{x})$ of the population may have much to recommend it. But we are unable to estimate the error of such an estimate, and for this reason, we generally* reject the method of sampling which provides such an estimate. We have, however, a sampling method which *does* permit determination of the error of estimate; somewhat unfortunately, one method for infinite populations and a different one for sampling without replacement from finite populations. Both are generally described as *random sampling* (Wadsworth and Bryan [192] call the latter *conditional* random sampling); both can be brought under one broad definition but the gain is merely semantic. We describe random sampling from an infinite population in Section 19–3 and random sampling from a finite population in Section 19–4.

In the remainder of this chapter, we therefore restrict our attention to the mathematical and operational nature of sampling "at random" from a population. In Parts IV and V, where we will be drawing inferences from samples of observed data as to the values of various characteristics of populations (and sampling is a principal means by which such values are estimated), only samples which at some stage are "random" will avail. The theory of probability which we have already introduced and the current theories of statistical inference to which we will come have little to say regarding the behavior of nonrandom samples, and therefore little to say regarding the confidence with which we can draw inferences from them.

19–2 Review. Let us first repeat certain material from Chapter 7. Consider k discrete random variables $\mathbf{x}_1, \mathbf{x}_2, \ldots, \mathbf{x}_k$ with joint density $f(x_1, x_2, \ldots, x_k)$,

$$p(\mathbf{x}_1 = x_1, \mathbf{x}_2 = x_2, \ldots, \mathbf{x}_k = x_k) = f(x_1, x_2, \ldots, x_k). \qquad (19\text{–}1)$$

If the k random variables are mutually independent, (19–1) becomes

$$f(x_1, x_2, \ldots, x_k) = f_1(x_1)f_2(x_2)\ldots f_k(x_k);$$

in addition, the k density functions may possibly be identical, that is,

$$f_1 = f_2 = \cdots = f_k. \qquad (19\text{–}2)$$

For k continuous random variables, we could write

$$p(x_1 \le \mathbf{x}_1 \le x_1 + dx_1, x_2 \le \mathbf{x}_2 \le x_2 + dx_2, \ldots, x_k \le \mathbf{x}_k \le x_k + dx_k)$$
$$= f(x_1, x_2, \ldots, x_k)\, dx_1\, dx_2 \ldots dx_k, \qquad (19\text{–}3)$$

---

* It may, however, be going too far to say that in the absence of a measure of error, we will always reject a method which we believe to be good.

which, with mutual independence, reduces to

$$f_1(x_1)\,dx_1 \cdot f_2(x_2)\,dx_2 \cdot \ldots \cdot f_k(x_k)\,dx_k, \tag{19–4}$$

moreover, the k density functions may possibly be identical, as before.

No concept of sampling, let alone random sampling, is involved here; $\mathbf{x}_1, \mathbf{x}_2, \ldots, \mathbf{x}_k$ may be a set of variables such as height, weight, . . . , intelligence, and we have merely described the possible joint density function of such a set of random variables. But the review will serve as a useful preliminary to the concept of random sampling from an infinite population.

19–3 Random sampling from infinite populations. Now, and apart from the review, we turn to an operational and a mathematical view of sampling. Consider a single random variable $\mathbf{x}$, discrete or continuous, with density function $f(x)$. Draw repeatedly samples of size n from the physical population described by $f(x)$, that is to say, replicate (make repeatedly) n trials on the random variable $\mathbf{x}$; the sample data of each trial will be of the form of an n-tuple:

$$\begin{array}{l} \text{Replication 1: } x_1',\ x_2',\ \ldots,\ x_n' \\ \text{Replication 2: } x_1'',\ x_2'',\ \ldots,\ x_n'' \\ \text{Replication 3: } x_1''',\ x_2''',\ \ldots,\ x_n''' \\ \quad\vdots \qquad\qquad\qquad \vdots \end{array} \tag{19–5}$$

Before continuing, note that all real populations are discrete either by nature or by practical limitation on the number of decimals we can carry. But we can often with advantage approximate the population by a continuous density function $f(x)$, for example, the population of human weights, measured in fact only to ounces, by a normal density function. This does not imply that we are engaged here in any such activity as drawing a random sample from a population containing an infinite number of units.

We now postulate that the outcome of any replication, say the n-tuple

$$x_1^{(i)}, x_2^{(i)}, \ldots, x_n^{(i)}$$

may be regarded as a value of the joint random variable

$$\mathbf{x}_1, \mathbf{x}_2, \ldots, \mathbf{x}_n,$$

with joint density function $g(x_1, x_2, \ldots, x_n)$. This is the critical postulate; it connects the outcomes of the physical process of sampling with the abstract theory of probability. Now, for the sampling operation which produces (19–5) to be *random* we impose two conditions; first, that the outcomes of the n trials be statistically independent of each other, by which we mean

$$g(x_1, x_2, \ldots, x_n) = g_1(x_1)g_2(x_2) \cdot \ldots \cdot g_n(x_n), \tag{19–6}$$

and, second, that all marginal distributions be identical with and equal to f, the density function of $\mathbf{x}$ in the population,

$$g_1 = g_2 = \cdots = g_n = f. \tag{19–7}$$

If (19–6) and (19–7) are satisfied, the sampling method is random; the joint density function $g(x_1, x_2, \ldots, x_n)$ of the sample variables is expressible in terms of the density function f of the population random variable $\mathbf{x}$. The sample is connected in probability to the population from which it came; its properties depend on the properties of f.

The mathematical formulae (19–6) and (19–7) involved here are identical with those shown in the review of material from Chapter 7, but the use to which they are put here is different.

As an example, in Chapter 10 we considered a binomial random variable which on a single trial took on one of two values, success or failure. From trial to trial the probability of success remained constant at p, the probability of success in the binomial population. The n trials were independent, by which was meant that the probability of success or failure in later trials was unaffected by the results of earlier trials (independence of later to earlier outcomes is, mathematically, equivalent to *mutual* independence); evidently the n Bernoulli trials of Chapter 10 constituted a random sample.

No set of precautions can guarantee that any *physical* operation of sampling will satisfy the *mathematical* conditions (19–6) and (19–7). We make every effort to draw the sample so that each unit drawn has no effect on any other, thus hoping to achieve independence in a probability sense, and we are careful that all units are in fact drawn from the population defined by $f(x)$. The first is generally the more difficult to assure.

19–4 Random sampling with and without replacement from a finite population. If samples are drawn *with replacement* from a finite population, the population is effectively infinite with respect to the sampling process in that its density as described by f is unaffected by sampling. Expressions (19–6) and (19–7) remain the conditions for random sampling, and we seek, operationally, to achieve the following: from the N units of the population draw n units in such a way that each of the N units has the constant population probability $1/N$ of being drawn. For example, if the population consists of 5 different objects A, B, C, D, E, a random sample of size 2 with replacement can be drawn by insuring that the first of the two places in the sample be any one of the 5 letters, with equal probabilities, and that the second of the two places in the sample be any one of the 5 letters with equal probabilities. Both (19–6) and (19–7) are satisfied. Note that the number of different ordered samples is N^n; in our example, 5^2 ordered samples:

$$\begin{array}{ccccc} AA & BA & CA & DA & EA \\ AB & BB & CB & DB & EB \\ AC & BC & CC & DC & EC \\ AD & BD & CD & DD & ED \\ AE & BE & CE & DE & EE \end{array}$$

and each has probability $(1/5)^2$ of being drawn. No such combinatorial calculation is possible, of course, when sampling from populations described by continuous density functions.

Sampling *without replacement* from a finite population, for example, from a population described by the hypergeometric distribution of Chapter 12, would be an example of nonrandom sampling. The appearance of success or failure in the later elements of a sample depends on the outcomes of earlier trials. In this situation, the withdrawal, via the sampling operation, of elements from the population affects the probability distribution of the population. A modification of the original definition of random sampling from infinite populations is necessary. The customary modification is the following: in sampling without replacement from a finite population, the statistician introduces "randomness" in such a way as will ensure that all possible sequences of n trials, that is, all possible samples of size n, have the *same* probability of occurrence; in a sense the population is a synthesis of all possible different samples of size n. For example, if the finite population consists of 5 objects A, B, C, D, E, a sample of size 2 drawn *without replacement* is called random if the sampling operation is such that the 20 possible sequences

$$\begin{array}{ccccc} AB & BA & CA & DA & EA \\ AC & BC & CB & DB & EB \\ AD & BD & CD & DC & EC \\ AE & BE & CE & DE & ED \end{array}$$

are equally likely to occur. One sampling operation which assures such equal likelihood of samples is the following: select the first element of the sample from the population in such a way that all N elements in the population have the same probability $1/N$ of being drawn. Once the first element is selected, the second element of the sample is drawn from the remaining $N - 1$ elements of the population in such a way that all $N - 1$ remaining elements in the population have the same probability $1/(N - 1)$ of being drawn. Each sample of size n evidently has probability

$$\frac{1}{N} \cdot \frac{1}{N-1} \cdot \cdots \cdot \frac{1}{N-(n-1)}$$

of being drawn. Samples so drawn, without replacement from a finite population, will be called random, and what is important, errors of inference based on the data in such samples can be determined.

In combinatorial symbols, the number of samples each of size n that can be obtained without replacement from a finite population of size N is P_n^N; in our example $P_2^5 = 20$. For the sampling process to be random, the probability associated with each sample is $1/P_n^N$; in our example $1/P_2^5 = \frac{1}{20}$. If order is neglected, the number of different samples is C_n^N; in our example $C_2^5 = 10$; an example of the set of samples neglecting order is

$$\begin{array}{llll} AB & & & \\ AC & BC & & \\ AD & BD & CD & \\ AE & BE & CE & DE \end{array}$$

and for the sampling process to be random, the probability associated with each sample must be $1/C_2^5 = \frac{1}{10}$.

19–5 General remarks. *Equal likelihood of all random samples.* The common remark, "all random samples of size n are equally likely," must be made with care. As we have seen, it is meaningful when sampling without replacement from a finite population. What it might mean when sampling from a population described by a normal distribution is hard to say; The random variable of a normal distribution is continuous and has an unlimited store of variates which permit an unlimited number and variety of random samples of size n.

And it does not mean that in sampling, for example, from a binomial population (with $p \neq \frac{1}{2}$) the eight (and only eight) possible outcomes in a sample of size $n = 3$

$$111 \quad 110 \quad 101 \quad 100 \quad 011 \quad 010 \quad 001 \quad 000$$

are equally likely to occur. The sampling process that produces these 3-tuples is random, but the 3-tuples are not equally likely.

It is true that all samples of size n contributing to a fixed number of successes are equally likely. Outcomes 2, 3, and 5 above yield 2 successes in a sample of size 3 and are equally likely; the probabilities are ppq, pqp, qpp. Equal likelihood of all samples of size n containing x successes follows here from independence and constant probability p, and the fact that if x is fixed, the number of p's and q's in each sample of size n is the same. This equal likelihood was explicitly used in Chapter 10 in the derivation of the binomial formulae for the probability of a fixed number x of successes. Similarly, and though the probability does *not* remain constant from trial to trial, n trials on the hypergeometric variable of Chapter 12 also lead to equally likely samples in this sense. This equal likelihood was used in Chapter 12 in the derivation of the hypergeometric formulae for the probability of a fixed number x of successes. For example, for samples of size $n = 3$ yielding 2 successes

$$110 \quad 101 \quad 011$$

drawn without replacement from a hypergeometric population containing A successes and B failures, $A + B = N$, the probabilities associated with the three outcomes are, respectively,

$$\frac{A}{N} \cdot \frac{A-1}{N} \cdot \frac{B}{N-2}, \qquad \frac{A}{N} \cdot \frac{B}{N-1} \cdot \frac{A-1}{N-2}, \qquad \frac{B}{N} \cdot \frac{A}{N-1} \cdot \frac{A-1}{N-2},$$

which are equal.

Note that if we imagine spreading out the *finite* number A of successes (1's) and B of failures (0's) in the hypergeometric population to *single* elements: 1, 1, 1, 1, 1, . . . , 0, 0, 0, 0, . . . , all elements in the population are easily seen to have the same probability $1/N$ of being drawn first, all remaining elements the same probability $1/(N - 1)$ of being drawn second, . . . ; this spreading will be used in Chapter 21.

Sampling, and not the sample, is random. Note that it is sampling, and not the sample, that should be called random; a sample obtained by random sampling may be a strange sample indeed. All we can do is attempt to realize in the sampling operation the mathematical conditions defining a random sample and put up (or not!) with the occasional nonrepresentative appearance of some of the samples that are drawn.

Tables of random numbers. In sampling from finite populations, tables of random numbers are often useful. These are digits produced by processes which offer maximum assurance of statistical independence among the digits. Assignment of different numbers to the N objects of a real finite population followed by selection of n objects for the random sample by use of a table of random numbers often facilitates and improves the random selection process. The RAND tables [162] are recent but well known; an excerpt from them is given on pages 418–419.

Practical sampling and partitioning. To repeat, the mathematical conditions for random sampling from infinite or finite populations are not automatically or uniquely translatable into operational rules. For example, to obtain, for a test of the proportion of clean wool, a random sample of a pound of wool from a frozen bale of compressed wool standing on an open wharf in Boston is not easy. But one must try, for it is only from such samples that we can effectively estimate the clean content of the bale and the error of that estimate, since it is to such samples that the probability calculus has been particularly successfully applied. As we shall see in Chapter 27, this is not quite as limiting as it sounds. To reduce the frequency of samples not representative of the populations from which they are drawn, populations may first be nonrandomly partitioned, with respect to sex, income, geographical region, or any other factor that is relevant (in our example, distance from the center of the bale is such a factor) and the sizes of the subsamples drawn from the partitions may be profitably decided by nonrandom criteria; but finally in the operation, random sampling of the ultimate partitions must take place if we are to estimate successfully the errors of our inferences.

Remaining chapters of Part III. In the chapters to follow in Part III we form and study functions of (and only of) the random variables of random samples. Such functions are called *statistics*; they are themselves random variables and their probability behavior will be uniquely determined by the distribution function $F(x)$ associated with the population from which the random samples are drawn. In Chapters 20 through 22 we deal with generally described populations, in Chapter 23 with a specific (normal) population. In all four chapters we deal with statistics which on the whole are simple, though in Chapter 23, some less intuitive statistics are considered.

Collateral Reading

H. D. Brunk, *An Introduction to Mathematical Statistics.* Boston, Ginn and Co., 1960.

D. A. S. Fraser, *Statistics: An Introduction.* New York, John Wiley and Sons, 1957.

M. G. Kendall and A. Stuart, *The Advanced Theory of Statistics, Volume* 1. London, Charles Griffin and Co., 1958. Includes a very good discussion of the problems of random sampling.

(The) RAND Corporation, *A Million Random Digits with 100,000 Normal Deviates.* Glencoe (Ill.), The Free Press, 1955.

G. P. Wadsworth and J. G. Bryan, *Introduction to Probability and Random Variables.* New York, McGraw-Hill Book Co., 1960. A brief but sharp discussion.

CHAPTER 20

CONVOLUTION

20–1 The problem, and solution by change of variable. We begin a discussion, to continue through Chapter 22, of sampling from quite generally described populations. In this chapter we consider a basic probability argument sometimes known as convolution. We consider only continuous random variables.

Let $\mathbf{x}$ and $\mathbf{y}$ be *independent* random variables with distribution functions $F(x)$, $G(y)$:

$$F(x) = \int_{-\infty}^{x} f(\xi)\, d\xi, \qquad G(y) = \int_{-\infty}^{y} g(\eta)\, d\eta.$$

Out of interest arising from its importance in statistical theory and applications, we seek the distribution function $H(u)$ [and the density function $h(u)$] of the linear sum $\mathbf{u} = \mathbf{x} + \mathbf{y}$; $\mathbf{u}$ is a function of random variables and is therefore itself a random variable. The problem may be solved by simple change of variable. We have

$$\mathbf{u} = \mathbf{x} + \mathbf{y} \quad \text{or} \quad \mathbf{x} = \mathbf{u} - \mathbf{y} = \mathbf{u} - \mathbf{v},$$

$$\mathbf{v} = \mathbf{y} \quad \text{or} \quad \mathbf{y} = \mathbf{v},$$

$$J = \begin{vmatrix} \dfrac{\partial x}{\partial u} & \dfrac{\partial x}{\partial v} \\[2ex] \dfrac{\partial y}{\partial u} & \dfrac{\partial y}{\partial v} \end{vmatrix} = \begin{vmatrix} 1 & -1 \\ 0 & 1 \end{vmatrix} = 1,$$

and the joint density function $\psi(u, v)$ of $\mathbf{u}$, $\mathbf{v}$ is

$$\psi(u, v) = f\{x(u, v)\} g\{y(u, v)\}\ J = f(u - v)g(v),$$

from which the density function $h(u)$ of $\mathbf{u}$ is found by integrating over v,

$$h(u) = \int_{-\infty}^{\infty} f(u - v)g(v)\, dv.$$

Had we made the dummy variable $\mathbf{v} = \mathbf{x}$ instead of $\mathbf{v} = \mathbf{y}$, we would have found

$$h(u) = \int_{-\infty}^{\infty} f(v)g(u - v)\, dv,$$

as the student may verify.

20–2 Convolution. We now reach the same results by a slightly different path:

$$H(u) = p[\mathbf{u} \le u] = p[\mathbf{x} + \mathbf{y} \le u] = \int_{x+y<u}\int f(x)g(y)\,dy\,dx.$$

The region $\mathbf{u} \le u$, that is, $\mathbf{x} + \mathbf{y} \le u$, is shown shaded in Fig. 20–1. From $\mathbf{x} + \mathbf{y} \le u$ the upper limit of $\mathbf{y}$ is $u - x$; we have

$$H(u) = \int_{-\infty}^{\infty}\int_{-\infty}^{u-x} f(x)g(y)\,dy\,dx = \int_{-\infty}^{\infty}\left[f(x)\,dx\int_{-\infty}^{u-x} g(y)\,dy\right]$$
$$= \int_{-\infty}^{\infty} G(u - x)f(x)\,dx. \tag{20–1}$$

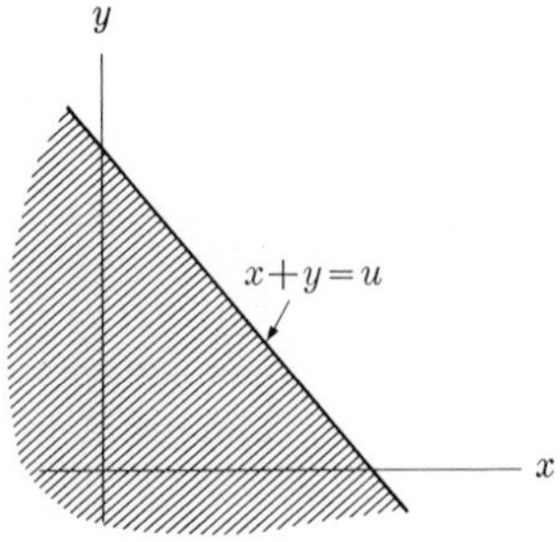

FIG. 20–1. The region $x + y \le u$.

If, for known G and f, the integration can be carried out, the problem is solved. Alternatively, with $u - y$ as the upper limit of $\mathbf{x}$, the distribution function $H(u)$ takes the form

$$H(u) = \int_{-\infty}^{\infty} F(u - y)g(y)\,dy. \tag{20–2}$$

If $H(u)$ is differentiable under the integral with respect to u, we also have, from (20–1) and (20–2),

$$h(u) = \frac{dH(u)}{du} = \int_{-\infty}^{\infty} g(u - x)f(x)\,dx = \int_{-\infty}^{\infty} f(u - y)g(y)\,dy, \tag{20–3}$$

the relationship between the density function of $\mathbf{x} + \mathbf{y}$ and the density functions of $\mathbf{x}$ and of $\mathbf{y}$, results already reached. The operations shown in (20–1) and (20–2), or in (20–3), are sometimes described as the convolution of $f(x)$ and $g(y)$ in $\mathbf{u} = \mathbf{x} + \mathbf{y}$. We describe (20–3) by saying that the density function of the sum of two independent continuous random variables is the convolution of the density functions of the random variables. An entirely similar result holds for discrete random variables; it will be needed in the solution of problem 20–2.

The argument of convolution is basic and therefore valuable, though in operation it may be intractable for other than simple probability distributions. For some distributions, it is simpler to obtain the probability behavior of $\mathbf{x} + \mathbf{y}$ by working with the moment generating function of $\mathbf{x} + \mathbf{y}$.

For n variables and $\mathbf{u} = \mathbf{x}_1 + \mathbf{x}_2 + \cdots + \mathbf{x}_n$, $H(u)$ and $h(u)$ may be obtained by repeated application of (20–1) and (20–2), first to $\mathbf{x}_1 + \mathbf{x}_2$, then to $(\mathbf{x}_1 + \mathbf{x}_2) + \mathbf{x}_3$, etc. The special case $\mathbf{u} + \mathbf{x}_1 + \mathbf{x}_2 + \cdots + \mathbf{x}_n$ in which $\mathbf{x}_1, \mathbf{x}_2, \ldots, \mathbf{x}_n$ are mutually independent with identical probability distributions is important. In this case, $\mathbf{x}_1, \mathbf{x}_2, \ldots, \mathbf{x}_n$ may be viewed as elements of a random sample, and we are discussing the probability behavior of the sum of the n observations of a sample drawn at random from a population. We have already dealt with a random variable which was the sum of n independent random variables all with the same probability distribution; in the binomial distribution, the number of successes $\mathbf{x}$ in n independent trials can be viewed as the sum of the number of successes $\mathbf{x}_1, \mathbf{x}_2, \ldots, \mathbf{x}_n$ in each of the n independent trials.

For the general problem, we write

$$\mathbf{u} = (\mathbf{x}_1 + \cdots + \mathbf{x}_{n-1}) + \mathbf{x}_n = \mathbf{y} + \mathbf{x}_n.$$

With subscripts on F, G, f, and g indicating the number of random variables involved, (20–1) and (20–2) become

$$H_n(u) = \int_{-\infty}^{\infty} G_{n-1}(u - x_n) f_1(x_n)\, dx_n = \int_{-\infty}^{\infty} F_1(u - y) g_{n-1}(y)\, dy, \tag{20–4}$$

and if $H_n(u)$ is differentiable under the integral, (20–3) becomes

$$h_n(u) = \int_{-\infty}^{\infty} g_{n-1}(u - x_n) f_1(x_n)\, dx_n = \int_{-\infty}^{\infty} f_1(u - y) g_{n-1}(y)\, dy \tag{20–5}$$

$$\left(\text{also} = \int_{-\infty}^{\infty} g_m(u - x) f_{n-m}(x)\, dx, \quad m < n\right).$$

Application: in the random assembly of mass-produced parts intended to be identical, the dimension of interest in the assembly is often the sum of a dimension of each of the parts; length is an example. Expressions (20–4) and (20–5) describe the probability distribution of the dimension in the assembly in terms of the probability distributions of the dimensions of the parts. As an example of the use of such a result (and one which plays a useful role in modern reliability theory), the proportion of "good" assemblies $p(u_1 < \mathbf{u} < u_2)$ can be expressed in terms of the quality of the parts.

20–3 Application to uniform random variables. We wish to find the probability distribution of the sum of n independent random variables each of which is uniformly distributed over the range 0 to 1. Or, in the framework of sampling, we wish to find the probability distribution of the sum of the elements of a sample of size n drawn at random from a population described by a uniform distribution of range 0 to 1.

While the distributions being convolved here are particularly simple, the fact, discussed in Chapter 8, that any continuous probability distribution can be transformed into a uniform distribution adds some significance to this example. In problems 20–3 through 20–5, other distributions are convolved.

We have

$$\mathbf{u} = (\mathbf{x}_1 + \cdots + \mathbf{x}_{n-1}) + \mathbf{x}_n = \mathbf{y} + \mathbf{x}_n.$$

Also

$$\begin{aligned} f_1(x_n) &= 1, \quad 0 \leq x_n \leq 1, \\ &= 0, \quad x_n \text{ otherwise,} \end{aligned}$$

or

$$\begin{aligned} f_1(u - y) &= 1, \quad 0 \leq u - y < 1, \\ &= 0, \quad u - y \text{ otherwise.} \end{aligned}$$

The second integral of (20–5) reduces to

$$h_n(u) = \int_{u-1}^{u} g_{n-1}(y)\, dy, \tag{20–6}$$

the range of y being $u - 1$ to u. Thus we find the density function of $\mathbf{u} = \mathbf{x}_1 + \mathbf{x}_2 + \cdots + \mathbf{x}_n$ from the density function of $\mathbf{x}_1 + \mathbf{x}_2 + \cdots + \mathbf{x}_{n-1}$, the density function of the latter found from an earlier convolution, and so on.

We illustrate in detail for $n = 2$. With $\mathbf{u} = \mathbf{x}_1 + \mathbf{x}_2$, the range of u extends from 0 to 2, and the lower and upper limits of the integral in (20–6) are -1 and 2. But the functional form of $g_1(=f_1)$ varies within the range $u = -1$ to $u = +2$. Integral (20–6) must therefore be split into parts within each of which the functional form of $f_1(y)$ is constant, in this example, 1 or 0. We consider first the integral (20–6) for $0 \leq u \leq 1$, then for $1 \leq u \leq 2$.

When $0 \leq u \leq 1$, the lower limit of the integral satisfies

$$-1 \leq u - 1 \leq 0.$$

Of this range, only $u - 1 = 0$ need be considered, since $f_1(y) = 0$ for all values of **y** less than 0. When $0 \leq u \leq 1$, the upper limit of the integral satisfies

$$0 \leq u \leq 1$$

for all values of which $f_1(y)$ is nonvanishing. In summary,

$$\int_{u-1}^{u} f_1(y)\,dy = \int_0^u dy = u, \quad 0 \leq u \leq 1.$$

When $1 \leq u \leq 2$, the lower limit of the integral (20–6) satisfies

$$0 \leq u - 1 \leq 1$$

for all values of which $f_1(y)$ is nonvanishing. When $1 \leq u \leq 2$, the upper limit of the integral satisfies

$$1 \leq u \leq 2.$$

Of this range, only $u = 1$ need be considered, since $f_1(y) = 0$ for all values of **y** greater than 1. In summary,

$$\int_{u-1}^{u} f_1(y)\,dy = \int_{u-1}^{1} dy = 2 - u, \quad 1 \leq u \leq 2.$$

The result is shown in Fig. 20–2; for values of u outside the range $0 \leq u \leq 2$, $h_2(u) = 0$.

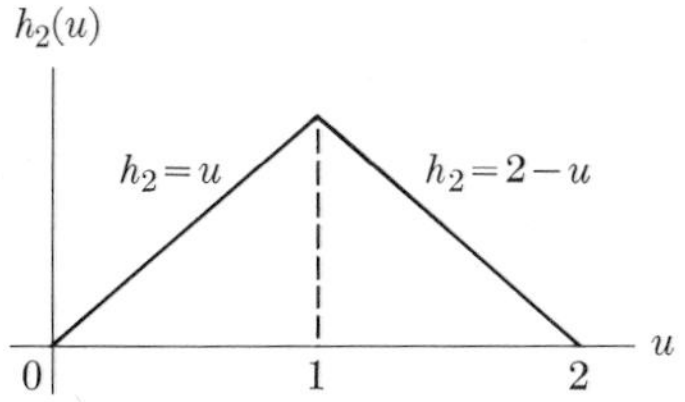

FIG. 20–2. Density function of the sum of the elements of a random sample of size 2 from a population described by the uniform distribution over the range (0, 1).

Collateral Reading

H. Cramér, *Mathematical Methods of Statistics.* Princeton, Princeton University Press, 1945.

W. Feller, *An Introduction to Probability and Its Applications.* 2nd ed., New York, John Wiley and Sons, 1957.

M. E. Munroe, *Theory of Probability.* New York, McGraw-Hill Book Co., 1951. A very good discussion.

E. Parzen, *Modern Probability Theory and Its Applications.* New York, John Wiley and Sons, 1960.

J. V. Uspensky, *Introduction to Mathematical Probability.* New York, McGraw-Hill Book Co., 1937.

P. M. Woodward, *Probability and Information Theory, with Applications to Radar.* New York, McGraw-Hill Book Co., 1953. An excellent reference. Includes a detailed example illustrating the discrete convolution process, convolution with application to the rectangular and the normal distributions, and a discussion of applications within the field of information theory.

Problems

20–1. The reader may check his understanding of the example of this chapter by convolving three uniform distributions each of range 0 to 1 and finding for $h_3(x)$, where $x = x_1 + x_2 + x_3$,

$$\begin{aligned} h_3(x) &= \tfrac{1}{2}x^2, & 0 \le x \le 1, \\ &= \tfrac{1}{2}(-2x^2 + 6x - 3), & 1 \le x \le 2, \\ &= \tfrac{1}{2}(3 - x)^2, & 2 \le x \le 3. \end{aligned}$$

Construct a graph of the result. In this problem the integral (20–6) will have to be split into three parts.

20–2. By convolution, show that the mean of the binomial random variable $\mathbf{x} = \mathbf{x}_1 + \mathbf{x}_2 + \cdots + \mathbf{x}_n$ is np and that its variance is npq.

20–3. By convolution, show that the mean $(\mathbf{x}_1 + \mathbf{x}_2)/2$ of two independent identically distributed Cauchy variables $\mathbf{x}_1$ and $\mathbf{x}_2$ has the (remarkable) property of having the probability distribution of each.

20–4. The sum of two independent normal variables is itself normal. The converse, that the sum of two independent variables is normal if each is normal, is also true; a brilliant result due to Lévy and to Cramér.

CHAPTER 21

SAMPLING FROM GENERAL POPULATIONS

In this chapter we continue the discussion of sampling from general populations. In Section 1, we sample at random from a population whose continuous probability distribution is quite generally described. In Section 2, the population has been partitioned into subpopulations from which random samples are drawn. In Section 3, we sample without replacement from a general discrete population. In all three sections the probability behavior of simple but important statistics is studied.

21–1 Sampling from a continuous population. *The sample mean.* A random sample $\mathbf{x}_1, \mathbf{x}_2, \ldots, \mathbf{x}_n$ is drawn from a population described by a continuous random variable $\mathbf{x}$ with density function $f(x)$. From the elements of the sample, we construct the *sample mean* $\bar{\mathbf{x}}$, a familiar measure of the average value of the sample variables defined by

$$\bar{\mathbf{x}} = \frac{\mathbf{x}_1 + \mathbf{x}_2 + \cdots + \mathbf{x}_n}{n}. \tag{21–1}$$

As $\bar{\mathbf{x}}$ is a function of random variables, $\bar{\mathbf{x}}$ itself is a random variable. Its moment generating function is

$$M_{\bar{\mathbf{x}}}(\theta) = M_{(\mathbf{x}_1+\mathbf{x}_2+\cdots+\mathbf{x}_n)/n}(\theta) = M_{\mathbf{x}_1+\mathbf{x}_2+\cdots+\mathbf{x}_n}\left(\frac{\theta}{n}\right) = M_{\mathbf{x}}^n\left(\frac{\theta}{n}\right), \tag{21–2}$$

the last expression on the right following from the expression to its left by the mutual independence of $\mathbf{x}_1, \mathbf{x}_2, \ldots, \mathbf{x}_n$ and the fact that each sample element $\mathbf{x}_i$ has density function $f(x)$; in a different framework, this result was reached in (7–9). Thus (21–2) connects the moment generating function of the sample mean and the moment generating function of the random variable of the population, and permits us to obtain the moments of $\bar{\mathbf{x}}$ in terms of the moments of $\mathbf{x}$.

Either by expansion or differentiation of (21–2) or, as below, directly from definition,

$$\begin{aligned} E(\bar{\mathbf{x}}) &= E\left(\frac{\mathbf{x}_1 + \mathbf{x}_2 + \cdots + \mathbf{x}_n}{n}\right) = \frac{1}{n}E(\mathbf{x}_1 + \mathbf{x}_2 + \cdots + \mathbf{x}_n) \\ &= \frac{nE(\mathbf{x})}{n} = E(\mathbf{x}); \end{aligned} \tag{21–3}$$

$$\begin{aligned} \sigma^2(\bar{\mathbf{x}}) &= E[\bar{\mathbf{x}} - E(\bar{\mathbf{x}})]^2 = E\left[\frac{\mathbf{x}_1 + \mathbf{x}_2 + \cdots + \mathbf{x}_n}{n} - E(\mathbf{x})\right]^2 \\ &= \frac{1}{n^2}E\{[\mathbf{x}_1 - E(\mathbf{x})] + [\mathbf{x}_2 - E(\mathbf{x})] + \cdots + [\mathbf{x}_n - E(\mathbf{x})]\}^2. \end{aligned}$$

As $\mathbf{x}_1, \mathbf{x}_2, \ldots, \mathbf{x}_n$ are mutually independent, the cross-product terms in the last expression vanish, and we have

$$\sigma^2(\bar{\mathbf{x}}) = \frac{\sigma^2(\mathbf{x})}{n}. \tag{21–4}$$

Both (21–3) and (21–4) follow immediately from (7–15) and (7–16) if in the latter $\mathbf{x}_1, \ldots, \mathbf{x}_n$ are described as n mutually independent random variables. Here this is also true, but $\{\mathbf{x}_1, \ldots, \mathbf{x}_n\}$ is now described as a random sample of size n.

The results (21–3) and (21–4) may be summarized as follows: the mean of a sample of size n drawn at random from a population of mean $E(\mathbf{x})$ and variance $\sigma^2(\mathbf{x})$ has mean $E(\mathbf{x})$ and variance $\sigma^2(\mathbf{x})/n$.

Equation (21–3) connects the mean of the (probability distribution of the) sample mean with the mean of the population (wherever the latter mean exists). Equation (21–4) connects the variance of the (probability distribution of the) sample mean with the variance of the population (wherever the latter variance exists). Equation (21–3) reveals that *on the average* the sample mean is equal to the population mean; (21–4) reveals that, *on the average*, the variability of the sample mean is, for $n > 1$, *less* than the variability of the random variable of the population. As estimators of population parameters which on the average coincide with the parameters are of interest in statistical inference, and as the error of an estimator is often judged by the variance of the estimator (the smaller the variance, the smaller the error), both (21–3) and (21–4) are of interest and will reappear frequently in the remainder of this text.

The combination of (21–4) and the Bienaymé-Chebychev inequality (3–12) yield an important theorem on the behavior of the sample mean $\bar{\mathbf{x}}$, as the sample size n becomes large. Applying (3–12) to the sample mean $\bar{\mathbf{x}}$, we have

$$p\left[|\bar{\mathbf{x}} - E(\bar{\mathbf{x}})| \geq k\frac{\sigma(\mathbf{x})}{\sqrt{n}}\right] \leq \frac{1}{k^2}.$$

Letting

$$\frac{k\sigma(\mathbf{x})}{\sqrt{n}} = b,$$

we find

$$p[|\bar{\mathbf{x}} - E(\mathbf{x})| \geq b] \leq \frac{\sigma^2(\mathbf{x})}{b^2 n}, \tag{21–5}$$

which, for each fixed $b > 0$, approaches 0 as n approaches ∞. For all $\mathbf{x}$ for which $\sigma^2(\mathbf{x})$ exists, the sample mean $\bar{\mathbf{x}}$ is said to *converge stochastically* to the population mean $E(\mathbf{x})$ as the sample size becomes indefinitely large. This is one of the forms of the "law of large numbers."

The sample variance. The *sample variance* $\mathbf{s}^2$ is a measure of the variability of the sample variables about their mean $\bar{\mathbf{x}}$. It is here defined by

$$\mathbf{s}^2 = \frac{(\mathbf{x}_1 - \bar{\mathbf{x}})^2 + (\mathbf{x}_2 - \bar{\mathbf{x}})^2 + \cdots + (\mathbf{x}_n - \bar{\mathbf{x}})^2}{n}; \tag{21-6}$$

$\mathbf{s}^2$ is a function of random variables and therefore is itself a random variable.

As for the relationship between $\mathbf{s}^2$ and the parameters of the population from which the random sample was drawn, we have

$$E(\mathbf{s}^2) = \frac{1}{n} E \sum_{i=1}^{n} (\mathbf{x}_i - \bar{\mathbf{x}})^2 = \frac{1}{n} E \sum_{i=1}^{n} [\mathbf{x}_i - E(\mathbf{x}) + E(\mathbf{x}) - \bar{\mathbf{x}}]^2$$

$$= \frac{1}{n} \sum_{i=1}^{n} \{E[\mathbf{x}_i - E(\mathbf{x})]^2 - E[\bar{\mathbf{x}} - E(\mathbf{x})]^2\}$$

$$= \frac{1}{n}\left[n\sigma^2(\mathbf{x}) - \frac{n\sigma^2(\mathbf{x})}{n}\right] = \frac{n-1}{n}\sigma^2(\mathbf{x}); \tag{21-7}$$

on the average, the sample variance is smaller than the population variance. The student may wish to show that

$$\sigma^2(\mathbf{s}^2) = \frac{\mu_4(\mathbf{x})}{n}\left(\frac{n-1}{n}\right)^2 - \frac{(n-3)(n-1)}{n^3}\sigma^4(\mathbf{x}),$$

which is of lesser value in inference. For while each of the results (21-3), (21-4), and (21-7) depends on *one* parameter of the population (and, therefore, the random variable involved in each case can, and later will be, utilized as an estimator of that parameter), the variance of the sample variance is seen to depend on *two* parameters, one of them the less informative fourth moment $\mu_4(\mathbf{x})$.

All four of the major results of this section to this point may be verified for discrete random variables.

The central limit theorem. We now prove the famous central limit theorem, but under somewhat stronger conditions than are needed. We have n independently distributed random variables $\mathbf{x}_1, \ldots, \mathbf{x}_n$ with common mean λ and with common (finite) variance σ^2. We form the sum

$$\mathbf{w} = \mathbf{x}_1 + \cdots + \mathbf{x}_n;$$

from earlier work,

$$E(\mathbf{w}) = n\lambda$$

and

$$\sigma^2(\mathbf{w}) = n\sigma^2.$$

The central limit theorem states that, for $n \to \infty$, the standardized sum

$$\mathbf{z} = \frac{\mathbf{w} - E(\mathbf{w})}{\sigma(\mathbf{w})}$$

is *normally* distributed with mean 0 and variance 1.

The proof is accomplished by showing that the moment generating function of $\mathbf{z}$, for n large, is the same as (16–13), the moment generating function of a normal random variable of mean 0 and variance 1. First, we express the moment generating function of $\mathbf{z}$ in terms of the moment generating function of $\bar{\mathbf{x}}$.

$$M_{\mathbf{z}}(\theta) = M_{(\bar{\mathbf{x}}-\lambda)/(\sigma/\sqrt{n})}(\theta) = \int_{-\infty}^{\infty} \exp \theta\left[\left\{\frac{\bar{x}-\lambda}{\sigma/\sqrt{n}}\right\}\right] f(\bar{x})\, d\bar{x}$$

$$= \exp\left[-\frac{\theta\lambda}{\sigma/\sqrt{n}}\right] M_{\bar{\mathbf{x}}}\left(\frac{\theta}{\sigma/\sqrt{n}}\right). \tag{21–8}$$

Now, repeating (21–2), we express the moment generating function of $\bar{\mathbf{x}}$ in (21–8) in terms of the moment generating function of $\mathbf{x}$:

$$M_{\bar{\mathbf{x}}}(\theta) = \left[M_{\mathbf{x}}\left(\frac{\theta}{n}\right)\right]^n,$$

or

$$M_{\bar{\mathbf{x}}}\left(\frac{\theta}{\sigma/\sqrt{n}}\right) = \left[M_{\mathbf{x}}\left(\frac{\theta}{\sigma\sqrt{n}}\right)\right]^n,$$

and (21–8) becomes

$$M_{\mathbf{z}}(\theta) = \exp\left[-\frac{\theta\lambda}{\sigma/\sqrt{n}}\right]\left[M_{\mathbf{x}}\left(\frac{\theta}{\sigma\sqrt{n}}\right)\right]^n. \tag{21–9}$$

Finally we express the moment generating function of $\mathbf{x}$ in terms of the moment generating function of $(\mathbf{x} - \lambda)/\sigma$:

$$M_{(\mathbf{x}-\lambda)/\sigma}(\theta) = \int_{-\infty}^{\infty} \exp\left[\theta\left(\frac{x-\lambda}{\sigma}\right)\right] f(x)\, dx = \exp\left[-\frac{\theta\lambda}{\sigma}\right] M_{\mathbf{x}}\left(\frac{\theta}{\sigma}\right),$$

or

$$M_{(\mathbf{x}-\lambda)/\sigma}\left(\frac{\theta}{\sigma\sqrt{n}}\right) = \exp\left[-\frac{\theta\lambda}{\sigma\sqrt{n}}\right] M_{\mathbf{x}}\left(\frac{\theta}{\sigma\sqrt{n}}\right),$$

which, introduced into (21–9), yields

$$M_{\mathbf{z}}(\theta) = \left[M_{(\mathbf{x}-\lambda)/\sigma}\left(\frac{\theta}{\sqrt{n}}\right)\right]^n.$$

Write $(\mathbf{x} - \lambda)/\sigma = \mathbf{t}$; we have

$$\left[M_{\mathbf{t}}\left(\frac{\theta}{\sqrt{n}}\right)\right]^n = \left[\int_{-\infty}^{\infty} \exp\left[\frac{\theta}{\sqrt{n}}\, t\right] f(t)\, dt\right]^n.$$

Introducing

$$\exp\left[\frac{\theta}{\sqrt{n}}\, t\right] = 1 + \frac{\theta t}{\sqrt{n}} + \frac{\theta^2 t^2}{2n} + \frac{\theta^3 t^3}{6n^{3/2}} + \cdots,$$

we find

$$\left[M_{\mathbf{t}}\left(\frac{\theta}{\sqrt{n}}\right)\right]^n = \left\{\int_{-\infty}^{\infty}\left[1 + \frac{\theta t}{\sqrt{n}} + \frac{\theta^2 t^2}{2n} + \frac{\theta^3 t^3}{6n^{3/2}} + \cdots\right] f(t)\, dt\right\}^n$$

$$= \left[\int_{-\infty}^{\infty} f(t)\, dt + \int_{-\infty}^{\infty} \frac{\theta}{\sqrt{n}}\, t f(t)\, dt + \int_{-\infty}^{\infty} \frac{\theta^2}{2n}\, t^2 f(t)\, dt + \int_{-\infty}^{\infty} \frac{\theta^3}{6n^{3/2}}\, t^3 f(t)\, dt + \cdots\right]^n$$

$$= \left[1 + \frac{\theta}{\sqrt{n}}\, \lambda_1(\mathbf{t}) + \frac{\theta^2}{2n}\, \lambda_2(\mathbf{t}) + \frac{\theta^3}{6n^{3/2}}\, \lambda_3(\mathbf{t}) + \cdots\right]^n$$

$$= \left[1 + 0 + \frac{\theta^2}{2n} + \text{terms in } n^{-3/2}, n^{-2}, \cdots\right]^n,$$

which, for $n \to \infty$, yields

$$M_{\mathbf{z}}(\theta) = e^{\theta^2/2}.$$

By comparison with (16–13), $\mathbf{z}$ is normal with mean 0 and variance 1. A special case of this general result was given in Section 16–4, where the random variables $\mathbf{x}_1, \ldots, \mathbf{x}_n$ were independently *binomially* distributed.

In this section we have considered only two statistics $\bar{\mathbf{x}}$ and $\mathbf{s}$. In Part IV we shall sometimes need to consider the probability behavior of other statistics, that is, of other functions of the variables of a random sample.

21–2 Sampling from partitioned populations. In this section we consider the probability behavior of a simple statistic formed from the variables of random samples drawn from the partitions or strata of a population described by a continuous random variable. In Chapter 27, we will see that adroit partitioning of a population before sampling from it generally leads to reduced errors in our estimates of characteristics of the population.

Consider a population π (personal incomes in the United States) described by the density function $f(x)$ of the random variable $\mathbf{x}$. Let this population consist of h mutually exclusive subpopulations or partitions or strata $\pi_1, \pi_2, \ldots, \pi_h$ (personal incomes in the fifty states) with respective density functions $f_1(x), f_2(x), \ldots, f_h(x)$ (distribution of personal income in each of the fifty states). For the subpopulations we may write

$$F_1(x) = \int_{-\infty}^{x} f_1(x)\,dx = \text{prob}\,[\mathbf{x} < x | \mathbf{x}\,c\,\pi_1],$$

$$F_2(x) = \int_{-\infty}^{x} f_2(x)\,dx = \text{prob}\,[\mathbf{x} < x | \mathbf{x}\,c\,\pi_2],$$

$$\vdots$$

$$F_h(x) = \int_{-\infty}^{x} f_h(x)\,dx = \text{prob}\,[\mathbf{x} < x | \mathbf{x}\,c\,\pi_h],$$

where prob $[\mathbf{x} < x | \mathbf{x}\,c\,\pi_2]$ stands for the probability that the value of the random variable $\mathbf{x}$ is a number smaller than x, *conditional* on $\mathbf{x}$ being a member of π_2 (the probability that a personal income is less than \$5000, given that the personal income belongs to a resident of Arizona).

Now to the relationship between the probability behavior of $\mathbf{x}$ in π and the probability behavior of $\mathbf{x}$ in $\pi_1, \pi_2, \ldots, \pi_h$,

$$\begin{aligned} F(x) = \int_{-\infty}^{x} f(x)\,dx &= \text{prob}\,[\mathbf{x} < x | \mathbf{x}\,c\,\pi] \\ &= \text{prob}\,[\mathbf{x} < x | \mathbf{x}\,c\,\pi_1] \cdot \text{prob}\,[\mathbf{x}\,c\,\pi_1] \\ &+ \text{prob}\,[\mathbf{x} < x | \mathbf{x}\,c\,\pi_2] \cdot \text{prob}\,[\mathbf{x}\,c\,\pi_2] + \cdots \\ &+ \text{prob}\,[\mathbf{x} < x | \mathbf{x}\,c\,\pi_h] \cdot \text{prob}\,[\mathbf{x}\,c\,\pi_h], \end{aligned}$$

or, writing $p_i = \text{prob}\,[\mathbf{x}\,c\,\pi_i]$, we have

$$F(x) = \sum_{i=1}^{h} p_i F_i(x) \tag{21–10}$$

and

$$\frac{dF(x)}{dx} = f(x) = \sum_{i=1}^{h} p_i \frac{dF_i(x)}{dx} = \sum_{i=1}^{h} p_i f_i(x), \tag{21–11}$$

basic relationships between the distribution and density functions of $\mathbf{x}$ in π and the distribution and density functions of $\mathbf{x}$ in $\pi_1, \pi_2, \ldots, \pi_h$. Note the important role played by the parameters p_i in both.

We now find relationships between the mean λ and the variance σ^2 of $\mathbf{x}$ in π on the one hand and the means $\lambda_1, \lambda_2, \ldots, \lambda_h$ and the variances $\sigma_1^2, \sigma_2^2, \ldots, \sigma_h^2$ of $\mathbf{x}$ in $\pi_1, \pi_2, \ldots, \pi_h$ on the other. No sampling is involved here; we seek relationships between the principal parameters of the population and the principal parameters of its partitions;

$$\lambda = E(\mathbf{x}) = \int_{-\infty}^{\infty} xf(x)\,dx = \int_{-\infty}^{\infty} \sum_{i=1}^{h} xp_i f_i(x)\,dx$$

$$= \sum_{i=1}^{h} p_i \int_{-\infty}^{\infty} xf_i(x)\,dx = \sum_{i=1}^{h} p_i\lambda_i; \tag{21–12}$$

$$\sigma^2 = E(\mathbf{x} - \lambda)^2 = \int_{-\infty}^{\infty} (x - \lambda)^2 f(x)\,dx$$

$$= \int_{-\infty}^{\infty} (x - \lambda)^2 \sum_{i=1}^{h} p_i f_i(x)\,dx.$$

The expression on the right may be written

$$\sum_{i=1}^{h} p_i \int_{-\infty}^{\infty} [(x - \lambda_i) + (\lambda_i - \lambda)]^2 f_i(x)\,dx,$$

which, carrying out the squaring and noting that the cross-product term vanishes, reduces to

$$\sigma^2 = \sum_{i=1}^{h} p_i[\sigma_i^2 + (\lambda_i - \lambda)^2], \tag{21–13}$$

and (21–13) will be recognized as a special case of (6–8): total variance = mean of the conditional variance + variance of the conditional mean. From (21–12), the population mean is a linear function of the subpopulation means; from (21–13), the population variance is a function both of the subpopulation variances and the spread of the subpopulations means about the population mean.

As already noted, (21–10) through (21–13) connect the population π and the subpopulations $\pi_1, \pi_2, \ldots, \pi_h$; they are not concerned with sampling. We now draw random samples from the subpopulations, a random sample of size n_1 from π_1, of size n_2 from $\pi_2, \ldots$, of size n_h from π_h, with

$$\sum_{i=1}^{h} n_i = n.$$

Let $\bar{\mathbf{x}}_1$ be the mean of the random sample drawn from π_1, $\bar{\mathbf{x}}_2$ the mean of the random sample from $\pi_2, \ldots, \bar{\mathbf{x}}_h$ the mean of the random sample from π_h. The overall mean $\bar{\mathbf{x}}_s$ of such a partitioned or stratified sample of total size n is, by definition, the sum of all elements in the sample divided by the number of these elements:

$$\bar{\mathbf{x}}_s = \frac{\sum_{i=1}^{n_1} \mathbf{x}_i + \sum_{i=1}^{n_2} \mathbf{x}_i + \cdots + \sum_{i=1}^{n_h} \mathbf{x}_i}{n}$$

(the sums referring respectively to samples from $\pi_1, \pi_2, \ldots, \pi_h$)

$$= \frac{n_1\bar{\mathbf{x}}_1}{n} + \frac{n_2\bar{\mathbf{x}}_2}{n} + \cdots + \frac{n_h\bar{\mathbf{x}}_h}{n}. \tag{21–14}$$

Again, as $\bar{\mathbf{x}}_s$ is a function of random variables, $\bar{\mathbf{x}}_s$ is a random variable.

A specific probability distribution of $\bar{\mathbf{x}}_s$ cannot be given when the subpopulations are so generally described as in this section, but we can find the mean and the variance of $\bar{\mathbf{x}}_s$ in terms of moments of the subpopulations:

$$E(\bar{\mathbf{x}}_s) = \frac{n_1}{n}\lambda_1 + \frac{n_2}{n}\lambda_2 + \cdots + \frac{n_h}{n}\lambda_h = \sum_{i=1}^{h} \frac{n_i}{n}\lambda_i. \tag{21–15}$$

Note from (21–15) and (21–12) that unless $n_i/n = p_i$ the mean $\bar{\mathbf{x}}_s$ is not, on the average, equal to the mean λ of the population for all λ_i. Or as we shall begin to say in Chapter 25, unless $n_i/n = p_i$, $\bar{\mathbf{x}}_s$ is not an *unbiased* estimator of λ.

$$\sigma^2(\bar{\mathbf{x}}_s) = E[\bar{\mathbf{x}}_s - E(\bar{\mathbf{x}}_s)]^2 = E\left[\sum_{i=1}^{h}\left(\frac{n_i}{n}\bar{\mathbf{x}}_i - \frac{n_i}{n}\lambda_i\right)\right]^2$$

$$= E\left[\sum_{i=1}^{h}\frac{n_i}{n}(\bar{\mathbf{x}}_i - \lambda_i)\right]^2.$$

Carrying out the squaring and noting that the cross-product terms vanish, we have

$$\sigma^2(\bar{\mathbf{x}}_s) = \frac{1}{n^2}\sum_{i=1}^{h} n_i^2 E(\bar{\mathbf{x}}_i - \lambda_i)^2 = \frac{1}{n^2}\sum_{i=1}^{h} n_i^2\sigma^2(\bar{\mathbf{x}}_i)$$

$$= \frac{1}{n}\sum_{i=1}^{h}\frac{n_i}{n}\sigma_i^2. \tag{21–16}$$

The variance $\sigma^2(\bar{\mathbf{x}}_s)$ of the mean of a partitioned sample, unlike the variance σ^2 of the population, does not involve $\lambda_i - \lambda$. Once again, (21–16) being essentially the variance of the sum of independent random variables, the answer follows directly from (7–16).

In Chapter 27 we will return to such results as (21–15) and (21–16) and show the generally smaller error of estimates of the parameters of a population that result from random sampling of the partitions of a carefully partitioned population. We will show the effect of wise partitioning (which determines the earlier parameters p_i) and we will also determine values of n_i which minimize $\sigma^2(\bar{\mathbf{x}}_s)$ of (21–16).

21–3 Sampling without replacement from discrete populations. In this section, we begin with a population described by a general discrete random variable $\mathbf{x}$, and we deal with the behavior of random samples drawn without replacement from such a generally described population. The example is from Wilks [206].

We have a population of N *different* elements, each characterized by a distinct number (which we need not identify); in this section we use *capital* letters for the elements of the population: X_i, $i = 1, 2, \ldots, N$. A sample $\mathbf{x}_1, \mathbf{x}_2, \ldots, \mathbf{x}_n$ is drawn with the proviso that not more than one element of the sample can fall on any one element of the population; thus if $\mathbf{x}_1 = X_3$, $\mathbf{x}_2 \neq X_3$, or generally, $p(\mathbf{x}_i = \mathbf{x}_j) = 0$, $i \neq j$; in short, the sampling is without replacement.

Without the proviso of nonreplacement, there would be N^n n-tuples in the event space, but with it the first element of the sample, say x_1, can be one of the N values of X, the second element, say x_2, can be one of only $N - 1$ values of $X, \ldots$, the nth element of the sample, say x_n, can be one of only $[N-(n-1)]$ values of X; there are $N!/(N-n)!$ points in the event space. We illustrate for $N = 4$ and $n = 3$; the number of points in the event space is 24, not 64 (Fig. 21–1).

Let all points in the event space have the same probability p; that is, let the sampling process be random. Then

$$p = \frac{(N-n)!}{N!}.$$

By definition, the population mean λ and variance σ^2 are

$$\lambda = \sum_{i=1}^{N} \frac{X_i}{N}, \qquad \sigma^2 = \sum_{i=1}^{N} \frac{[X_i - E(X)]^2}{N}.$$

Now consider the probability behavior of the mean $\bar{\mathbf{x}}$ of a (random) sample of size n drawn without replacement from this population. We have

$$E(\bar{\mathbf{x}}) = \frac{1}{n} E[\mathbf{x}_1 + \mathbf{x}_2 + \cdots + \mathbf{x}_n] = \frac{1}{n}[E(\mathbf{x}_1) + E(\mathbf{x}_2) + \cdots + E(\mathbf{x}_n)], \tag{21–17}$$

the right-hand term of (21–17) following from the term to the left by

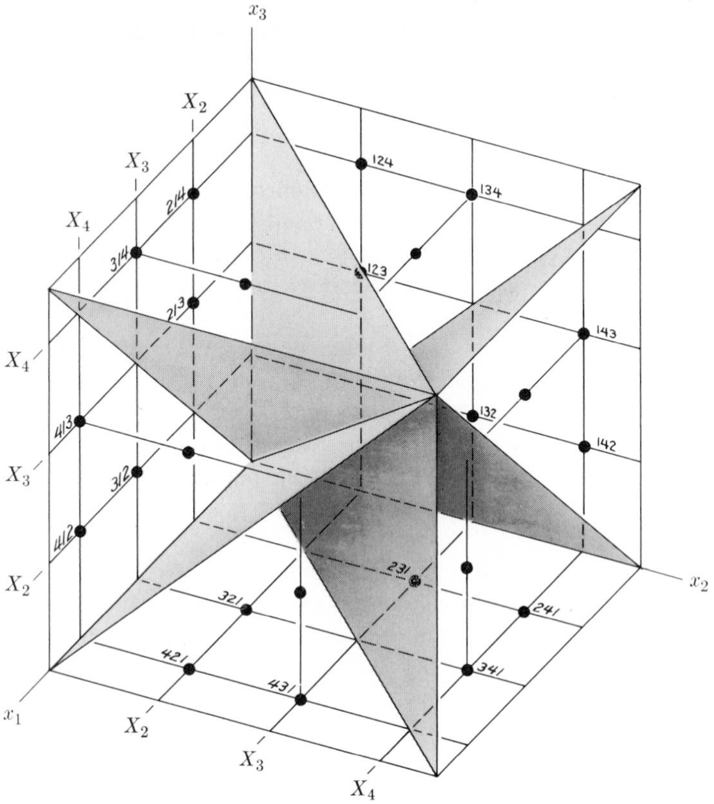

FIG. 21–1. Sampling without replacement. Planes of zero probability for $N = 4, n = 3$.

(7–15) which holds for all random variables, independent or [as in (21–17)] not. To evaluate (21–17), consider one of the expectations in (21–17), say $E(\mathbf{x}_2)$. By definition, this expectation is given by multiplying each possible value taken on by $\mathbf{x}_2$ by the probability of its occurrence and summing this product over all possible values of $\mathbf{x}_2$:

$$E(\mathbf{x}_2) = X_1 \cdot \text{prob}\,(\mathbf{x}_2 = X_1) + X_2 \cdot \text{prob}\,(\mathbf{x}_2 = X_2) + \cdots + X_n \cdot \text{prob}\,(\mathbf{x}_2 = X_n). \quad (21\text{–}18)$$

Consider the first term to the right of the equality sign. How many points are there in the event space? $N!/(N - n)!$ How many of these points have $\mathbf{x}_2 = X_1$? Only $(1)(N - 1)(N - 2) \ldots (N - [n - 1]) = (N - 1)!/(N - n)!$; for if $\mathbf{x}_2$ is to be X_1, $\mathbf{x}_2$ can assume only 1 population value (X_1 itself), $\mathbf{x}_1$ can be only one of the remaining $N - 1$ population values,

$\mathbf{x}_3$ one of the remaining $N - 2$ population values, etc. Therefore,

$$\text{prob}\,(\mathbf{x}_2 = X_1) = \frac{(N-1)!}{(N-n)!}\bigg/\frac{N!}{(N-n)!} = \frac{1}{N}\cdot$$

One may also note from the evident symmetry under permutation that all $\mathbf{x}_i$ have the same marginal distribution. For the same reason $(\mathbf{x}_i, \mathbf{x}_j)$ have the same joint distribution for any i, j, etc.

Similarly, all probabilities in (21–18) are $1/N$; we get $E(\mathbf{x}_2) = \lambda$ and, finally from (21–17)

$$E(\bar{\mathbf{x}}) = \lambda. \tag{21–19}$$

On the average, the mean of a sample drawn without replacement is equal to the mean of the finite population.

Now to obtain the variance of $\bar{\mathbf{x}}$; we have

$$\begin{aligned}\sigma^2(\bar{\mathbf{x}}) &= E(\bar{\mathbf{x}} - \lambda)^2 = E\left[\frac{\mathbf{x}_1 + \mathbf{x}_2 + \cdots + \mathbf{x}_n}{n} - \lambda\right]^2 \\ &= \frac{1}{n^2} E[(\mathbf{x}_1 - \lambda) + (\mathbf{x}_2 - \lambda) + \cdots + (\mathbf{x}_n - \lambda)]^2.\end{aligned}$$

Carrying out the squaring and noting that here the cross-product terms do *not* vanish, we may write the result in compressed form:

$$\sigma^2(\bar{\mathbf{x}}) = \frac{1}{n^2}\sum_{i,j=1}^{n} \rho_{ij}\sigma_i\sigma_j, \tag{21–20}$$

where σ_i^2 is the variance of $\mathbf{x}_i$, σ_j^2 is the variance of $\mathbf{x}_j$, and ρ_{ij} is the correlation coefficient between $\mathbf{x}_i$ and $\mathbf{x}_j$.

To evaluate (21–20) in terms of the parameters of the population, let us first consider $i = j$; the right-hand term of (21–20) reduces to $\sum_{i=1}^{n} \sigma_i^2/n$. Consider one of these variances, say σ_2^2. By definition,

$$\sigma_2^2 = E(\mathbf{x}_2 - \lambda)^2.$$

This expectation is found by multiplying each possible value taken on by $(\mathbf{x}_2 - \lambda)^2$ by its probability of occurrence and summing these products over all possible values of $(\mathbf{x}_2 - \lambda)^2$:

$$\sigma_2^2 = E(\mathbf{x}_2 - \lambda)^2 = \sum_{i=1}^{N} (X_i - \lambda)^2 \cdot \frac{1}{N} = \sigma^2, \tag{21–21}$$

and similarly for all other variances $\sigma_1^2, \ldots, \sigma_n^2$. Again, we can in fact regard $\mathbf{x}_2$ simply as a sample of size 1; its variance is, by definition, σ^2. The extended argument is provided because of the fact that random sampling here is achieved via sampling without replacement.

For $i \neq j$, consider a typical term of (21–20), say $i = 1, j = 2$. We have

$$\rho_{12}\sigma_1\sigma_2 = E(\mathbf{x}_1 - \lambda)(\mathbf{x}_2 - \lambda).$$

The expectation is given by multiplying each possible value taken on by $\mathbf{x}_1 - \lambda$ and $\mathbf{x}_2 - \lambda$ by the joint probability of that pair of values, and summing these products over all possible pairs of values of $\mathbf{x}_1 - \lambda$ and $\mathbf{x}_2 - \lambda$:

$$\begin{aligned}\rho_{12}\sigma_1\sigma_2 &= E(\mathbf{x}_1 - \lambda)(\mathbf{x}_2 - \lambda)\\ &= \sum_{\substack{i,j=1\\ i\neq j}}^{N} (X_i - \lambda)(X_j - \lambda)\,\frac{1}{N}\cdot\frac{1}{N-1}\\ &= \frac{1}{N(N-1)}\left[\sum_{i=1}^{N}(X_i - \lambda)\sum_{j=1}^{N}(X_j - \lambda) - \sum_{i=1}^{N}(X_i - \lambda)^2\right].\end{aligned}$$

But the product of the sums vanishes, and we have left

$$\rho_{12}\sigma_1\sigma_2 = -\frac{1}{N-1}\,\frac{\sum_{i=1}^{N}(X_i - \lambda)^2}{N} = -\frac{\sigma^2}{N-1}\cdot \tag{21–22}$$

Since $\sigma_1 = \sigma_2 = \sigma$, the correlation coefficient ρ_{ij} between the sample variables $\mathbf{x}_i$ and $\mathbf{x}_j$, $i \neq j$ is $-1/(N-1)$ (negative!).

Similarly for other $i \neq j$; placing in (21–20) the results found in (21–21) and (21–22), we have

$$\sigma^2(\bar{\mathbf{x}}) = \frac{1}{n^2}\left[n\sigma^2 - n(n-1)\frac{\sigma^2}{N-1}\right]$$

[noting that, in the sum in (21–20), such terms as $\rho_{12}\sigma_1\sigma_2$ and $\rho_{21}\sigma_2\sigma_1$, though equal, must be counted separately]. Finally,

$$\sigma^2(\bar{\mathbf{x}}) = \frac{\sigma^2}{n}\left(\frac{N-n}{N-1}\right)\cdot \tag{21–23}$$

Note three features of (21–23). First, the variance of the mean, sampling without replacement, is never greater than the variance of the mean, sampling with replacement, the latter given in (21–4). Second, as $N \to \infty$, (21–23) $\to$ (21–4). Third, as $n \to \infty$, (21–23) $\to 0$.

Results (21–19) and (21–23) should be compatible with those obtained in Chapter 12, where we dealt with sampling without replacement from the finite hypergeometric distribution. There the population was dichotomous; the N single population elements $X_1, X_2, \ldots, X_N$ of this section were in Chapter 12 contracted to A of X_1 (successes) and $N - A$ of X_2

(failures). In Chapter 12 we studied the random variable $\mathbf{x}$, the *number* of successes in a sample of size n drawn without replacement from a population of A successes and $N - A$ failures. We found

$$E(\mathbf{x}) = np, \qquad \sigma^2(\mathbf{x}) = npq\frac{N-n}{N-1},$$

where $p = A/N$. In this section we studied the random variable $\bar{\mathbf{x}}$, the *mean* of a sample of size n drawn without replacement from a finite population of N different elements of mean λ. We found

$$E(\bar{\mathbf{x}}) = \lambda, \qquad \sigma^2(\bar{\mathbf{x}}) = \frac{\sigma^2}{n}\frac{N-n}{N-1},$$

where $p = (N - n)!/N!$. The results are compatible; the sample mean of this section corresponds to $\mathbf{x}/n$ of Chapter 12.

Collateral Reading

S. S. Wilks, *Mathematical Statistics*. Princeton, Princeton University Press, 1943. Also, with many associated topics, in the 1962 Wiley volume.

CHAPTER 22

ORDER STATISTICS

22–1 The basic theorem. By *order statistics* we mean a generic term to cover the variables of a random sample arranged in order of magnitude, as well as functions of such ordered variables. The probability behavior of order statistics was in large part developed and unified by S. S. Wilks [206], [207] and his associates and students at Princeton. The basic theorem and several examples are given in this chapter.

Let a random sample of size n be drawn from a population described by the continuous random variable $\mathbf{x}$ with density function $f(x)$. Let the n sample variables be arranged in order of magnitude. Write $\mathbf{x}_1$ for the smallest, $\mathbf{x}_2$ for the next larger, . . . , $\mathbf{x}_n$ for the largest. Each of these elements of the sample remains a random variable as is the set of all of them.

We could find the joint density function $h(x_1, x_2, \ldots, x_n)$ of the now-ordered sample $\mathbf{x}_1, \mathbf{x}_2, \ldots, \mathbf{x}_n$ or, to clarify the later exposition, the probability element

$$h(x_1, x_2, \ldots, x_n)\, dx_1\, dx_2 \ldots dx_n, \tag{22–1}$$

but it suits our later purposes to answer a somewhat more general question. Write down a set of integers $r_1, r_2, \ldots, r_k$ with the property

$$1 \le r_1 < r_2 \cdots < r_k \le n,$$

and instead of (22–1), find the probability element

$$h(x_{r_1}, x_{r_2}, \ldots, x_{r_k})\, dx_{r_1}\, dx_{r_2} \ldots dx_{r_k} \tag{22–2}$$

of the order statistics $\mathbf{x}_{r_1}, \mathbf{x}_{r_2}, \ldots, \mathbf{x}_{r_k}$. To illustrate, for $n = 8$, $r_1 = 3$, $r_2 = r_k = 6$, we are seeking

$$h(x_3, x_6)\, dx_3\, dx_6,$$

the joint probability element of the third and sixth ordered $\mathbf{x}_i$. Note that the necessary conditions on the remaining 6 of the 8 sample variables are: $\mathbf{x}_1$ and $\mathbf{x}_2$ smaller than x_3, $\mathbf{x}_4$ and $\mathbf{x}_5$ between x_3 and x_6, $\mathbf{x}_7$ and $\mathbf{x}_8$

larger than x_6. In general, (22–2) requires the following allocation of the n sample variables:

Number of variables	$r_1 - 1$	1	$r_2 - r_1 - 1$	1	$\dots$	$r_k - r_{k-1} - 1$	1	$n - r_k$
Value of variables	$-\infty$ $\quad x_{r_1}$	$x_{r_1}+dx_{r_1}$	x_{r_2}	$x_{r_2}+dx_{r_2}$	$\dots$	x_{r_k}	$x_{r_k}+dx_{r_k}$	$+\infty$

There is one sample variable at each of $x_{r_1}, x_{r_2}, \dots, x_{r_k}$ with the following conditions on the remaining sample variables: $r_1 - 1$ variables less than x_{r_1}, $r_2 - 1$ variables less than x_{r_2} [that is, $(r_2 - 1) - (r_1 - 1) - 1$ variables between $x_{r_1} + dx_{r_1}$ and x_{r_2}) $\dots$, and finally, $n - r_k$ variables greater than $x_{r_k} + dx_{r_k}$.

The probability of *one* such allocation is

$$\left[\int_{x_{r_1}}^{x_{r_1}+dx_{r_1}} f(x)\,dx \int_{x_{r_2}}^{x_{r_2}+dx_{r_2}} f(x)\,dx \dots \int_{x_{r_k}}^{x_{r_k}+dx_{r_k}} f(x)\,dx\right] \times$$

$$\left[\left\{\int_{-\infty}^{x_{r_1}} f(x)\,dx\right\}^{r_1-1} \left\{\int_{x_{r_1}+dx_{r_1}}^{x_{r_2}} f(x)\,dx\right\}^{r_2-r_1-1} \cdots \left\{\int_{x_{r_k}+dx_{r_k}}^{\infty} f(x)\,dx\right\}^{n-r_k}\right],$$

the square bracket in the first line showing the probability of one sample variable at each of $x_{r_1}, x_{r_2}, \dots, x_{r_k}$, and the square bracket in the second line showing the probability of $r_1 - 1$ variables in the indicated cell (less than x_{r_1}), $r_2 - r_1 - 1$ variables in the indicated cell (between $x_{r_1} + dx_{r_1}$ and x_{r_2}) $\dots$, $n - r_k$ variables in the indicated cell (greater than $x_{r_k} + dx_{r_k}$). But n variables taken n at a time can be so allocated in

$$\frac{n!}{1!1!\dots 1!(r_1 - 1)!(r_2 - r_1 - 1)!\dots(n - r_k)!}$$

different and equally likely ways; the required probability element (22–2) is

$$\frac{n!}{(r_1 - 1)!(r_2 - r_1 - 1)!\dots(n - r_k)!}$$

$$\times \left\{\int_{-\infty}^{x_{r_1}} f(x)\,dx\right\}^{r_1-1} \left\{\int_{x_{r_1}}^{x_{r_2}} f(x)\,dx\right\}^{r_2-r_1-1} \dots \left\{\int_{x_{r_k}}^{\infty} f(x)\,dx\right\}^{n-r_k}$$

$$\times f(x_{r_1})f(x_{r_2})\dots f(x_{r_k})\,dx_{r_1}\,dx_{r_2}\dots dx_{r_k}, \qquad (22\text{–}3)$$

where

$$\int_{x_{r_1}}^{x_{r_1}+dx_{r_1}} f(x)\,dx \qquad \text{has been replaced by} \qquad f(x_{r_1})\,dx_{r_1},$$

$$\int_{x_{r_1}+dx_{r_1}}^{x_{r_2}} f(x)\,dx \qquad \text{has been replaced by} \qquad \int_{x_{r_1}}^{x_{r_2}} f(x)\,dx,$$

etc. Here and throughout this chapter, note that all f are strictly the same; f refers to the density function of the univariate population being sampled. Note also that (22–3) might be simply and perhaps profitably expressed in terms of the *distribution* function $F(x)$. Furthermore, note that (22–3) may be viewed as a term of a multinomial joint density function of the random variables $\mathbf{x}_{r_1}, \mathbf{x}_{r_2}, \ldots, \mathbf{x}_{r_k}$; the n independent trials of Chapter 14 are here a random sample of size n.

Expression (22–3) is the basic result in the probability theory of order statistics. Using it, we can find the probability distributions of various important order statistics such as $\mathbf{x}_n$, the largest variable in the sample, $\mathbf{x}_1$, the smallest variable in the sample, $\tilde{\mathbf{x}}$, the middle variable of the *ordered* sample (the *median*), and $\mathbf{x}_n - \mathbf{x}_1$, the *range* of the sample, as special cases of (22–3).

22–2 Distribution of the ordered sample. As the first special case, the probability element of the ordered sample $\mathbf{x}_1, \mathbf{x}_2, \ldots, \mathbf{x}_n$ itself is seen to be

$$n!f(x_1)f(x_2)\ldots f(x_n)\,dx_1\,dx_2\ldots dx_n, \tag{22–4}$$

for here $r_1 = 1$, $r_2 = 2, \ldots, r_k = n$, and the general allocation of the n sample variables reduces to

	0		1		0		1		. . .		1		0	
$-\infty$		x_1		$x_1 + dx_1$		x_2		$x_2 + dx_2$		$\cdots\ x_n$		$x_n + dx_n$		$+\infty$

Expression (22–4) could have been obtained directly, without recourse to (22–3).

22–3 Distribution of the sample maximum. As a second and important special case of (22–3), consider the probability behavior of $\mathbf{x}_n$, the largest variable in the sample. The general allocation of the n sample variables here reduces to

	$n - 1$		1		0	
$-\infty$		x_n		$x_n + dx_n$		$+\infty$

and (22–3) reduces to

$$n\left[\int_{-\infty}^{x_n} f(x)\,dx\right]^{n-1} f(x_n)\,dx_n, \tag{22–5}$$

a function of the random variable $\mathbf{x}_n$ and the sample size n. If the population density function is specified, (22–5) can be further developed.

Expression (22–5) could also have been obtained directly. Write $F(x)$ for the distribution function of $\mathbf{x}$. The probability that any element of a random sample from $F(x)$ is less than or equal to a number x_n is $F(x_n)$, and since the variables of a random sample are independent, the probability that all n sample variables are less than or equal to x_n is $F^n(x_n)$. But this is identical with the probability $G(x_n)$ that the largest variable $\mathbf{x}_n$ in the sample is less than or equal to x_n, so $G(x_n) = F^n(x)$. The derivative of $G(x_n)$ with respect to x_n is the density function $g(x_n)$ of $\mathbf{x}_n$; this derivative is identical with (22–5).

The probability distribution of the maximum variable of a sample has been extensively applied to floods and dam construction (and the minimum variable of a sample to droughts and irrigation). In such applications the random variable often is annual rainfall and the sample size n is the number of years for which annual rainfall data are available. Many other applications are found in Gumbel [79].

22–4 Distribution of the sample median. As a third example, consider the distribution of the sample median $\tilde{\mathbf{x}}$, the central value of the ordered random variables of the sample, $\mathbf{x}_1, \mathbf{x}_2, \ldots, \mathbf{x}_n$. Considering only n odd, write $n = 2s + 1$. The general allocation reduces to

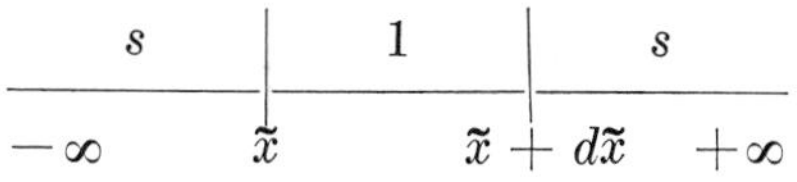

for which (22–3) reduces to

$$p(\tilde{x})\,d\tilde{x} = \frac{(2s+1)!}{s!s!}\left\{\int_{-\infty}^{\tilde{x}} f(x)\,dx\right\}^{s}\left\{\int_{\tilde{x}}^{\infty} f(x)\,dx\right\}^{s} f(\tilde{x})\,d\tilde{x}. \tag{22–6}$$

If the form f of the population density function is specified, (22–6) can be evaluated. Also, it is not difficult to show that, for n large, (22–6) leads to $\tilde{\mathbf{x}}$ normally distributed with mean $\tilde{X}$ and variance

$$\frac{1}{4n[f(\tilde{X})]^2},$$

where f is the density function of the population being sampled and $\tilde{X}$

is the population median. For examples of the latter,

(a) for $\mathbf{x}$ normal with mean (median) λ and variance σ^2, $\tilde{\mathbf{x}}$ in large samples is normal with mean λ and variance

$$\sigma^2(\tilde{\mathbf{x}}) = \frac{1}{4n(1/\sigma\sqrt{2\pi})^2} = \frac{\pi\sigma^2}{2n}.$$

As we shall later be interested in the variance of the median, in large random samples drawn from populations described by (b) a Cauchy distribution and (c) a Laplace distribution, we record here, for large n,

(b) for $\mathbf{x}$ Cauchy, $\quad \sigma^2(\tilde{\mathbf{x}}) = \dfrac{1}{4n(1/\pi)^2} = \dfrac{\pi^2}{4n},$

(c) for $\mathbf{x}$ Laplace, $\quad \sigma^2(\tilde{\mathbf{x}}) = \dfrac{1}{4n(1/2)^2} = \dfrac{1}{n}.$

22–5 Distribution of the sample range. A fourth example of considerable practical interest is the distribution of the sample range $\mathbf{x}_n - \mathbf{x}_1$; the sample range is frequently used to estimate the variability of a population.

The general allocation of the n sample variables here reduces to

0	1	$n-2$	1	0
$-\infty$ to x_1	x_1 to x_1+dx_1	x_1+dx_1 to x_n	x_n to x_n+dx_n	x_n+dx_n to $+\infty$

and (22–3) reduces to the joint probability element of $\mathbf{x}_1$, $\mathbf{x}_n$,

$$h(x_1, x_n)\,dx_1\,dx_n = n(n-1)\left(\int_{x_1}^{x_n} f(x)\,dx\right)^{n-2} f(x_1)f(x_n)\,dx_1\,dx_n. \tag{22–7}$$

As we want the distribution of a specific function of $\mathbf{x}_1$, $\mathbf{x}_n$, namely $\mathbf{x}_n - \mathbf{x}_1$, we write

$$\mathbf{r} = \mathbf{x}_n - \mathbf{x}_1, \qquad \mathbf{s} = \mathbf{x}_1.$$

We find, changing from the random variables $\mathbf{x}_1$, $\mathbf{x}_n$ to the random variables $\mathbf{r}$, $\mathbf{s}$ by the method of Chapter 8, the joint probability element of $\mathbf{r}$ and $\mathbf{s}$,

$$n(n-1)\left(\int_s^{s+r} f(x)\,dx\right)^{n-2} f(r+s)f(s)\,dr\,ds, \tag{22–8}$$

from which the required probability element of $\mathbf{r}$ is obtained by integrating over $-\infty < s < \infty$. The result cannot, however, be put into practical form unless the form of f is specified. For $\mathbf{x}$ normal, the distribution of $\mathbf{r}$ was developed by L. H. C. Tippett; for $\mathbf{x}$ normal and $n = 2(1)20$, tables of the distribution function

$$G(w) = \int_0^w g(w)\,dw,$$

where

$$\mathbf{w} = \frac{\mathbf{r}}{\sigma},$$

have been prepared by Pearson and Hartley [151]. The integral gives the probability that the range of a random sample of size n drawn from a normal population of standard deviation σ is less than a given multiple of σ. An excerpt is shown on pages 420–421.

Here we illustrate for the simpler case of the uniform population $f(x) = 1/\theta$, $0 \leq x \leq \theta$; (22–8) becomes

$$n(n-1)\left(\int_{x_1}^{x_n} \frac{1}{\theta}\,dx\right)^{n-2} \frac{1}{\theta^2}\,dx_1\,dx_n = n(n-1)\,\frac{1}{\theta^n}\,(x_n - x_1)^{n-2}\,dx_1\,dx_n,$$

$$= n(n-1)\,\frac{1}{\theta^n}\,r^{n-2}\,dr\,ds,$$

the Jacobian of the transformation from $\mathbf{x}_1$, $\mathbf{x}_n$ to $\mathbf{r}$, $\mathbf{s}$ being 1. Integrating on $\mathbf{s}$ from 0 to $\theta - r$ (not to θ, since $s = x_1 = x_n - r$ and the largest value of $\mathbf{x}_n$ is θ), we have

$$g(r)\,dr = n(n-1)\,\frac{1}{\theta^n}\,r^{n-2}(\theta - r)\,dr.$$

Without loss of generality let $\theta = 1$. We have

$$g(r)\,dr = n(n-1)r^{n-2}(1-r)\,dr,$$

the probability element of the range $\mathbf{r}$ in random samples of size n from a population described by the uniform random variable over 0 to 1.

22–6 Wilks' tolerance theorem. Wilks has obtained a remarkable result by a simple change of the variables in (22–7). Change from the random variables $\mathbf{x}_1$, $\mathbf{x}_n$ to the random variables $\mathbf{u}$, $\mathbf{v}$, where

$$u = \int_{-\infty}^{x_1} f(x)\,dx, \qquad v = \int_{x_1}^{x_n} f(x)\,dx.$$

The probability distribution of $\mathbf{v}$ (which is seen to be the proportion of the population lying between the smallest and largest elements in the sample) turns out to be independent of the density function $f(x)$ describing the population. To show this, recall from Chapter 8 that

$$p(u, v)\, du\, dv = p[u(x_1, x_n), v(x_1, x_n)] \begin{vmatrix} \dfrac{\partial u}{\partial x_1} & \dfrac{\partial v}{\partial x_1} \\ \dfrac{\partial u}{\partial x_n} & \dfrac{\partial v}{\partial x_n} \end{vmatrix}^{-1} dx_1\, dx_n,$$

the inverse Jacobian

$$\left|\frac{\partial(u, v)}{\partial(x_1, x_n)}\right|^{-1}$$

being preferred to its equal

$$\left|\frac{\partial(x_1, x_n)}{\partial(u, v)}\right|,$$

for here the partial derivatives in the inverse Jacobian are directly obtained. We have

$$\frac{\partial u}{\partial x_1} = f(x_1), \qquad \frac{\partial u}{\partial x_n} = 0, \qquad \frac{\partial v}{\partial x_1} = -f(x_1), \qquad \frac{\partial v}{\partial x_n} = f(x_n),$$

and the Jacobian of the transformation from $\mathbf{x}_1$, $\mathbf{x}_n$ to $\mathbf{u}$, $\mathbf{v}$ is $1/f(x_1)f(x_n)$. Thus $f(x_1)f(x_n)$ in (22–7) is exactly cancelled, and we get

$$p(u, v)\, du\, dv = n(n - 1)v^{n-1}\, du\, dv.$$

Integrating out $\mathbf{u}$, the lower limit of $\mathbf{u}$ being 0 and the upper limit $1 - v$, we get the probability element of $\mathbf{v}$,

$$q(v)\, dv = n(n - 1)v^{n-2}(1 - v)\, dv, \tag{22–9}$$

depending on n *but not on* $f(x)$.

In practical application of (22–9) we usually wish to determine n so that the probability is α (high) that at least proportion β (high) of the population lies within the range of the sample. We want to find the value of n which satisfies

$$\alpha = \int_\beta^1 n(n - 1)v^{n-2}(1 - v)\, dv,$$

which reduces to a solution for n of the transcendental equation

$$n\beta^{n-1} + (n - 1)\beta^n = 1 - \alpha. \tag{22–10}$$

Several approximate solutions of (22–10) for n have been proposed. Tables have been prepared by Dion [47]; his major table is shown on page 422. For example,

β	α	n
0.95	0.95	93
0.95	0.99	130
0.99	0.95	473
0.99	0.99	661

Collateral Reading

H. Cramér, *Mathematical Methods of Statistics.* Princeton, Princeton University Press, 1946.

R. A. Fisher, *Contributions to Mathematical Statistics.* New York, John Wiley and Sons, New York, 1950.

E. J. Gumbel, *Statistics of Extremes.* New York, Columbia University Press, 1958. Theory and applications of the probability behavior of sample maxima and minima by a pioneer in this area. Includes such varied applications as flood flow, strength of material, precipitation, wind speed and pressures, various meteoric phenomena, gust velocity in relation to aircraft design, old age, breakdown voltage, wave height and impact of pitch and roll, cobble and gravel size, and study of stream deposits. The bibliography fills twenty-three pages.

A. M. Mood, *Introduction to the Theory of Statistics.* New York, McGraw-Hill Book Co., 1950. The discussion here serves as an introduction to the important area of nonparametric tests (not discussed in this text, see [208]).

S. S. Wilks, *Mathematical Statistics.* Princeton, Princeton University Press, 1943. Or the 1962 Wiley volume. By far the best general reference.

S. S. Wilks, "Order statistics," *Bull. Amer. Math. Soc.*, Vol. 54 (1948), pp. 6–50. The basic reference.

Problems

22–1. Find the distribution of the minimum variable of a random sample.

22–2. Find the distribution of the range in random samples of size n from a population described by $f(x) = \theta e^{-\theta x}$, $x \geq 0$.

22–3. Let $\mathbf{x}$ be normal with variance σ^2. Find the expected value of the larger of the variables in a random sample of size $n = 2$.

22–4. The solution of Wilks' tolerance problem was

$$\alpha = \int_{\beta}^{1} n(n-1)v^{n-2}(1-v)\,dv,$$

an incomplete beta-function. Recall the relationship (10–11) between the binomial sum and the incomplete beta-function:

$$\sum_{j=0}^{k} C_j^n p^j q^{n-j} = 1 - \frac{n!}{k!(n-k-1)!} \int_0^p (p-\theta)^k (q+\theta)^{n-k-1}\,d\theta.$$

Is the relationship of the tolerance problem to binomial probabilities fortuitous or natural?

22–5. It was shown (page 73) that $F(\mathbf{x})$ was uniformly distributed. So if $\mathbf{x}_1, \ldots, \mathbf{x}_n$ are independent ordered random variables, $F(\mathbf{x}_1), \ldots, F(\mathbf{x}_n)$ are independent ordered uniform random variables. Ordering $\mathbf{x}_1, \ldots, \mathbf{x}_n$ orders $F(\mathbf{x}_1), \ldots, F(\mathbf{x}_n)$. Can you develop Wilks' tolerance theorem along these lines?

CHAPTER 23

SAMPLING FROM A NORMAL POPULATION

23–1 Motivation. In this chapter we consider the behavior of functions of the variables of random samples drawn from populations described by normal distributions. We are interested in such statistics, for many populations in the real world are normal, or nearly normal, or by simple change of variable may become normal. This fact, to which must be added the relative simplicity of theory in sampling from normal populations and the remarkable simplicity of many of the results, motivates the discussion of this chapter.

As earlier, we must decide what statistics of the random sample to study. Some statistics such as the sample mean have sufficient intuitive appeal to warrant study of their probability behavior without further defense (though one will be provided). But some of the statistics studied in this chapter are less intuitive functions of the sample variables. Their selection will be justified in Chapter 28, where it will be seen that in studying hypotheses on the values of the unknown parameters of one or more normal distributions, the statistics whose probability behavior we discuss here generally provide superior tests of such hypotheses. That is, statistics whose probability behavior we obtain here in what sometimes appear merely to be mathematical exercises will reappear in Chapter 28 as statistics best fitted to test certain hypotheses. These statistics will also appear, with optimal properties, in Chapter 29, where we attempt to establish limits within which the unknown value of a population parameter is likely to lie.

23–2 Moment generating function of a normal variable. First, we describe the parent normal population (uniquely) by its moment generating function. Beginning with the random variable $\mathbf{x}$ with normal density function

$$f(x) = \frac{1}{\sigma\sqrt{2\pi}} \exp\left[-\frac{1}{2}\left(\frac{x-\lambda}{\sigma}\right)^2\right], \quad -\infty < x < \infty,$$

we found in (16–3) and (16–4)

$$M_{\mathbf{x}-\lambda}(\theta) = e^{\theta^2\sigma^2/2}, \tag{23–1}$$

the moment generating function of the normal variable $\mathbf{x} - \lambda$ of mean 0 and variance σ^2. For *any* random variable whose moment generating function exists, repeating (3–19),

$$M_{\mathbf{x}-\lambda}(\theta) = e^{-\theta\lambda}M_{\mathbf{x}}(\theta).$$

Therefore, for the normal variable $\mathbf{x}$ of mean λ and variance σ^2,

$$M_{\mathbf{x}}(\theta) = e^{\theta\lambda}e^{\theta^2\sigma^2/2}. \qquad (23\text{–}2)$$

23–3 Distribution of the sample mean. From (21–2) and (23–2) we have, for the moment generating function of the mean of a random sample of size n from a normal population of mean λ and variance σ^2,

$$M_{\bar{\mathbf{x}}}(\theta) = e^{\theta\lambda}e^{\theta^2\sigma^2/2n} \qquad (23\text{–}3)$$

or

$$M_{\bar{\mathbf{x}}-\lambda}(\theta) = e^{\theta^2\sigma^2/2n}, \qquad (23\text{–}4)$$

which with respect to θ is identical with (23–1). It differs from (23–1) only in constants multiplying θ; the term σ^2/n replaces the term σ^2. Therefore, on expansion or differentiation with respect to θ, (23–4) will yield moments of $\bar{\mathbf{x}} - \lambda$ which, with σ^2/n replacing σ^2, are the same as those given for the normal variable $\mathbf{x} - \lambda$ in (23–1). From the uniqueness theorem on moment generating functions outlined on page 33, we conclude that in random samples from a normal population (and, though we do not prove it, *only* from a normal population), $\bar{\mathbf{x}}$ is normally distributed, and from the argument above, with moments 0 and σ^2 of the random variable $\mathbf{x}$ replaced by 0 and σ^2/n of the random variable $\bar{\mathbf{x}}$. Theoretical, as well as somewhat conflicting empirical, evidence on the distribution of $\bar{\mathbf{x}}$ in random samples from a wide variety of nonnormal populations suggests that in samples of $n \geq 30$ from such populations $\bar{\mathbf{x}}$ is nearly normally distributed.

More formally, for the mean and variance of $\bar{\mathbf{x}}$ (which completely fix the normal distribution of $\bar{\mathbf{x}}$), we have

$$E(\bar{\mathbf{x}} - \lambda) = \left.\frac{\partial M_{\bar{\mathbf{x}}-\lambda}(\theta)}{\partial\theta}\right|_{\theta=0} = 0, \qquad E(\bar{\mathbf{x}}) = \lambda, \qquad (23\text{–}5)$$

$$E(\bar{\mathbf{x}} - \lambda)^2 = \left.\frac{\partial^2 M_{\bar{\mathbf{x}}-\lambda}(\theta)}{\partial\theta^2}\right|_{\theta=0} = \frac{\sigma^2}{n}, \qquad \sigma^2(\bar{\mathbf{x}}) = \frac{\sigma^2}{n}; \qquad (23\text{–}6)$$

two results obtained here only for the normal variable $\bar{\mathbf{x}}$ but true, as we have seen in Chapter 21, for the mean of a random sample from *any* population (whose mean and variance exist).

The results contained in the last three equations may be assembled and brought to bear on the *standardized* random variable

$$\mathbf{u} = \frac{\bar{\mathbf{x}} - \lambda}{\sigma/\sqrt{n}} \tag{23–7}$$

as follows: in random samples of size n from a normal population of mean λ and variance σ^2, $\mathbf{u}$ is normally distributed with mean 0 and variance 1.

In Chapter 28 we shall use the statistic $\bar{\mathbf{x}}$ in tests of hypotheses on the value of the mean of a normal distribution whose variance is known. The same statistic will also appear in Chapter 29 in the determination of limits within which the unknown mean of a normal distribution of known variance is likely to lie.

23–4 Distribution of the difference of two independent sample means. By definition, the moment generating function of the difference $\bar{\mathbf{x}} - \bar{\mathbf{y}}$ between the means of two random samples of size n_x and n_y, respectively, is

$$M_{\bar{\mathbf{x}}-\bar{\mathbf{y}}}(\theta) = \int_{-\infty}^{\infty}\int_{-\infty}^{\infty} e^{\theta(\bar{x}-\bar{y})} f(x, y)\, dx\, dy, \tag{23–8}$$

which, for $\bar{\mathbf{x}}$, $\bar{\mathbf{y}}$ independent,

$$= \int_{-\infty}^{\infty} e^{\theta\bar{x}} g(x)\, dx \int_{-\infty}^{\infty} e^{-\theta\bar{y}} h(y)\, dy = M_{\bar{\mathbf{x}}}(\theta) M_{\bar{\mathbf{y}}}(-\theta),$$

and which, replacing the general moment generating functions by expressions of the form (23–3), becomes

$$M_{\bar{\mathbf{x}}-\bar{\mathbf{y}}}(\theta) = \exp\{\theta[E(\mathbf{x}) - E(\mathbf{y})]\} \exp\left[\frac{\theta^2}{2}\left(\frac{\sigma^2(\mathbf{x})}{n_x} + \frac{\sigma^2(\mathbf{y})}{n_y}\right)\right]. \tag{23–9}$$

With respect to θ, (23–9) is identical with (23–2); it differs only in constants multiplying θ. Again, by the uniqueness theorem of page 33, we conclude that in independent random samples from normal populations, $\bar{\mathbf{x}} - \bar{\mathbf{y}}$ is normally distributed. And again, expanding or differentiating the moment generating function (23–9), or by noting that λ and σ^2 of (23–2) are replaced in (23–9) by

$$E(\mathbf{x}) - E(\mathbf{y}) \qquad \text{and} \qquad \frac{\sigma^2(\mathbf{x})}{n} + \frac{\sigma^2(\mathbf{y})}{n},$$

we have

$$E(\bar{\mathbf{x}} - \bar{\mathbf{y}}) = E(\bar{\mathbf{x}}) - E(\bar{\mathbf{y}}) \tag{23–10}$$

$$\sigma^2(\bar{\mathbf{x}} - \bar{\mathbf{y}}) = \frac{\sigma^2(\mathbf{x})}{n_x} + \frac{\sigma^2(\mathbf{y})}{n_y}, \tag{23–11}$$

two results obtained here only for the normal variable $\bar{\mathbf{x}} - \bar{\mathbf{y}}$ but obtainable, as we saw in Chapter 21, by elementary operations on sample means for the difference of the means of independent random samples from *any* populations whose means and variances exist.

This theorem on the normality of $\bar{\mathbf{x}} - \bar{\mathbf{y}}$ is a special case of the more general theorem on the probability behavior of the sum $a\mathbf{x} + b\mathbf{y}$ of two independent normal variables $\mathbf{x}$ and $\mathbf{y}$, which the student has already considered in problem 16–3.

Again, the results contained in the last three equations may be assembled and brought to bear on the *standardized* random variable

$$\mathbf{v} = \frac{(\bar{\mathbf{x}} - \bar{\mathbf{y}}) - [E(\mathbf{x}) - E(\mathbf{y})]}{\sqrt{[\sigma^2(\mathbf{x})/n_x] + [\sigma^2(\mathbf{y})/n_y]}} \tag{23–12}$$

as follows: in independent random samples of sizes n_x and n_y from normal populations of means $E(\mathbf{x})$ and $E(\mathbf{y})$ and variances $\sigma^2(\mathbf{x})$ and $\sigma^2(\mathbf{y})$, $\mathbf{v}$ is normally distributed with mean 0 and variance 1.

In Chapter 28 we shall use the statistic $\bar{\mathbf{x}} - \bar{\mathbf{y}}$ in tests of hypotheses on the value of the difference of the means of two normal distributions whose variances are known. This statistic will also appear in Chapter 29 in the determination of limits within which the unknown difference of the means of two normal distributions of known variances is likely to lie.

23–5 Chi-square. We interrupt the orderly development of this chapter for the following result which will immediately prove to be useful in describing the probability behavior of certain functions of the variables of random samples drawn from normal populations. Consider a continuous random variable $\mathbf{x}$ with density function

$$f(x) = \frac{1}{2\Gamma(b/2)}\left(\frac{x}{2}\right)^{(b/2)-1} e^{-(1/2)x}, \quad 0 \leq x < \infty, \tag{23–13}$$

where b is an integral constant and $\Gamma(b/2)$ is the gamma function of $b/2$, defined by (9–5), also a constant. We seek the moment generating function of the random variable $\mathbf{x}$ of (23–13). After an elementary change of variable, (9–5) becomes

$$\frac{1}{a^k}\Gamma(k) = \int_0^\infty e^{-au}u^{k-1}\,du,$$

and the moment generating function of the random variable $\mathbf{x}$ is

$$M_{\mathbf{x}}(\theta) = \int_0^\infty e^{\theta x} f(x)\,dx = (1 - 2\theta)^{-b/2}. \tag{23–14}$$

The random variable $\mathbf{x}$ of (23–13) is often called χ^2 (*chi-square*) and the constant b is often called the number of *degrees of freedom*. Reasons underlying the latter terminology will appear in the final section of this chapter; b will reappear as the number of independent random variables minus the number of constraints on them. Several tables of areas under the χ^2 density function are available; those by Catherine Thompson [186] are shown on pages 423–424.

Several important functions of the variables of random samples from normal populations will now be seen to follow the χ^2 distribution.

23–6 Distribution of the sum of squares. By definition, the moment generating function of the sum of squares of the sample variables $\mathbf{x}_1, \mathbf{x}_2, \ldots, \mathbf{x}_n$ from a population described by the general continuous density function $f(x)$ is

$$M_{\sum_{i=1}^n \mathbf{x}_i^2}(\theta) = \int_{-\infty}^{\infty}\int_{-\infty}^{\infty}\cdots\int_{-\infty}^{\infty}\exp\left[\theta\sum_{i=1}^n x_i^2\right] \times f(x_1)f(x_2)\ldots f(x_n)\,dx_1\,dx_2\ldots dx_n. \tag{23–15}$$

Let $f(x)$ be the density function of the standardized normal random variable (mean 0 and variance 1); we have

$$f(x_1)f(x_2)\ldots f(x_n) = \frac{1}{(2\pi)^{n/2}}\exp\left[\frac{1}{2}\sum_{i=1}^n x_i^2\right],$$

and (23–15) becomes

$$\frac{1}{(2\pi)^{n/2}}\int_{-\infty}^{\infty}\int_{-\infty}^{\infty}\cdots\int_{-\infty}^{\infty}\exp\left[\theta\sum_{i=1}^n x_i^2\right]\exp\left[-\frac{1}{2}\sum_{i=1}^n x_i^2\right]dx_1\,dx_2\ldots dx_n$$

$$= \frac{1}{(2\pi)^{n/2}}\prod_{i=1}^n\int_{-\infty}^{\infty}\exp\left[\theta x_i^2\right]\exp\left[-\frac{1}{2}x_i^2\right]dx_i. \tag{23–16}$$

Let

$$(1-2\theta)^{-1/2}x_i = u_i, \qquad dx_i = \frac{du_i}{(1-2\theta)^{1/2}};$$

the integral reduces to that evaluated on page 138. We finally obtain for the moment generating function of $\sum_{i=1}^n \mathbf{x}_i^2$,

$$M_{\sum_{i=1}^n \mathbf{x}_i^2}(\theta) = \frac{1}{(2\pi)^{n/2}}\frac{1}{(\frac{1}{2}-\theta)^{n/2}}\prod_{i=1}^n\int_{-\infty}^{\infty}e^{-u_i^2}\,du_i = (1-2\theta)^{-n/2}. \tag{23–17}$$

Comparing (23–17) with (23–14), we have that $\mathbf{x}_1^2 + \mathbf{x}_2^2 + \cdots + \mathbf{x}_n^2$ in random samples from a standardized normal population is distributed as χ^2 with n degrees of freedom. The student should verify that if the normal population has mean λ and variance σ^2, the corresponding sum of squares

$$\sum_{i=1}^{n} \left(\frac{\mathbf{x}_i - \lambda}{\sigma}\right)^2 \tag{23–18}$$

is distributed as χ^2 with n degrees of freedom, a result which we shall immediately use.

23–7 Distribution of the sample variance. The probability behavior of the variance $\mathbf{s}^2$,

$$\mathbf{s}^2 = \frac{(\mathbf{x}_1 - \bar{\mathbf{x}})^2 + (\mathbf{x}_2 - \bar{\mathbf{x}})^2 + \cdots + (\mathbf{x}_n - \bar{\mathbf{x}})^2}{n},$$

in random samples from a normal population of mean λ and variance σ^2 differs fundamentally if not always practically from that of (23–18), for in the sample variance, the quantity $\bar{\mathbf{x}}$ as well as $\mathbf{x}_1, \mathbf{x}_2, \ldots, \mathbf{x}_n$ is a random variable. We express $\mathbf{s}^2$ in terms of deviations about the population mean λ in order to bring to bear the theorem already obtained on (23–18).

Since

$$ns^2 = \sum_{i=1}^{n} (x_i - \bar{x})^2 = \sum_{i=1}^{n} [(x_i - \lambda) + (\lambda - \bar{x})]^2$$

$$= \sum_{i=1}^{n} (x_i - \lambda)^2 - n(\bar{x} - \lambda)^2,$$

or better, as we shall see,

$$\frac{ns^2}{\sigma^2} = \sum_{i=1}^{n} \left(\frac{x_i - \lambda}{\sigma}\right)^2 - \left(\frac{\bar{x} - \lambda}{\sigma/\sqrt{n}}\right)^2, \tag{23–19}$$

we find the probability distribution of $n\mathbf{s}^2/\sigma^2$ (which is $\mathbf{s}^2$ multiplied by a constant) by studying its moment generating function.

For (23–19) write $A = B - C$. Then $B = A + C$ and $M_B(\theta) = M_{A+C}(\theta)$; from the critical theorem that A and C are statistically independent, which we shall prove in Section 23–11, we have

$$M_B(\theta) = M_A(\theta) M_C(\theta), \qquad M_A(\theta) = \frac{M_B(\theta)}{M_C(\theta)}. \tag{23–20}$$

First we consider the numerator of (23–20). From (23–18) we have

$$M_B(\theta) = (1 - 2\theta)^{-n/2}.$$

Now we consider the denominator of (23–20), noting that C is a single term, not a sum. But (23–18) also applies to C, for $\bar{\mathbf{x}}$ in samples from a normal population of mean λ and variance σ^2 is normally distributed with mean λ and variance σ^2/n. That is,

$$M_C(\theta) = (1 - 2\theta)^{-1/2},$$

and finally,

$$M_A(\theta) = (1 - 2\theta)^{-(n-1)/2}; \tag{23–21}$$

$n\mathbf{s}^2/\sigma^2$ is distributed as χ^2 with $n - 1$ degrees of freedom, a result due to the astronomer and geodesist, Helmert in 1876 and independently to K. Pearson [154] in 1900.

The statistic $\mathbf{s}^2$ will be used in Chapter 28 in tests of hypotheses on the unknown value of the variance of a normal population, and in Chapter 29 in the determination of the limits within which the unknown value of the variance of a normal population is likely to lie.

23–8 Distribution of the sum of independent chi-squares. Consider k independent variables $\chi_1^2, \chi_2^2, \ldots, \chi_k^2$ distributed as χ^2 with $b_1, b_2, \ldots, b_k$ degrees of freedom, respectively. The moment generating function of $\chi_1^2 + \chi_2^2 + \cdots + \chi_k^2$ is

$$M_{\sum_{i=1}^k \chi_i^2}(\theta) = \int_0^\infty \int_0^\infty \cdots \int_0^\infty \exp\left[\theta \sum_{i=1}^k \chi_i^2\right] \times h(\chi_1^2, \chi_2^2, \ldots, \chi_k^2)\, d\chi_1^2\, d\chi_2^2 \cdots d\chi_k^2,$$

which by virtue of the mutual independence of $\chi_1^2, \chi_2^2, \ldots, \chi_k^2$ reduces to

$$\prod_{i=1}^k \int_0^\infty e^{\theta\chi_i^2} f(\chi_i^2)\, d\chi_i^2 = (1 - 2\theta)^{-b/2}, \tag{23–22}$$

where $b = b_1 + b_2 + \cdots + b_k$. From (23–14) it follows that

$$\chi_1^2 + \chi_2^2 + \cdots + \chi_k^2$$

is distributed as χ^2 with $b = b_1 + b_2 + \cdots + b_k$ degrees of freedom. This result also follows from the characterization of χ^2 as the sum of squares of independent standardized normal variables.

Example: If independent random samples of sizes $n_1, n_2, \ldots, n_k$ are drawn from normal populations of *common* variance σ^2, the quantity

$$\frac{n_1\mathbf{s}_1^2}{\sigma^2} + \frac{n_2\mathbf{s}_2^2}{\sigma^2} + \cdots + \frac{n_k\mathbf{s}_k^2}{\sigma^2} \tag{23–23}$$

is distributed as χ^2 with $n_1 + n_2 + \cdots + n_k - k$ degrees of freedom. If k samples of sizes $n_1, n_2, \ldots, n_k$ are available for testing a hypothesis on the variance σ^2 of a normal distribution, the statistic $n_1\mathbf{s}_1^2 + n_2\mathbf{s}_2^2 + \cdots + n_k\mathbf{s}_k^2$ will be seen in Chapter 28 to provide an efficient test of the hypothesis. Alternatively, in Chapter 29, the same statistic will enable us to determine the limits within which the unknown value of the variance of a normal distribution is likely to lie.

23–9 Distribution of Student's t. Let $\mathbf{x}$ and $\mathbf{y}$ be independent. Let $\mathbf{x}$ be distributed as χ^2 with m degrees of freedom and $\mathbf{y}$ be normally distributed with mean 0 and variance 1.

$$g(x) = \frac{1}{2\Gamma(m/2)}\left(\frac{x}{2}\right)^{(m/2)-1} e^{-(1/2)x}, \quad 0 \le x < \infty,$$

$$h(y) = \frac{1}{\sqrt{2\pi}} e^{-(1/2)y^2}, \qquad -\infty < y < \infty.$$

Since $\mathbf{x}$ and $\mathbf{y}$ are independently distributed, the joint probability element factors,

$$f(x, y)\,dx\,dy = g(x)\,dx\,h(y)\,dy.$$

We want the distribution of the random variable

$$\mathbf{t} = \frac{\mathbf{y}}{\sqrt{\mathbf{x}/m}}, \quad -\infty < t < \infty, \tag{23–24}$$

a quantity which like certain other quantities studied in this chapter will reappear more meaningfully in Chapters 28 and 29.

We shift from the random variables $\mathbf{x}$ and $\mathbf{y}$ to the random variables $\mathbf{t}$ and $\mathbf{u}(=\mathbf{y})$:

$$f(x, y)\,dx\,dy = f[x(t, u), y(t, u)]\,\frac{\partial(x, y)}{\partial(t, u)}\,dt\,du.$$

The Jacobian of the transformation from $\mathbf{x}$, $\mathbf{y}$ to $\mathbf{t}$, $\mathbf{u}$ is $\sqrt{u/m}$. Integrating out u, which, over $0, \infty$, is a gamma function, we have the probability element of the random variable $\mathbf{t}$,

$$f(t)\,dt = \frac{1}{\sqrt{m\pi}}\,\frac{[(m-1)/2]!}{[(m-2)/2]!}\left(\frac{t^2}{m} + 1\right)^{-(m+1)/2} dt, \quad -\infty < t < \infty, \tag{23–25}$$

due, by a somewhat more restricted argument, to "Student" [180]. We speak of a random variable being distributed as Student's $\mathbf{t}$ or having Student's distribution with m degrees of freedom. Several independent tables of areas under the density function of $\mathbf{t}$ are available; those by Maxine Merrington [122] are shown on page 425.

Example 1. From a normal population of mean λ and variance σ^2, a random sample of size n is drawn. From (23-21) we know that $n\mathbf{s}^2/\sigma^2$ is distributed as χ^2 with $n - 1$ degrees of freedom and from (23-7) we know that $(\bar{\mathbf{x}} - \lambda)/\sigma/\sqrt{n}$ is (independently, as shown in Section 23-11) normally distributed with mean 0 and variance 1. Therefore,

$$\mathbf{t} = \frac{\bar{\mathbf{x}} - \lambda}{\mathbf{s}/\sqrt{n-1}} \tag{23-26}$$

has Student's distribution with $n - 1$ degrees of freedom.

Note that the variance σ^2 has disappeared and that (23-26) depends only on one population parameter (λ), and on three quantities (n, $\bar{x}$, and s) all the latter known from the sample. The quantity (23-26) will play the central role in Chapter 28 in testing hypotheses on the mean λ of a normal distribution whose variance is *unknown*, and, in Chapter 29, in estimating the limits within which the unknown mean λ of a normal distribution of unknown variance is likely to lie.

Example 2. If, from two normal populations of means $E(\mathbf{x})$ and $E(\mathbf{y})$ and common variance $\sigma^2(\mathbf{x}) = \sigma^2(\mathbf{y}) = \sigma^2$, independent random samples of sizes n_x and n_y are drawn, we have from (23-23) that

$$\frac{n_x\mathbf{s}_x^2 + n_y\mathbf{s}_y^2}{\sigma^2}$$

is distributed as χ^2 with $n_x + n_y - 2$ degrees of freedom. We have from (23-12) that

$$\frac{(\bar{\mathbf{x}} - \bar{\mathbf{y}}) - [E(\mathbf{x}) - E(\mathbf{y})]}{\sigma\sqrt{(1/n_x) + (1/n_y)}}$$

is (independently, as shown in Section 23-11) normally distributed with mean 0 and variance 1. Therefore,

$$\mathbf{t} = \frac{(\bar{\mathbf{x}} - \bar{\mathbf{y}}) - [E(\mathbf{x}) - E(\mathbf{y})]}{\sqrt{(n_x\mathbf{s}_x^2 + n_y\mathbf{s}_y^2)/(n_x + n_y - 2)}\,\sqrt{(1/n_x) + (1/n_y)}} \tag{23-27}$$

has Student's distribution with $n_x + n_y - 2$ degrees of freedom.

Again (the common) σ^2 has disappeared; $\mathbf{t}$ depends on the parameter difference $[E(\mathbf{x}) - E(\mathbf{y})]$, and on six quantities $(n_x, n_y, \bar{x}, \bar{y}, s_x, s_y)$ known from the sample. The quantity (23–27) will reappear in Chapter 28 in connection with testing hypotheses on the difference of the means $E(\mathbf{x}) - E(\mathbf{y})$ of two normal populations of equal but unknown variances, and in Chapter 29 in connection with estimating the limits within which the unknown difference of population means $E(\mathbf{x}) - E(\mathbf{y})$ of normal distributions of unknown but equal variances is likely to lie.

Other examples, most of them due to Fisher and showing the remarkable range and diversity of application of Student's $\mathbf{t}$, will appear in Chapter 30.

23–10 Distribution of Fisher's F. Let $\mathbf{x}$ and $\mathbf{y}$ be independently distributed as χ^2 with a and b degrees of freedom, respectively:

$$f(x, y)\,dx\,dy = g(x)\,dx \cdot h(y)\,dy$$
$$= \frac{1}{4\Gamma(a/2)\Gamma(b/2)}\left(\frac{1}{2}\right)^{(a/2)+(b/2)-2} x^{(a/2)-1}y^{(b/2)-1}e^{-1/2(x+y)}\,dx\,dy.$$

We shall want the distribution of the random variable

$$\mathbf{F} = \frac{\mathbf{x}/a}{\mathbf{y}/b}. \tag{23–28}$$

Replace the random variables $\mathbf{x}$ and $\mathbf{y}$ by the random variables $\mathbf{F}$ and $\mathbf{v} = \mathbf{y}$.* The Jacobian of the transformation from $\mathbf{x}$, $\mathbf{y}$ to $\mathbf{F}$, $\mathbf{v}$ is $(a/b)v$. We have

$$g(F, v)\,dF\,dv = \frac{1}{4\Gamma(a/2)\Gamma(b/2)}$$
$$\times\left(\frac{1}{2}\right)^{[(a+b)/2]-2}\left(\frac{a}{b}\right)^{a/2} F^{(a/2)-1}v^{[(a+b)/2]-1}e^{-(1/2)v[(aF/b)+1]}dF\,dv.$$

To find the distribution of $\mathbf{F}$, we need to integrate out v. The reader should verify that

$$\int_0^\infty \left(\frac{v}{2}\right)^{[(a+b)/2]-1} e^{-(1/2)v[(aF/b)+1]}\,dv = \frac{2\Gamma(a+b)/2}{[(aF/b)+1]^{(a+b)/2}}.$$

Hence

$$h(F)\,dF = \frac{\Gamma(a+b)/2}{\Gamma(a/2)\Gamma(b/2)}\left(\frac{a}{b}\right)^{a/2}\left(\frac{a}{b}F + 1\right)^{-(a+b)/2} F^{(a/2)-1}\,dF,$$
$$0 \le F < \infty, \tag{23–29}$$

---

* The less natural $\mathbf{v} = \mathbf{x} + \mathbf{y}$ is better; J is less simple but we are lead to the separable joint density function $g(F, v) = h(F)p(v)$.

due to Fisher [60]. The distribution (23–29) depends, but not symmetrically, on the two parameters a and b. We speak of a quantity being distributed as **F** or having Fisher's distribution with a and b degrees of freedom. Excerpts from the tables of **F**, by Merrington and Thompson [123], are shown on pages 426–429.

Example 1. If, from normal populations of variances $\sigma_{\mathbf{x}}^2$ and $\sigma_{\mathbf{y}}^2$ independent random samples of sizes n_x and n_y are drawn, we have from (23–21) that $n_x\mathbf{s}_x^2/\sigma_{\mathbf{x}}^2$ is distributed as χ^2 with $n_x - 1$ degrees of freedom, and that $n_y\mathbf{s}_y^2/\sigma_{\mathbf{y}}^2$ is *independently* (it is a function of different random variables) distributed as χ^2 with $n_y - 1$ degrees of freedom. The ratio

$$F = \frac{n_x\mathbf{s}_x^2/\sigma_{\mathbf{x}}^2}{n_x - 1} \Big/ \frac{n_y\mathbf{s}_y^2/\sigma_{\mathbf{y}}^2}{n_y - 1} \tag{23–30}$$

has Fisher's **F** distribution with $n_x - 1$ and $n_y - 1$ degrees of freedom. The ratio (23–30) will reappear in Chapter 28 in testing hypotheses on the equality of the variances of two normal populations [all quantities in (23–30) other than these variances are known from the sample] and in Chapter 29 in determining the limits within which the unknown difference in the variances of normal populations likely lies.

Example 2. (Introduction to analysis of variance.) Though artificial here, this example is an important application of the F distribution; F as involved here will be seen in Chapter 30 to be central in efficient tests of important linear hypotheses on the means of several normal populations.

From a normal population of variance σ^2 and mean λ a random sample $x_{11}, \ldots, x_{kn}$ of size kn is drawn. Place the kn variables of the sample in a columnar array of n variables per column, k columns, as shown in Table 23–1. First we consider a purely algebraic property of the elements

TABLE 23–1

ALLOCATION OF kn NORMAL RANDOM VARIABLES WITHIN A COLUMNAR DESIGN

x_{11}	x_{21}	$\cdots$	x_{k1}	
x_{12}	x_{22}	$\cdots$	x_{k2}	
		x_{ij}		
$\vdots$	$\vdots$		$\vdots$	
x_{1n}	x_{2n}	$\cdots$	x_{kn}	
$\bar{x}_1$	$\bar{x}_2$	$\bar{x}_i$	$\bar{x}_k$	$\bar{x}$

of the array in Table 23–1. Write

$$\sum_{i=1}^{k}\sum_{j=1}^{n}(x_{ij}-\bar{x})^2=\sum_{i=1}^{k}\sum_{j=1}^{n}[(x_{ij}-\bar{x}_i)+(\bar{x}_i-\bar{x})]^2; \qquad (23\text{–}31)$$

the cross-product term vanishes and (23–31) reduces to

$$\sum_{i=1}^{k}\sum_{j=1}^{n}(x_{ij}-\bar{x}_i)^2+n\sum_{i=1}^{k}(\bar{x}_i-\bar{x})^2$$

$$=(ns_1^2+ns_2^2+\cdots+ns_k^2)+nks_{\bar{x}}^2, \qquad (23\text{–}32)$$

which we write $U+V$.

In Section 23–11 we will show that U and V are independent. From (23–23), U/σ^2 is distributed as χ^2 with $nk-k$ degrees of freedom. Now consider V. As in a random sample of n observations $\mathbf{x}_1, \mathbf{x}_2, \ldots, \mathbf{x}_n$ from a normal population of variance σ^2, the quantity $n\mathbf{s}^2/\sigma^2$ is distributed as χ^2 with $n-1$ degrees of freedom, so in a random sample of k *means* $\bar{\mathbf{x}}_1, \bar{\mathbf{x}}_2, \ldots, \bar{\mathbf{x}}_k$ from a normal population of variance $\sigma_{\bar{\mathbf{x}}}^2$, the quantity $k\mathbf{s}_{\bar{x}}^2/\sigma_{\bar{\mathbf{x}}}^2$ is distributed as χ^2 with $k-1$ degrees of freedom (remember that $\bar{\mathbf{x}}$ is normal if $\mathbf{x}$ is normal). But

$$\frac{k\mathbf{s}_{\bar{x}}^2}{\sigma_{\bar{\mathbf{x}}}^2}=\frac{V}{n\sigma_{\bar{\mathbf{x}}}^2}=\frac{V}{\sigma^2}.$$

Therefore

$$\frac{V/(k-1)}{U/(nk-k)} \qquad (23\text{–}33)$$

is distributed as $\mathbf{F}$ with $k-1$ and $nk-k$ degrees of freedom. These results are often arranged in an "analysis-of-variance" table as shown in Table 23–2.

This, and similar but more general results, play a central role in testing the linear hypotheses so frequently considered in modern experimental design. The present example will reappear more meaningfully, that is, less as a mathematical exercise, in Chapter 30. There the columnar array of Table 23–1 will arise as a simple experimental design in the test of an interesting hypothesis on the means of k normal populations. In the present chapter, all k columns of data in Table 23–1 are known to come from k normal populations of known common mean λ and known common variance σ^2 (from one normal population of known mean λ and known variance σ^2). In Chapter 30 we continue the assumption of common variance σ^2 and we study the *hypothesis* that the k normal populations have common means $\lambda_1=\lambda_2=\cdots=\lambda_k$. If the hypothesis is true,

TABLE 23-2

ANALYSIS-OF-VARIANCE TABLE

Sum of squares	Degrees of freedom	Mean square	F
$n\sum_{i=1}^{k}(\bar{x}_i - \bar{x})^2 \quad (V)$	$k - 1$	$\frac{V}{k-1}$	$\frac{V/(k-1)}{U/(nk-k)}$
$\sum_{i=1}^{k}\sum_{j=1}^{n}(x_{ij} - \bar{x}_i)^2 \quad (U)$	$nk - k$	$\frac{U}{nk-k}$	
$\sum_{i=1}^{k}\sum_{j=1}^{n}(x_{ij} - \bar{x})^2$	$nk - 1$		

the probability behavior of the statistic **F** reduces to that studied here; if the hypothesis is false, the observed mean square $V/(k-1)$ will be *larger* and the computed F will be larger. Thus by studying the magnitude of observed F we will be able to study the validity of the hypothesis $\lambda_1 = \lambda_2 = \cdots = \lambda_k$.

23-11 Linear transformations; independence of $\bar{\mathbf{x}}$ and s. While studying the behavior of random samples drawn from a normal population of mean λ and variance σ^2, we wrote (23-19)

$$\frac{ns^2}{\sigma^2} = \sum_{i=1}^{n}\left(\frac{x_i - \lambda}{\sigma}\right)^2 - \left(\frac{\bar{x} - \lambda}{\sigma/\sqrt{n}}\right)^2$$

or

$$A = B - C, \qquad B = A + C,$$

and (23-20) as

$$M_B(\theta) = M_{A+C}(\theta) = M_A(\theta)M_C(\theta),$$

which assumes that A and C (essentially, the second and the first powers of normal variables) are statistically independent. This now needs to be proved.

Rather than work with the moment generating functions (23-20) themselves, we use a method of proof due to Fisher [62]. This method enables us to introduce the concepts of linear and orthogonal transformations, concepts which are generally valuable in statistical theory, in particular, in experimental design.

First consider k variables $\mathbf{y}_1, \ldots, \mathbf{y}_k$ which are linear functions of k random variables $\mathbf{x}_1, \ldots, \mathbf{x}_k$:

$$y_i = c_{i1}x_1 + \cdots + c_{ik}x_k, \qquad i = 1, \ldots, k, \tag{23–34}$$

where $\mathbf{x}_1, \ldots, \mathbf{x}_k$ are mutually independent, normally distributed random variables with common means 0 and common variances σ^2. We readily obtain

$$E(\mathbf{y}_i) = c_{i1}E(\mathbf{x}_1) + \cdots + c_{ik}E(\mathbf{x}_k) = 0, \qquad i = 1, \ldots, k, \tag{23–35}$$

$$\sigma^2(\mathbf{y}_i) = c_{i1}^2\sigma^2(\mathbf{x}_1) + \cdots + c_{ik}^2\sigma^2(\mathbf{x}_k) = \sigma^2 \sum_{j=1}^{k} c_{ij}^2, \quad i = 1, \ldots, k. \tag{23–36}$$

What restrictions on $\{c_{ij}\}$ are required for $\mathbf{y}_1, \ldots, \mathbf{y}_k$ to have exactly the same probability behavior as $\mathbf{x}_1, \ldots, \mathbf{x}_k$? That is, we want $\mathbf{y}_1, \ldots, \mathbf{y}_k$ to be mutually independent, normal (random) variables with common means 0 and common variances σ^2.

Normality is already assured; $\mathbf{y}_i$ is a linear sum of independent normal variables and therefore, by problem 16–3, $\mathbf{y}_1, \ldots, \mathbf{y}_k$ are normal. Common zero means of $\mathbf{y}_1, \ldots, \mathbf{y}_k$ were already shown in (23–35). For common variances σ^2 of $\mathbf{y}_1, \ldots, \mathbf{y}_k$, we must require, noting (23–36),

$$\sum_{j=1}^{k} c_{ij}^2 = 1, \quad i = 1, \ldots, k. \tag{23–37}$$

As for independence, pairwise zero correlation is sufficient for independence of *normal* random variables (page 151). Since the $\mathbf{y}_i$ have zero means, the numerator of the correlation coefficient $\rho(\mathbf{y}_i, \mathbf{y}_m)$, $i \neq$ m reduces to

$$E(\mathbf{y}_i\mathbf{y}_m) = (c_{i1}c_{m1} + \cdots + c_{ik}c_{mk})\sigma^2, \quad i \neq m, \quad i, m = 1, \ldots, k,$$

and we must require

$$c_{i1}c_{m1} + \cdots + c_{ik}c_{mk} = 0, \quad i \neq m, \quad i, m = 1, \ldots, k. \tag{23–38}$$

Conditions (23–37) and (23–38) may be written

$$\sum_{j=1}^{k} c_{ij}c_{mj} = 1, \quad i = m; \qquad \sum_{j=1}^{k} c_{ij}c_{mj} = 0, \quad i \neq m \tag{23–39}$$

with $i, m = 1, \ldots, k$.

If we began with $p < k$ variables $\mathbf{y}_1, \ldots, \mathbf{y}_p$, the linear functions (23-24) would be

$$y_i = c_{i1}x_1 + \cdots + c_{ik}x_k, \quad i = 1, \ldots, p,$$

and the two conditions (23-39) would become

$$\sum_{j=1}^{k} c_{ij}c_{mj} = 1, \quad i = m; \qquad \sum_{j=1}^{k} c_{ij}c_{mj} = 0, \quad i \neq m, \tag{23-40}$$

with $i, m = 1, \ldots, p$. Also, it is important to note that to the constants

$$\begin{matrix} c_{11} & \ldots & c_{1k} \\ \vdots & & \vdots \\ c_{p1} & \ldots & c_{pk} \end{matrix}$$

on which (23-40) hold, we can always add $k - p$ rows

$$\begin{matrix} c_{p+1,1} & \ldots & c_{p+1,k} \\ \vdots & & \vdots \\ c_{k1} & \ldots & c_{kk} \end{matrix}$$

so that (23-39) are satisfied. The student is asked to show that additional constants $c_{p+1,1}, \ldots, c_{kk}$ can be found to assure this.

Now consider the sum of squares $\mathbf{y}_1^2 + \cdots + \mathbf{y}_k^2$ as functions of $\mathbf{x}_1, \ldots, \mathbf{x}_k$. We have

$$y_i^2 = c_{i1}^2x_1^2 + \cdots + c_{ik}^2x_k^2 + c_{i1}c_{i2}x_{i1}x_{i2} + \cdots, \quad i = 1, \ldots, k.$$

and

$$y_1^2 + \cdots + y_k^2 = x_1^2(c_{11}^2 + \cdots + c_{k1}^2) + \cdots + x_k^2(c_{1k}^2 + \cdots + c_{kk}^2)$$
$$+ x_1x_2(c_{11}c_{12} + \cdots + c_{k1}c_{k2}) + \cdots. \tag{23-41}$$

What additional conditions must be imposed on $\{c_{ij}\}$ for $\mathbf{y}_1^2 + \cdots + \mathbf{y}_k^2$ to be equal to $\mathbf{x}_1^2 + \cdots + \mathbf{x}_k^2$? Evidently, from (23-41), we need to require

$$\sum_{i=1}^{k} c_{ij}^2 = 1, \quad j = 1, \ldots, k, \tag{23-42}$$

as well as

$$\sum_{i=1}^{k} c_{ir}c_{is} = 0, \quad r \neq s, \quad r, s = 1, \ldots, k. \tag{23-43}$$

If (23-42) and (23-43) hold and if (23-39) also hold, then $\mathbf{y}_1, \ldots, \mathbf{y}_k$

are independent normal random variables with means 0 and variances σ^2 and

$$\sum_{i=1}^{k} y_i^2 = \sum_{i=1}^{k} x_i^2.$$

Note also that as $\mathbf{y}_i$ are $N(0, \sigma^2)$ random variables, the quantity

$$\sum_{i=1}^{k} \mathbf{y}_i^2/\sigma^2$$

follows the χ^2 distribution with k degrees of freedom.

The four required conditions may be written as follows:

$$\sum_{j=1}^{k} c_{ij}c_{mj} = 0, \quad i \neq m; \qquad \sum_{j=1}^{k} c_{ij}c_{mj} = 1, \quad i = m; \quad i, m = 1, \ldots, k, \tag{23-44}$$

$$\sum_{i=1}^{k} c_{ij}c_{im} = 0, \quad j \neq m; \qquad \sum_{i=1}^{k} c_{ij}c_{im} = 1, \quad j = m; \quad j, m = 1, \ldots, k,$$

and (23–44) may be described as the conditions for an *orthogonal* transformation of $\mathbf{x}_1, \ldots, \mathbf{x}_k$ to $\mathbf{y}_1, \ldots, \mathbf{y}_k$; sometimes this term is confined to the conditions (23–39).

The geometry of an orthogonal transformation, a subject with which we do not deal, may well be instructive to the reader; Fraser [70] contains a particularly effective discussion.

We complete the general argument by forming

$$Q = \mathbf{x}_1^2 + \cdots + \mathbf{x}_k^2 - \mathbf{y}_1^2 - \cdots - \mathbf{y}_p^2 = \mathbf{y}_{p+1}^2 + \cdots + \mathbf{y}_k^2, \tag{23-45}$$

and noting (what was known earlier) that Q/σ^2 is distributed as χ^2 with $k - p$ degrees of freedom and, what is most important for us here, that Q evidently is independent of $\mathbf{y}_1, \ldots, \mathbf{y}_p$. Now return to the problem of the independence of $\bar{\mathbf{x}}$ and $\mathbf{s}$ in random samples from a normal population of mean λ and variance σ^2. To correspond to the discussion of this section, let $\lambda = 0$ in (23–19); this is without loss of generality, for the distribution of $\mathbf{s}^2$ does not involve λ and the distribution of $\bar{\mathbf{x}}$ is affected only by an additive constant. Thus (23–19) reduces to

$$ns^2 = \sum_{i=1}^{n} x_i^2 - n\bar{x}^2. \tag{23-46}$$

But

$$n\bar{x}^2 = (\sqrt{n}\,\bar{x})^2 = \left(\frac{x_1}{\sqrt{n}} + \cdots + \frac{x_n}{\sqrt{n}}\right)^2 = \text{say } y_1^2,$$

where $\mathbf{y}_1^2$ is the square of a linear transformation of $\mathbf{x}_1, \ldots, \mathbf{x}_n$, with

$$c_1 = \cdots = c_n = \frac{1}{\sqrt{n}} \qquad \text{and} \qquad c_1^2 + \cdots + c_n^2 = 1,$$

which satisfy (23–44); $\mathbf{y}_1$ is an *orthogonal* transformation of $\mathbf{x}_1, \ldots, \mathbf{x}_n$. From (23–45) we conclude both that

$$\frac{n\mathbf{s}^2}{\sigma^2} = \frac{\mathbf{x}_1^2 + \cdots + \mathbf{x}_n^2 - \mathbf{y}_1^2}{\sigma^2}$$

follows the χ^2 distribution with $n - 1$ degrees of freedom [already obtained in (23–21)] and (the important new result) that the distribution of $\mathbf{s}^2$ is independent of that of $\mathbf{y}_1$, that is, of that of $\bar{\mathbf{x}}$.

This example of orthogonality is important in itself; it justifies the independence argument used earlier in (23–20) and in Section 23–9 on the behavior of Student's **t**. The same example can also be brought to bear, with far-reaching impact, on the basic statistical model underlying modern experimental design. On page 212, we argued that

$$\frac{V/(k - 1)}{U/(nk - k)}$$

followed the **F** distribution. This implied that

$$V = n \sum_{i=1}^{k} (\bar{x}_i - \bar{x})^2,$$

$$U = \sum_{i=1}^{k} \sum_{j=1}^{n} (x_{ij} - \bar{x}_i)^2 = ns_1^2 + \cdots + ns_k^2$$

are statistically independent. This we can now prove; V is evidently a function only of $\bar{\mathbf{x}}_1, \ldots, \bar{\mathbf{x}}_k$ (since $\bar{\mathbf{x}}$ is also a function only of $\bar{\mathbf{x}}_1, \ldots, \bar{\mathbf{x}}_k$). Examining U, $\mathbf{s}_i^2$ is independent of $\bar{\mathbf{x}}_m$, $i = m$, as we have just shown; moreover $\mathbf{s}_i^2$ is independent of $\bar{\mathbf{x}}_m$, $i \neq m$, since $\mathbf{s}_i^2$ and $\bar{\mathbf{x}}_m$, $i \neq m$, are functions of different (independent) random variables and are thereby (by problem 4–3) themselves independent. So V and U are independent and the argument on page 212 is justified.

Tables

C. M. Thompson, "Tables of percentage points of the χ^2 distribution," *Biometrika,* Vol. 32 (1941), pp. 188–189.

M. Merrington, "Tables of percentage points of the *t* distribution," *Biometrika,* Vol. 32 (1941–1942), p. 300. See also E. T. Federighi, "Extended tables of the percentage points of Student's **t** distribution," *Journ. Amer. Stat. Assoc.*, Vol. 54 (1959), pp. 683–688. The latter are tables of upper percentage points of *t*, for degrees of freedom 1(1)30(5)60(10)100, 200, 500, 1000, 2000, 10,000, and probabilities 0.0005, 0.00025, 0.0001, 0.000025, 0.00001.

M. Merrington and C. M. Thompson, "Tables of percentage points of the inverted beta (**F**) distribution," *Biometrika,* Vol. 33 (1943), pp. 73–88.

Collateral Reading

H. D. Brunk, *An Introduction to Mathematical Statistics.* Boston, Ginn and Co., 1960.

H. Cramér, *Mathematical Methods of Statistics.* Princeton, Princeton University Press, 1950.

S. S. Wilks, *Mathematical Statistics.* Princeton, Princeton University Press, 1943. Or the 1962 Wiley volume.

Problems

23–1. Find the mean and the variance of χ^2.

23–2. Find the mean of **F**.

23–3. Prove that for 1 degree of freedom, Student's **t** has a Cauchy distribution.

23–4. Prove that for 1 and n degrees of freedom, $\mathbf{F} = \mathbf{t}^2$.

Part IV

Statistical Inference

In Part IV we finally reach problems of statistical inference. A population will be characterized by information in samples drawn from the population. We will consider two categories of problems. In the first, we estimate values of the parameters of populations from the data in samples drawn from them; in the second, we test hypotheses on values of the parameters of populations, using in the tests the data in samples drawn from them. In both categories, additional axioms beyond those of Part I are required, and the division of chapters in Part IV in large part reflects the current choice of axioms as to what constitutes a "good" estimate or a "good" test of a hypothesis.

CHAPTER 24

BAYES' RULE

24–1 Bayes' rule. In this chapter we begin a discussion of the problem of estimating the values of parameters of probability distributions. A particular problem and a particular solution are discussed here; a more general discussion of estimation is deferred to Chapter 25.

The content of this chapter is somewhat "classical" in character. It is concerned only with the role of Bayes' rule in the estimation of a parameter; it does not consider the resurgent role of the Bayesian principle in more general decision-making processes.

For the simple case in which each independent trial produces a success or a failure, we wish to estimate p, the probability of a success. This is a strikingly different problem from any encountered so far. In the discussion to this point we have always assumed knowledge of the value of the parameter (here p) and have gone on to calculate probabilities of various functions of the variables of a random sample, for example, of x successes in a random sample of size n. Now the situation is reversed; the data of the sample are known to us and p is unknown. Since we do not (and in fact cannot) know p, we wish to estimate it.

This latter is obviously the practical real-world problem; it is a problem not in deductive probability, from population to sample, but in statistical inference, from sample to population. The laws of probability which rest on the three axioms of Chapters 1 and 2 and which enabled us to describe the behavior of samples drawn from populations of known parameters will again be necessary in our current discussion, but they will not be sufficient. The problem of estimation requires an additional axiom or principle as to what constitutes a *good* estimator. Quite different from the axiomatic situation in probability theory, there is little agreement as to the best axiom or principle to adopt in estimation, and therefore we shall have to consider several.

In this chapter we consider the pioneering and controversial system of estimation due to the Reverend Thomas Bayes, 1763. In this system, the estimation of p is accomplished by combining *a priori* knowledge of p with information contained in a finite number n of current independent trials. One may also think of Bayes' rule as a system for modifying historical information on a parameter in the light of current data. An extension of the logic and the method of Bayes' rule to parameters other than p is possible but this is not discussed here.

In estimating the value of p among all possible values between 0 and 1, the literature often uses the more general language of choosing a hypothesis among all possible hypotheses on p, or of selecting a cause system among all possible cause systems. One may then speak of the probability of a hypothesis, the probability of a cause system, inverse probability, etc.

The procedure associated with the name of Bayes is the following. Though p is a parameter, the argument of Bayes requires us to regard it as a random variable with initial (*a priori*, with respect to the current experiment) density function $a(p)$; if a population characterized by p can itself be regarded as a sample from a larger population, such a model may be reasonable. Now let the observed number of successes in n independent trials (or in the language of sampling, in a random sample of size n) be m; it is assumed that the sample comes from one population of parameter p. Write $f(m|n, p)$ for the probability of m successes given the number of trials n and given p. We combine *a priori* $[a(p)]$ and current $[m, n]$ information by forming the product

$$a(p) \cdot f(m|n, p)$$

and the final (*a posteriori*) probabilities $z(p)$ [which might better have been written $z(p|m, n)$] of various p are proportional to this product, the proportionality arranged so that the final probabilities over all possible values of p add to 1.

A numerical example may clarify the argument. We have three objects, each of which is good or bad. The proportion p of bad objects among the three objects is unknown. Our *a priori* knowledge (more likely, as Bayes proposed, without conviction for most of us, our ignorance) may suggest that all values of p possible here are equally likely $[a(p) = \frac{1}{4}]$. A random sample of size $1 (n = 1)$ is drawn; it yields a bad object $(m = 1)$. The calculations leading to $z(p)$ are shown in Table 24–1; if one value of p is to be selected, $p = 1$ (all objects are bad) is the posterior mode.

TABLE 24–1

CALCULATION OF BAYES' FINAL PROBABILITIES, $n = 1, m = 1, a(p) = \frac{1}{4}$

p	$a(p)$	$f(1\|1, p)$	$a(p) \cdot f(1\|1, p)$	$z(p)$
0	$\frac{1}{4}$	0	0	0
$\frac{1}{3}$	$\frac{1}{4}$	$\frac{1}{3}$	$\frac{1}{12}$	$\frac{1}{6}$
$\frac{2}{3}$	$\frac{1}{4}$	$\frac{2}{3}$	$\frac{2}{12}$	$\frac{2}{6}$
1	$\frac{1}{4}$	1	$\frac{3}{12}$	$\frac{3}{6}$

Evidently the general expression for the final probability of p is

$$z(p_i) = \frac{a(p_i)f(m|n, p_i)}{\Sigma_i a(p_i)f(m|n, p_i)}, \tag{24-1}$$

which, for $a(p_i)$ constant (as in the numerical example and, in fact, the only case considered by Bayes), reduces to

$$z(p_i) = \frac{f(m|n, p_i)}{\Sigma_i f(m|n, p_i)}.$$

Expression (24-1) also follows from, or better, is identical with the definition of conditional probability. (For economy in symbolism, we now use the single letter f to represent different density functions.) Repeating (4-24), we had for two *random* variables θ and $\mathbf{x}$,

$$f(\theta, x) = f(\theta)f(x|\theta) = f(x)f(\theta|x),$$

or

$$f(\theta|x) = \frac{f(\theta)f(x|\theta)}{f(x)}. \tag{24-2}$$

But $f(x)$ is a marginal distribution

$$f(x) = \Sigma_\theta f(\theta, x) = \Sigma_\theta f(\theta)f(x|\theta),$$

which, placed in (24-2), gives

$$f(\theta|x) = \frac{f(\theta)f(x|\theta)}{\Sigma_\theta f(\theta)f(x|\theta)},$$

a result different in appearance but identical in interpretation to (24-1). This view of Bayes' rule (as a simple and uncontroversial restatement of conditional probability) is common in modern probability if not in modern statistics. The expression of one conditional probability in terms of the other conditional probability and one marginal distribution is mathematically incontrovertible. It is the use of the expression for statistical inference by means of prior probabilities that is controversial.

Regarding θ as continuous, and using an integral rather than a sum, we could write (24-2) as

$$f(\theta|x)\,d\theta = \frac{f(\theta)f(x|\theta)\,d\theta}{\int f(\theta)f(x|\theta)\,d\theta}, \tag{24-3}$$

and in this form a weakness of Bayes' rule (noted by R. A. Fisher) is readily exhibited. If ignorance (or knowledge) of θ is interpreted to mean that all values of θ are *a priori* equally likely [$f(\theta)$ = constant], (24–3) becomes

$$f(\theta|x)\,d\theta = \frac{f(x|\theta)\,d\theta}{\int f(x|\theta)\,d\theta}.$$

But the parameter θ may be no more natural than the parameter θ^2 (for example, σ as against σ^2), and if ignorance (or knowledge) is interpreted to mean that all values of θ^2 are *a priori* equally likely [$f(\theta^2)$ = constant], (24–3) becomes

$$f(\theta|x)\,d\theta = \frac{\theta f(x|\theta)\,d\theta}{\int \theta f(x|\theta)\,d\theta},$$

a result which differs from that which precedes it.

24–2 Bayes' rule specialized to binomial p. If m follows a binomial distribution [which may be as hard to justify, in some practical problems, as a particular form for $a(p)$], we have

$$f(m|n, p_i) = C_m^n p_i^m (1 - p_i)^{n-m}.$$

We find (Laplace, 1774)

$$z(p_i) = \frac{a(p_i) p_i^m (1 - p_i)^{n-m}}{\Sigma_i a(p_i) p_i^m (1 - p_i)^{n-m}}, \tag{24–4}$$

and, for $a(p_i)$ constant (Bayes, 1763),

$$z(p_i) = \frac{p_i^m (1 - p_i)^{n-m}}{\Sigma_i p_i^m (1 - p_i)^{n-m}}. \tag{24–5}$$

Since p is continuous over its range 0 to 1, we may write

$$z(p_1 \le p \le p_2) = \frac{\int_{p_1}^{p_2} a(p) p^m (1 - p)^{n-m}\,dp}{\int_0^1 a(p) p^m (1 - p)^{n-m}\,dp}, \tag{24–6}$$

which, for $a(p)$ constant, reduces to the ratio of the incomplete to the complete beta-function, tabled by Karl Pearson [157]. The single value of p_i at which $z(p_i)$ in (24–5) is maximum is an example of *point* estimation of a parameter; *one* estimator is formed of p. On the other hand, (24–6) is an example of an *interval* estimator of p; limits p_1 and p_2 are found within which the unknown parameter p has a certain probability of falling. Point estimators in general are the subject of Chapters 25 to 27, interval estimators are the subject of Chapter 29.

The following special example assumes that the number of successes m in a sample of size n follows a binomial distribution: given that a random sample of size n_1 contains m_1 successes, find the probability that a random sample of size n_2 from the same population will contain m_2 successes.

From (24-6) the final Bayes probability of p, given the facts (n_1, m_1) of the first sample, is

$$\frac{a(p)p^{m_1}(1 - p)^{n_1-m_1}}{\int_0^1 a(p)p^{m_1}(1 - p)^{n_1-m_1}\,dp}.$$

By the binomial theorem, the probability of m_2 successes in a sample of size n_2, given p, is

$$C_{m_2}^{n_2}p^{m_2}(1 - p)^{n_2-m_2}.$$

The product, integrated over all possible values of p (since in this problem it does not matter what values of p give rise to m_1 and m_2 successes), answers the question. For $a(p)$ constant, the integrated product reduces to

$$z(m_2|n_2, n_1, m_1) = C_{m_2}^{n_2}\frac{\int_0^1 p^{m_1+m_2}(1 - p)^{n_1+n_2-m_1-m_2}\,dp}{\int_0^1 p^{m_1}(1 - p)^{n_1-m_1}\,dp}, \qquad (24\text{–}7)$$

where the integrals are complete beta-functions. For given m_1, n_1, m_2, n_2, this probability may be read from Karl Pearson's tables [157].

24-3 General remarks. Bayes' suggestion that ignorance or insufficient reason guarantees equal *a priori* probabilities is probably the weakest link in the chain. For while the parameter p may indeed sometimes be properly regarded as a random variable (the current lot may, for example, sometimes be properly regarded as itself a sample from a production process whose proportion defective we seek to estimate), much less frequently, on any basis, is the assumption of uniform *a priori* distribution of p reasonable. Equal *a priori* probability may sometimes be compatible with probability as a measure of attitude (even there the restriction would be severely limiting), but if we regard (and we do) probability as the limit of relative frequency, this may, as M. Kendall has remarked, be asking too much of the universe. If no more plausible, the assumption of uniform distribution of p is agreeable if n is *large*. In fact, if n is large enough, it does not matter what we assume for $a(p)$ (see von Mises [126]); the current sample will dominate the determination of the final probabilities.

But what of $a(p)$ in conjunction with modest n? It is surely true that in some areas of application (e.g., Mendelian genetics, meteorology, sporting events, and industrial quality control) *a priori* probabilities have evident meaning; in Mendelian problems, for example, not only meaning but clear (and unequal) values. To reject them outright (say in problem 24–2) and to form estimates exclusively from the data of the current sample is questionable judgment. If we cannot say that *a priori* probabilities are equal by ignorance, we may be able to say that they are unequal by knowledge and it may often be well to seek the knowledge. The path will likely be a rocky one; in this writer's experience it has not been often that the miscellaneous past probability history of any phenomenon can be convincingly assembled in the form of a known $a(p)$. Yet such assembly may be better for the purpose at hand than to attempt to escape the difficulty by using a method of doubtful relevance in which other arbitrary choices are also required.

Collateral Reading

Of the large supporting and critical literature on Bayes' Rule, the following are of particular value.

T. Bayes, "An essay toward solving a problem in the doctrine of chances," *Phil. Trans. Roy. Soc., London,* Vol. 53 (1763), pp. 370–418. Reproduced, no date given (with a second paper), by The Graduate School, U.S. Dept. of Agriculture, with an editorial preface and remarks by W. E. Deming and comments by E. C. Molina.

F. David, *Probability Theory for Statistical Method.* Cambridge (Eng.), Cambridge University Press, 1949. Includes a good example from genetics.

A. Fisher, *The Mathematical Theory of Probabilities,* 2nd ed., New York, Macmillan Co., 1922.

W. Feller, *An Introduction to Probability and Its Applications,* 2nd ed., New York, John Wiley and Sons, 1957. A brief but sharp presentation of the probabilitist's position.

T. C. Fry, *Probability and Its Engineering Uses.* New York, D. Van Nostrand Co., 1928. With good examples.

R. T. Cox, *The Algebra of Probable Inference.* Baltimore, The Johns Hopkins Press, 1961. Interesting discussion of Laplace's law of succession (problem 24–1).

L. Hogben, *Chance and Choice by Cardpack and Chessboard, Volumes 1 and 2.* New York, Chanticleer Press, 1950. A particularly interesting discussion of the assumption of continuity of θ.

H. Jeffreys, *Theory of Probability,* 3rd ed., New York, Oxford University Press, 1961.

M. G. Kendall and A. Stuart, *The Advanced Theory of Statistics, Volumes* 1 *and* 2. London, Charles Griffin and Co., 1958 and 1961. Very good discussion.

W. Kneale, *Probability and Induction.* Oxford, Clarendon Press, 1949. A good discussion of the continuity question and its impact on the form of the parameter (θ or θ^2) and on such problems as 24–1.

R. von Mises, *Notes on Mathematical Theory of Probability and Statistics.* Cambridge (Mass.), Harvard University, Graduate School of Engineering, Special Publication No. 1, 1946.

R. von Mises, "On the correct use of Bayes' formula," *Ann. Math. Stat.*, Vol. 13 (1942), pp. 156–165. Von Mises (a) argues that the existence of $a(p)$ does not imply that p is a random variable, (b) gives a simple set-theoretic proof (more general than his original proof in 1919) that, for large n, $a(p)$ does not affect $z(p)$, (c) discusses Bayes estimates versus confidence intervals, the latter a system of estimation independent of *a priori* considerations, to which we turn in Chapter 29.

R. von Mises, *Probability, Statistics, and Truth.* New York, Macmillan Co. 2nd revised English ed., 1957.

E. Parzen, *Modern Probability Theory and Its Applications.* New York, John Wiley and Sons, 1960. Admirable examples.

"Studies in the history of probability and statistics, IX. Bayes' essay towards solving a problem in the doctrine of chance," *Biometrika*, Vol. 45, 1958, pp. 293–315. A reprint of the original paper by Bayes and a biographical note by G. A. Barnard.

I. Todhunter, *A History of the Mathematical Theory of Probability from the Time of Pascal to that of Laplace.* London, Macmillan Co., 1865. Reprinted by Chelsea Pub. Co., 1949. Chaotic but interesting.

J. V. Uspensky, *Introduction to Modern Probability.* New York, McGraw-Hill Book Co., 1937.

P. M. Woodward, *Probability and Information Theory, with Applications to Radar.* London, Pergamon Press, 1955. Application to theory of reception in such observational systems as radar.

Problems

24–1. Let $a(p)$ be constant. An event occurs b times in n independent trials. Show that the Bayes probability of its occurrence in the next trial is $(b+1)/(n+2)$. This is the "law of succession" of Laplace, 1774. Does this result support the idea that the more days a man has lived, the more likely he is to live another? Or that if Fermat's last theorem is once disproved, the probability is $2/(n+2)$ that it will be disproved again?

24–2. Fry [72]. The historical record* of the quality of a product is

Percentage defective in lot	0	1	2	3	4	5	6
Percentage of lots	78	17	3.4	0.9	0.5	0.2	0.0

Standard quality is 2 percent defective or less. In a random sample of 50 from a lot of 1200, 6 defectives were found. Have standards been maintained?

* Here we do not mean in time sequence, though an appropriate model including prior probabilities need not be indifferent to time.

CHAPTER 25

POINT ESTIMATION

25–1 Introduction. In the preceding chapter we began a discussion of estimation. The parameter estimated there was p and the particular solution advanced there involved both an *a priori* probability distribution of p (and complete knowledge of the same) and the data in a current random sample from the population characterized by p. We found both point and interval estimators of p with probabilities attached to both.

The problem discussed in the preceding chapter is but one of many estimation problems, and the solution advanced there is but one of many solutions. We may wish to estimate other parameters, or we may wish to estimate *entire* probability distributions. We might hope to find a method or methods which, applied to whatever *a priori* information we have and to the current data, always reveal the true value of the parameter, parameters, or the distribution function. None exist. Instead, we have a number of methods each with errors of inference peculiar to it; some methods guarantee that the *maximum* error in inferring from the current sample and prior information to the population will be held to a minimum, others guarantee that *on the average* the estimate will equal the true value, still others have excellent properties when the current sample size is large. Among these and many other methods we cannot often make an easy choice since we seldom have any firm guide as to the kind of errors of inference we should avoid.

In the discussion of estimation to follow, we will often restrict ourselves to estimation of parameters (generally to a single parameter) of probability distributions of *known* form, that is to say, the form of f in the density function $f(x; \theta)$ is known. We will also exclude methods which require detailed *a priori* knowledge. While there can be no doubt that the experimenter often brings to his estimation problem *a priori* information which should be considered along with whatever data his current experiment produces, most current systems of estimation which utilize *a priori* information require it to be in form and of rigidity which he may not often be able to satisfy. Therefore, from this point on in this introductory account, all estimates are made on the basis of information contained in the current sample. Moreover, this sample will always be fixed in size; there is no discussion of recently developed powerful methods of estimation in which the sample size itself is a random variable, depending on the current data themselves as they unfold. Even with these restrictions, there is no shortage of properties and associated methods from which we must choose.

It will be convenient to separate discussions of point and interval estimation. In point estimation a single function of the sample data is constructed to estimate the value of a parameter. Properties of such point estimators will be the subject of this chapter; two methods of point estimation which have several of the desirable properties discussed in this chapter are the subjects of the two following chapters (Chapters 26 and 27). Point estimators are particularly required when the estimate itself must be inserted as a number in a larger mathematical structure, as is common in, for example, modern physics and economics. In interval estimation, generally *two* functions of the sample data are constructed within which, in a certain probability sense, the unknown value of a parameter lies. This is the subject of Chapter 29; the discussion there is coupled with Chapter 28 on tests of statistical hypotheses, for the particular method of interval estimation of parameters discussed in Chapter 29 is in many respects the complement of methods used in Chapter 28 to test statistical hypotheses on those parameters.

25–2 Unbiasedness. The first property of a point estimator of a parameter of a probability distribution of known form which we discuss is *unbiasedness.* We have a population described by the density function $f(x; \theta)$, where f is known and the value of the parameter θ is unknown. A random sample $\mathbf{x}_1, \mathbf{x}_2, \ldots, \mathbf{x}_n$ is drawn from this population. The statistic $\mathbf{t}(\mathbf{x}_1, \mathbf{x}_2, \ldots, \mathbf{x}_n)$ is an *unbiased estimator* of the parameter θ if

$$E(\mathbf{t}) = \theta \tag{25–1}$$

for all n and for any possible value of θ. Otherwise $\mathbf{t}$ is *biased.*

Note that it is the mean value rather than say the median value of $\mathbf{t}$ which is set equal to θ in (25–1); this is a matter of definition (and in this instance, one resting largely on mathematical convenience). Note also that the suggestive language "unbiased" is merely language. Note that once we find $\mathbf{t} = t$, we do not imply that $E(\theta) = \theta = t$. Note that no *a priori* concepts are implied here; unbiasedness is a property of a statistic constructed from the current random sample. Finally note that (25–1) is required to hold for *all* sample sizes *n;* we will sometimes deal with statistics which are merely *asymptotically unbiased,* that is, statistics for which, roughly, $E(\mathbf{t}) = \theta$ only when $n \to \infty$.

Is it important that the estimator of a parameter be unbiased? If a parameter is to be repeatedly estimated, in situations in which it always has the same value, and significant decisions based on its estimated value, the answer is probably yes; any estimating process used repeatedly and which on the average (mean) is not equal to the parameter leads to a sure cumulation of error in one direction. Moreover unbiasedness is particularly attractive (and absence of it is particularly unattractive) if several experimenters are reporting on the same parameter, for unbiased estimators may be readily combined (and biased estimators of unknown bias may not); see, for example, problem 25–1. But there surely are situations, perhaps particularly arising in occasional estimation, in which a slightly biased estimator which has small variability from sample to sample and is therefore generally close to the mark is preferable to an estimator which on the average is equal to the parameter but which has high variability from sample to sample.

25–3 Examples of unbiased estimators. We now give examples of unbiased estimators. Certain additional features of such estimators will emerge in the examples; these include (a) the nonuniqueness of an unbiased estimator, (b) the different variances of different unbiased estimators of the same parameter, (c) the possibility of converting biased to unbiased estimators.

(a) From (21–3) we have

$$E(\mathbf{x}_1) = \lambda, \qquad E(\bar{\mathbf{x}}) = \lambda, \tag{25–2}$$

if the mean λ exists. Both a single random observation $\mathbf{x}_1$ and the mean $\bar{\mathbf{x}}$ of a random sample of n observations from a population of mean λ are unbiased estimators of λ. Their respective variances are σ^2 and σ^2/n. If both are available, we will, for $n > 1$, generally prefer $\bar{\mathbf{x}}$ to $\mathbf{x}_1$; this preference, based on lower variance, will be discussed in Section 25–9.

(b) From (21–3) we have

$$E(\bar{\mathbf{x}} - \bar{\mathbf{y}}) = E(\mathbf{x}) - E(\mathbf{y}); \tag{25–3}$$

the difference of the means of two independent random samples from populations of means $E(\mathbf{x})$ and $E(\mathbf{y})$ is an unbiased estimator of the difference of the population means, if the latter exist.

(c) From (21–7) we have

$$E(\mathbf{s}^2) = \frac{n-1}{n}\sigma^2 \tag{25–4}$$

if the variance σ^2 exists; $\mathbf{s}^2$ is evidently a biased estimator of σ^2. But here, as often, the bias can be removed by multiplying the random variable (here $\mathbf{s}^2$) by a suitable function of n. Thus

$$E\left(\frac{n\mathbf{s}^2}{n-1}\right) = \sigma^2, \tag{25–5}$$

and $n\mathbf{s}^2/(n-1)$ is an unbiased estimator of σ^2. A quite general method, developed by M. H. Quenouille, for correcting bias in estimators is described by Kendall and Stuart, [104], Volume 2.

(d) The examples given to this point hold for certain parameters of the distributions of random variables with *any* density function $f(x;\theta)$. The form f of the density function need not be specified in the arguments that lead to results (25–2) through (25–5). Merely then as an exercise in integration (though other use will be made of it shortly), we indicate that (25–4) which holds for all $f(x;\sigma^2)$ holds for f normal. We want to evaluate

$$E(\mathbf{s}^2) = \int_0^\infty s^2 f(s^2)\, ds^2, \tag{25–6}$$

where $\mathbf{s}^2$ is now the variance in random samples of size n from a *normal* population of variance σ^2. From (23–18), $n\mathbf{s}^2/\sigma^2$ in random samples from a normal population of variance σ^2 is distributed as χ^2 with $n - 1$ degrees of freedom; the distribution is shown in (23–13). By change of variable from $n\mathbf{s}^2/\sigma^2$ to $\mathbf{s}^2$ we obtain from (23–13) the density function

$$f(s^2) = \frac{(n/2\sigma)^{(n-1)/2}}{\Gamma(n-1)/2}\, e^{-ns^2/\sigma^2}(s^2)^{(n-3)/2}. \tag{25–7}$$

Placing (25–7) in (25–6) and carrying out the integration, we get (25–4).

(e) Some unbiased estimators can be established only for parameters of *particular* probability distributions. For example, and unlike (25–4) which was readily shown on page 181 to hold for all $f(x;\sigma^2)$, the expectation

$$E(\mathbf{s}) = E\sqrt{\frac{\Sigma(\mathbf{x} - \bar{\mathbf{x}})^2}{n}} \tag{25–8}$$

cannot be evaluated for all $f(x;\sigma^2)$; the integral operation on the root is often intractable. In such cases we must be content with unbiased estimators obtained only for particular obliging f, and such examples, being less general, are less valuable. For example, for $f(x;\sigma^2)$ normal, the density function of $\mathbf{s}^2$ was given by (25–7). By change of variable from $\mathbf{s}^2$ to $\mathbf{s}$, the student will find that the density function of $\mathbf{s}$ is

$$f(s) = \frac{2(n/2\sigma)^{(n-1)/2}}{\Gamma(n-1)/2}\, e^{-ns^2/2\sigma^2} s^{n-2} \tag{25–9}$$

and, now holding only for $\mathbf{x}$ normal,

$$E(\mathbf{s}) = \int_0^\infty s f(s)\, ds = \sqrt{\frac{2}{n}}\,\frac{[(n-2)/2]!}{[(n-3)/2]!}\,\sigma, \tag{25–10}$$

from which an unbiased estimator of σ is immediately obtained.

We give three further examples of unbiased estimators which hold for the parameters of *particular* probability distributions.

(f) For $\mathbf{x}$ binomial,

$$E\left(\frac{\mathbf{x}}{n}\right) = \frac{1}{n}E(\mathbf{x}) = \frac{np}{n} = p; \tag{25–11}$$

the proportion $\mathbf{x}/n$ of successes in a random sample of size n from a binomial population is an unbiased estimator of the proportion of successes p in the population.

(g) Consider the sample *midrange*

$$\mathbf{w} = \frac{\mathbf{x}_{\min} + \mathbf{x}_{\max}}{2}$$

as an estimator of the center (θ) of a uniform distribution extending from $\theta - \frac{1}{2}$ to $\theta + \frac{1}{2}$. Writing $\mathbf{x}_{\min} = \mathbf{u}$ and $\mathbf{x}_{\max} = \mathbf{v}$, we have, repeating a result from Chapter 22,

$$f(u, v) = n(n-1)(v-u)^{n-2},$$

and, with the limits of integration following from

$$\frac{2\theta - 1}{2} \leq \frac{\mathbf{u} + \mathbf{v}}{2} \leq \frac{2\theta + 1}{2},$$

we have

$$E(\mathbf{w}) = E\left(\frac{\mathbf{u}+\mathbf{v}}{2}\right)$$

$$= \int_{\theta-(1/2)}^{\theta+(1/2)} \int_0^{\theta+(1/2)-u} \frac{u+v}{2} n(n-1)(v-u)^{n-2}\, dv\, du, \tag{25–12}$$

which readily integrates out to θ; $\mathbf{w}$ is an unbiased estimator of θ. Alternatively, $\theta - \mathbf{x}_{\min}$ and $\mathbf{x}_{\max} - \theta$ have, by symmetry, the same distribution, so

$$0 = E\{(\theta - \mathbf{x}_{\min}) - (\mathbf{x}_{\max} - \theta)\} = 2\theta - E\{\mathbf{x}_{\min} + \mathbf{x}_{\max}\}$$

and (25–12) follows.

(h) Consider the mean absolute deviation

$$\mathbf{v} = \sum_{i=1}^{n} |\mathbf{x}_i - \lambda|/n$$

as an estimator of the standard deviation σ of a normal distribution of known mean λ. Writing $\mathbf{x}_1 - \lambda = \mathbf{t}_i$, we have

$$E(\mathbf{v}) = \frac{1}{n}E\sum_{i=1}^{n}|\mathbf{t}_i| = \frac{1}{n}\{E|\mathbf{t}_1| + E|\mathbf{t}_2| + \cdots + E|\mathbf{t}_n|\}.$$

Writing $\mathbf{t}$ for any $\mathbf{t}_i$, we have

$$E(\mathbf{v}) = E|\mathbf{t}| = \int_{-\infty}^{\infty} |t|N(t)\,dt = \int_{-\infty}^{0} -tN(t)\,dt + \int_{0}^{\infty} tN(t)\,dt,$$

where $N(t)$ is written for the density function of a normal random variable $\mathbf{t}$ of mean 0 and variance σ^2. Continuing,

$$E(\mathbf{v}) = 2\int_0^{\infty} t\frac{1}{\sigma\sqrt{2\pi}}e^{-(1/2)(t/\sigma)^2}\,dt = \sqrt{\frac{2}{\pi}}\,\sigma; \qquad (25\text{–}13)$$

$\mathbf{v}$ is biased but

$$\mathbf{u} = \sqrt{\frac{\pi}{2}}\sum_{i=1}^{n}|\mathbf{x}_i - \lambda|/n \qquad (25\text{–}14)$$

is an unbiased estimator of σ.

As we shall later need the variance of the unbiased estimator $\mathbf{u}$, we find it here. Compactly,

$$\begin{aligned}\sigma^2(\mathbf{u}) &= \sigma^2\left\{\sqrt{\frac{\pi}{2}}\sum_{i=1}^{n}\mathbf{t}_i/n\right\} = \frac{1}{n}\sigma^2\left\{\sqrt{\frac{\pi}{2}}\,\mathbf{t}\right\}\\ &= \frac{\pi}{2n}\sigma^2(\mathbf{t}) = \frac{\pi}{2n}[E(\mathbf{t}^2) - \{E(\mathbf{t})\}^2] \qquad (25\text{–}15)\\ &= \frac{\pi}{2n}\left(\sigma^2 - \frac{2}{\pi}\sigma^2\right) = \frac{\sigma^2}{2n}(\pi - 2).\end{aligned}$$

We do not consider situations in which (a) *no* unbiased estimator exists, (b) an unbiased estimator gives absurd results over certain or over all ranges of the parameter. For both, see Kendall and Stuart [104], Volume 2.

25–4 Invariance. If we estimate θ and θ^2 from the same data by the same method, we may feel that the estimate of θ^2 should be the square of the estimate of θ. If our estimate of θ is t, our estimate of $\theta + k$ and of $k\theta$ should be $t + k$ and kt, respectively. A shift from ounces to (equally acceptable) pounds, from Centigrade to Fahrenheit, from σ to σ^2, should merely alter "accordingly" our estimator of the transformed parameter. We call a method of estimation *invariant* under transformation of a parameter if, when the method leads to $\mathbf{t}$ as the estimator of θ, the method also leads to $g(\mathbf{t})$ as the estimator of $g(\theta)$. We can speak of $\mathbf{t}$ as an invariant estimator for a certain class of transformations g if, when the parameter θ is transformed by g to $g(\theta)$, the estimator $\mathbf{t}$ is transformed to $g(\mathbf{t})$.

In our examples of unbiased estimators, note that invariance is not satisfied; given the unbiased estimator (25–5) of σ^2, the square root of

(25–5) is *not* an unbiased estimator of σ. Even simpler, consider the estimators $\bar{\mathbf{x}}$ and $\bar{\mathbf{x}}^2$. We have

$$E(\bar{\mathbf{x}}) = \lambda,$$

but

$$E(\bar{\mathbf{x}}^2) = E\left(\frac{\mathbf{x}_1 + \cdots + \mathbf{x}_n}{n}\right)^2 = \frac{1}{n^2} E(\mathbf{x}_1 + \cdots + \mathbf{x}_n)^2.$$

Expanding and recalling (3–9), we obtain

$$E(\bar{\mathbf{x}}^2) = \lambda^2 + \frac{\sigma^2}{n},$$

which is larger than λ^2. Unbiasedness and invariance cannot always be simultaneously satisfied; for parameters θ and θ^2, they can *never* be simultaneously satisfied [unless $\sigma^2(\mathbf{t}) = 0$].

Although not always an easy property to handle mathematically, invariance is an important and desirable property of an estimator. We seldom have any meaningful way of deciding whether it is θ or θ^k or $k\theta$ which we are really estimating; so much greater then is the need that an estimator be invariant under transformation of a parameter.

25–5 Consistency. We define consistency as follows: $\mathbf{w}$ is a consistent estimator of θ if, for arbitrarily small given positive numbers a and b, a value N of n exists such that if $n > N$,

$$p\{|\mathbf{w} - \theta| < a\} > 1 - b. \tag{25–16}$$

Roughly, as $n \to \infty$, $\mathbf{w} \to \theta$ in a certain sense (other senses are possible).

We prove that if

$$E(\mathbf{w}) \to \theta \quad \text{and} \quad \sigma^2(\mathbf{w}) \to 0 \quad \text{as} \quad n \to \infty, \tag{25–17}$$

then $\mathbf{w}$ is a consistent estimator of θ.

From the Bienaymé-Chebychev inequality

$$p\{|\mathbf{w} - E(\mathbf{w})| > k\sigma(\mathbf{w})\} < \frac{1}{k^2},$$

or

$$p\{|\mathbf{w} - E(\mathbf{w})| < k\sigma(\mathbf{w})\} > 1 - \frac{1}{k^2}, \quad \text{for any } k.$$

Writing $k\sigma(\mathbf{w}) = c$, we have

$$p\{|\mathbf{w} - E(\mathbf{w})| < c\} > 1 - \frac{\sigma^2(\mathbf{w})}{c^2}, \quad \text{for any } c.$$

When

$$|\mathbf{w} - E(\mathbf{w})| < c, \qquad |\mathbf{w} - \theta| < c + |\theta - E(\mathbf{w})|,$$

and we have

$$p\{|\mathbf{w} - \theta| < c + |\theta - E(\mathbf{w})|\} > 1 - \frac{\sigma^2(\mathbf{w})}{c^2}. \tag{25-18}$$

But (25–17) implies that there exists $n = N$ such that if $n > N$

$$|\theta - E(\mathbf{w})| < d, \qquad \sigma^2(\mathbf{w}) < e$$

for d, e arbitrary positive numbers. Thus (25–18) becomes, for $n > N$,

$$\begin{aligned} p\{|\mathbf{w} - \theta| < c + d\} &> p\{|\mathbf{w} - \theta| < c + |\theta - E(\mathbf{w})|\} \\ &> 1 - \frac{\sigma^2(\mathbf{w})}{c^2} \\ &> 1 - \frac{e}{c^2}, \end{aligned}$$

where c, d, and e are arbitrary. Therefore, we can satisfy (25–16) for $n > N$ by choosing $c + d = a$ and then $e = bc^2$.

Many estimators are consistent; of them we may often prefer that one with most probability mass around θ as measured by variance. But it is well to keep in mind that consistency is an asymptotic property; it may not be of great interest to those engaged in estimating parameters from small samples.

We give two examples of consistent estimators.

(a) The mean of a random sample is a consistent estimator of the population mean if the latter exists. For $E(\bar{\mathbf{x}}) = \lambda$ and $\sigma^2(\bar{\mathbf{x}}) = (\sigma^2/n) \to 0$ as $n \to \infty$; conditions (25–17) are satisfied. Or, consistency follows directly from the central limit theorem (page 183).

(b) The variance $\mathbf{s}^2$ of a random sample from a normal population of variance σ^2 is a consistent estimator of σ^2. For

$$\mathbf{s}^2 = \frac{(\mathbf{x}_1 - \bar{\mathbf{x}})^2 + (\mathbf{x}_2 - \bar{\mathbf{x}})^2 + \cdots + (\mathbf{x}_n - \bar{\mathbf{x}})^2}{n - 1} \cdot \frac{n-1}{n}$$

is a constant multiplied by the mean of what may be regarded as *independent* random variables $(\mathbf{x}_1 - \bar{\mathbf{x}})^2, (\mathbf{x}_2 - \bar{\mathbf{x}})^2, \ldots, (\mathbf{x}_n - \bar{\mathbf{x}})^2$, as considered in (23–23). Therefore, $\mathbf{s}^2$ satisfies the conditions of the central limit theorem and is consistent. Or, substituting (16–5) in the expression following (21–9), we could find $\sigma^2(\mathbf{s}^2)$ in normal samples and show that $\sigma^2(\mathbf{s}^2) \to 0$ as $n \to \infty$. This coupled with the fact that from (25–4) $\mathbf{s}^2$

is an asymptotically unbiased estimator of σ^2 would lead to consistency of $\mathbf{s}^2$.

Another demonstration might follow along these lines:

$$\mathbf{s}^2 = \frac{1}{n}\sum_{i=1}^{n} \mathbf{x}_i^2 - \overline{\mathbf{x}}^2;$$

$$\frac{1}{n}\sum_{i=1}^{n} \mathbf{x}_i^2 \text{ is a consistent estimator of } \sigma^2 + \lambda^2,$$

$$\overline{\mathbf{x}} \text{ is a consistent estimator of } \lambda,$$

from which it can be shown that $\mathbf{s}^2$ is a consistent estimator of σ^2 for any population with finite μ_4. This is a special case of a general theorem (for proof, see Brunk [23]) that the kth sample moment about the origin,

$$\frac{1}{n}\sum_{i=1}^{n} \mathbf{x}_i^k$$

is a consistent estimator of the corresponding population moment about the origin $E\mathbf{x}^k$ if $E(\mathbf{x}^k)$ and $\sigma^2(\mathbf{x}^k)$ exist.

25–6 Sufficiency. Consider the parameter θ and two statistics $\mathbf{u}$ and $\mathbf{w}$; the joint probability density of $\mathbf{u}$ and $\mathbf{w}$ is $f(u, w; \theta)$. If given u, the conditional density function $g(w|u; \theta)$ of $\mathbf{w}$ for any statistic $\mathbf{w}$ does not depend on θ, the statistic $\mathbf{u}$ will be called *sufficient* for θ. Intuitively we argue that if, given u, the conditional density function of $\mathbf{w}$ does not involve θ, $\mathbf{w}$ can contain no further information (beyond that in $\mathbf{u}$) on θ. Also intuitively, if $\mathbf{u}$ is a sufficient statistic for θ with respect to any other statistic $\mathbf{w}$, $\mathbf{u}$ (or a statistic based on $\mathbf{u}$) is preferable; the reduction of the total information in the sample with respect to θ to the single statistic $\mathbf{u}$ has been without loss of information on θ.

Example. Let $\mathbf{x}_1, \mathbf{x}_2, \ldots, \mathbf{x}_n$ be the outcomes of n Bernoulli trials, with $p(\mathbf{x}_i = \text{success}) = p$, $p(\mathbf{x}_i = \text{failure}) = 1 - p$, for all i. The total number of successes $\mathbf{t} = \sum_{i=1}^{n} \mathbf{x}_i$ in n trials is a sufficient statistic for p. For *given* $\mathbf{t} = t$, there are now exactly C_t^n possible outcomes $x_1, x_2, \ldots, x_n$ of n Bernoulli trials. Each has probability

$$\frac{p^t(1-p)^{n-t}}{C_t^n p^t(1-p)^{n-t}} = \frac{1}{C_t^n},$$

a quantity *independent of p*. Given the statistic $\mathbf{t}$, the sample data $x_1, x_2, \ldots, x_n$ contain no further information on the parameter p; $\mathbf{t}$ is sufficient for p.

Which sufficient statistic should we use? The sample $x_1, x_2, \ldots, x_n$ itself is sufficient, but we prefer a smaller set. We will sometimes call a sufficient statistic *minimal* if it is a function of every other sufficient statistic, and we will prefer a minimal sufficient statistic. For one of the purposes of a sufficient statistic is to reduce a complex of probabilities of the multivariables $\mathbf{x}_1, \mathbf{x}_2, \ldots, \mathbf{x}_n$ with respect to a parameter, and we should generally prefer to reduce this complex as much as possible.

25–7 The Fisher-Neyman factorization theorem. Finding sufficient statistics directly from their definition is cumbersome; the example just given is one of the few that can be readily produced. But a result due to Fisher and Neyman (proof of sufficiency by Fisher [59], [61], proof of necessity by Neyman [136]) provides us with a practical method of obtaining sufficient statistics if they exist: $\mathbf{t}$ is sufficient if and only if the density function of the sample can be factored as follows:

$$f(x_1, \ldots, x_n; \theta) = g(t; \theta)h(x_1, \ldots, x_n), \tag{25–19}$$

where g depends on the statistic $\mathbf{t}$ and the parameter θ, and h is independent of θ. The equivalence of (25–19) and the definition of a sufficient statistic is intuitively reasonable, for both describe the exhaustion of the information in the sample on θ by the single statistic $\mathbf{t}$, but the formal equivalence under general conditions has only recently been settled by Halmos and Savage [83].

We verify, for continuous random variables, that the Fisher-Neyman factorization criterion is sufficient for sufficiency. We show that, given the factorization (25–19), where $g(t; \theta)$ is the density function of $\mathbf{t}$, $\mathbf{t}$ is sufficient for θ.

For symmetry in the following argument, write $\mathbf{t} = \mathbf{t}_1$. Consider the change of variable from $\mathbf{x}_1, \ldots, \mathbf{x}_n$ to $\mathbf{t}_1, \ldots, \mathbf{t}_n$:

$$t_i = t_i(x_1, \ldots, x_n), \quad i = 1, \ldots, n,$$

with inverses

$$x_i = x_i(t_1, \ldots, t_n), \quad i = 1, \ldots, n,$$

and Jacobian J. Thus (25–19) becomes

$$f[x_1(t_1, \ldots, t_n), \ldots, x_n(t_1, \ldots, t_n)]J$$
$$= g(t_1; \theta)h[x_1(t_1, \ldots, t_n), \ldots, x_n(t_1, \ldots, t_n)]J; \tag{25–20}$$

the left side is the joint density function of $\mathbf{t}_1, \ldots, \mathbf{t}_n$, and $g(t_1; \theta)$ is the density function of $\mathbf{t}_1$. From (7–10), their ratio (which is hJ) is the conditional joint density function of $\mathbf{t}_2, \ldots, \mathbf{t}_n$, given $\mathbf{t}_1$. But h does not depend on θ and J does not introduce θ. Therefore the conditional joint density function of $\mathbf{t}_2, \ldots, \mathbf{t}_n$ given $\mathbf{t}_1$ does not depend on θ, and $\mathbf{t}_1$ is a sufficient statistic for θ. The argument for necessity of (25–19) is similar in character and we omit it.

Using (25–19), we give two examples of sufficient statistics; both would be lengthier projects from definition.

(a) Consider a normal random variable has unknown mean λ and known variance σ^2. Is the statistic $\bar{\mathbf{x}}$ a sufficient statistic for λ? We have

$$f(x_1, x_2, \ldots, x_n; \theta) = \frac{1}{(\sigma/\sqrt{2\pi})^n} \exp\left[-\frac{1}{2\sigma^2}\sum_{i=1}^{n}(x_i - \lambda)^2\right]$$

$$= \frac{1}{(\sigma/\sqrt{2\pi})^n} \exp\left[-\frac{1}{2\sigma^2}\left\{\sum_{i=1}^{n}(x_i - \bar{x})^2 + n(\bar{x} - \lambda)^2\right\}\right],$$

which factors according to (25–19); $\bar{\mathbf{x}}$ is a sufficient statistic for (and a good estimator of) λ.

(b) A Poisson random variable has parameter θ. Is $\mathbf{x}_1 + \mathbf{x}_2 + \cdots + \mathbf{x}_n$ a sufficient statistic for θ? We have

$$f(x_1, x_2, \ldots, x_n; \theta) = \frac{\theta^{x_1+x_2+\cdots+x_n} e^{-\theta n}}{x_1!x_2!\ldots x_n!}.$$

It is evident that $\mathbf{x}_1 + \mathbf{x}_2 + \cdots + \mathbf{x}_n$ is a sufficient statistic for (and *not* a good estimator of!) θ.

25–8 A property of certain sufficient statistics. Consider two statistics $\mathbf{u}$ and $\mathbf{w}$, both normally distributed with means equal to the parameter θ, with variances $\sigma_{\mathbf{u}}^2$ and $\sigma_{\mathbf{w}}^2$, and linearly correlated with correlation coefficient ρ;

$$f(u, w) = \frac{1}{2\pi\sigma_{\mathbf{u}}\sigma_{\mathbf{w}}\sqrt{1-\rho^2}} \times \exp\left\{-\frac{1}{2(1-\rho^2)}\left[\left(\frac{u-\theta}{\sigma_{\mathbf{u}}}\right)^2 - \frac{2\rho(u-\theta)(w-\theta)}{\sigma_{\mathbf{u}}\sigma_{\mathbf{w}}} + \left(\frac{w-\theta}{\sigma_{\mathbf{w}}}\right)^2\right]\right\}. \tag{25–21}$$

Under what conditions is $\mathbf{u}$ a sufficient statistic for θ with respect to $\mathbf{w}$? We find $h(w|u)$ and determine the conditions under which this conditional density function does not depend on θ:

$$h(w|u) = \frac{f(u, w)}{g(u)} = \frac{f(u, w)}{\int_{-\infty}^{\infty} f(u, w)\, dw},$$

which, repeating (17–18) for $f(u, w)$ bivariate normal, reduces to

$$h(w|u) = \frac{\exp\left\{-\frac{1}{2}\left[\frac{w - \{\theta + \rho(\sigma_{\mathbf{w}}/\sigma_{\mathbf{u}})(u - \theta)\}}{\sigma_{\mathbf{w}}\sqrt{1 - \rho^2}}\right]^2\right\}}{\sigma_{\mathbf{w}}\sqrt{2\pi}\,\sqrt{1 - \rho^2}}. \tag{25–22}$$

The conditional distribution of **w** is normal with mean

$$\theta + \rho\frac{\sigma_{\mathbf{w}}}{\sigma_{\mathbf{u}}}(u - \theta)$$

and standard deviation $\sigma_{\mathbf{w}}\sqrt{1 - \rho^2}$. Evidently $h(w|u)$ involves θ; for θ to vanish we need

$$1 = \rho\frac{\sigma_{\mathbf{w}}}{\sigma_{\mathbf{u}}} \quad \text{or} \quad \frac{\sigma_{\mathbf{w}}}{\sigma_{\mathbf{u}}} = \frac{1}{\rho}.$$

But $\sigma_{\mathbf{w}}/\sigma_{\mathbf{u}}$ is positive, so ρ must be positive. Moreover, if ρ is positive, we must have $0 \leq \rho \leq 1$, from which $\sigma_{\mathbf{u}} \leq \sigma_{\mathbf{w}}$. If the joint normal density function (25–21) arises, as is often likely in *large* samples, we could conclude that within the class of asymptotically unbiased and normally distributed statistics, sufficient statistics have asymptotically least variance.

We will soon come to an important relationship, holding for *all* sample sizes n, between sufficient statistics and minimum-variance estimators. Finally we note here two important properties of sufficient statistics. First, if **t** is sufficient for θ so is any single-valued function of **t** with a single-valued inverse; for example, if **t** is sufficient for θ, so is $a + b$**t**. Second, if **t** is sufficient for θ with respect to a statistic **u**, then **t** is sufficient for θ with respect to *every* other statistic. Both properties are discussed in Hogg and Craig [91].

25–9 Risk functions; variance and relative efficiency. There is no function of the random variables $\mathbf{x}_1, \mathbf{x}_2, \ldots, \mathbf{x}_n$ which is equal to the parameter θ for all θ. Accordingly, in any estimation procedure in which the estimator depends on random variables $\mathbf{x}_1, \mathbf{x}_2, \ldots, \mathbf{x}_n$, some risk of error of inference must be run. The amount of error depends on the particular estimator chosen; roughly speaking, a good estimator will generally be closer to θ than a poor estimator. The risk is also generally a function of θ.

As the estimator **t** depends on random variables, it is a random variable; its use involves risk but we cannot usefully define risk in terms of a random variable. Risk is often defined in terms of *average loss* over all values of the random variable **t**. Thinking of the risk associated with a particular value of **t** as measured by $r(t; \theta)$, the risk R associated with **t** as an estimator of θ is often defined by

$$R = \int_{-\infty}^{\infty} r(t; \theta)f(t)\, dt. \tag{25–23}$$

Precisely what risk function of θ we should seek to minimize is seldom suggested by the facts surrounding any statistical estimation problem, and systems of estimation based on different minimized risk functions are available. Minimizing the *maximum* risk, a conservative strategy which seems far from relevant in most real estimation problems, leads to *minimax* estimation. Minimizing average risk relative to an *a priori* distribution of θ leads to *Bayes' estimation.* In the customary absence of usable prior information, we are often persuaded to agree to an arbitrary settlement of the risk function to be minimized. By far the most common settlement, from the time of Gauss to the present, is the minimization, as a function of θ, of the (nonnegative) variance of a (generally unbiased) estimator **t**:

$$\sigma^2(\mathbf{t}) = \int_{-\infty}^{\infty} (t - \theta)^2 f(t; \theta)\, dt.$$

If estimators are normally distributed about θ, the variance is clearly the parameter of choice in measuring the variability of each estimator, and minimization of variance is of great interest; the normally distributed estimator of smaller variance is almost surely preferable to the normally distributed estimator of larger variance. If one estimator is normally distributed and another not, or if both are nonnormally distributed, variance as a measure of risk is somewhat less convincing.

We begin by considering the variances of certain pairs of estimators of the same parameter.

(a) Consider $\mathbf{x}_1$ and $\bar{\mathbf{x}}$. Both are unbiased estimators of the mean λ, if λ exists. The respective variances are σ^2 and σ^2/n. For **x** normal both estimators are normally distributed; for **x** normal and $n > 1$, $\bar{\mathbf{x}}$ is surely preferable. For populations described by other than the normal distributions, neither is normally distributed for finite n, and comparison of their variances is less decisive. But the fact that $\bar{\mathbf{x}}$ is consistent while $\mathbf{x}_1$ is not lends support to $\bar{\mathbf{x}}$ as the better estimator.

(b) Consider two estimators **s** and **t** of the mean θ of a uniform distribution,

$$\mathbf{s} = \frac{\mathbf{x}_{\text{min}} + \mathbf{x}_{\text{max}}}{2}, \quad \mathbf{t} = \bar{\mathbf{x}}.$$

Both **s** and **t** have already been shown to be unbiased estimators of θ. To calculate their variances, we take, without loss of generality, $0 \leq x \leq 1$, that is, $\theta = \frac{1}{2}$.

First consider $\mathbf{t} = \bar{\mathbf{x}}$. For uniform **x** over the range $0 \leq x \leq 1$,

$$\sigma^2(\mathbf{t}) = \sigma^2(\bar{\mathbf{x}}) = \sigma^2/n = 1/12n.$$

Now consider **s**. Write $\mathbf{u} = \mathbf{x}_{\min}$, $\mathbf{v} = \mathbf{x}_{\max}$, and we have

$$\sigma^2\left(\frac{\mathbf{u}+\mathbf{v}}{2}\right) = \frac{E(\mathbf{u}^2) - 2E(\mathbf{uv}) + E(\mathbf{v}^2)}{4} - \left(\frac{1}{2}\right)^2.$$

But $f(u, v)$ was found on page 197. We have

$$E(\mathbf{u}^2) = \int_0^1 \int_u^1 u^2 n(n-1)(v-u)^{n-2}\,du\,dv = \frac{2}{(n+1)(n+2)},$$

$$E(\mathbf{uv}) = \int_0^1 \int_u^1 uvn(n-1)(v-u)^{n-2}\,du\,dv = \frac{1}{n+2},$$

$$E(\mathbf{v}^2) = \int_0^1 \int_u^1 v^2 n(n-1)(v-u)^{n-2}\,du\,dv = \frac{n}{n+2},$$

from which

$$\sigma^2\left(\frac{\mathbf{u}+\mathbf{v}}{2}\right) = \frac{1}{2(n+1)(n+2)}.$$

It is evident that for $n \geq 3$, $\sigma^2(\mathbf{s}) < \sigma^2(\mathbf{t})$; the unbiased estimator **s** which is indifferent to the behavior of the $n - 2$ sample observations between $x_{\min}$ and $x_{\max}$ has lower variance than $\bar{\mathbf{x}}$ which takes careful account of all sample observations; this obtains, of course, only for the problem at hand (estimating the center of a uniform distribution), and not generally. Moreover, even here the comparison is not decisive; neither **s** nor **t** is normally distributed for finite n, and of the two only **t** is normal for large n.

If we introduce the term *relative efficiency* of a statistic to stand for the ratio of its variance to that of another statistic, the relative efficiency of **t** with respect to **s** is

$$\frac{6n}{(n+1)(n+2)},$$

which is evidently less than 1 for $n \geq 3$.

The problem and its solution are found in R. J. Brookner, "A note on the mean as a poor estimate of central tendency," *Jour. Amer. Stat. Assoc.*, Vol. 36 (1941), pp. 410–412.

25–10 Minimum variance. The discussion of Section 25–9 leads naturally to the concept of a statistic with *minimum* variance. For certain classes of estimators such statistics often exist; we seek ways and means of finding them. In some cases the result is immediate. For the important class of unbiased estimators all of which are *linear* functions of the sample variables,

$$\mathbf{t} = c_1\mathbf{x}_1 + c_2\mathbf{x}_2 + \cdots + c_n\mathbf{x}_n, \quad E(\mathbf{t}) = \theta,$$

we can for a certain class of parameters θ readily determine $c_1, c_2, \ldots, c_n$ so that $\sigma^2(\mathbf{t})$ is minimum [with $\mathbf{t}$ subject to the desirable condition $E(\mathbf{t}) = \theta$]. This class of estimators is the subject of Chapter 27. For a class of estimators which are asymptotically normal and asymptotically unbiased, we can find asymptotically minimum-variance estimators; these are the subject of Chapter 26.

The concept of relative efficiency used in Section 25–9 can be extended to judge estimators in relation to estimators which have *minimum* variance. Interest in statistics with larger-than-minimum variance implies that there may be reasons (and there are) for choosing an estimator whose variance is not minimum. Certain minimum-variance estimators may in one sense or another be uneconomic; for example, they may be difficult to construct.

We will call $\mathbf{t}$ an *efficient* estimator of θ if $\sigma^2(\mathbf{t}) \leq \sigma^2(\mathbf{w})$ for every other estimator $\mathbf{w}$ of θ. The definition of an efficient estimator is usually restricted to estimators $\mathbf{t}$ and $\mathbf{w}$ which are unbiased, and so it will be here. Thus, if $\mathbf{t}$ is an unbiased estimator of θ with minimum variance V among unbiased estimators of θ (note that $\mathbf{t}$ may not be unique), the efficiency of any unbiased estimator $\mathbf{w}$ of θ is $V/\sigma^2(\mathbf{w})$; the estimator $\mathbf{w}$ is efficient only if its efficiency is equal to 1.

The definition of efficiency has sometimes been restricted to large-sample statistics which are asymptotically unbiased and asymptotically normal; some texts speak only of *asymptotic efficiency*. This has the merit of confining the comparison of estimators by their variances to normally distributed statistics (where variance is decisive) at the cost of restricting comparison to values of n not of ordinary interest.

Apart from such special classes in which we can in effect investigate every possible statistic of the class, the practical question of finding an unbiased estimator of minimum variance remains. Two general theorems are relevant here, each remarkable in its own right. We consider them now.

25–11 The Cramér-Rao bound. Cramér [35] and independently Rao [163] have shown that within the general class of all unbiased estimators of θ, and under fairly broad conditions, no estimator $\mathbf{u}$ can have variance which is less than a quantity which depends only on $f(x; \theta)$ and n. This important result which we now derive is

$$\sigma^2(\mathbf{u}) \geq \frac{1}{nE[(\partial/\partial\theta) \log f(x; \theta)]^2}. \qquad (25\text{–}24)^*$$

Let $\mathbf{x}_1, \mathbf{x}_2, \ldots, \mathbf{x}_n$ be a random sample from $f(x; \theta)$, and let

$$\mathbf{u}(\mathbf{x}_1, \mathbf{x}_2, \ldots, \mathbf{x}_n)$$

be an unbiased estimator of θ. By definition,

$$\begin{aligned}\theta &= \int_{-\infty}^{\infty}\int_{-\infty}^{\infty}\cdots\int_{-\infty}^{\infty} u(x_1, x_2, \ldots, x_n) f(x_1, x_2, \ldots, x_n; \theta)\, dx_1\, dx_2 \ldots dx_n \\ &= \int_{-\infty}^{\infty}\int_{-\infty}^{\infty}\cdots\int_{-\infty}^{\infty} u(x_1, x_2, \ldots, x_n) \\ &\qquad \times f(x_1; \theta)\, f(x_2; \theta) \ldots f(x_n; \theta)\, dx_1\, dx_2 \ldots dx_n.\end{aligned}$$

If differentiation under the integral (with respect to θ) is possible, we find

$$\begin{aligned}1 &= \int_{-\infty}^{\infty}\int_{-\infty}^{\infty}\cdots\int_{-\infty}^{\infty} u(x_1, x_2, \ldots, x_n; \theta) \\ &\times \sum_{i=1}^{n}\left[\frac{1}{f(x_i; \theta)}\frac{\partial}{\partial\theta} f(x_i; \theta)\right] f(x_1, x_2, \ldots, x_n; \theta)\, dx_1\, dx_2 \ldots dx_n, \qquad (25\text{–}25)\end{aligned}$$

where the term $f(x_i; \theta)$ is introduced into the denominator to complete the sequence of terms under the summation in the numerator. Thus (25–25) may be written

$$1 = E\left[\mathbf{u} \sum_{i=1}^{n} \frac{\partial}{\partial\theta} \log f(x_i; \theta)\right].$$

Now let $u = 1$; assuming differentiation under the integral (with respect to θ) is possible, we differentiate the identity

$$E(1) = \int_{-\infty}^{\infty}\int_{-\infty}^{\infty}\cdots\int_{-\infty}^{\infty} f(x_1; \theta) f(x_2; \theta) \ldots f(x_n; \theta)\, dx_1\, dx_2 \ldots dx_n = 1$$

and obtain

$$E \sum_{i=1}^{n} \frac{\partial}{\partial\theta} \log f(x_i; \theta) = 0.$$

Writing $\mathbf{t}$ for the sum, we have found

$$E(\mathbf{ut}) = 1, \qquad E(\mathbf{t}) = 0.$$

---

* This important theorem can also be credited to Fisher, Frechet, Darmois, Dugué, Aitken and Silverstone.

Since the covariance of $\mathbf{u}$ and $\mathbf{t}$ is

$$\text{cov}\,(\mathbf{u}, \mathbf{t}) = E(\mathbf{ut}) - E(\mathbf{u})E(\mathbf{t}) = 1,$$

we have

$$\frac{1}{\sigma^2(\mathbf{u})\sigma^2(\mathbf{t})} = \rho^2(\mathbf{u}, \mathbf{t}) \leq 1,$$

where $\rho(\mathbf{u}, \mathbf{t})$ is the linear correlation coefficient of $\mathbf{u}$ and $\mathbf{t}$. We find

$$\sigma^2(\mathbf{u}) \geq \frac{1}{\sigma^2(\mathbf{t})}$$

and since $E(\mathbf{t}) = 0$,

$$\sigma^2(\mathbf{u}) \geq \frac{1}{E\{\sum_{i=1}^n (\partial/\partial\theta) \log f(x_i; \theta)\}^2}\cdot \tag{25–26}$$

We now find an expression for the right-hand side of (25–26) which involves only simple operations on the original density function $f(x; \theta)$. Writing

$$t_i = \frac{\partial}{\partial\theta} \log f(x_i; \theta),$$

(25–26) becomes

$$\sigma^2(\mathbf{u}) \geq \frac{1}{E\{\sum_{i=1}^n \mathbf{t}_i\}^2}\cdot$$

We have

$$\begin{aligned} E(\mathbf{t}_1 + \cdots + \mathbf{t}_n)^2 &= E(\mathbf{t}_1^2) + \cdots + E(\mathbf{t}_n^2) + E(\text{cross-product terms}) \\ &= nE\mathbf{t}^2 + E(\text{cross-product terms}). \end{aligned}$$

Consider a cross-product term,

$$E\frac{\partial}{\partial\theta}\log f(x_i; \theta)\cdot\frac{\partial}{\partial\theta}\log f(x_j; \theta), \quad i \neq j,$$

$$= E\frac{\partial}{\partial\theta}\log f(x_i; \theta) E\frac{\partial}{\partial\theta}\log f(x_j; \theta), \quad i \neq j.$$

But

$$\begin{aligned} E\frac{\partial}{\partial\theta}\log f(x; \theta) &= \int_{-\infty}^{\infty} \frac{\partial}{\partial\theta}\log f(x; \theta)\cdot f(x; \theta)\,dx \\ &= \int_{-\infty}^{\infty}\frac{\partial}{\partial\theta} f(x; \theta)\,dx. \end{aligned}$$

We know that

$$\int_{-\infty}^{\infty} f(x; \theta)\,dx = 1;$$

if we may differentiate (with respect to θ) under the integral and if the limits of $\mathbf{x}$ do not depend on θ, we will have

$$\int_{-\infty}^{\infty} \frac{\partial}{\partial\theta} f(x;\theta)\,dx = 0,$$

and finally,

$$\sigma^2(u) \geq \frac{1}{nE\{(\partial/\partial\theta)\log f(x;\theta)\}^2},$$

the Cramér-Rao inequality or bound. A valuable alternative form of this inequality is found as follows:

$$0 = \int_{-\infty}^{\infty} \frac{\partial}{\partial\theta} f(x;\theta)\,dx = \int_{-\infty}^{\infty} \left\{\frac{1}{f(x;\theta)}\left[\frac{\partial}{\partial\theta} f(x;\theta)\right]\right\} f(x;\theta)\,dx.$$

Abbreviating the notation and differentiating under the integral, we obtain

$$\begin{aligned} 0 &= \int_{-\infty}^{\infty} \left\{\frac{1}{f}\frac{\partial f}{\partial\theta}\frac{\partial f}{\partial\theta} + f\frac{\partial}{\partial\theta}\left[\frac{\partial}{\partial\theta}\log f\right]\right\} dx \\ &= \int_{-\infty}^{\infty} \left\{\left(\frac{1}{f}\frac{\partial f}{\partial\theta}\right)^2 + \frac{\partial^2 \log f}{\partial\theta^2}\right\} f\,dx \\ &= \int_{-\infty}^{\infty} \left\{\left(\frac{\partial \log f}{\partial\theta}\right)^2 + \frac{\partial^2 \log f}{\partial\theta^2}\right\} f\,dx, \end{aligned}$$

from which we obtain an alternative form of the Cramér-Rao inequality involving the *second* derivative of the density function,

$$\sigma^2(\mathbf{u}) \geq \frac{1}{-nE\{\partial^2 \log f(x;\theta)/\partial\theta^2\}}. \tag{25–27}$$

The denominator of (25–27) has been called I, the "amount of information" on θ in $x_1, \ldots, x_n$. Note that I depends only on f and n, and not on any particular method of estimation.

In some discussions, the efficiency of an estimator $\mathbf{t}$ is defined relative to the Cramér-Rao bound [the right-hand side of (25–24)]. This has one disadvantage. In many situations it may not be possible for *any* statistic to reach the Cramér-Rao bound; a relatively small number of distributions $f(x;\theta)$ admit statistics for θ which have variance equal to the Cramér-Rao bound. These restrictions, as well as other lower bounds on the variance due principally to Bhattacharyya and to Kiefer, are discussed in detail in Kendall and Stuart [104], Volume 2.

But whether we define efficiency in terms of this bound or not, the Cramér-Rao result provides a method of identifying certain minimum-variance unbiased statistics: in any point estimation problem find the Cramér-Rao bound; if an unbiased estimator which has this variance is known, the search is ended.

Note that an estimator having the variance of the Cramér-Rao lower bound is consistent; such an estimator is unbiased and, noting (25–24), its variance goes to 0 as n becomes indefinitely large.

We give three examples, leaving it to the student to verify the Cramér-Rao bound in each example.

(a) For $\mathbf{x}$ normal, unknown mean λ, known variance σ^2, the CR bound is σ^2/n. The unbiased estimator $\bar{\mathbf{x}}$ has this variance. Therefore $\bar{\mathbf{x}}$ is a minimum-variance unbiased estimator of λ.

(b) For $\mathbf{x}$ normal, known mean λ, unknown variance σ^2, the CR bound is $2\sigma^4/n$. The unbiased estimator $\sum_{i=1}^n (\mathbf{x}_i - \lambda)^2/n$ has this variance. Therefore $\sum_{i=1}^n (\mathbf{x}_i - \lambda)^2/n$ is a minimum-variance unbiased estimator of σ^2.

(c) For $\mathbf{x}$ normal, unknown mean λ, unknown variance σ^2, consider the (best) unbiased estimator $\sum_{i=1}^n (\mathbf{x}_i - \bar{\mathbf{x}})^2/(n - 1)$. It has variance $2\sigma^4/(n-1)$, larger than the CR bound which, extending the Cramér-Rao theorem to two parameters, is $2\sigma^4/n$. No unbiased estimator with variance as small as the CR bound is available.

25–12 The Blackwell-Rao theorem. A more powerful method of finding minimum-variance unbiased estimators is due to Blackwell and, independently, Rao; see [15]. They show that if both a sufficient statistic for θ and an unbiased estimator of θ exist, the two statistics can be combined to produce one estimator which is sufficient, unbiased, and has smaller variance than the original unbiased estimator. The resulting estimator is often unique, in which case it is best among unbiased estimators. The following simplified proof is due to Fraser [70].

Let $\mathbf{x}_1, \mathbf{x}_2, \ldots, \mathbf{x}_n$ be a random sample from $f(x; \theta)$. Let $\mathbf{t}$ be a sufficient statistic for θ. Let $\mathbf{u}(\mathbf{x}_1, \mathbf{x}_2, \ldots, \mathbf{x}_n)$ be an unbiased estimator of θ, with $\mathbf{u}$ not a function of $\mathbf{t}$ alone. Then $E(\mathbf{u}|t)$, the conditional mean of $\mathbf{u}$, given t, is a statistic which is a function only of the sufficient statistic $\mathbf{t}$ and, as we shall now show, $E(\mathbf{u}|t)$ is an unbiased estimator of θ and has smaller variance than $\mathbf{u}$. If a sufficient statistic $\mathbf{t}$ exists, it is therefore sensible to confine our search to functions of $\mathbf{t}$, for if we begin with an unbiased estimator $\mathbf{u}$ (not a function of $\mathbf{t}$ alone) we can always improve $\mathbf{u}$ by computing $E(\mathbf{u}|t)$, which will also be unbiased and have smaller (not necessarily minimum) variance than $\mathbf{u}$.

Consider the joint density function $f(u, t; \theta)$. The marginal distribution of $\mathbf{t}$ is

$$h(t; \theta) = \int_{-\infty}^{\infty} f(u, t; \theta)\, du,$$

and the conditional distribution of **u**, given **t**, is

$$g(u|t) = \frac{f(u, t; \theta)}{h(t; \theta)},$$

which, since **t** is sufficient, does not depend on θ. Now **u** is unbiased,

$$E(\mathbf{u}) = \int_{-\infty}^{\infty} \int_{-\infty}^{\infty} uf(u, t)\, du\, dt = \theta,$$

and we propose to improve on **u** by averaging **u** over fixed t, that is, we form the estimator $\mathbf{v}(\mathbf{t})$:

$$\mathbf{v}(\mathbf{t}) = E(\mathbf{u}|t) = \int_{-\infty}^{\infty} ug(u|t)\, du,$$

which is a statistic; it depends only on **t** (and not on θ).

In the first place $\mathbf{v}(\mathbf{t})$ is unbiased:

$$\begin{aligned} E[\mathbf{v}(\mathbf{t})] &= \int_{-\infty}^{\infty} v(t)h(t; \theta)\, dt \\ &= \int_{-\infty}^{\infty} \int_{-\infty}^{\infty} ug(u|t)h(t; \theta)\, du\, dt \\ &= \int_{-\infty}^{\infty} \int_{-\infty}^{\infty} uf(u, t; \theta)\, du\, dt = \theta. \end{aligned} \tag{25–28}$$

In the second place, $\sigma^2(\mathbf{v}) \leq \sigma^2(\mathbf{u})$; consider

$$\mathbf{u} - \theta = [\mathbf{u} - \mathbf{v}(\mathbf{t})] + [\mathbf{v}(\mathbf{t}) - \theta].$$

$E_{u,t}(\mathbf{u} - \theta)^2 = E_{u,t}\{\mathbf{u} - \mathbf{v}(\mathbf{t})\}^2 + E_{u,t}\{\mathbf{v}(\mathbf{t}) - \theta\}^2 +$ a cross-product term, that is,

$$\sigma^2(\mathbf{u}) = \text{nonnegative term} + \sigma^2(\mathbf{v}) + \text{a cross-product term}, \tag{25–29}$$

from which

$$\sigma^2(\mathbf{v}) \leq \sigma^2(\mathbf{u}),$$

for the cross-product term of (25–29) vanishes, as follows:

$$\begin{aligned} E(\mathbf{u} - \mathbf{v})(\mathbf{v} - \theta) &= \int_{-\infty}^{\infty} \int_{-\infty}^{\infty} [u - v(t)][v(t) - \theta] f(u, t; \theta)\, du\, dt \\ &= \int_t \left\{ \int_u [u - v(t)] g(u|t)\, du \right\} \cdot [v(t) - \theta] h(t; \theta)\, dt, \end{aligned}$$

and by the definition of $\mathbf{v}(\mathbf{t})$, the expression in braces vanishes.

These two results may be obtained more compactly:

$$E\{\mathbf{v}(\mathbf{t})\} = E[E(\mathbf{u}|\mathbf{t})] = E(\mathbf{u}) = \theta,$$

$$\sigma^2(\mathbf{u}) = \sigma^2\{E(\mathbf{u}|\mathbf{t})\} + E\{\sigma^2(\mathbf{u}|\mathbf{t})\}$$

$$\geq \sigma^2\{E(\mathbf{u}|\mathbf{t})\}$$

the latter only if $\sigma^2(\mathbf{u}|\mathbf{t}) = 0$, that is, only if

$$\mathbf{u} \equiv \{E\mathbf{u}|\mathbf{t}\} \equiv \mathbf{v}(\mathbf{t}).$$

Roughly, the class of distributions in which a sufficient statistic for θ exists is limited and is identical with the class of distributions which admit estimators of θ with variance reaching the Cramér-Rao bound. See Hogg and Craig [91] or Kendall-Stuart [104], Volume 2.

25–13 Possible uniqueness of the improved unbiased estimator. The improved (smaller) unbiased estimator $\mathbf{v}$ found in Section 25–12 is often unique. Let $\mathbf{t}$ be a sufficient estimator of θ, with density function $g(t; \theta)$. Let both $\mathbf{v}_1(\mathbf{t})$ and $\mathbf{v}_2(\mathbf{t})$ be unbiased estimators of θ depending on $\mathbf{t}$,

$$E\{\mathbf{v}_1(\mathbf{t})\} = \int_{-\infty}^{\infty} v_1(t)g(t; \theta)\, dt = \theta,$$

$$E\{\mathbf{v}_2(\mathbf{t})\} = \int_{-\infty}^{\infty} v_2(t)g(t; \theta)\, dt = \theta.$$

Write $\mathbf{v}(\mathbf{t}) = \mathbf{v}_1(\mathbf{t}) - \mathbf{v}_2(\mathbf{t})$;

$$E\{\mathbf{v}(\mathbf{t})\} = 0 = \int_{-\infty}^{\infty} v(t)g(t; \theta)\, d\theta. \tag{25–30}$$

Now for a large class of density functions $g(t; \theta)$, the only statistic $\mathbf{v}(\mathbf{t})$ which satisfies (25–30) is $\mathbf{v}(\mathbf{t}) = 0$, implying

$$\mathbf{v}_1(\mathbf{t}) = \mathbf{v}_2(\mathbf{t}),$$

and the improved unbiased estimator is unique. In this case there exists only one unbiased estimator based on the sufficient statistic; it has *minimum* variance.

Formally, we call $g(t; \theta)$ *complete* if $v(t) = 0$ at all points t for which there is a value of θ at which $g(t; \theta) > 0$. In this case, there is only one (unique) continuous function of $\mathbf{t}$ which is an unbiased estimator of θ. It has minimum variance.

We give three examples of complete distributions.

(a) $h(t;\theta) = \dfrac{1}{\theta}, \quad 0 < t < \theta, \quad \theta > 0,$

$$= 0 \text{ otherwise,}$$

$$0 = \int_{-\infty}^{\infty} s(t)h(t;\theta)\,dt = \int_0^{\theta} s(t)\,\frac{1}{\theta}\,dt = \int_0^{\theta} s(t)\,dt$$

for all θ. Write

$$S(\theta) = \int_0^{\theta} s(t)\,dt = 0, \quad \text{for all } \theta > 0,$$

$$\frac{\partial S(\theta)}{d\theta} = 0 = s(\theta), \quad \text{for all } \theta > 0,$$

$$0 = s(t), \quad \text{for all } t > 0,$$

for all θ for which $h(t;\theta) \neq 0$.

(b) $h(t;\theta) = \theta^t(1-\theta)^{1-t}, \quad t = 0, 1, \quad 0 \le \theta \le 1,$

$$0 = \sum_t s(t)h(t;\theta) = \sum_{t=0}^{1} s(t)\theta^t(1-\theta)^{1-t} = s(0)\cdot(1-\theta) + s(1)\cdot\theta,$$

for all $0 \le \theta \le 1$. That is, $a\theta + b = 0$ for all θ, where $a = s(1) - s(0)$ and $b = s(0)$. This requires $a = 0$, $b = 0$, from which

$$s(0) = 0, \qquad s(1) = 0,$$

for all θ for which $h(t;\theta) \neq 0$.

(c) Consider the Poisson population

$$f(x;\mu) = e^{-\mu}\mu^x/x!, \quad x = 0, 1, \ldots.$$

We have shown that a sufficient statistic for μ is $\mathbf{t} = \mathbf{x}_1 + \cdots + \mathbf{x}_n$. We know from problem 11–1 that $\mathbf{t}$ itself follows a Poisson distribution,

$$h(t;\lambda) = e^{-\lambda}\lambda^t/t!, \quad t = 0, 1, \ldots,$$

with $\lambda = n\mu$. Is $h(t;\lambda)$ complete? We have

$$0 = \sum_{t=0}^{\infty} s(t)e^{-\lambda}\lambda^t/t!$$

$$= s(0)e^{-\lambda} + s(1)e^{-\lambda}\lambda + s(2)e^{-\lambda}\frac{\lambda^2}{2!} + s(3)e^{-\lambda}\frac{\lambda^3}{3!} + \cdots$$

for all λ, or

$$s(0) + \lambda s(1) + \frac{\lambda^2}{2!} s(2) + \frac{\lambda^3}{3!} s(3) + \cdots = 0 \quad \text{for all } \lambda,$$

which requires

$$s(0) = s(1) = s(2) = s(3) = \cdots = 0,$$

where $h(t; \lambda) \neq 0$ for some λ. So there is only one function of **t** which is a (minimum-variance) unbiased estimator of μ. As for its identity, since

$$E(\mathbf{x}_1 + \cdots + \mathbf{x}_n) = n\mu = E(\mathbf{t}),$$

we have

$$\frac{1}{n} E(\mathbf{t}) = \mu;$$

$\mathbf{t}/n$ is the minimum-variance unbiased estimator of μ.

25–14 Remaining chapters on point estimation. In the next two chapters we consider two systems of point estimation which have certain attractive principles of their own and which enjoy certain of the desirable properties described in this chapter.

The first is *maximum likelihood*. The density associated with the random sample $\mathbf{x}_1, \mathbf{x}_2, \ldots, \mathbf{x}_n$ drawn from a population described by $f(x; \theta)$ (where the form f is known and the value of the parameter θ is unknown) is

$$g(x_1, x_2, \ldots, x_n; \theta) = f(x_1; \theta) f(x_2; \theta) \ldots f(x_n; \theta). \qquad (25\text{–}31)$$

If we regard $x_1, x_2, \ldots, x_n$ in (25–31) as *given* sample data, the maximum likelihood principle suggests that we select as estimator of θ that function T of the sample data which, when θ is replaced by T, maximizes (25–31); roughly, T is determined so that the "probability" associated with the observed data $x_1, x_2, \ldots, x_n$ is maximized.

The second is *least squares*. We shall consider in detail, but only within the class of *linear* estimators

$$\mathbf{t} = c_1\mathbf{x}_1 + c_2\mathbf{x}_2 + \cdots + c_n\mathbf{x}_n,$$

those estimators which have the desirable properties

$$E(\mathbf{t}) = \theta \quad \text{and} \quad E(\mathbf{t} - \theta)^2 \text{ minimum};$$

the least squares estimator **t** will be unbiased and will have minimum variance.

Collateral Reading

N. Arley and K. R. Buch, *Introduction to the Theory of Probability and Statistics*. New York, John Wiley and Sons, 1950.

D. Blackwell, "Conditional expectation and unbiased sequential estimation," *Ann. Math. Stat.*, Vol. 18 (1947), pp. 105–110.

J. T. Chu and H. Hotelling, "The moments of the sample median," *Ann. Math. Stat.*, Vol. 26 (1955), pp. 593–606. Also A. E. Sarhan, "Estimation of the mean and standard deviation by order statistics," *Ann. Math. Stat.*, Vol. 25 (1954), pp. 317–328. The variance of a Laplace variable is 2 (page 134); the variance of the mean of a random sample of size n from a Laplace population is therefore $2/n$. Chu and Hotelling have shown that the sample median has smaller variance than the sample mean (both of which are unbiased estimators of the Laplace parameter) for $n \geq 7$ (for $n \to \infty$ this is clearly the case; we had, page 196, for n large, $\sigma^2(\tilde{x}) = 1/8n$ which is less than $2/n$). Sarhan extends the Chu-Hotelling result to $n = 2, 3, 4, 5$.

H. Cramér, "Contributions to the theory of statistical estimation," *Skand. Aktuarietidskrift*, Vol. 29 (1946), pp. 85–94.

———, *Mathematical Methods of Statistics*. Princeton, Princeton University Press, 1955.

D. Duncan, *Statistical Theory II*, Mimeographed notes. Chapel Hill, University of North Carolina.

R. A. Fisher, "On the mathematical foundations of theoretical statistics," *Phil. Trans. Royal Soc. London*, Series A, Vol. 222 (1922), pp. 309–368.

———, "Theory of statistical estimation," *Proc. Cambridge Phil. Soc.*, Vol. 22 (1925), pp. 700–725. Both reprinted in R. A. Fisher, *Contributions to Mathematical Statistics*. New York, John Wiley and Sons, 1950.

M. G. Kendall and A. Stuart, *The Advanced Theory of Statistics, Vol. 2*. London, Charles Griffin and Co., 1961.

J. Kiefer, "On minimum variance estimators," *Ann. Math. Stat.*, Vol. 23 (1952), pp. 627–629.

E. Lehmann, *Notes on the Theory of Estimation*, Chapters I to V. Mimeographed. Berkeley, University of California, Associated Students Store, 1950. The best discussion.

J. Neyman, "Su un teorema concernente de cosidette statistiche sufficienti," *Giorn. Ist. Ital. Att.*, Vol. 6 (1935), pp. 320–334.

C. R. Rao, "Information and the accuracy attainable in the estimation of statistical parameters," *Bull. Calcutta Math. Soc.*, Vol. 37 (1945), pp. 81–91.

———, *Advanced Statistical Methods in Biometric Research*. New York, John Wiley and Sons, 1952. An excellent source of information.

B. L. Van Der Waerden, *Mathematische Statistik*. Berlin, Springer-Verlag, 1957. A good reference.

S. S. Wilks, *Mathematical Statistics*. New York, John Wiley and Sons, 1962. Also the 1943 volume of the same title.

Problems

25–1. Show that the weighted mean of uncorrelated unbiased estimators of the same parameter θ is unbiased if the sum of the weights is 1. In particular, if $\mathbf{u}$ and $\mathbf{v}$ are uncorrelated unbiased estimators of θ with variances $\sigma^2(\mathbf{u})$ and $\sigma^2(\mathbf{v})$, show that the weighted function

$$w\mathbf{u} + (1 - w)\mathbf{v}$$

is unbiased and find the weights w and $1 - w$ which minimize the variance of this estimator.

25–2. Show that the sample median is an unbiased estimator of the parameter of the Cauchy distribution.

25–3. The radius of a circle is measured with an error of observation which is normally distributed about zero, with unknown variance σ^2. Given n independent measurements of the radius, find *an* unbiased estimator of the area of the circle.

25–4. Given that $\mathbf{x}$ is normal with variance σ^2; let $\mathbf{s}$ be the sample standard deviation. Determine $E(\mathbf{s})$ [verifying (25–10)].

25–5. Show that $f(x; \theta) = 1/2\theta$, $-\theta \leq x \leq \theta$, is not complete.

CHAPTER 26

MAXIMUM LIKELIHOOD

26–1 The maximum likelihood principle. Again, in this chapter, we are interested in a single (point) estimate of the unknown value of a parameter of a probability distribution. The discussion here will center on a particular method of point estimation known as *maximum likelihood;* this method has several of the desirable properties of point estimators discussed in Chapter 25.

We begin with a random variable $\mathbf{x}$ with density function $f(x; \theta)$; the form of f is known but the value of the constant parameter θ is unknown. A random sample $\mathbf{E}$: $\mathbf{x}_1, \mathbf{x}_2, \ldots, \mathbf{x}_n$ is drawn. The probability density of $\mathbf{E}$ is written $p(\mathbf{E}|\theta)$. For given (observed) sample data $\mathbf{E} = E$: $x_1, x_2, \ldots, x_n$, the density $p(\mathbf{E}|\theta)$ evaluated at $\mathbf{E} = E$ will be a function of the unknown θ.

We wish to estimate θ by forming a statistic $T(E)$, a function of and only of the sample data $x_1, x_2, \ldots, x_n$. As we already know, many functions are possible; here we consider the following principle suggested by Gauss and systematically developed by R. A. Fisher: the best T is that which, when T is substituted for θ, maximizes for the observed E the density $p(\mathbf{E}|\theta)$ evaluated at $\mathbf{E} = E$. Roughly, the best value to give the unknown parameter θ is that which maximizes the probability associated with the sample point E which has in fact been observed. Such a value often exists.

Fisher's principle must be added to those of Chapter 25. It is intuitively appealing, the operational procedure for finding the estimator is practical, and the estimator enjoys several of the properties discussed in Chapter 25.

We have

$$p(\mathbf{E}|\theta) = p(\mathbf{x}_1, \mathbf{x}_2, \ldots, \mathbf{x}_n|\theta) = f(x_1; \theta)f(x_2; \theta) \ldots f(x_n; \theta). \qquad (26\text{–}1)$$

We want to maximize $p(\mathbf{E}|\theta)$ for *given* E: $x_1, x_2, \ldots, x_n$; the $x_1, x_2, \ldots, x_n$ in (26–1) are numbers, not random variables. [This need not prevent us from regarding $\mathbf{T}$ as, in fact, a random variable and later considering such functions as $f(T)$, $E(\mathbf{T})$ and $\sigma^2(\mathbf{T})$.]

We require

$$f(x_1, \mathbf{T})f(x_2, \mathbf{T}) \ldots f(x_n, \mathbf{T}) > f(x_1, \mathbf{U})f(x_2, \mathbf{U}) \ldots f(x_n, \mathbf{U}), \qquad (26\text{–}2)$$

where $\mathbf{U}$ is any estimator of θ other than the maximum likelihood estimator $\mathbf{T}$; the requirement is similar if we have several parameters θ_1, $\theta_2, \ldots, \theta_k$ with maximum likelihood estimators respectively $\mathbf{T}_1, \mathbf{T}_2, \ldots, \mathbf{T}_k$ and other estimators $\mathbf{U}_1, \mathbf{U}_2, \ldots, \mathbf{U}_k$. If θ is continuous, (26–2) can be replaced by

$$\frac{dp(E|\theta)}{d\theta} = 0 \quad \text{and} \quad \frac{d^2p(E|\theta)}{d\theta^2} < 0, \tag{26–3}$$

or since $\log p(E|\theta)$ (often written L) and $p(E|\theta)$ reach their maximum values at the same value of θ, the first condition of (26–3) may be written

$$\frac{d \log p(E|\theta)}{d\theta} = \frac{d}{d\theta} \sum_{i=1}^{n} \log f(x_i; \theta) = 0, \tag{26–4}$$

which can often be solved for θ. Relation (26–4) is usually preferable to (26–3), for it is easier to differentiate a sum than a product. While we may speak of "solving" (26–4) for θ, note that the solution does not reveal the true value of θ. It merely produces an equation for θ in terms of observed values $x_1, x_2, \ldots, x_n$ of the random variables $\mathbf{x}_1, \mathbf{x}_2, \ldots, \mathbf{x}_n$.

We consider only those roots of (26–4) which depend on observed values of the random variables $\mathbf{x}_1, \mathbf{x}_2, \ldots, \mathbf{x}_n$. Any "solution" $\theta = T =$ constant is disregarded; for while the estimate $T = 17$ is without competition if θ is in fact 17, it will be unimpressive if θ is 1700.

If a root (often unique) of (26–4) corresponds to the maximum of (26–1), we call it the *maximum likelihood* estimator of θ.

For random samples of size n drawn with replacement from a *discrete* population, with x_1 appearing n_1 times, x_2 appearing n_2 times, $\ldots$, x_h appearing n_h times, $n_1 + n_2 + \ldots + n_h = n$, the function to be maximized is

$$[f(x_1; \theta)]^{n_1}[f(x_2; \theta)]^{n_2} \ldots [f(x_h; \theta)]^{n_h}, \tag{26–5}$$

with $x_1, x_2, \ldots, x_h, n_1, n_2, \ldots, n_h$ given numbers.

26–2 Examples of maximum likelihood estimators.

We now consider examples of maximum likelihood estimators; each one will bring out some useful detail.

(a) *Binomial parameter p.* The binomial variable has the density function

$$f(x; p) = p^x(1 - p)^{1-x}, \quad x = 0, 1, \tag{26–6}$$

and, repeating (10–4), we have

$$p(E|p) = C_x^n p^x (1 - p)^{n-x}, \quad x = 0, 1, \ldots, n. \tag{26–7}$$

Note that (26–6) and not the more familiar (26–7) is the $f(x;\theta)$ of (26–4).

$$L = \log p(E|p) = \log C_x^n + x \log p + (n - x) \log (1 - p),$$

$$\frac{dL}{dp} = \frac{x}{p} - \frac{n - x}{1 - p} = 0 \quad \text{with unique solution} \quad \mathbf{T} = \frac{x}{n}. \tag{26–8}$$

The proportion of successes in a random sample from a binomial population is the maximum likelihood estimator of the probability of success in the population. As we already know, this estimator is also unbiased, consistent, sufficient, has minimum variance among unbiased estimators and is asymptotically normal.

(b) *Parameter of the Laplace distribution.*

$$f(x;\theta) = e^{-|x-\theta|}, \quad -\infty < x < \infty, \tag{26–9}$$

$$p(E|\theta) = \exp\left[-\sum_{i=1}^{n} |x_i - \theta|\right], \quad L = -\sum_{i=1}^{n} |x_i - \theta|;$$

in order to find the value of θ which maximizes L, we might consider arranging $x_1, x_2, \ldots, x_n$ in (ascending) order of magnitude:

$$x_1, x_2, \ldots, x_k \text{ below } \theta,$$

$$x_{k+1}, x_{k+2}, \ldots, x_n \text{ above } \theta,$$

with k unknown. We would have

$$L = -(\theta - x_1) - (\theta - x_2) - \cdots - (\theta - x_k) - (x_{k+1} - \theta) - (x_{k+2} - \theta) - \cdots - (x_n - \theta),$$

$$\frac{dL}{d\theta} = -k + (n - k) = 0, \qquad k = \frac{n}{2} \quad \text{and} \quad \mathbf{T} = \tilde{x}, \tag{26–10}$$

the sample median. But $d^2L/d\theta^2 = 0$, so we have no guarantee that $\tilde{x}$ maximizes L. However,

$$M = \sum_{i=1}^{n} |x_i - \theta| = |x_1 - \theta| + \cdots + |x_k - \theta| + \cdots + |x_n - \theta|$$

is minimized when the *number* of ordered x's below and above θ is the same, that is, when $\theta = \tilde{x}$. Any shift in θ away from $\tilde{x}$ increases M. If $\theta = \theta' = \tilde{x} + \Delta$ (positive), each term involving x_i less than θ' is reduced by Δ, and each term involving x_i greater than θ' is increased by Δ; but there are more of the latter terms. Similarly if $\theta = \theta' = \tilde{x} - \Delta$ (positive), each term involving x_i less than θ' is increased by Δ, and each term involving x_i

greater than θ' is decreased by Δ; but there are more of the former terms. Thus M is minimum (and L is maximum) at $\theta = \tilde{x}$.

(c) *Parameter N of the hypergeometric distribution.* In the example on page 115 dealing with the size of a zoological population, the probability distribution involved was the hypergeometric; the probability $p(a; n, N, A)$ that a random sample of size n drawn from a population of size N containing A successes and $N - A$ failures contains a successes and $n - a$ failures is, repeating (12–5),

$$p(a; n, N, A) = \frac{C_a^A C_{n-a}^{N-A}}{C_n^N}. \tag{26–11}$$

The unknown parameter was N, the number of fish in the pond. As N is discrete, we repeat problem 12–2 to determine the value of N at which (26–11) is maximum. Form

$$\frac{p_N}{p_{N-1}} = \frac{(N - n)(N - A)}{(N - n - A + a)N},$$

which is greater than 1 if $Na < nA$ and less than 1 if $Na > nA$. So the probability $p(a; n, N, A)$ in (26–11) is maximum when N is the largest integer less than nA/a, that is, when N is approximately equal to nA/a. For the example of page 115, the maximum likelihood estimate of N is 10,000 fish.

(d) *Parameter of the Cauchy distribution.*

$$f(x; \theta) = \frac{1}{\pi} \cdot \frac{1}{1 + (x - \theta)^2}, \quad -\infty < x < \infty, \tag{26–12}$$

$$p(E|\theta) = \frac{1}{\pi^n} \prod_{i=1}^{n} \frac{1}{1 + (x_i - \theta)^2},$$

$$L = \log p(E|\theta) = \log \frac{1}{\pi^n} \sum_{i=1}^{n} \log\,[1 + (x_i - \theta)]^2,$$

$$\frac{dL}{d\theta} = \sum_{i=1}^{n} \frac{2(x_i - \theta)}{1 + (x_i - \theta)^2} = 0, \tag{26–13}$$

which can be solved for θ only by approximation. Not always does (26–4) lead to an explicit solution for the maximum likelihood estimator. Details of approximations to maximum likelihood estimators in intractable situations, including the present example, are discussed by Kendall and Stuart [104], Volume 2.

(e) *The variance or standard deviation of a normal variable, mean known.* We have

$$f(x; \lambda, V) = \frac{1}{\sqrt{2\pi V}} \exp\left[-\frac{1}{2V}(x-\lambda)^2\right], \tag{26-14}$$

$$p(E|\lambda, V) = \frac{1}{(2\pi)^{n/2} V^{n/2}} \exp\left[-\frac{1}{2V}\sum_{i=1}^{n}(x_i-\lambda)^2\right],$$

$$L = \log p(E|\lambda, V) = \log \frac{1}{(2\pi)^{n/2}} - \frac{n}{2}\log V - \frac{1}{2V}\sum_{i=1}^{n}(x_i-\lambda)^2,$$

$$\frac{dL}{dV} = -\frac{n}{2V} + \frac{1}{2V^2}\sum_{i=1}^{n}(x_i-\lambda)^2 = 0,$$

$$\mathbf{T} = \frac{\sum_{i=1}^{n}(x_i-\lambda)^2}{n}, \tag{26-15}$$

also the minimum-variance estimator of V.

Had we considered the standard deviation σ[the (positive) square root of V] rather than V as the unknown parameter, we would have had

$$\frac{d \log p(E|\lambda, \sigma)}{d\sigma} = -\frac{n}{\sigma} + \frac{1}{\sigma^3}\sum_{i=1}^{n}(x_i-\lambda)^2 = 0,$$

$$\mathbf{T} = \sqrt{\frac{\sum_{i=1}^{n}(x_i-\lambda)^2}{n}}, \tag{26-16}$$

the square root of the maximum likelihood estimator of V. This (desirable) invariance was to be expected. For $V = \sigma^2$ is a single-valued function of σ with (as σ is positive) a unique inverse. Wherever in the estimation of V we had the symbol V, we now have

$$V(\sigma)\frac{dV}{d\sigma} = 2\sigma^3$$

and invariance follows.

26-3 Invariance of the maximum likelihood estimator. Quite generally, if $\mathbf{T}$ is the maximum likelihood estimator of θ, $u(\mathbf{T})$ is the maximum likelihood estimator of $u(\theta)$; maximum likelihood estimation is invariant under a transformation of parameters. Let $\mathbf{T}$ be the maximum likelihood estimator of θ, let $\phi = u(\theta)$ be a *single-valued* function of θ, and let the inverse of $\phi = u(\theta)$ be $\theta = v(\phi)$. As $\mathbf{T}$ is the maximum likelihood estimator of θ, the likelihood $L(\theta)$ is maximum at $\theta = \mathbf{T}$, or $L(\theta) = L\{v(\phi)\}$ is maximum at $\theta = \mathbf{T} = v(\phi)$, that is, maximum at $\phi = u(\mathbf{T})$.

26–4 Further examples. (a) *Parameter of the uniform distribution.*

$$\begin{aligned} f(x;\theta) &= \frac{1}{\theta}, \quad 0 \le x < \theta, \\ p(E|\theta) &= \frac{1}{\theta^n}. \end{aligned} \tag{26–17}$$

Here differentiation would fail to produce maximum L (in fact, if applied formally, it would produce an apparent minimum, for $d^2L/d\theta^2$ is positive!), for L does not have zero slope at its maximum. But (26–17) is evidently maximum when θ is minimum, and the smallest θ, calculated from the sample and compatible with θ being the range of $\mathbf{x}$, is the largest variable in the sample. Therefore $\mathbf{T} = x_{\max}$, all other sample variables being neglected.

Substituting $f(x;\theta)$ of (26–17) into (22–5), we get the density function of the largest variable $\mathbf{T} = \mathbf{x}_{\max}$ in a random sample from a uniform population:

$$f(T)\,dT = \theta^{-n}nT^{n-1}, \quad 0 \le T < \theta, \tag{26–18}$$

and

$$E(\mathbf{T}) = \frac{n}{n+1}\,\theta$$

follows; the maximum likelihood estimator $\mathbf{T} = x_{\max}$ is biased.

(b) *Simultaneous estimation of two parameters; normal distribution.* We give an example of simultaneous or joint maximum likelihood estimation of two parameters. For $f(x;\theta_1,\theta_2)$, θ_1 and θ_2 continuous, we have

$$p(E|\theta_1,\theta_2) = f(x_1;\theta_1,\theta_2)f(x_2;\theta_1,\theta_2)\ldots f(x_n;\theta_1,\theta_2),$$

$$L = \log p(E|\theta_1,\theta_2) = \sum_{i=1}^{n} \log f(x_i;\theta_1,\theta_2).$$

Since L involves both parameters θ_1 and θ_2, we need

$$\frac{\partial L}{\partial \theta_1} = 0, \qquad \frac{\partial L}{\partial \theta_2} = 0, \tag{26–19}$$

which are to be solved wherever possible for

$$\theta_1 = \mathbf{T}_1 \qquad \text{and} \qquad \theta_2 = \mathbf{T}_2.$$

Any pair of roots of (26–19) which correspond to maximum L constitute the joint maximum likelihood estimators of θ_1 and θ_2.

For the two parameters of a normal distribution

$$f(x;\lambda,\sigma) = \frac{1}{\sigma\sqrt{2\pi}} \exp\left[-\frac{1}{2}\left(\frac{x-\lambda}{\sigma}\right)^2\right],$$

$$p(E|\lambda,\sigma) = \frac{1}{\sigma^n(2\pi)^{n/2}} \exp\left[-\frac{1}{2}\sum_{i=1}^{n}\left(\frac{x_i-\lambda}{\sigma}\right)^2\right],$$

$$L = \log p(E|\lambda,\sigma) = \log\frac{1}{(2\pi)^{n/2}} - n\log\sigma - \frac{1}{2\sigma^2}\sum_{i=1}^{n}(x_i-\lambda)^2,$$

$$\frac{\partial L}{\partial\lambda} = \frac{1}{\sigma^2}\sum_{i=1}^{n}(x_i-\lambda) = 0, \qquad \frac{\partial L}{\partial\sigma} = \frac{n}{\sigma} + \frac{1}{\sigma^3}\sum_{i=1}^{n}(x_i-\lambda)^2 = 0,$$

$$\mathbf{T}_1 = \bar{x} \quad \text{and} \quad \mathbf{T}_2 = \sqrt{\frac{\sum_{i=1}^{n}(x_i-\bar{x})^2}{n}}, \tag{26–20}$$

$\mathbf{T}_1$ unbiased, $\mathbf{T}_2$ biased. A lengthy home problem but one which is sufficiently valuable to warrant note in the text is: the maximum likelihood estimators of the five parameters $E(\mathbf{x})$, $E(\mathbf{y})$, $\sigma^2(\mathbf{x})$, $\sigma^2(\mathbf{y})$, $\rho(\mathbf{x},\mathbf{y})$ of the bivariate normal probability distribution are respectively the sample means $\bar{x}$, $\bar{y}$, the sample variances s_x^2, s_y^2, and the sample correlation coefficient

$$r_{xy} = \sum_{i=1}^{n}(x_i-\bar{x})(y_i-\bar{y})/ns_xs_y.$$

26–5 Asymptotic variance of the maximum likelihood estimator. We now find an expression for the variance, in large samples, of the maximum likelihood estimator. The following argument is entirely heuristic, but it can be made rigorous (Cramér [36], Kendall and Stuart [104]).

Expanding the right-hand side of the maximum likelihood equation

$$0 = \sum_{i=1}^{n}\frac{\partial}{\partial\theta}\log f(x_i;\theta) \tag{26–21}$$

in a Taylor series in the vicinity of the true value of θ, we have

$$0 = \sum_{i=1}^{n}\left.\frac{\partial\log f(x_i;\theta)}{\partial\theta}\right|_{\theta=T} = \sum_{i=1}^{n}\frac{\partial\log f(x_i;\theta)}{\partial\theta} + (T-\theta)\sum_{i=1}^{n}\frac{\partial^2\log f(x_i;\theta)}{\partial\theta^2} + \cdots,$$

from which, heuristically, and neglecting $+\cdots$,

$$\sqrt{n}(T-\theta) = -\sqrt{n}\,\frac{\displaystyle\sum_{i=1}^{n}\frac{\partial\log f(x_i;\theta)}{\partial\theta}}{n} \qquad E\,\frac{\partial^2\log f(x_i;\theta)}{\partial\theta^2}$$

The numerator of the right-hand side is the sample mean of n identically distributed random variables

$$y_i = \frac{\partial}{\partial \theta} \log f(x_i; \theta)\Big|_{\theta=\theta_0};$$

by the central limit theorem, this mean is, for n large, normally distributed.

Now recalling the argument used in reaching the Cramér-Rao bound, we had

$$E\left\{\frac{\partial \log f(x; \theta)}{\partial \theta}\right\} = 0,$$

and, recalling the argument used in reaching the alternative form of the Cramér-Rao bound,

$$E\left\{\frac{\partial \log f(x; \theta)}{\partial \theta}\right\}^2 = -E\left\{\frac{\partial^2 \log f(x; \theta)}{\partial \theta^2}\right\}.$$

From these results, we have, heuristically, for n large, that $\mathbf{T}$ is normally distributed with mean 0 and variance

$$E(\mathbf{T} - \theta)^2 = \sigma^2(\mathbf{T}) = \frac{E\left(\frac{\partial \log f(x; \theta)}{\partial \theta}\right)^2}{\left[E\left(\frac{\partial^2 \log f(x; \theta)}{\partial \theta^2}\right)\right]^2} = -\frac{1}{nE\,\frac{\partial^2 \log f(x; \theta)}{\partial \theta^2}}. \tag{26–22}$$

Note that for n large, the variance of maximum likelihood estimator reaches the Cramér-Rao bound; the maximum likelihood estimator has minimum asymptotic variance.

26–6 Asymptotic variance of the maximum likelihood estimator; examples. (a) *Asymptotic variance of the maximum likelihood estimator of the variance V of a normal distribution of known mean.* The maximum likelihood estimator of V is given by (26–15). To find the variance, in large samples, of this estimator,

$$f(x; \lambda, V) = \frac{1}{\sqrt{2\pi V}} \exp\left[-\frac{1}{2V}(x-\lambda)^2\right],$$

$$\frac{\partial^2 \log f(x; \lambda, V)}{\partial V^2} = \frac{1}{2V^2} - \frac{1}{V^3}(x-\lambda)^2,$$

$$E\,\frac{\partial^2 \log f(x; \lambda, V)}{\partial V^2} = \frac{1}{2V^2} - \frac{V}{V^3} = -\frac{1}{2V^2}, \qquad \sigma^2(\mathbf{T}) = \frac{2V^2}{n}. \tag{26–23}$$

(b) *Asymptotic variance of the maximum likelihood estimator of the binomial p.* The maximum likelihood estimator of p is given by (26–8). To find the variance, in large samples, of this estimator,

$$f(x; p) = p^x(1-p)^{1-x},$$

$$\frac{\partial^2 \log f(x; p)}{\partial p^2} = -\frac{x}{p^2} - \frac{1-x}{(1-p)^2} = -\frac{1}{pq}, \qquad \sigma^2(\mathrm{T}) = \frac{pq}{n}, \tag{26–24}$$

which equals the exact (for all n) variance already found in (10–9).

(c) *Asymptotic variance of the maximum likelihood estimator of the Cauchy parameter.* While equation (26–13) for the maximum likelihood extimator of the Cauchy parameter was found to be intractable, the asymptotic variance of the maximum likelihood estimator is readily found:

$$\log f(x; \theta) = \log \frac{1}{\pi} - \log [1 + (x - \theta)^2],$$

$$\frac{\partial^2}{\partial \theta^2} \log f(x; \theta) = \frac{-2 + 2(x + \theta)^2}{[1 + (x - \theta)^2]^2},$$

$$E \frac{\partial^2}{\partial \theta^2} \log f(x; \theta) = \int_{-\infty}^{\infty} \frac{-2 + 2(x - \theta)^2}{[1 + (x - \theta)^2]^2} \cdot \frac{1}{\pi} \frac{1}{1 + (x - \theta)^2}\, dx$$

$$= -\frac{2}{\pi} \int_{-\infty}^{\infty} \frac{1 - u^2}{(1 + u^2)^3}\, du,$$

which, as shown in problem 26–5, reduces to $\frac{1}{2}$. Finally,

$$\sigma^2(\mathrm{T}) = \frac{2}{n}\cdot \tag{26–25}$$

Thus the maximum likelihood estimator of the Cauchy parameter (which can only be approximately determined) has asymptotic variance $2/n$, while the much more readily found sample median which, as noted in problem 25–2, is an unbiased estimator of the Cauchy parameter, has asymptotic variance $\pi^2/4n$, as noted on page 196. The asymptotic efficiency of the sample median here is $8/\pi^2$, about 88 percent.

26–7 Maximum likelihood estimators and sufficiency. Maximum likelihood estimators are functions of sufficient statistics. If a sufficient statistic exists, it can always be found by the method of maximum likelihood; any nontrivial solution (if there is one) of the maximum likelihood equation (26–4) will be a function of the sufficient statistic. For if a sufficient statistic **t** exists, **t** must satisfy (25–19),

$$f(x_1, x_2, \ldots, x_n; \theta) = g(t; \theta)h(x_1, x_2, \ldots, x_n).$$

But the maximum likelihood equation is

$$\frac{\partial}{\partial\theta} f(x_1, x_2, \ldots, x_n; \theta) = 0,$$

which, using (25–19), becomes

$$\frac{\partial}{\partial\theta} g(t; \theta) = 0,$$

and any solution for θ depends only on the sufficient statistic **t**.

Note that the examples of sufficient statistics in Chapter 25 are functions of maximum likelihood estimators; $\mathbf{x}/n$ was a sufficient statistic for binomial p; it is also the maximum likelihood estimator of p. Also, $\mathbf{x}_1 + \mathbf{x}_2 + \cdots + \mathbf{x}_n$ was a sufficient statistic for the Poisson parameter λ; $\bar{\mathbf{x}}$ is the maximum likelihood estimator of λ. But keep in mind that dependence on a sufficient statistic does not guarantee that the maximum likelihood estimator will always be unbiased or that it will have variance reaching the Cramér-Rao bound. It may or it may not. For example, $\sum_{i=1}^{n} (x_i - \lambda)^2/n$ is an unbiased estimator of σ^2 and has variance equal to the CR bound; it is also the maximum likelihood estimator of σ^2. Also, $\sum_{i=1}^{n} (\mathbf{x}_i - \bar{\mathbf{x}})^2/n$ is the maximum likelihood estimator of σ^2, but is biased and does not have variance reaching the CR bound. In the estimation of the parameter of the Laplace distribution, the maximum likelihood estimator (the sample median) of the parameter has larger variance than another estimator for *all* finite n (see Tukey [189]). But for *large* n, all (normally distributed) maximum likelihood estimators are unbiased, and, as we found in Section 26–5, have asymptotic variance less than or equal to that of any other asymptotically unbiased and asymptotic normal estimator.

Collateral Reading

H. Cramér, *Mathematical Methods of Statistics*. Princeton, Princeton University Press, 1946.

J. Durbin and M. G. Kendall, "The geometry of estimation," *Biometrika*, Vol. 38 (1951), pp. 150–158. A valuable geometrical exposition of the earlier material of this chapter, and of the basic argument of the following chapter.

R. A. Fisher, "On the mathematical foundations of theoretical statistics," *Phil. Trans. Roy. Soc. London*, Series A, Vol. 222 (1922), pp. 309–368.

————, "Theory of statistical estimation," *Proc. Camb. Phil. Soc.*, Vol. 22 (1925), pp. 700–725. The two pioneering papers. These and others by Fisher on maximum likelihood estimation are reprinted in his *Contributions to Mathematical Statistics*. New York, John Wiley and Sons, 1950.

————, *Statistical Theory of Estimation*. Calcutta, University of Calcutta, 1938. A good summary of Fisher's major contributions to maximum likelihood estimation.

D. A. S. Fraser, *Statistics: an Introduction*. New York, John Wiley and Sons, 1958.

R. V. Hogg and A. T. Craig, *Introduction to Mathematical Statistics*. New York, Macmillan Co., 1959.

M. G. Kendall and A. Stuart, *The Advanced Theory of Statistics, Vol. 2*. London, Charles Griffin and Co., 1961. The best reference.

A. M. Mood, *Introduction to the Theory of Statistics*. New York, McGraw-Hill Book Co., 1950.

J. W. Tukey, in "Questions and answers." *The American Statistician*, October, 1949, p. 12.

S. S. Wilks, *Mathematical Statistics*. New York, John Wiley and Sons, 1962. See also the 1943 volume. Also *Lectures on the Theory of Statistical Inference*. Ann Arbor, Edwards Bros., 1937.

Separate References on Special Topics

NORMALITY, FOR LARGE n

D. Dugué, *Traité de Statistique Théorique et Appliquée*. Paris, Masson et Cie, 1958.

CONSISTENCY

R. R. Bahadur, "Examples of inconsistency of maximum likelihood estimates," *Sankhyā*, Vol. 20 (1958), 207–210.

J. L. Doob, "Probability and statistics," *Trans. Amer. Math. Soc.*, Vol. 36 (1934), pp. 759–775.

D. Dugué, *Traité de Statistique Théorique et Appliquée*. Paris, Masson et Cie, 1958.

J. Neyman and E. L. Scott, "Consistent estimates based on partially consistent observations," *Econometrica*, Vol. 16 (1948), pp. 1–32.

A. Wald, "Note on the consistency of the maximum likelihood estimate," *Ann. Math. Stat.*, Vol. 20 (1949), pp. 595–601. The basic proof.

SUFFICIENCY

D. E. Barton, "A class of distributions for which the maximum likelihood estimate is unbiased and of minimum variance for all sample sizes," *Biometrika*, Vol. 43 (1956), pp. 200–202. A valuable method for finding best unbiased sufficient statistics, when the Blackwell-Rao theorem of Chapter 25 is intractable.

R. A. Fisher, "Theory of statistical estimation," *Proc. Camb. Phil. Soc.*, Vol. 22 (1925), pp. 702–725.

ASYMPTOTIC EFFICIENCY

H. CRAMÉR, *Mathematical Methods of Statistics.* Princeton, Princeton University Press, 1950.

RELATIONSHIP TO BAYES' RULE

R. VON MISES, *Notes on Mathematical Theory of Probability and Statistics.* Spec. Pub. No. 1, Harvard University, Graduate School of Engineering, Cambridge, 1946, Chapter IX. Von Mises argues that initial and final probabilities are involved in maximum likelihood, that Fisher implies constant initial probabilities in maximum likelihood. Supported by Jeffreys [93] who writes "accurate *a priori* probabilities are not needed if n is large so maximum likelihood is practically indistinguishable from inverse probability in estimation problems."

PROBLEMS

26–1. Find the maximum likelihood estimator of the parameter of a Poisson distribution. Find the variance of the estimator.

26–2. Find the maximum likelihood estimator of the mean of a normal distribution, variance known.

26–3. Find the maximum likelihood estimator of the parameter θ in $f(x; \theta) = \theta e^{-\theta x}$, $x \geq 0$. Note that $\theta = 0$ is a (trivial) solution of the maximum likelihood equation.

26–4. Show that the expected value of maximum likelihood estimator nA/a of the parameter N (page 256) is ∞ (!).

26–5. Show that the variance of the maximum likelihood estimator of the Cauchy parameter is $2/n$.

CHAPTER 27

BEST LINEAR UNBIASED ESTIMATORS

27–1 The problem and the solution. We continue the discussion of point estimation, turning in this chapter to a certain restricted class of point estimators which have a number of the properties described in Chapter 25 and which have wide utility in the statistical models frequently met in modern experimentation, in modern econometrics, in sampling human populations, and elsewhere.

Again we have a parameter θ. The value of θ is unknown and is to be estimated by a statistic $\mathbf{T}$ which is to depend on and only on the values of n independent* random variables $\mathbf{x}_1, \ldots, \mathbf{x}_n$. In this chapter, these need not be the variables of a random sample from a single population described by $f(x; \theta)$. They may be, more generally, n independent random variables with known and possibly different means $\lambda_1, \ldots, \lambda_n$ and known and possibly different variances $\sigma_1^2, \ldots, \sigma_n^2$. The variate x_2 can be thought of as the outcome of a random sample of size 1 drawn from a population π_2 whose mean is λ_2 and variance σ_2^2.

We want the estimator $\mathbf{T}$ to be unbiased,

$$E(\mathbf{T}) = \theta, \tag{27–1}$$

and to have minimum mean square error,

$$\sigma^2(\mathbf{T}) \text{ minimum.} \tag{27–2}$$

An estimator $\mathbf{T}$ with properties (27–1) and (27–2) is often called "best," though it is well to keep in mind that unbiasedness is not always critically important in an estimator and that minimum variance in an estimator may not always be equivalent to highest concentration of density about the parameter.

The problem is to determine $\mathbf{T}$. The argument which follows has been extended to random variables $\mathbf{x}_1, \ldots, \mathbf{x}_n$ which are not mutually uncor-

---

* Pairwise uncorrelated $\mathbf{x}_1, \ldots, \mathbf{x}_n$ are sufficient. We will need, essentially,

$$E(\mathbf{x}_1 \cdot \cdots \cdot \mathbf{x}_n) = E(\mathbf{x}_1) \cdots E(\mathbf{x}_n),$$

and for this, pairwise uncorrelated $\mathbf{x}_1, \ldots, \mathbf{x}_n$ are sufficient, as shown in the discussion of problem 5–4.

related [1], but this important extension is not considered here. Moreover, and critically, we confine ourselves to *linear** estimators, that is, to estimators which are linear in the random variables $\mathbf{x}_1, \ldots, \mathbf{x}_n$, for it is only in comparison with other estimators within this restricted class that we can always find estimators with properties (27–1) and (27–2). We have

$$\mathbf{T} = c_1\mathbf{x}_1 + \cdots + c_n\mathbf{x}_n,$$

the $\{c_i\}$ to be determined. The quantities (27–1) and (27–2) become

$$E(\mathbf{T}) = c_1\lambda_1 + \cdots + c_n\lambda_n = \theta, \tag{27–3}$$

$$\sigma^2(\mathbf{T}) = c_1^2\sigma_1^2 + \cdots + c_n^2\sigma_n^2 \tag{27–4}$$

to be minimum. We shall call (27–3) and (27–4) the first and second condition, respectively.

Noting (27–3), any solution for $\{c_i\}$ contains the unknown θ. But under certain conditions, θ can be eliminated from the argument. If the means $\{\lambda_i\}$ and the parameter θ are linearly related, or somewhat more generally, if we can express $\{\lambda_i\}$ and θ as linear functions of k unknown parameters $z_1, \ldots, z_k$,

$$\begin{aligned} \lambda_1 &= a_{11}z_1 + \cdots + a_{1k}z_k \\ &\vdots \\ \lambda_n &= a_{n1}a_1 + \cdots + a_{nk}z_k \\ \theta &= b_1z_1 + \cdots + b_kz_k, \end{aligned} \tag{27–5}$$

where $\{a_{ij}\}$ and $\{b_j\}$ are known constants, (27–3) becomes

$$E(\mathbf{T}) = c_1 \sum_{j=1}^{k} a_{ij}z_j + \cdots + c_n \sum_{j=1}^{k} a_{ij}z_j = \sum_{j=1}^{k} b_jz_j,$$

or, grouping terms in z_j,

$$z_1(c_1a_{11} + \cdots + c_na_{n1}) + \cdots + z_k(c_1a_{1k} + \cdots + c_na_{nk}) = z_1b_1 + \cdots + z_kb_k. \tag{27–6}$$

But $\{\mathbf{z}_j\}$ are *arbitrary* parameters and (27–6) must hold for any value of each $\mathbf{z}_j$; for each j, the coefficient of $\mathbf{z}_j$ on the left of (27–6) must equal the

* For best *nonlinear* unbiased estimators, see, for example, P. R. Halmos [82] and P. L. Hsu [92]. It should also be noted that best linear unbiased estimators often are best unbiased estimators.

coefficient of $\mathbf{z}_j$ on the right of (27-6). That is, we must have

$$\begin{aligned} c_1 a_{11} + \cdots + c_n a_{n1} &= b_1 \\ &\vdots \\ c_1 a_{1k} + \cdots + c_n a_{nk} &= b_k, \end{aligned} \tag{27-7}$$

which can often be solved for $c_1, \ldots, c_n$.

Before continuing, we must consider certain necessary relationships between n (the number of random variables $\mathbf{x}_1, \ldots, \mathbf{x}_n$) and k (the number of parameters $z_1, \ldots, z_k$). To this end we will use one elementary result from matrix algebra. First, for $z_1, \ldots, z_k$ to exist, the rank of the matrix A of the coefficients $a_{11}, \ldots, a_{nk}$ and the rank of the augmented matrix B of the coefficients $a_{11}, \ldots, a_{nk}, \lambda_1, \ldots, \lambda_n$

$$A = \left\| \begin{matrix} a_{11} & a_{12} & \ldots & a_{1k} \\ a_{21} & a_{22} & \ldots & a_{2k} \\ \vdots & \vdots & & \vdots \\ a_{n1} & a_{n2} & \ldots & a_{nk} \end{matrix} \right\|, \qquad B = \left\| \begin{matrix} a_{11} & a_{12} & \ldots & a_{1k} & \lambda_1 \\ a_{21} & a_{22} & \ldots & a_{2k} & \lambda_2 \\ \vdots & \vdots & & \vdots & \vdots \\ a_{n1} & a_{n2} & \ldots & a_{nk} & \lambda_n \end{matrix} \right\|$$

must be the same. Second, for $z_1, \ldots, z_k$ to be unique, $k \leq n$; if $k > n$, many values of $z_1, \ldots, z_k$ will satisfy (27-6). Third, the common rank of A and B cannot be less than k, otherwise many values of $z_1, \ldots, z_k$ will satisfy (27-6); so the rank of both matrices A and B is $k \leq n$. Finally the augmented matrix C of (27-7)

$$C = \left\| \begin{matrix} a_{11} & a_{21} & \ldots & a_{n1} & b_1 \\ a_{12} & a_{22} & \ldots & a_{n2} & b_2 \\ \vdots & \vdots & & \vdots & \vdots \\ a_{1k} & a_{2k} & \ldots & a_{nk} & b_k \end{matrix} \right\|$$

must have rank k for solutions $c_1, \ldots, c_n$ to exist. As $k \leq n$, there may be many solutions for $c_1, \ldots, c_n$; we select that solution for which $\sigma^2(\mathbf{T})$ is minimum.

The solution of (27-7) for $c_1, \ldots, c_n$ may be found directly, using elementary methods. As we have k equations, place k unknowns $c_1, \ldots, c_k$ (choosing a nonsingular portion of $||A||$) on the left:

$$\begin{aligned} c_1 a_{11} + \cdots + c_k a_{k1} &= b_1 - c_{k+1} a_{k+1,1} - \cdots - c_n a_{n1} \\ &\vdots \\ c_1 a_{1k} + \cdots + c_k a_{kk} &= b_k - c_{k+1} a_{k+1,k} - \cdots - c_n a_{nk}. \end{aligned} \tag{27-8}$$

Now solve (by Cramer's rule) for $c_1, \ldots, c_k$ in terms of $c_{k+1}, \ldots, c_n$ and substitute these values for $c_1, \ldots, c_k$ into (27–4); $\sigma^2(\mathbf{T})$ then becomes a function of the $n - k$ unknown $c_{k+1}, \ldots, c_n$. To find the minimum of $\sigma^2(\mathbf{T})$, differentiate (27–4) partially with respect to each of these $n - k$ unknowns and set each of the $n - k$ partial derivatives equal to zero. Solve (uniquely) for the $n - k$ unknowns $c_{k+1}, \ldots, c_n$, put these values back into (27–8), and find $c_1, \ldots, c_k$. Thus we can find all required quantities $c_1, \ldots, c_n$.

We shall obtain **T** by two other methods, but the student may wish to carry out the foregoing in detail and see if his solution for **T** agrees with those given later.

27–2 Example. Given that θ is the unknown parameter, $\mathbf{x}_1$, $\mathbf{x}_2$ are two independent random variables with means λ_1, λ_2, variances σ_1^2, σ_2^2. We have

$$\mathbf{T} = c_1\mathbf{x}_1 + c_2\mathbf{x}_2$$

$$E(\mathbf{T}) = c_1\lambda_1 + c_2\lambda_2 = \theta, \qquad \sigma^2(\mathbf{T}) = c_1^2\sigma_1^2 + c_2^2\sigma_2^2$$

to be minimum. We assume

$$\lambda_1 = a_{11}z_1, \qquad \lambda_2 = a_{21}z_1, \qquad \theta = b_1z_1.$$

Note that λ_1 and λ_2 are linearly related to θ. The condition $E(\mathbf{T}) = \theta$ yields

$$c_1a_{11}z_1 + c_2a_{21}z_1 = b_1z_1, \qquad c_1 = \frac{b_1 - c_2a_{21}}{a_{11}}. \tag{27–9}$$

The condition $\sigma^2(\mathbf{T})$ minimum yields

$$\frac{\partial}{\partial c_2}\sigma^2(\mathbf{T}) = \frac{\partial}{\partial c_2}\left[\left(\frac{b_1 - c_2a_{21}}{a_{11}}\right)^2\sigma_1^2 + c_2^2\sigma_2^2\right] = 0,$$

that is,

$$c_2 = \frac{b_1a_{21}\sigma_1^2}{a_{11}^2\sigma_2^2 + a_{21}^2\sigma_1^2},$$

and from (27–9),

$$c_1 = \frac{b_1a_{11}\sigma_2^2}{a_{11}^2\sigma_2^2 + a_{21}^2\sigma_1^2}.$$

Finally,

$$\mathbf{T} = \frac{b_1a_{11}\sigma_2^2\mathbf{x}_1 + b_1a_{21}\sigma_1^2\mathbf{x}_2}{a_{11}^2\sigma_2^2 + a_{21}^2\sigma_1^2}. \tag{27–10}$$

Note the inverse weighting (by variances) of the random variables $\mathbf{x}_1$ and $\mathbf{x}_2$ in the estimator **T**. The example is due to A. T. Craig [33].

Applications, in which practical sources of the constants $\{a_{ij}\}$ and $\{b_j\}$ will be indicated, will appear later in this chapter.

27-3 Undetermined multipliers of Lagrange. In the preceding section we found values of c_i in the estimator $\mathbf{T}$ which minimize $\sigma^2(\mathbf{T})$, with $\mathbf{T}$ subject to the condition $E(\mathbf{T}) = \theta$. The problem of determining maxima or minima of functions, given side conditions on the variables in the functions, is met in many areas of applied mathematics. Therefore we include here a brief example illustrating a systematic and elegant method of solution, due to Lagrange.

The method suggested in Section 27-1 for solving this problem was one of reduction; the number of equations to be solved was reduced from n to $n - k$ by substituting the k conditions on $\mathbf{T}$ into $\sigma^2(\mathbf{T})$, the function to be minimized. Thus $\sigma^2(\mathbf{T})$ became a function of $n - k$ *independent* variables and the ordinary methods were applicable. In the method of Lagrange, the number of equations to be solved is increased from n to $n + k$ by the introduction of new parameters, and the larger set of equations is solved.

For example, to minimize $H(w, x, y, z)$ subject to the side conditions $f(w, x, y, z) = 0$, $g(w, x, y, z) = 0$, with $f \not\equiv g$, we have

$$\begin{aligned} dH &= \frac{\partial H}{\partial w}\,dw + \frac{\partial H}{\partial x}\,dx + \frac{\partial H}{\partial y}\,dy + \frac{\partial H}{\partial z}\,dz, \\ df &= \frac{\partial f}{\partial w}\,dw + \frac{\partial f}{\partial x}\,dx + \frac{\partial f}{\partial y}\,dy + \frac{\partial f}{\partial z}\,dz = 0, \qquad (27\text{-}11) \\ dg &= \frac{\partial g}{\partial w}\,dw + \frac{\partial g}{\partial x}\,dx + \frac{\partial g}{\partial y}\,dy + \frac{\partial g}{\partial z}\,dz = 0. \end{aligned}$$

For the two side conditions to be different, at least one of the six Jacobians

$$\left|\frac{\partial(f, g)}{\partial(w, x)}\right|, \quad \left|\frac{\partial(f, g)}{\partial(w, y)}\right|, \quad \left|\frac{\partial(f, g)}{\partial(w, z)}\right|,$$

$$\left|\frac{\partial(f, g)}{\partial(x, y)}\right|, \quad \left|\frac{\partial(f, g)}{\partial(x, z)}\right|, \quad \left|\frac{\partial(f, g)}{\partial(y, z)}\right|$$

must not vanish. Let the last Jacobian not vanish. Then we can always find two constants (undetermined multipliers) a, b, so that [adding the last two columns in (27-11)]

$$\frac{\partial H}{\partial y}\,dy + a\,\frac{\partial f}{\partial y}\,dy + b\,\frac{\partial g}{\partial y}\,dy = 0, \qquad (27\text{-}12)$$

$$\frac{\partial H}{\partial z}\,dz + a\,\frac{\partial f}{\partial z}\,dz + b\,\frac{\partial g}{\partial z}\,dz = 0, \qquad (27\text{-}13)$$

which gives us *four* equations (the two side conditions and the two equations just produced) in *six* unknowns w, x, y, z, a, b. The final two equations are now produced, from properties not yet used. Multiply df by a, dg by b, and form $dH + adf + bdg$. Adding the first two columns in and remembering that for an extreme value of H to exist, $dH = 0$, we get

$$\left(\frac{\partial H}{\partial w} + a\frac{\partial f}{\partial w} + b\frac{\partial g}{\partial w}\right) dw + \left(\frac{\partial H}{\partial x} + a\frac{\partial f}{\partial x} + b\frac{\partial g}{\partial x}\right) dx = 0.$$

But w and x are *independent* variables, as, therefore, are dw and dx, and *each* parenthesis must be zero. The final two equations are

$$\frac{\partial H}{\partial w} + a\frac{\partial f}{\partial w} + b\frac{\partial g}{\partial w} = 0, \qquad \frac{\partial H}{\partial x} + a\frac{\partial f}{\partial x} + b\frac{\partial g}{\partial x} = 0, \tag{27–14}$$

and we can solve for the six unknowns. The six equations we end with are seen to be those that are obtained by minimizing

$$H = af + bg$$

(four equations) plus the two side conditions (two equations).

Example. Find the minimum value of $H = x^2 + y^2 + z^2$, with condition $x + y + z = 3k$. First, we consider the solution by reduction to two independent variables and two equations:

$$H = x^2 + y^2 + (3k - x - y)^2,$$

$$\frac{\partial H}{\partial x} = 2x - 2(3k - x - y) = 0, \qquad \frac{\partial H}{\partial y} = 2y - 2(3k - x - y) = 0,$$

from which $x = k$, $y = k$, and finally from the condition equation, $z = k$. As the second derivatives are positive, $H = 3k^2$ is the value of minimum H.

In the method of Lagrange, we increase to four variables and four equations. In our example, the four equations are

$$x + y + z = 3k, \qquad 2z + a = 0, \qquad 2x + a = 0, \qquad 2y + a = 0,$$

with the same result.

The general representation of the method of Lagrange is evident. To minimize or maximize

$$H(x_1, x_2, \ldots, x_n), \tag{27–15}$$

where the n variables are subject to m different conditions $(m < n)$

$$\phi_i(x_1, x_2, \ldots, x_n) = 0, \quad i = 1, 2, \ldots, m, \tag{27–16}$$

form

$$\frac{\partial H}{\partial x_i} + a_1 \frac{\partial \phi_1}{\partial x_i} + a_2 \frac{\partial \phi_2}{\partial x_i} + \cdots + a_m \frac{\partial \phi_m}{\partial x_i} = 0, \quad i = 1, 2, \ldots, n. \tag{27–17}$$

There are $n + m$ unknowns (n variables x_i and m multipliers a_i) and $n + m$ equations [n differential equations (27–17), m condition equations (27–16)], and a unique solution may exist.

27–4 The best linear unbiased estimator, using the method of multipliers. Returning to the general problem of Section 25–1, we want to determine $c_1, \ldots, c_n$ in (27–3),

$$E(\mathbf{T}) = \sum_{i=1}^{n} c_i \lambda_i = \theta,$$

with (27–4):

$$\sigma^2(\mathbf{T}) = \sum_{i=1}^{n} c_i^2 \sigma_i^2$$

minimum. Condition (27–3) was found to be equivalent to the k side conditions,

$$\sum_{i=1}^{n} c_i a_{ij} - b_j = 0, \quad j = 1, 2, \ldots, k. \tag{27–18}$$

Using Lagrange multipliers, we form

$$\sum_{i=1}^{n} c_i^2 \sigma_i^2 + e_1 \left[\sum_{i=1}^{n} c_i a_{i1} - b_1 \right] + \cdots + e_k \left[\sum_{i=1}^{n} c_i a_{ik} - b_k \right].$$

Differentiating partially with respect to c_i, we have

$$2 c_i \sigma_i^2 + e_1 a_{i1} + \cdots + e_k a_{ik} = 0, \quad i = 1, 2, \ldots, n,$$

$$c_i = -\frac{1}{2\sigma_i^2} \sum_{j=1}^{k} e_j a_{ij}, \quad i = 1, 2, \ldots, n,$$

which, placed in (27–18), yield

$$-\frac{1}{2} \sum_{i=1}^{n} \left[\frac{1}{\sigma_i^2} \sum_{j=1}^{k} e_j a_{ij} \right] a_{ij} = b_j, \quad j = 1, 2, \ldots, k,$$

$$e_1 \sum_{i=1}^{n} -\frac{a_{i1} a_{ij}}{2\sigma_i^2} + \cdots + e_k \sum_{i=1}^{n} -\frac{a_{ik} a_{ij}}{2\sigma_i^2} = b_j, \quad j = 1, 2, \ldots, k, \tag{27–19}$$

k equations in k unknown multipliers, $e_1, \ldots, e_k$. The solution is unique if $\Delta \neq 0$, where

$$\Delta = \begin{vmatrix} \sum_{i=1}^{n} \frac{a_{i1}^2}{-2\sigma_i^2} & \cdots & \sum_{i=1}^{n} \frac{a_{ik}a_{i1}}{-2\sigma_i^2} \\ \vdots & & \vdots \\ \sum_{i=1}^{n} \frac{a_{i1}a_{ik}}{-2\sigma_i^2} & \cdots & \sum_{i=1}^{n} \frac{a_{ik}^2}{-2\sigma_i^2} \end{vmatrix}.$$

But since we assumed that $\lambda_1, \ldots, \lambda_n$ could be expressed in terms of $\{z_i\}$, we must have

$$\begin{vmatrix} a_{11} & \cdots & a_{n1} \\ \vdots & & \vdots \\ a_{1k} & \cdots & a_{nk} \end{vmatrix} \neq 0, \qquad \text{and since } \sigma_i > 0, \qquad \begin{vmatrix} \frac{a_{11}}{\sigma_1} & \cdots & \frac{a_{n1}}{\sigma_n} \\ \vdots & & \vdots \\ \frac{a_{1k}}{\sigma_1} & \cdots & \frac{a_{nk}}{\sigma_n} \end{vmatrix} \neq 0,$$

and since each element of Δ is the sum of squares of elements in the right-hand determinant directly above, Δ cannot be zero.

Writing out (27–19), we obtain

$$e_1 \sum_{i=1}^{n} \frac{a_{i1}^2}{-2\sigma_i^2} + \cdots + e_k \sum_{i=1}^{n} \frac{a_{ik}a_{i1}}{-2\sigma_i^2} = b_1,$$

$$\vdots$$

$$e_1 \sum_{i=1}^{n} \frac{a_{i1}a_{ik}}{-2\sigma_i^2} + \cdots + e_k \sum_{i=1}^{n} \frac{a_{ik}^2}{-2\sigma_i^2} = b_k,$$

and

$$e_m = \frac{\sum_{j=1}^{k} b_j \, \Delta_{jm}}{\Delta}, \quad m = 1, 2, \ldots, k,$$

where Δ_{jm} is the minor of the jmth element of Δ. Finally,

$$c_i = -\frac{1}{2\sigma_i^2} \sum_{m=1}^{k} e_m a_{im} = -\frac{1}{2\sigma_i^2} \sum_{m=1}^{k} \sum_{j=1}^{k} \frac{a_{im} b_j \, \Delta_{jm}}{\Delta}, \quad i = 1, 2, \ldots, n,$$

and

$$T = \sum_{i=1}^{n} c_i x_i = -\frac{1}{2\Delta} \sum_{i=1}^{n} \frac{x_i}{\sigma_i^2} \sum_{m=1}^{k} a_{im} \sum_{j=1}^{k} b_j \, \Delta_{jm}, \qquad (27\text{–}20)$$

a ponderous expression but special cases of which have, as we shall see, impressive possibilities in application.

27–5 The Gauss-Markov solution by least squares. Gauss and later Markov* [121] were able to find the best unbiased estimator $\mathbf{T}$ by a different (and simpler) argument. Write

$$G = \sum_{i=1}^{n} \left(\frac{\mathbf{x}_i - \lambda_i}{\sigma_i}\right)^2; \tag{27–21}$$

the standardized sum of squares of the random variables $\mathbf{x}_1, \ldots, \mathbf{x}_n$,

$$= \sum_{i=1}^{n} \left(\frac{x_i - a_{i1}z_1 \cdots - a_{ik}z_k}{\sigma_i}\right)^2.$$

Minimize G; form the k conditions

$$\frac{\partial G}{\partial z_1} = \sum_{i=1}^{n} \left(\frac{x_i - a_{i1}z_1 - \cdots - a_{ik}z_k}{\sigma_i}\right)\frac{a_{i1}}{\sigma_i} = 0,$$

$$\vdots$$

$$\frac{\partial G}{\partial z_k} = \sum_{i=1}^{n} \left(\frac{x_i - a_{i1}z_1 - \cdots - a_{ik}z_k}{\sigma_i}\right)\frac{a_{ik}}{\sigma_i} = 0,$$

which may be written in the form

$$\sum_{i=1}^{n} \frac{a_{i1}x_i}{\sigma_i^2} = z_1 \sum_{i=1}^{n} \frac{a_{i1}^2}{\sigma_i^2} + \cdots + z_k \sum_{i=1}^{n} \frac{a_{ik}a_{i1}}{\sigma_i^2},$$

$$\vdots$$

$$\sum_{i=1}^{n} \frac{a_{ik}x_k}{\sigma_i^2} = z_1 \sum_{i=1}^{n} \frac{a_{i1}a_{ik}}{\sigma_i^2} + \cdots + z_k \sum_{i=1}^{n} \frac{a_{ik}^2}{\sigma_i^2}.$$

We have k equations in k parameters z_i. Solving for z_t, we have

$$z_t = \sum_{j=1}^{k} \sum_{i=1}^{n} \frac{a_{ij}x_i}{\sigma_i^2} D_{jt} \Big/ D,$$

where D_{jt} is the minor of the $j$tth element of D, and $D = -2\,\Delta$; this "least-squares" solution for z_t placed in

$$\theta = \sum_{i=1}^{n} b_i z_i$$

yields T of (27–20).

---

* See, however, the historical note by Plackett [159]; this note also contains an elegant proof of the Gauss-Markov theorem.

What has been shown is this. The least-squares principle (27–21) (due to Legendre, 1805 and Gauss, 1809), that is, the set of values $\{z_t\}$ which minimize G, yield, as Markov explicitly showed, the best linear unbiased estimator of θ. This is perhaps the strongest justification we have for the extensive use of the method of least squares in statistical inference, better, probably, than the classical justification which required that $\{\mathbf{x}_i\}$ be normally distributed. As we shall see in this chapter and in Chapter 30, least squares has the further advantage of being operationally simple and open to great systematization in tests of linear hypotheses.

We apply the Gauss-Markov argument to the example of Section 27–2:

$$G = \sum_{i=1}^{2} \left(\frac{x_i - \lambda_i}{\sigma_i}\right)^2 = \sum_{i=1}^{2} \left(\frac{x_i - a_{i1}z_1}{\sigma_i}\right)^2.$$

From $\partial G/\partial z_1 = 0$, we have

$$\sum_{i=1}^{2} \frac{a_{i1}x_i}{\sigma_i^2} = z_1 \sum_{i=1}^{2} \frac{a_{i1}^2}{\sigma_i^2},$$

from which

$$z_1 = \frac{a_{11}\sigma_2^2 x_1 + a_{21}\sigma_1^2 x_2}{a_{11}^2\sigma_2^2 + a_{21}^2\sigma_1^2},$$

which, placed in $\theta = b_1 z_1$, yields (27–10).

27–6 Best linear unbiased estimators and stratified random sampling.* First we review certain results from Chapter 22. In Section 22–2 we considered a population π described by a random variable $\mathbf{x}$ with mean λ and variance σ^2. This population was partitioned or stratified into subpopulations $\pi_1, \ldots, \pi_k$ with means $\lambda_1, \ldots, \lambda_k$ and variances $\sigma_1^2, \ldots, \sigma_k^2$, respectively. We wrote prob $[\mathbf{x} \in \pi_i] = p_i$, and we found, repeating (21–12) and (21–13),

$$\lambda = \sum_{i=1}^{k} p_i\lambda_i, \qquad \sigma^2 = \sum_{i=1}^{k} p_i[\sigma_i^2 + (\lambda_i - \lambda)^2],$$

relationships between the parameters of the population π and parameters of the subpopulations $\pi_1, \ldots, \pi_k$.

We wrote $\bar{\mathbf{x}}_s$ for the mean of a *stratified* random sample (a random sample of size n_1 from $\pi_1, \ldots,$ a random sample of size n_k from π_k, with $\sum_{i=1}^{k} n_i = n$),

$$\bar{\mathbf{x}}_s = \frac{n_1\bar{\mathbf{x}}_1 + \cdots + n_k\bar{\mathbf{x}}_k}{n}, \tag{27–22}$$

---

* Section 27–6 may be omitted.

and we found

$$E(\bar{\mathbf{x}}_s) = \sum_{i=1}^{k} \frac{n_i}{n} \lambda_i, \qquad \sigma^2(\bar{\mathbf{x}}_s) = \frac{1}{n^2} \sum_{i=1}^{k} n_i \sigma_i^2,$$

relationships between the statistic $\bar{\mathbf{x}}_s$ and parameters of the subpopulations. No discussion of the usefulness of $\bar{\mathbf{x}}_s$ was pursued in Chapter 22; we merely determined by deductive mathematical argument the mean and variance of a particular statistic $\bar{\mathbf{x}}_s$. If now, with statistical inference in mind, we impose on $\bar{\mathbf{x}}_s$ the first (desirable) condition

$$E(\bar{\mathbf{x}}_s) = \lambda, \tag{27-23}$$

we have at once

$$\sum_{i=1}^{k} \frac{n_i}{n} \lambda_i = \sum_{i=1}^{k} p_i \lambda_i. \tag{27-24}$$

Since (27-24) should hold whatever the values of λ_i, we must have

$$p_i = n_i/n;$$

the quantities p_i associated with π_i are thus set equal to the proportions of the total sample coming from π_i. Such a stratified sample could be called a *proportional* sample. For this case we write $\bar{\mathbf{x}}_s = \bar{\mathbf{x}}_p$, and for the mean of a proportional sample, we find

$$\sigma^2(\bar{\mathbf{x}}_p) = \frac{1}{n^2} \sum_{i=1}^{k} n_i \sigma_i^2 = \frac{n}{n^2} \sum_{i=1}^{k} p_i \sigma_i^2 = \frac{1}{n} [\sigma^2 - (\lambda_i - \lambda)^2]$$

$$= \frac{\sigma^2}{n} - \frac{\sum_{i=1}^{k} p_i (\lambda_i - \lambda)^2}{n}. \tag{27-25}$$

Since $\sigma^2(\bar{\mathbf{x}}) = \sigma^2/n$ is the variance of the mean of a *random* sample of size n drawn from π, we have

$$\sigma^2(\bar{\mathbf{x}}_p) \le \sigma^2(\bar{\mathbf{x}}); \tag{27-26}$$

the mean of a proportional sample never has larger variance than the mean of a random sample. This result has important implications for efficient sampling of real populations.

Now let us go further. Similar to stratified sampling in which the statistic is (27-22), let us estimate the mean λ of π by a weighted mean of the means $\bar{x}_1, \ldots, \bar{x}_k$ of random samples from the partitions or strata or subpopulations $\pi_1, \ldots, \pi_k$, but this time, and unlike (27-22), with *weights* c_i not involving n_i and with $\sum_{i=1}^{k} c_i = 1$. We call this a "Markov sample" and write its mean $\bar{x}_M$,

$$\mathbf{T} = \bar{\mathbf{x}}_M = c_1 \bar{\mathbf{x}}_1 + \cdots + c_k \bar{\mathbf{x}}_k. \tag{27-27}$$

The first condition $E(\bar{\mathbf{x}}_M) = \theta$ applied to (27–27) yields

$$\sum_{i=1}^{k} c_i\lambda_i = \sum_{i=1}^{k} p_i\lambda_i,$$

and since this should hold whatever the values of λ_i, we must have $c_i = p_i$. The weights c_i in our estimator (27–27) are now fixed, but the $n_1, \ldots, n_k$ implicit in $\bar{x}_1, \ldots, \bar{x}_k$ of (27–27) are not yet fixed. We settle them by requiring the second condition to hold:

$$\sigma^2(\mathbf{T}) = \sigma^2(\bar{\mathbf{x}}_M) = \sum_{i=1}^{k} c_i^2\sigma^2(\bar{\mathbf{x}}_i) = \sum_{i=1}^{k} c_i^2 \frac{\sigma_i^2}{n_i}, \tag{27–28}$$

which, with $c_i = p_i$, is to be minimum. The values of n_i which, subject to the side condition $\sum_{i=1}^{k} n_i = n$, minimize (27–28), may be obtained using the method of undetermined multipliers (regarding n_i as continuous). Forming

$$G = \sigma^2(\bar{\mathbf{x}}_M) + a\left(\sum_{i=1}^{k} n_i - n\right),$$

we have

$$\frac{\partial G}{\partial n_i} = \frac{\partial}{\partial n_i}\sigma^2(\bar{\mathbf{x}}_M) + \frac{\partial}{\partial n_i} a\left(\sum_{i=1}^{k} n_i - n\right) = -\frac{p_i^2\sigma_i^2}{n_i^2} + a = 0,$$

$$\frac{\partial G}{\partial a} = \sum_{i=1}^{k} n_i - n = 0,$$

from which

$$n_i^2 = \frac{p_i^2\sigma_i^2}{a}, \qquad n_i = \frac{p_i\sigma_i}{\sqrt{a}},$$

neglecting the negative root. Also

$$\sum_{i=1}^{k} n_i = \frac{1}{\sqrt{a}}\sum_{i=1}^{k} p_i\sigma_i = n,$$

from which

$$\sqrt{a} = \frac{\sum_{i=1}^{k} p_i\sigma_i}{n},$$

and finally,

$$n_i = \frac{np_i\sigma_i}{\sum_{i=1}^{k} p_i\sigma_i}. \tag{27–29}$$

The sizes n_i of the random samples from the subpopulations π_i should be proportional to the product of the probabilities p_i and the standard deviations σ_i in the subpopulations.

Putting (27–29) into (27–28), we have

$$\sigma^2(\bar{\mathbf{x}}_M) = \frac{1}{n}\left(\sum_{i=1}^{k} p_i\sigma_i\right)^2, \tag{27–30}$$

and repeating (27–25) for the mean $\bar{\mathbf{x}}_p$ of a proportional sample, we have

$$\sigma^2(\bar{\mathbf{x}}_p) = \frac{1}{n}\sum_{i=1}^{k} p_i\sigma_i^2. \tag{27–31}$$

Is (27–30) smaller than (27–31)? Recall the Schwarz inequality of problem 5–3,

$$\left(\sum_{i=1}^{k} x_iy_i\right)^2 \leq \sum_{i=1}^{k} x_i^2 \sum_{i=1}^{k} y_i^2,$$

and write

$$x_i = \sqrt{p_i}, \qquad y_i = \sqrt{p_i}\,\sigma_i.$$

We have at once

$$\sigma^2(\bar{\mathbf{x}}_M) \leq \sigma^2(\bar{\mathbf{x}}_p), \tag{27–32}$$

the equality holding only when $\sigma_1 = \cdots = \sigma_k$.

We close the application to sampling by searching for the best unbiased estimator of a parameter θ which is a linear sum, such as income in the United States. We use the language of this area of application:

$$\theta = N_1\lambda_1 + \cdots + N_k\lambda_k,$$

where $N_1, \ldots, N_k$ are the number of income receivers in the subpopulations (say, the $k = 50$ states) $\pi_1, \ldots, \pi_k$. We consider an estimator $\mathbf{T}$ of θ which permits weighting of individual incomes:

$$\begin{aligned}\mathbf{T} &= \underbrace{c_{11}x_{11} + \cdots + c_{1n_1}x_{1n_1}}_{\substack{\text{from}\\ \text{subpopulation 1}}} + \cdots + \underbrace{c_{k1}x_{k1} + \cdots + c_{kn_k}x_{kn_k}}_{\substack{\text{from}\\ \text{subpopulation } k}} \\ &= \sum_{i=1}^{k}\sum_{j=1}^{n_i} c_{ij}x_{ij}. \end{aligned} \tag{27–33}$$

We require $E(\mathbf{T}) = \theta$ and $\sigma^2(\mathbf{T})$ minimum. The first condition yields

$$E\sum_{i=1}^{k}\sum_{j=1}^{n_i} c_{ij}x_{ij} = \sum_{i=1}^{k} N_i\lambda_i, \qquad \text{or} \qquad E\sum_{i=1}^{k}\left(\sum_{j=1}^{n_i} c_{ij}x_{ij} - N_i\lambda_i\right) = 0.$$

We find

$$\sum_{i=1}^{k} E\left(\sum_{j=1}^{n_i} c_{ij}x_{ij} - N_i\lambda_i\right) = 0, \qquad \sum_{i=1}^{k}\left(\sum_{j=1}^{n_i} c_{ij}\lambda_i - N_i\lambda_i\right) = 0,$$

or

$$\sum_{i=1}^{k} \lambda_i\left(\sum_{j=1}^{n_i} c_{ij} - N_i\right) = 0. \tag{27–34}$$

Since (27–34) must hold for all λ_i, we have

$$\sum_{j=1}^{n_i} c_{ij} = N_i, \quad i = 1, \ldots, k,$$

which is satisfied by many weights c_{ij}. The sum of the still undetermined weights c_{ij} over all income receivers in subpopulation i must equal the number of income receivers in subpopulation i.

The second condition yields

$$\begin{aligned}
\sigma^2(\mathbf{T}) &= E(\mathbf{T} - \theta)^2 = E\left[\sum_{i=1}^{k}\sum_{j=1}^{n_i} c_{ij}x_{ij} - \sum_{i=1}^{k} N_i\lambda_i\right]^2 \\
&= E\sum_{i=1}^{k}\left[\sum_{j=1}^{n_i} c_{ij}x_{ij} - N_i\lambda_i\right]^2 \\
&= E\sum_{i=1}^{k}\left[\left(\sum_{j=1}^{n_i} c_{ij}x_{ij}\right)^2 - 2N_i\lambda_i\sum_{j=1}^{n_i} c_{ij}x_{ij} + N_i^2\lambda_i^2\right] \\
&= \sum_{i=1}^{k} E\left(\sum_{j=1}^{n_i} c_{ij}x_{ij}\right)^2 - 2N_i\lambda_i\sum_{j=1}^{n_i} c_{ij}\lambda_i + N_i^2\lambda_i^2.
\end{aligned} \tag{27–35}$$

If from each subpopulation we draw *without* replacement (and this is the usual practice), (27–35) finally reduces to

$$= \sum_{i=1}^{k} \sigma_i^2\left[n_i\frac{N_i - n_i}{N_i - 1}c_i^2 + \frac{N_i}{N_i - 1}\sum_{j=1}^{n_i}(c_{ij} - c_i)^2\right],$$

where c_i is the mean of c_{ij} summed over j.

Without need for Lagrange multipliers, $\sigma^2(\mathbf{T})$ is seen to be minimum when $c_{ij} = c_i$, that is, when

$$c_{ij} = c_i = \frac{1}{n_i}\sum_{j=1}^{n_i} c_{ij} = \frac{N_i}{n_i},$$

which satisfies the first condition. We get

$$\mathbf{T} = \sum_{i=1}^{k} \frac{N_i}{n_i} \sum_{j=1}^{n_i} x_{ij} = \sum_{i=1}^{k} \frac{N_i}{n_i} n_i \bar{x}_i = \sum_{i=1}^{k} N_i \bar{x}_i,$$

$$\sigma^2(\mathbf{T}) = \sum_{i=1}^{k} \sigma_i^2 \cdot n_i \cdot \frac{N_i - n_i}{N_i - 1} c_i^2 = \sum_{i=1}^{k} \sigma_i^2 \left(\frac{N_i - n_i}{N_i - 1} \frac{N_i^2}{n_i} \right). \tag{27–36}$$

Note that all weights, of $\bar{x}_i$ in $\mathbf{T}$ and of σ_i^2 in $\sigma^2(\mathbf{T})$, are constant within each subpopulation π_i, and that as $n_i \to N_i$, $\sigma^2(\mathbf{T}) \to 0$. Note also that while a minimum of $\sigma^2(\mathbf{T})$ has been found, $\sigma^2(\mathbf{T})$ is a function of n_i which still remains unfixed. We now determine the n_i so as to reach the minimum of $\sigma^2(\mathbf{T})$ with respect to n_i. To this end Neyman has suggested writing (27–36) in the following illuminating form, which permits us to view the minimization of $\sigma^2(\mathbf{T})$ in two stages. If we write

$$N = \sum_{i=1}^{k} N_i, \qquad n = \sum_{i=1}^{k} n_i,$$

and introduce the symbol

$$S_i^2 = \frac{N_i \sigma_i^2}{N_i - 1},$$

(27–36) becomes

$$\sigma^2(\mathbf{T}) = \frac{N-n}{n} \sum_{i=1}^{k} N_i S_i^2 + \sum_{i=1}^{k} n_i \left[\frac{N_i S_i}{n_i} - \frac{\sum_{i=1}^{k} N_i S_i}{n} \right]^2$$
$$- \frac{N}{n} \sum_{i=1}^{k} N_i \left[S_i - \frac{\sum_{i=1}^{k} N_i S_i}{N} \right]^2, \tag{27–37}$$

where the middle term is the only term depending on n_i. If in this term, n_i are chosen proportional to N_i, that is, if

$$\frac{n_i}{n} = \frac{N_i}{N},$$

the last two terms of (27–37) cancel each other as follows:

$$\sum_{i=1}^{k} \frac{N_i}{N} n \left[\frac{S_i^2 N^2}{n^2} - \frac{2 S_i N}{n^2} \sum_{i=1}^{k} N_i S_i + \frac{(\sum_{i=1}^{k} N_i S_i)^2}{n^2} \right]$$
$$- \frac{N}{n} \sum_{i=1}^{k} \left[N_i \left(S_i^2 - \frac{2 S_i}{N} \right) \sum_{i=1}^{k} N_i S_i + \frac{(\sum_{i=1}^{k} N_i S_i)^2}{N^2} \right].$$

When we expand to six separate terms, complete cancellation becomes obvious. We have

$$\mathbf{T} = \sum_{i=1}^{k} N_i\bar{x}_i, \qquad \text{with} \qquad n_i = \frac{N_i}{N}\,n, \tag{27–38}$$

and

$$\sigma^2(\mathbf{T}) = \frac{N-n}{n}\sum_{i=1}^{k} N_iS_i^2 = \frac{N-n}{n}\sum_{i=1}^{k}\frac{N_i^2\sigma_i^2}{N_i-1}. \tag{27–39}$$

But a superior minimum, and a useful result in practical sampling, can be found if only the middle term of (27–33) vanishes, since the final (nonnegative) term is preceded by a negative sign and reduces the variance as it stands. This clearly requires

$$\frac{N_iS_i}{n_i} = \frac{\sum_{i=1}^{k} N_iS_i}{n} \qquad \text{or} \qquad n_i = \frac{nN_iS_i}{\sum_{i=1}^{k} N_iS_i}, \tag{27–40}$$

with $\sum_{i=1}^{k} n_i = n$; this result could, of course, have been directly obtained from (27–36) by the use of Lagrange multipliers, with n_i regarded as continuous. For minimum variance of the estimator, the optimal sample sizes should be proportional to the product of the subpopulation sizes and the subpopulation standard deviations. As a practical procedure, inasmuch as σ_i are often unknown, one may first select n_i proportional to N_i as in (27–38), then adjust n_i on obtaining sample standard deviations, for the latter are estimates of σ_i. Detailed theory and technique along such lines have been developed and can be found in Deming [45].

Much of the systematic application of the Gauss-Markov theorem to stratified random sampling is due to Neyman [135], [139]. Many results were known earlier, but Neyman unified this area of application.

27–7 Best linear unbiased estimators and linear regression. Consider the relationship of the content of this chapter to linear regression. Let $\mathbf{x}$ be a random variable with expectation defined at each value of t by the linear regression function

$$E(\mathbf{x}|t) = \alpha + \beta t. \tag{27–41}$$

The correspondence of (27–41) to the general model

$$\theta = b_1z_1 + \cdots + b_kz_k$$

is immediate; we put $\theta = E(\mathbf{x}|t)$, $k = 2$, $b_1 = 1$, $b_2 = t$, $z_1 = \alpha$, $z_2 = \beta$.

At each value t_i of t, the mean value of $\mathbf{x}$ is

$$E(\mathbf{x}_i) = \alpha + \beta t_i, \tag{27–42}$$

so the correspondence of (27–42) to the general model

$$\lambda_i = a_{i1}z_1 + \cdots + a_{ik}z_k, \quad i = 1, \ldots, n$$

is also immediate; we put $E(\mathbf{x}_i) = \lambda_i$, $k = 2$, $a_{i1} = 1$, $a_{i2} = t_i$, $z_1 = \alpha$, $z_2 = \beta$. Placing these values in (27–20), we find the best unbiased estimator of the linear regression function $\alpha + \beta t$. In particular, for

$$\sigma_1 = \sigma_2 = \cdots = \sigma_n$$

(sometimes described as homoscedasticity), (27–20) becomes, after some algebraic reduction,

$$T = \bar{x} + r_{xt}\frac{s_x}{s_t}(t - \bar{t}), \tag{27–43}$$

the best linear least-squares estimator of the linear regression function $\alpha + \beta t$. Thus, the conventional least-squares operational technique on observed data $(x_1, t_1), \ldots, (x_n, t_n)$, yields the best linear unbiased estimator of $\alpha + \beta t$. As already noted, this is probably the strongest justification for the extensive use of least squares in regression analysis and in analysis of the models underlying modern experimental design. No assumptions of normality are involved.

By similar, though lengthier, demonstrations which the student might undertake, the whole least-squares apparatus for estimating multiple linear regression functions of the form (7–11) is obtained. No assumptions of normality are involved here. But, if in addition to finding best unbiased estimators of parameters, we want to find limits within which the parameters probably lie, then, as we shall see in Chapter 30, further assumptions will be needed.

27–8 Least squares and maximum likelihood. The relationship between the two systems of point estimation, maximum likelihood and least squares, is of interest. Consider a random variable $\mathbf{x}$ with unknown mean λ and known variance σ^2. If $\mathbf{x}$ is *normally* distributed, the likelihood L employed in maximum likelihood estimation of λ is

$$L = -\frac{n}{2}\log 2\pi - \frac{n}{2}\log \sigma^2 - \frac{1}{2\sigma^2}\sum_{i=1}^{n}(x_i - \lambda)^2,$$

but the necessary condition $dL/d\lambda = 0$ for maximum L is evidently identical with the necessary condition

$$\frac{d}{d\lambda}\sum_{i=1}^{n}(x_i - \lambda)^2 = 0$$

for least squares estimation of λ.

Collateral Reading

A. C. Aitken, "On least squares and linear combination of observations," *Proc. Roy. Soc. Edinburgh,* Vol. 55 (1935), pp. 42–48.

H. D. Brunk, *An Introduction to Mathematical Statistics.* Boston, Ginn and Co., 1960.

A. T. Craig, "On the mathematics of the representative method of sampling," *Ann. Math. Stat.*, Vol. 10 (1939), pp. 26–34.

F. N. David, *Probability Theory for Statistical Methods.* Chapters XIII and XIV. Cambridge (Eng.), Cambridge University Press, 1949. Particularly good for application to regression.

F. N. David and J. Neyman, "Extension of the Markoff theorem on least squares," *Statistical Research Memoirs,* Vol. II (1938), pp. 105–116.

W. E. Deming, *Some Theory of Sampling.* New York, John Wiley and Sons, 1950.

P. R. Halmos, "The theory of unbiased estimation," *Ann. Math. Stat.*, Vol. 17 (1946), pp. 34–43.

P. L. Hsu, "On the best unbiassed quadratic estimate of the variance," *Statistical Research Memoirs,* Vol. II (1938), pp. 91–104.

A. A. Markov, *Wahrscheinlichkeitsrechnung.* Teubner, Leipzig and Berlin, 1912.

J. Neyman, "On two different aspects of the representative method: the method of stratified sampling and the method of purposive selection," *Jour. Roy. Stat. Soc.*, Vol. 97 (1934), pp. 558–625.

————, "Contribution to the theory of sampling human populations," *Jour. Amer. Stat. Assoc.*, Vol. 33 (1938), pp. 101–116.

————, *Lectures and Conferences on Mathematical Statistics.* Washington, Graduate School of the U.S. Dept. of Agriculture, 2nd ed., revised, 1952.

R. L. Plackett, "A historical note on the method of least squares," *Biometrika,* Vol. 36 (1949), pp. 458–460.

S. S. Wilks, *Mathematical Statistics.* New York, John Wiley and Sons, 1962. Also the earlier volume of the same title, Princeton, Princeton University Press, 1943.

CHAPTER 28

TESTING STATISTICAL HYPOTHESES

28–1 Basic concepts of the Neyman-Pearson theory. Alternative to estimating the values of one or more parameters of a probability distribution, as was the objective in Chapters 25 through 27, we may test hypotheses on the values of such parameters. Both estimation and testing hypotheses may be profitably viewed as different aspects of the more general problem of reaching decisions on the basis of data, that is, of the general problem of statistical inference, though we do not take this broad view here. Considered separately, there is some question as to which of these two closely related branches of statistical inference is of greater practical import. Surely in some areas of pure and applied science, the testing of statistical hypotheses is of interest. Let us first fix some of the statistical ideas involved in such tests, and then consider a few of the broader problems as well as the details of some of the tests themselves.

Some of the ideas in the theory of testing statistical hypotheses are to be found in the work of R. A. Fisher [65], but explicit formulation as well as important basic concepts themselves are due to J. Neyman and E. S. Pearson [143]; the latter are surely the pioneers in this area of statistical inference.

Let **x** be a random variable with density function $f(x; \theta)$. In the class of problems most thoroughly investigated to date, the form f of the density function is (as generally in estimation) assumed to be known, with the value of the constant parameter θ unknown. Assumption of knowledge of the form f places, of course, a severe limitation on the practicality of any results we may reach. More general hypotheses in which the form of the probability distribution is also unknown is the subject matter of nonparametric statistics (see Fraser [69], Siegel [178]); such hypotheses are not considered here.

For the moment, consider two single-valued hypotheses H_0 and H_1 on the true value of the only parameter θ:

$H_0 : \theta = \theta_0$, sometimes called the *null* hypothesis,

$H_1 : \theta = \theta_1$, sometimes called the *alternative* hypothesis.

While acknowledging that the true value of θ may be different from either, we agree to decide between H_0 and H_1 after examining the outcome $E : x_1, x_2, \ldots, x_n$ of a random sample of fixed size n on the variable

x. In so deciding, we may reject H_0 when H_0 is true; such an error is called a Type I error. Or we may accept H_0 when H_0 is false (when H_1 is true); such an error is called a Type II error. Write

$$\alpha = \text{probability of a Type I error,}$$

$$\beta = \text{probability of a Type II error.}$$

The consequences of this decision-making process are shown in Table 28–1.

TABLE 28–1

CONSEQUENCES OF A TWO-DECISION PROCESS

Truth \ Decision	H_0	H_1
H_0	Correct	Type I error
H_1	Type II error	Correct

The decision-making process itself is the following: the random sample data E constitute a point in an n-dimensional Euclidean space. We separate this space into two regions. Let R be a region, sometimes called the *critical region* in this space such that if E falls in R, we reject H_0 (equivalent here to accepting H_1), while if E falls outside R, we do not reject (we accept) H_0. Prescribing a region R is thus equivalent to prescribing a test of a hypothesis. A sample point E in a critical region R is illustrated in Fig. 28–1. We have

$$p(E \in R|\theta_0) = \alpha, \tag{28–1}$$

where α (preferably small) may be described as the probability of rejecting

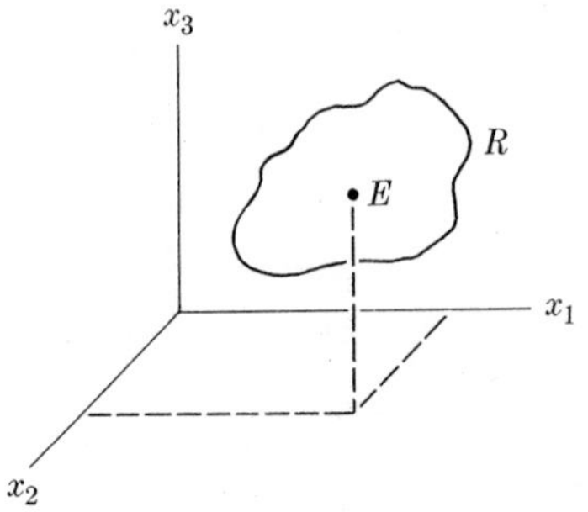

FIG. 28–1. A sample point E in a critical region R.

$H_0: \theta = \theta_0$ when true or the probability of an error of the first kind or the *size* of the region R, and

$$p(E \in R|\theta_1) = 1 - \beta, \tag{28-2}$$

where $1 - \beta$ (β preferably small) is called the probability of rejecting $H_0: \theta = \theta_0$ when $H_1: \theta = \theta_1$ is true or the probability of *not* making an error of the second kind or the *power* of the region R with respect to $\theta = \theta_1$.

Each region R has a power function $p(E \in R|\theta)$ (the probability of rejecting H_0 when the parameter has value θ) with respect to all possible values of θ, not merely with respect to $\theta = \theta_1$. The power function is valuable, for within the framework of the model in which it is defined [that a random sample E has been drawn from a population described by $f(x; \theta)$] the power function completely describes the performance characteristics of a test. A typical power function is shown in Fig. 28–2.

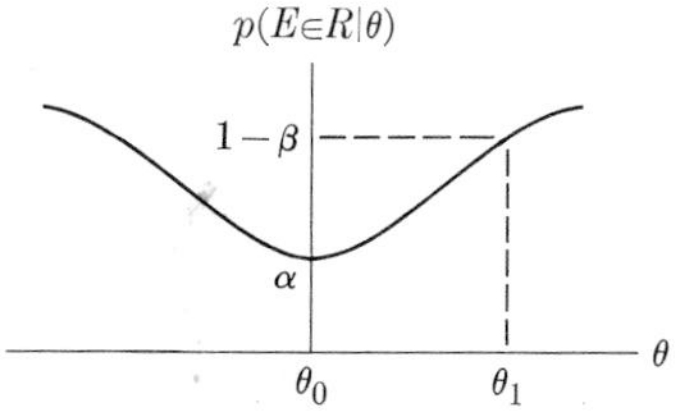

FIG. 28–2. Power function of a region R.

The power function suggests a way to choose the region R and therefore a way to choose a test: of all regions, choose that one which *minimizes* some function of α and β. Neyman and Pearson (who introduced this critical concept) have suggested that, restricting ourselves to regions of fixed size α, we should when possible choose that region which has greatest power with respect to *all* possible alternative hypotheses on θ. If such a region exists (seldom indeed) the test of the hypothesis H_0 using that region is called *uniformly most powerful.*

We note here that in certain discussions, in particular, in the field of sampling inspection, the expression "operating characteristic" (OC or OC curve) of a region or test is often met; it is the probability of accepting θ_0 when θ is true. Clearly for each possible value of θ, OC $= 1 -$ power.

28–2 Various problems in testing statistical hypotheses. First, we make a few brief remarks on risks. The proper specification of α and β, or more generally of the power function, is easier in such areas as sampling inspection where the cost of errors (rejecting good product and accepting bad product) is more likely to be measurable. In physical, biological, and social science, the difficulty of assessing the costs of wrong decisions is great (consider estimating the social consequences of errors of inference) and specification of risks has often been at best a murky enterprise. It may also be noted that while no explicit *a priori* concepts are present in the Neyman-Pearson theory of testing hypotheses, *a priori* preferences can, in fact, be introduced by appropriate choice of values of α and β. As an example, science generally favors, and for good historical reasons, simpler and/or established hypotheses; the probability of rejecting such hypotheses when true may be set very low.

The few ideas which have been introduced in this chapter (which will shortly be developed in some detail) form the foundation of a valuable branch of modern statistical inference. But it is well to note that these ideas are primarily concerned with statistical aspects of the testing of hypotheses and that other problems remain, e.g., the origin of hypotheses themselves. Simple hypotheses essentially testing the relevance of the past or testing indifference (that current quality is up to recent standards, that the established value of a parameter in one area of science is valid in another, that the public is evenly divided between two candidates) are easy to imagine, formulate, and test, and on the whole we confine ourselves to them. Some (many, if interpreted literally) may be simple to the point of being straw men; a test of the hypothesis that the linear correlation between income and savings is 0 would interest few economists indeed. But with more complex phenomena come more complex hypotheses (hypotheses involving several parameters and their interrelationship). Apart from the difficulty of finding efficient statistical tests of such hypotheses, their construction is often a matter of genius.

As we have seen, probability alone, unfortunately, cannot uniquely solve the problem of estimating parameters or deciding among hypotheses; some notion of good estimators and good tests must be adopted by the experimenter. It is often for this reason that the concept of an alternative hypothesis (and power, which need not mention a *specific* alternative hypothesis) is valuable, to provide a sensible basis for selecting, among regions of equal size, one which in some sense is superior. For without any alternative hypothesis in mind, one need not draw *any* data. A well-mixed box of one thousand chips of which one is red is apparatus enough to provide a technique which will guarantee a comfortably small probability of 0.001 of rejecting *any* hypothesis when true! Such a box will of course be of little comfort if and when, at unpredictable intervals, the hypothesis is false.

The earlier discussion with respect to alternative hypotheses may appear to be more limiting than it is. It seems that we are forced to choose between two hypotheses θ_0 and θ_1, while the truth on θ may be far from either. But we may, and this often makes good sense in science, focus all our attention on θ_0 alone and reject it or accept it as E falls in R or not. For while the concept of an alternative hypothesis is valuable, and even critical, its presence should not force us to accept it if the null hypothesis is rejected. We may reject H_0 because H_1 is more compatible with the data, but without detailed acceptance of H_1. Science does imply action; the correspondence of decision-making between two hypotheses and taking action is natural, and, in such areas as industrial sampling inspection, usually immediate. Acceptance or rejection of a hypothesis is usually followed by (or better, is usually equivalent to) acceptance or rejection of a lot. But even here the rejection of one hypothesis (that a lot is bad) need not imply acceptance of a particular alternative as to the precise quality of the accepted lot.

In certain earlier discussions of testing statistical hypotheses, one often finds no mention of alternative hypotheses; but it will generally be noted that in the selection of a method of testing the hypothesis at hand, a class of alternative hypotheses was implicit.

The hypotheses, both null and alternative, with which we begin are *simple;* a simple hypothesis completely specifies the probability distribution. Or, to put it differently, since the form of the probability distribution is assumed known, all parameters in a simple hypothesis are specified; in the parameter space, the hypothesis constitutes a point. More common and more difficult to handle are *composite hypotheses.* Here the probability distribution is not completely specified (some parameters are not specified in the hypothesis). A *composite hypothesis* is a region rather than a point in the parameter space. For example, a single-valued hypothesis on the mean λ of a normal distribution of *known* variance σ^2 is simple; it is a point in the two-dimensional λ, σ^2 parameter space. A single-valued hypothesis on the mean of a normal distribution of *unknown* variance σ^2 is composite; it is a line in the λ, σ^2 parameter space. The hypothesis $\theta = \theta_0$, other parameters specified, is simple; the more interesting hypothesis $\theta_1 \leq \theta \leq \theta_2$, other parameters specified or unspecified, is composite. A common situation is a simple null hypothesis $\theta = \theta_0$ with composite alternatives $\theta \neq \theta_0$. Some of our theorems follow from consideration of simple null and simple alternative hypotheses, but in particular cases these theorems may hold, as we shall see, for composite alternative hypotheses.

Finally let us admit that many areas of physical, biological, and social science have made great progress while paying remarkably little attention to the formal apparatus of hypothesis testing. In fact, advances may sometimes have been missed by clinging to statistically well-supported hypotheses. Moreover, not all in science is yet quantitative. Nor can everything be settled by random sampling; some of the most profound hypotheses of science (Schrödinger's wave theory, for example) have, as E. Bright Wilson [209] has remarked, not been open to test by random experiments and have been tested (and in this example supported) by deduction. Remember that no hypothesis can ever be "proved" by observed data; in fact, there is practically always a hypothesis often suggested by the data and different from the one under test which is more compatible with the observed data. Remember, too, that if α is fixed, any hypothesis will fall when the sample size n is large enough (it hardly follows that testing at small n is pointless, for α should sensibly be reduced if one uses large n). But despite all such important reservations, the method of testing statistical hypotheses by appeal to relevant random data is one of the best of all methods of verification available to the scientist.

28–3 Two examples. To illustrate the discussion, we give two simple examples; the first is due to Wald [195]. We have 4 bearings of unknown quality, each known only to be good or bad. Consider the statistical hypothesis H_0: 2 good, 2 bad. To test H_0 we decide to draw *with replacement* a random sample of size $n = 2$. The entire event space consists of four points, GG, GB, BG, BB. Or, transforming to the joint random variable "number of good bearings in the first, second bearing drawn," the event space becomes the four real number pairs shown in Fig. 28–3.

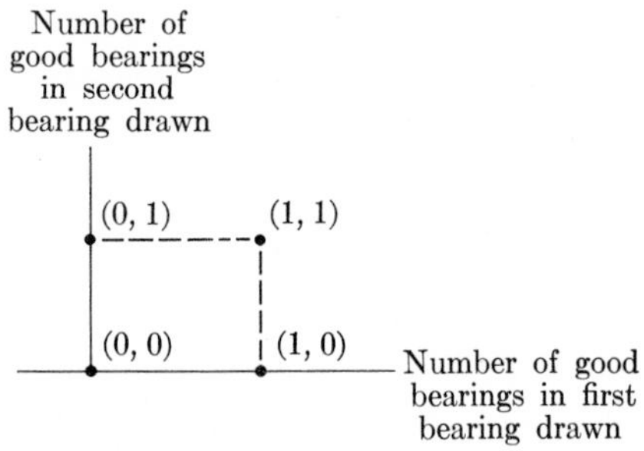

FIG. 28–3. An event space of four real number pairs.

Consider the three following regions R_1, R_2, R_3 in the event space:

R_1: GB or BG; reject H_0 if the sample yields GB or BG;

R_2: GG or BB; reject H_0 if the sample yields GG or BB;

R_3: BG or BB; reject H_0 if the sample yields BG or BB.

Table 28–2 shows, for each one of all possible hypotheses relevant here (the null and each of the four possible alternatives), the probability that a sample point $\mathbf{E}$ will fall into each of these regions. Note that the size of each region is $\frac{1}{2}$; if H_0 is true, the probability that $\mathbf{E}$ will fall into each of

Table 28–2

Power Functions of R_1, R_2, and R_3 of Page 288

Region \ Number of good bearings in the population	0	1	(H_0) 2	3	4
R_1	0	$\frac{6}{16}$	$\frac{8}{16}$	$\frac{6}{16}$	0
R_2	1	$\frac{10}{16}$	$\frac{8}{16}$	$\frac{10}{16}$	1
R_3	1	$\frac{12}{16}$	$\frac{8}{16}$	$\frac{4}{16}$	0

these regions is $\frac{1}{2}$. The regions R_1, R_2, and R_3 are equally effective (or ineffective!) in rejecting H_0 when H_0 is true.

Region R_1 (intuitively a poor region) is evidently a poor region. As there exists another region R_2 which is as good as or better than R_1 (has as high as or higher power than R_1 with respect to *all* alternative hypotheses), R_1 is called *inadmissable.* Neither of the better regions R_2 or R_3 is better with respect to *all* alternative hypotheses; R_2 is evidently better with respect to certain alternatives, R_3 is better with respect to others. If the class of alternatives to H_0 can be confidently specified, either by knowledge or by interest, then the choice between R_2 and R_3 can be readily made; if not, R_2 is probably the better compromise.

A second example will bring out further points in testing statistical hypotheses. Let a population be described by a normal random variable $\mathbf{x}$ with $\sigma = 1$ and with unknown mean λ. Let the null hypothesis H_0 be $\lambda = \lambda_0$. A random sample $\mathbf{E}$: $\mathbf{x}_1, \mathbf{x}_2, \ldots, \mathbf{x}_n$ is to be drawn. It is proposed to accept or reject H_0 on the basis of the observed value of the sample mean $\bar{\mathbf{x}}$.

If $H_0: \lambda = \lambda_0$ is true, we know from (23–7) that the probability of

$$\text{(a)}\quad \bar{\mathbf{x}} \leq \lambda_0 - 1.96\,\sigma(\bar{\mathbf{x}}) \qquad \text{and} \qquad \bar{\mathbf{x}} \geq \lambda_0 + 1.96\,\sigma(\bar{\mathbf{x}})$$

is 0.05. If we reject H_0 when $\bar{\mathbf{x}}$ lies outside the two indicated limits, the probability of rejecting H_0 when H_0 is true is 0.05. But many other ranges on the random variable $\bar{\mathbf{x}}$ yield $\alpha = 0.05$; two among them are

$$\text{(b)}\quad \bar{\mathbf{x}} \geq \lambda_0 + 1.65\,\sigma(\bar{\mathbf{x}}),$$

$$\text{(c)}\quad \lambda_0 - 0.065\,\sigma(\bar{\mathbf{x}}) \leq \bar{\mathbf{x}} \leq \lambda_0 + 0.065\,\sigma(\bar{\mathbf{x}}).$$

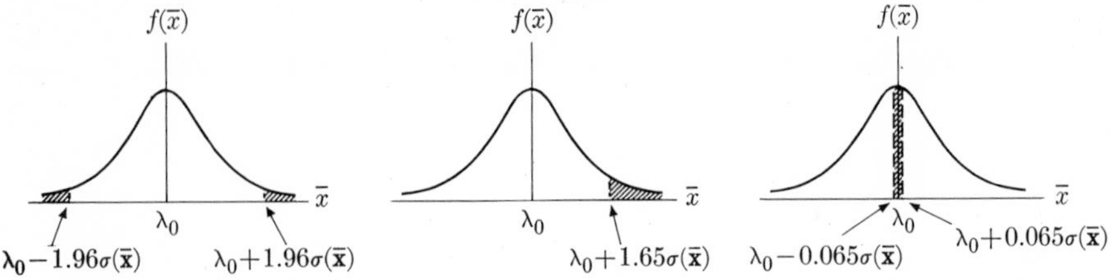

(a) (b) (c)

FIG. 28–4. Three ranges on $\bar{\mathbf{x}}$ in samples of size n from a normal population; all have size $\alpha = 0.05$.

The three ranges on $\bar{\mathbf{x}}$ and the corresponding $\alpha = 0.05$ (shaded) are shown in Fig. 28–4.

Among the practical questions which arise and to which the Neyman-Pearson theory of testing hypotheses suggests answers are: which of these three ranges on $\bar{\mathbf{x}}$, if any, should we use for rejecting H_0; why use the statistic $\bar{\mathbf{x}}$ in the first place; why incur a risk as high as 0.05 of a Type I error?

Note that, corresponding to given α, a particular range on a statistic corresponds to a particular region in the event space, though it may sometimes be difficult to show to just what region the range on a statistic corresponds. To illustrate the correspondence, we adapt the three ranges (a), (b), and (c) to a simpler example: $H_0 : \lambda = \lambda_0 = 0$, with $\sigma = 1$ and $n = 2$. Recalling that $\sigma(\bar{\mathbf{x}}) = \sigma/\sqrt{n}$, the three ranges on $\bar{\mathbf{x}}$ which we are considering (shown in Fig 28–4 with $\lambda_0 = 0$, $\sigma = 1$, and $n = 2$) become

$$(\text{a}')\ \bar{\mathbf{x}} \le \frac{-1.96}{\sqrt{2}} \quad \text{and} \quad \bar{\mathbf{x}} \ge \frac{1.96}{\sqrt{2}},$$

$$(\text{b}')\ \bar{\mathbf{x}} \ge \frac{1.65}{\sqrt{2}}, \qquad (\text{c}')\ -\frac{0.065}{\sqrt{2}} \le \bar{\mathbf{x}} \le \frac{0.065}{\sqrt{2}}.$$

These correspond respectively to the regions R_1', R_2', R_3' in the two-dimensional event space of x_1 and x_2 shown in Fig. 28–5. The precise position of these regions in the plane of x_1, x_2 can be determined from (a′), (b′), and (c′) by replacing $\bar{x}$ by $(x_1 + x_2)/2$. A similar correspondence, though often harder to visualize or graph, holds for all ranges on all statistics and at all values of n.

The power functions of R_1, R_2, and R_3 corresponding to the ranges (a), (b), (c) on $\bar{\mathbf{x}}$ are easily constructed for any sample size; we need only

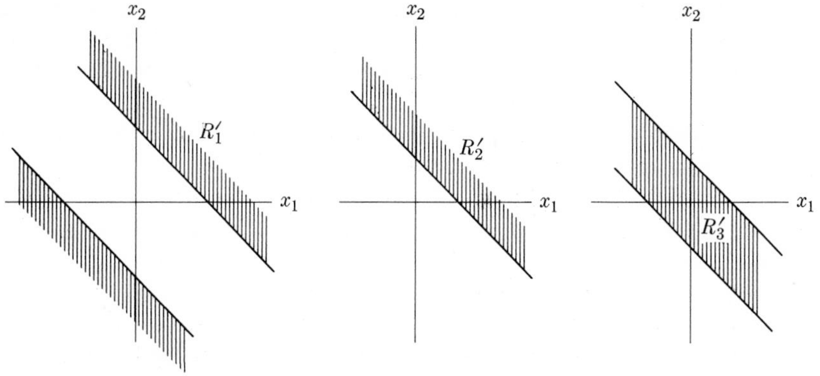

FIG. 28–5. Three regions in the event space of x_1, x_2 corresponding to the three ranges on $\bar{x}$ shown in (a′), (b′), (c′).

know the probability distribution of $\bar{x}$ in random samples of size n from normal populations of $\sigma = 1$ and of any mean λ, and this we know. The student should construct in detail the power function of each of the three ranges (a), (b), and (c) for, say, $n = 4$ and $n = 9$. The results will appear roughly as in Fig. 28–6, with $1 - \beta$ plotted against $\lambda - \lambda_0$; the dotted lines are for $n = 4$, the solid lines for $n = 9$.

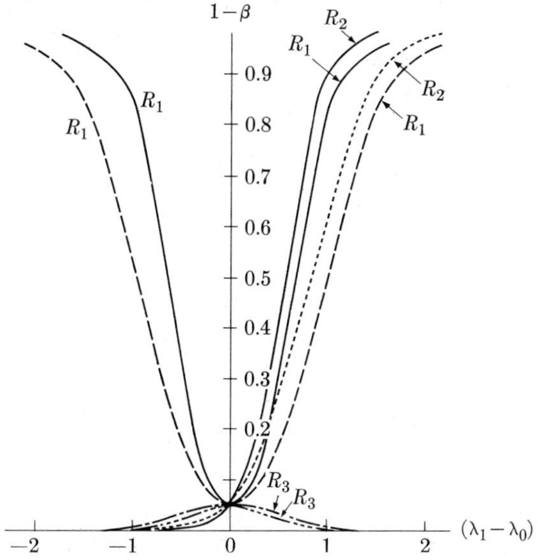

FIG. 28–6. Power functions, $n = 4$ and $n = 9$ corresponding to the three regions R_1, R_2, and R_3.

Once again, the choice of region depends on the experimenter's interest in and knowledge of alternative values of λ; in the absence of such interest of knowledge, R_1 is perhaps the best compromise. Is R_3 nonadmissible?

28–4 Further discussion of the second example. Using the statistic $\bar{\mathbf{x}}$ and any one of the regions R_1 or R_2 or R_3, the probability of rejecting H_0 when H_0 is true is $\alpha = 0.05$. But which of these three regions, if any, minimizes the probability β of accepting H_0 when H_1 is true? While general principles and methods can and will be brought to bear on this and similar problems later in this chapter, principles which will also enable us to decide if $\bar{\mathbf{x}}$ is the best statistic in the first place, and while the question is in fact already answered in Fig. 28–6, it is instructive to view the present example as one in elementary calculus. Shifting from $\bar{\mathbf{x}}$ to the standardized variable $\mathbf{u}$, we have

$$\mathbf{u} = \frac{\bar{\mathbf{x}} - \lambda_0}{1/\sqrt{n}}, \tag{28–3}$$

which, when $H_0 : \lambda = \lambda_0$ is true, has mean 0 and variance 1. The integral

$$\int_{u_1}^{u_2} \frac{1}{\sqrt{2\pi}} e^{-(1/2)u^2} \, du = 1 - \alpha = 0.95 \tag{28–4}$$

describes the probability of accepting H_0 when true. We ask what values of u_1 and u_2 in (28–4) will minimize the probability of accepting H_0 when H_1 is true, that is, minimize

$$\beta = p(u_1 \leq \mathbf{u} \leq u_2 | \lambda_1). \tag{28–5}$$

To answer, we evidently need to know the sampling behavior of $\mathbf{u}$ when H_1 is true (when $\lambda = \lambda_1 \neq \lambda_0$). If $\lambda = \lambda_1$ is true, the standardized variable $\mathbf{v}$

$$\mathbf{v} = \frac{\bar{\mathbf{x}} - \lambda_1}{1/\sqrt{n}}$$

is normal with mean 0 and variance 1. But

$$\mathbf{u} = \frac{\bar{\mathbf{x}} - \lambda_0}{1/\sqrt{n}} = \mathbf{v} + \sqrt{n}\,(\lambda_1 - \lambda_0);$$

$\mathbf{u}$ is a *linear* function of $\mathbf{v}$. So if $\lambda = \lambda_1$, the random variable of interest $\mathbf{u}$ (by which a decision between H_0 and H_1 is to be made) is normal with variance 1 [as in (28–3)] but with mean shifted from 0 to $\sqrt{n}\,(\lambda_1 - \lambda_0)$. (We note, in contrast, that in many other problems, the probability behavior of the appropriate statistic is difficult to determine when the alternative hypothesis is true.) Thus (28–5) becomes

$$\beta = \int_{u_1}^{u_2} \frac{1}{\sqrt{2\pi}} \exp\left\{-\frac{1}{2}\frac{[u - \sqrt{n}(\lambda_1 - \lambda_0)]^2}{1}\right\} du, \tag{28–6}$$

a function of u_1 and u_2. As already indicated, we want to determine u_1 and u_2, with both subject to condition (28–4), so that (28–6) will be minimum. The problem is solved here, somewhat crudely but instructively, as follows: of the two unknowns, regard u_1 as independent and u_2 as a function of u_1. If we write $B(u)$ for the integrand, (28–6) becomes

$$\beta = \int_0^{u_2} B(u)\, du - \int_0^{u_1} B(u)\, du. \tag{28–7}$$

To minimize β, we differentiate (28–7) with respect to u_1:

$$\frac{d\beta}{du_1} = B(u_2)\frac{du_2}{du_1} - B(u_1). \tag{28–8}$$

If we write $A(u)$ for the integrand of the side condition, (28–4) becomes

$$\int_0^{u_2} A(u)\, du - \int_0^{u_1} A(u)\, du = 0.95, \tag{28–9}$$

and, differentiating (28–9) with respect to u_1 to obtain du_2/du_1 required in (28–8), we have

$$A(u_2)\frac{du_2}{du_1} - A(u_1) = 0. \tag{28–10}$$

Inserting (28–10) in (28–8), we have

$$\frac{d\beta}{du_1} = B(u_2)\frac{A(u_1)}{A(u_2)} - B(u_1). \tag{28–11}$$

If we replace all terms in (28–11) by the original functions, and write $k = \sqrt{n}(\lambda_1 - \lambda_0)$, (28–11) reduces, after some algebra, to

$$\frac{d\beta}{du_1} = \frac{1}{\sqrt{2\pi}} e^{-(1/2)(u_1^2+k^2)}[e^{u_2 k} - e^{u_1 k}]$$

$$= \text{positive quantity} \cdot [e^{u_2 k} - e^{u_1 k}].$$

If $k > 0$, that is, if $\lambda_1 > \lambda_0$,

$$\frac{d\beta}{du_1} = \text{positive quantity (larger positive quantity} - \text{smaller positive quantity)}$$

$$= \text{positive quantity.}$$

The parentheses can be held to a minimum (thereby holding β to a minimum) by making u_1 as small as possible $(-\infty)$ and u_2 whatever is necessary for (28–4) to be satisfied. This is illustrated in the left-hand diagram of Fig. 28–7.

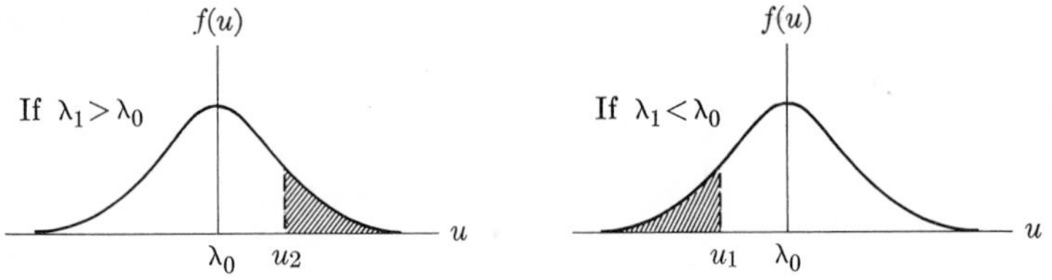

FIG. 28–7. Ranges of u which minimize errors of the second kind for the indicated alternative hypotheses.

If $k < 0$, that is, if $\lambda_1 < \lambda_0$,

$$\frac{d\beta}{du_1} = \text{positive quantity (smaller positive quantity} - \text{larger positive quantity)}$$

$$= \text{negative quantity.}$$

The parentheses can be held to smallest *absolute* value (thereby holding β to a minimum) by making u_2 as large as possible (∞) and u_1 whatever is necessary for (28–4) to be satisfied. This is illustrated in the right-hand diagram of Fig. 28–7.

Regions of the type R_2' of page 291 emerge if one can be confident that the alternatives to $\lambda = \lambda_0$ are greater than λ_0 (left diagram above) or (not "and") are less than λ_0 (right diagram above).

28–5 A basic theorem of Neyman and Pearson. We seek a region R which is most powerful with respect to an alternative hypothesis H_1. In the following demonstration, which is restricted to continuous density functions and, more important, to simple null and simple alternative hypotheses, it is proved that if R and a positive constant k exist satisfying

$$\frac{p(E|H_0)}{p(E|H_1)} \leq k \text{ for each } E \text{ inside } R,$$
$$\frac{p(E|H_0)}{p(E|H_1)} \geq k \text{ for each } E \text{ outside } R, \tag{28–12}$$

then R is most powerful with respect to H_1. The size of R is determined by k.

It is intuitive that (28–12) should lead to a good test. Remember that H_0 is rejected when E falls in R. Now there are E's for which the first inequality holds; for these E's the case for H_0 relative to H_1 is unimpressive. If all such E's can be assembled *inside* R, the test will be good, for each will (and should) lead to rejection of H_0. On the other hand, there are E's for which the second inequality holds; for these E's the case for H_0 relative to H_1 is impressive. If all such E's can be assembled *outside* R, the test will be good, for each will (and should) lead to acceptance of H_0.

The formal proof is straightforward. Consider two regions R and S of equal size,

$$\int_R p(E|H_0)\,dE = \alpha, \qquad \int_S p(E|H_0)\,dE = \alpha,$$

which are more compactly written

$$\int_R p_0(E)\,dE = \alpha, \qquad \int_S p_0(E)\,dE = \alpha.$$

The regions R and S may (but need not) overlap, as shown in Fig. 28–8.

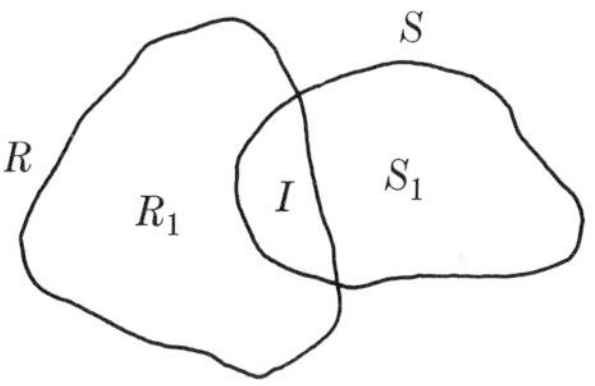

FIG. 28–8. Overlapping regions in the event space.

We have

$$\int_{R_1} p_0(E)\,dE = \int_R p_0(E)\,dE - \int_I p_0(E)\,dE$$
$$= \int_S p_0(E)\,dE - \int_I p_0(E)\,dE = \int_{S_1} p_0(E)\,dE.$$

As R_1 is inside R, apply the first condition of (28–12) to R_1; compactly,

$$\int_{R_1} p_1(E)\,dE \geq \frac{1}{k}\int_{R_1} p_0(E)\,dE. \tag{28–13}$$

As S_1 is outside R, apply the second condition of (28–12) to S_1; compactly,

$$\int_{S_1} p_1(E)\,dE \leq \frac{1}{k}\int_{S_1} p_0(E)\,dE. \tag{28–14}$$

From the two inequalities (28–13) and (28–14), we easily get an inequality on $\int_R p_1(E)\,dE$, the power of R with respect to the simple alternative H_1

and $\int_S p_1(E)\,dE$, the power of S with respect to the simple alternative H_1, as follows:

$$\begin{aligned}\int_R p_1(E)\,dE &= \int_{R_1} p_1(E)\,dE + \int_I p_1(E)\,dE \\ &\geq \frac{1}{k}\int_{R_1} p_0(E)\,dE + \int_I p_1(E)\,dE \qquad \text{from (28–13)} \\ &\geq \int_{S_1} p_1(E)\,dE + \int_I p_1(E)\,dE \qquad \text{from (28–14)} \\ &\geq \int_S p_1(E)\,dE. \qquad (28\text{–}15)\end{aligned}$$

Thus R is more powerful than S with respect to H_1, and since S is *any* other region of size α, R is *uniformly most powerful* with respect to all alternative hypotheses.

If the random variable $\mathbf{x}$ and the parameter θ are thought of as vectors (in Fig. 28–8, $\mathbf{x}$ is, in fact, two-dimensional, and one may note that the number of parameters does not enter the preceding proof) the Neyman-Pearson theorem can be extended to several random variables and several parameters.

28–6 Application of the Neyman-Pearson theorem. We illustrate the Neyman-Pearson theorem by applying it to an example discussed earlier in this chapter. Let $\mathbf{x}$ be normal with $\sigma = 1$ and unknown mean λ. Consider $H_0\colon \lambda = \lambda_0$ and $H_1\colon \lambda = \lambda_1$; we want to find the most powerful region (test) R. We have

$$p_0(E) = \frac{1}{(2\pi)^{n/2}} \exp\left[-\frac{1}{2}\sum_{i=1}^{n} (x_i - \lambda_0)^2\right],$$

$$p_1(E) = \frac{1}{(2\pi)^{n/2}} \exp\left[-\frac{1}{2}\sum_{i=1}^{n} (x_i - \lambda_1)^2\right],$$

$$\frac{p_0(E)}{p_1(E)} = \exp\left\{-\frac{1}{2}\left[\sum_{i=1}^{n} (x_i - \lambda_0)^2 - \sum_{i=1}^{n} (x_i - \lambda_1)^2\right]\right\}.$$

The square terms in x_i disappear, and (28–12) reduces to

$$\exp\left[-\frac{n}{2}(\lambda_0^2 - \lambda_1^2) + (\lambda_0 - \lambda_1)\sum_{i=1}^{n} x_i\right] \begin{matrix}\leq k \text{ inside } R, \\ \geq k \text{ outside } R.\end{matrix} \qquad (28\text{–}16)$$

Now

$$\exp\left[-\frac{n}{2}(\lambda_0^2 - \lambda_1^2)\right]$$

is constant and positive, so (28–16) becomes

$$(\lambda_0 - \lambda_1) \sum_{i=1}^{n} x_i \quad \begin{matrix} \leq k_1 \text{ inside } R, \\ \geq k_1 \text{ outside } R, \end{matrix}$$

or, replacing $\sum_{i=1}^{n} x_i$ by $n\bar{x}$,

$$(\lambda_0 - \lambda_1)\bar{x} \quad \begin{matrix} \leq k_2 \text{ inside } R, \\ \geq k_2 \text{ outside } R, \end{matrix} \tag{28–17}$$

where k_2 may be positive or negative. The linear function $\bar{\mathbf{x}}$ emerges as the statistic of choice.

Evidently there is no *one* most powerful region R against *all* alternatives to λ_0. For if $\lambda_1 > \lambda_0$, (28–17) become

$$\bar{x} \geq k_3 \text{ inside } R \qquad (\text{reject } \lambda = \lambda_0 \text{ if } \bar{x} \geq k_3),$$

$$\bar{x} \leq k_3 \text{ outside } R \qquad (\text{accept } \lambda = \lambda_0 \text{ if } \bar{x} \leq k_3),$$

while if $\lambda_1 < \lambda_0$, (28–17) become

$$\bar{x} \leq k_4 \text{ inside } R \qquad (\text{reject } \lambda = \lambda_0 \text{ if } \bar{x} \leq k_4),$$

$$\bar{x} \geq k_4 \text{ outside } R \qquad (\text{accept } \lambda = \lambda_0 \text{ if } \bar{x} \geq k_4),$$

where k_3 and k_4, which may be positive or negative, fix the size α of the region R. The results correspond precisely to those already found by *ad hoc* procedures but now have the value of following from a general principle of merit.

This example reveals that there is no one most powerful region R against all alternatives to $H_0 : \lambda = \lambda_0$. This is true fairly generally. Relative to *any* $H_0 : \theta = \theta_0$ and under quite general conditions, there exists no uniformly most powerful test of H_0 against alternatives which include *both* positive and negative values of $\theta - \theta_0$ (Kendall and Stuart [104]).

In the proof of the Neyman-Pearson theorem, H_0 and H_1 were simple hypotheses. But in the example to which we applied the theorem, the alternative hypothesis H_1 entered (28–17) in such a way that we were able to imply that the right-hand range of $\bar{\mathbf{x}}$ (left diagram of Fig. 28–7) is most powerful for *any* λ_1 greater than λ_0, and that the left-hand range of $\bar{\mathbf{x}}$ (right diagram of Fig. 28–7) is most powerful for *any* λ_1 less than λ_0. When the particular application of the theorem permits such extension to composite alternative hypotheses the results have added interest.

28–7 The Neyman-Pearson theorem and sufficient statistics. Note that in this example the Neyman-Pearson theorem lead to a sufficient statistic ($\bar{\mathbf{x}}$). This is generally true. If a sufficient statistic $\mathbf{t}$ exists, we have, repeating (25–19),

$$f(x_1, x_2, \ldots, x_n; \theta) = g(t; \theta) \cdot h(x_1, x_2, \ldots, x_n),$$

but the Neyman-Pearson ratio (28–12) then reduces to

$$\frac{f(x_1, x_2, \ldots, x_n; \theta_0)}{f(x_1, x_2, \ldots, x_n; \theta_1)} = \frac{g(t; \theta_0)}{g(t; \theta_1)}, \tag{28–18}$$

a function of $\mathbf{x}_1, \mathbf{x}_2, \ldots, \mathbf{x}_n$ only through the sufficient statistic $\mathbf{t}$. If a sufficient statistic $\mathbf{t}$ exists, the most powerful test against any alternative is based on $\mathbf{t}$.

28–8 Unbiased tests. Uniformly most powerful tests seldom exist, and particularly, as noted, if the alternatives to the hypothesis $H_0: \theta = \theta_0$ lie on both sides of the value of θ specified in H_0. We therefore must seek tests which are most powerful within a more restricted class of tests, for example, among tests enjoying certain valuable properties such as invariance or unbiasedness. Before discussing a general principle and technique which often yields good tests, we note briefly two types of *unbiased* tests, both suggested by Neyman and Pearson [144].

(a) A uniformly most powerful *unbiased* test of $H_0: \theta = \theta_0$ sensibly has minimum power at $\theta = \theta_0$, a property described by the term "unbiased," and is more powerful than any other *unbiased* test. Generally, at some values of θ, such a test will not be as powerful as a uniformly most powerful test. It is defined by a region R of size α satisfying

$$p[E \in R|\theta_0] = p[E \in R'|\theta_0] = \alpha \qquad \text{and} \qquad p[E \in R|\theta] \geq p[E \in R'|\theta], \tag{28–19}$$

where R' is any other unbiased region of size α. Such a test sometimes exists. For example, for a normal variable of variance 1, the uniformly most powerful unbiased test of $H_0: \lambda = 0$ against *all* $H_1: \lambda \neq 0$ leads to the two-sided range

(1) $$|\bar{x}| \geq c \tag{28–20}$$

[shown in Fig. 28–4(a)], as against the tests

(2) $$\bar{x} \geq c_1 \quad \text{for } \lambda > 0$$

and

(3) $$\bar{x} \leq c_2 \quad \text{for } \lambda < 0$$

(shown in Fig. 28–7), which are uniformly most powerful against the alternatives indicated. The power functions of (1), (2), and (3) are sketched in Fig. 28–9.

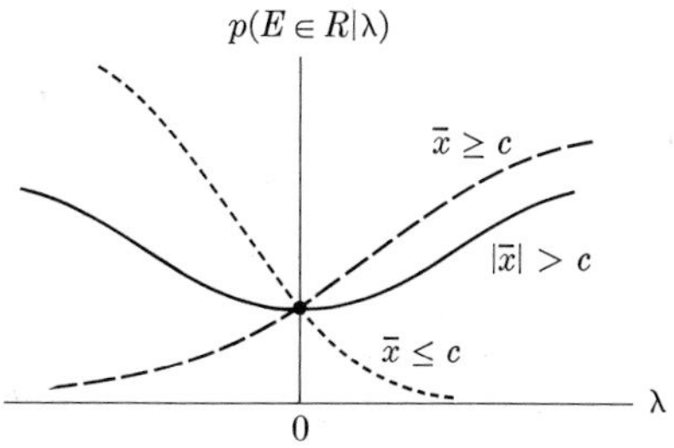

Fig. 28–9. Power functions of three ranges (1), (2), (3) on $\bar{\mathbf{x}}$.

(b) We now consider an unbiased test of $H_0: \theta = \theta_0$ which, of all unbiased tests of the same size, has the highest rate of increase of power in the vicinity of θ_0. Such a property has obvious merit if the alternative hypotheses on θ are close to θ_0 and, in many inquiries in mathematically well-structured science, this is the case. Such a test is defined by a region R of size α satisfying

$$\frac{\partial}{\partial \theta} p[E \in R|\theta]_{\theta=\theta_0} = 0, \tag{28–21}$$

which describes a property which might be called local unbiasedness, *and*

$$\frac{\partial^2}{\partial \theta^2} p[E \in R|\theta]_{\theta=\theta_0} \geq \frac{\partial^2}{\partial \theta^2} p[E \in R'|\theta]_{\theta=\theta_0}, \tag{28–22}$$

where R' is any other unbiased region of size α. Power functions satisfying (28–21) and both (28–21) and (28–22) are shown in Fig. 28–10.

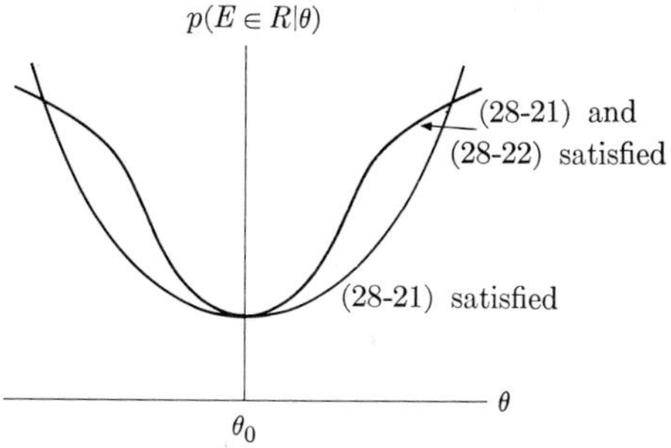

Fig. 28–10. Power functions of an unbiased and a steepest unbiased region.

A few examples of unbiased tests are given in Fraser [69] and Kendall and Stuart [104]. The sparsity of examples suggests the need for a more widely applicable method of testing statistical hypotheses, and to this we turn now.

28–9 The likelihood ratio. A practical procedure for testing simple and composite statistical hypotheses is now described. It centers on the *likelihood ratio* and is due to Neyman [137]. It has good general properties, as we shall see; among them: if a sufficient statistic for a parameter exists, the likelihood ratio produces a test based on it, and if a uniformly most powerful test exists, the likelihood ratio often leads to it.

We begin with the one-parameter case. We have a population described by a random variable $\mathbf{x}$ with density function $f(x;\ \theta)$; the value of the parameter θ is unknown. As usual, the observed value $E\colon x_1, x_2, \ldots, x_n$ of a random sample $\mathbf{E}\colon \mathbf{x}_1, \mathbf{x}_2, \ldots, \mathbf{x}_n$ is to be the basis of the decision on H_0.

Two preliminary remarks are necessary. For the given (observed) sample point E, $p(E|\theta)$ depends only on θ. The functional nature of the dependence might possibly differ from one observed value of E to another. Also, for any one observed E, certain values of θ lead to larger values of $p(\mathbf{E}|\theta)$ evaluated at $\mathbf{E} = E$ than others. Both of these facts as well as maximum values of $p(\mathbf{E}|\theta)$ evaluated at E_1 and E_2 are illustrated in Fig. 28–11.

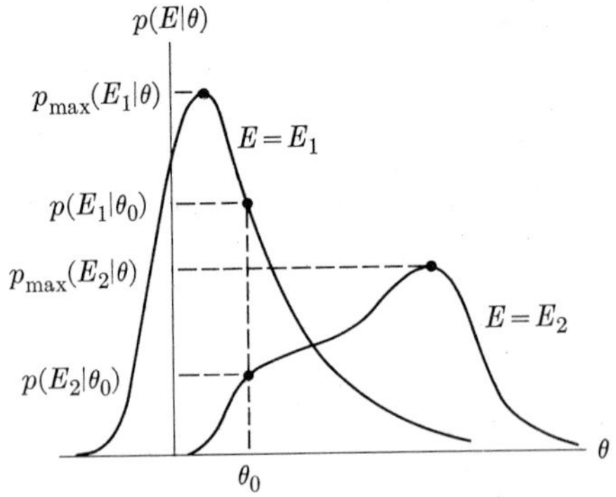

FIG. 28–11. Various characteristics of $p(\mathbf{E}|\theta)$ as a function of θ.

Consider the *likelihood ratio* L:

$$L = \frac{p_{\max}(E|\omega)}{p_{\max}(E|\Omega)}, \tag{28–23}$$

the numerator being the maximum value of the probability $p(\mathbf{E}|\theta)$ evaluated at $\mathbf{E} = E$ when the parameter θ is restricted to that part ω of the parameter space specified by the hypothesis H_0 (not necessarily simple), and the denominator being the maximum value of the probability $p(\mathbf{E}|\theta)$ evaluated at $\mathbf{E} = E$ when θ ranges without restriction over the entire admissible parameter space Ω.

For tests of hypotheses on several parameters, the statistic is similar. We test the hypothesis that values of the pair of parameters θ_1, θ_2 lie in the region ω, with the alternative hypothesis that the values lie in the region $\Omega - \omega$. Again we form (28–23), where now the numerator is the maximum of $p(\mathbf{E}|\omega)$ evaluated at $\mathbf{E} = E$ with respect to θ_1 and θ_2 subject to the condition that θ_1, θ_2 lie in ω, and the denominator is the unrestricted maximum of $p(\mathbf{E}|\Omega)$, evaluated at $\mathbf{E} = E$.

The similarity in spirit to the maximum likelihood criterion (26–2) is evident. The similarity to the Neyman-Pearson criterion (28–12) is also evident; for simple hypotheses, (28–23) is (28–12) with parameter(s) replaced by maximum likelihood estimator(s). In fact, the likelihood ratio may be regarded as an intuitive extension, via the method of maximum likelihood, of the Neyman-Pearson criterion (28–12), with the simple null and alternative hypotheses of (28–12) now replaced by simple *or* composite null and alternative hypotheses. For any observed E, the two maxima in (28–23) are achieved by replacing the parameter θ by its maximum likelihood estimator, within ω in the numerator of (28–23) and within Ω in the denominator of (28–23). Notice that $\mathbf{E}$ is in fact a random variable; so therefore is L. Clearly L cannot be less than 0 nor greater than 1. If for a particular observed E the value of L is near 1, that is, if $p_{\max}(E|H_0)$ is nearly equal to $p_{\max}(E|\theta)$, the hypothesis H_0 warrants support for the maximum probability density associated with the set of θ in the region ω cannot be much increased by shifting from them to other values of θ in Ω. If L is near 0, the contrary is true; the hypothesis H_0 does not warrant support. The distribution of L, for H_0 true, is illustrated in Fig. 28–12.

Using the L test, the probability α of an error of the first kind is evidently

$$\int_0^{L\alpha} h(L|\omega)\, dL = \alpha,$$

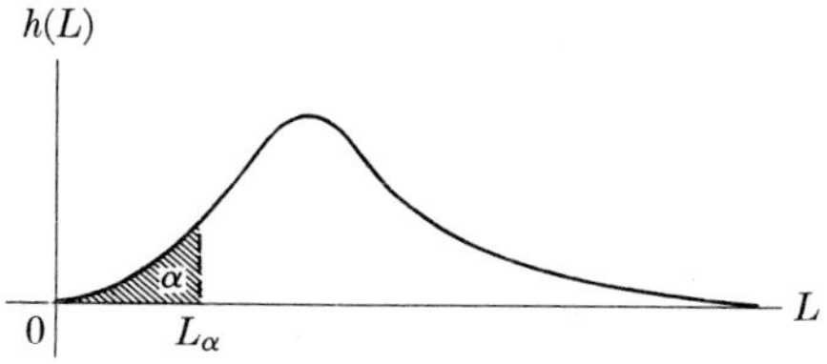

FIG. 28–12. The distribution of L for H_0 true.

while the power of the L test is

$$\int_0^{L\alpha} h(L|\theta)\, dL,$$

which is evidently a function of the possible values of θ.

For large n, an approximation to the distribution of L is available. For $H_0\colon \theta = \theta_0$ simple and true and with modest continuity restrictions on $f(x;\theta)$, the quantity $-2 \log L$ follows the χ^2 distribution with 1 degree of freedom. In the multiparameter case with similar continuity restrictions on the density function $f(x;\theta_1, \ldots, \theta_k)$, and for the composite hypothesis

$$H_0 : \theta_1 = \theta_1^0, \ldots, \theta_r = \theta_r^0, \quad r \leq k,$$

the quantity $-2 \log L$ follows, for H_0 true, the χ^2 distribution with $k - r$ degrees of freedom. These important asymptotic results are due to Wilks [205]; we shall not need them in our simple examples.

28–10 Two examples of the likelihood ratio test. We give two examples. First consider a simple hypothesis. Let $\mathbf{x}$ be normal with $\sigma = 1$ and unknown mean λ; consider $H_0 : \lambda = \lambda_0$. For the numerator of L, we have

$$p(E|\lambda_0) = \left(\frac{1}{\sqrt{2\pi}}\right)^n \exp\left[-\frac{1}{2}\sum_{i=1}^{n} (x_i - \lambda_0)^2\right] = p_{\max}(E|\lambda_0). \qquad (28\text{–}24)$$

In this example there are no unspecified parameters in H_0 (σ is known to be 1 and the remaining parameter λ is specified by H_0 to be λ_0). For an observed E, $p(E|\lambda_0)$ in (28–24) is already at its only, and therefore, maximum value; the subspace ω of the parameter space Ω is a point.

Now we consider the denominator of L. We have

$$p(E|\lambda) = \left(\frac{1}{\sqrt{2\pi}}\right)^n \exp\left[-\frac{1}{2}\sum_{i=1}^{n} (x_i - \lambda)^2\right].$$

To find $p_{\max}(E|\lambda)$ for an observed E, we have the (maximum likelihood) equation

$$\frac{\partial}{\partial\lambda}\, p(E|\lambda) = \text{constant}\left\{\exp\left[-\frac{1}{2}\sum_{i=1}^{n} (x_i - \lambda)^2\right]\right\}\sum_{i=1}^{n} (x_i - \lambda) = 0,$$

which, as we already know from problem 26–2, is satisfied in Ω by $\lambda = \bar{x}$. Thus (28–23) becomes

$$L = \frac{(1/\sqrt{2\pi})^n \exp\left[-\frac{1}{2}\sum_{i=1}^{n} (x_i - \lambda_0)^2\right]}{(1/\sqrt{2\pi})^n \exp\left[-\frac{1}{2}\sum_{i=1}^{n} (x_i - \bar{x})^2\right]} = \exp\left[-\frac{n}{2}(\bar{x} - \lambda_0)^2\right], \qquad (28\text{–}25)$$

a function only of the random variable $\bar{x}$. Finally we need the critical value L_α satisfying, say, $\alpha = 0.05$,

$$\int_0^{L_{0.05}} h(L|H_0)\, dL = 0.05;$$

from which, substituting in (28–25), we can find the (two) critical values of the statistic $\bar{\mathbf{x}}$ for the problem at hand. As already indicated, the distribution of L for H_0 simple and true is known in general only for *large n*. But in the present example, L is a function only of $\bar{\mathbf{x}}$ and as the distribution of $\bar{\mathbf{x}}$ in random samples from normal populations is known for *any n*, we can here reach a solution valid for any n. Moreover, in the present example L_α need not in fact be found; from (28–25) it is evident that for each critical value of L there are two critical values of $\bar{\mathbf{x}}$ symmetrically placed around λ_0 and at the *tails* of the density function of $\bar{\mathbf{x}}$. The 0.05 level of L evidently corresponds to $\bar{x}_l$, $\bar{x}_u$, the two 0.025 levels of $\bar{x}$ as illustrated in Fig. 28–13, and the region for rejection of $H_0\colon \lambda = \lambda_0$ is seen to be identical with that produced by the uniformly most powerful *unbiased* test of $\lambda = \lambda_0$, namely,

$$\bar{x} \leq \lambda_0 - \frac{1.96}{\sqrt{n}} \text{ and } \bar{x} \geq \lambda_0 + \frac{1.96}{\sqrt{n}}. \tag{28–26}$$

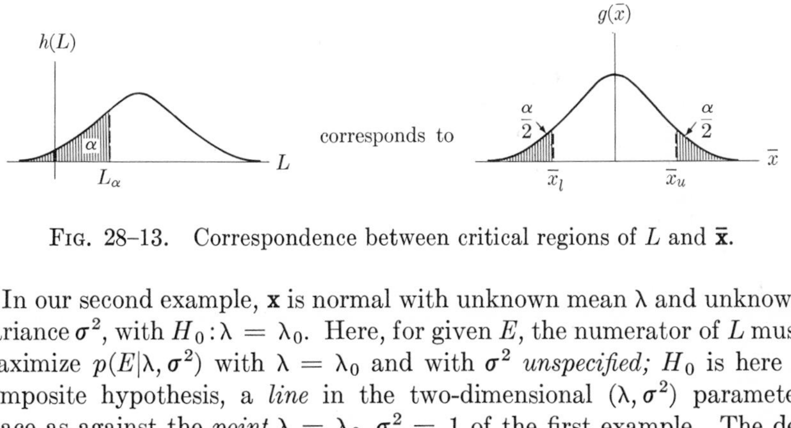

FIG. 28–13. Correspondence between critical regions of L and $\bar{\mathbf{x}}$.

In our second example, $\mathbf{x}$ is normal with unknown mean λ and unknown variance σ^2, with $H_0\colon \lambda = \lambda_0$. Here, for given E, the numerator of L must maximize $p(E|\lambda, \sigma^2)$ with $\lambda = \lambda_0$ and with σ^2 *unspecified;* H_0 is here a composite hypothesis, a *line* in the two-dimensional (λ, σ^2) parameter space as against the *point* $\lambda = \lambda_0$, $\sigma^2 = 1$ of the first example. The denominator must maximize $p(E|\lambda, \sigma^2)$ without restriction on λ *or* σ^2.

The numerator requires maximization only with respect to σ^2, for in the numerator, λ is again fixed at λ_0. As we know, the maximum likelihood estimator $\mathbf{T}_2$ of σ^2, for λ fixed at λ_0, is

$$\mathbf{T}_2 = \frac{1}{n}\sum_{i=1}^{n} (x_i - \lambda_0)^2,$$

and the numerator of L is

$$p_{\max}(E|\lambda_0, \sigma^2) = \left[\frac{n}{2\pi\sum_{i=1}^{n}(x_i - \lambda_0)^2}\right]^{n/2} e^{-n/2}.$$

The denominator requires maximization with respect to λ and σ^2,

$$\frac{\partial p(E|\lambda, \sigma)}{\partial \lambda} = 0, \qquad \frac{\partial p(E|\lambda, \sigma^2)}{\partial \sigma^2} = 0.$$

As we already know, the joint maximum likelihood estimators of λ and σ^2 are

$$\mathbf{T}_1 = \bar{x} \qquad \text{and} \qquad \mathbf{T}_2 = \frac{\sum_{i=1}^{n}(x_i - \bar{x})^2}{n},$$

and the denominator of L is

$$p_{\max}(E|\lambda, \sigma^2) = \left[\frac{n}{2\pi\sum_{i=1}^{n}(x_i - \bar{x})^2}\right]^{n/2} e^{-n/2}.$$

From these results, we have

$$L = \left\{\frac{\sum_{i=1}^{n}(x_i - \bar{x})^2}{\sum_{i=1}^{n}(x_i - \lambda_0)^2}\right\}^{n/2}. \tag{28–27}$$

Expanding the denominator of L, we have

$$\sum_{i=1}^{n}(x_i - \lambda_0)^2 = \sum_{i=1}^{n}(x_i - \bar{x})^2 + n(\bar{x} - \lambda_0)^2,$$

and recalling that in random samples from a normal population of mean λ_0 and unknown variance σ^2,

$$\mathbf{t} = \frac{\bar{\mathbf{x}} - \lambda_0}{\sqrt{\sum_{i=1}^{n}(\mathbf{x}_i - \bar{\mathbf{x}})^2/n(n-1)}} \tag{28–28}$$

has Student's distribution with $n - 1$ degrees of freedom, we get, after some reduction,

$$L = \left[\frac{1}{1 + t^2/(n-1)}\right]^{n/2}. \tag{28–29}$$

The statistic L whose distribution we know generally only for large n is here expressed as a function only of Student's $\mathbf{t}$ whose distribution we know for any n. As in the earlier example, we see from (28–29) that the critical value L_α for rejecting the hypothesis H_0 corresponds to the two extremities of the symmetrical tails, each of area $\alpha/2$, of the **t** distribution; this result holds for all n.

28–11 The likelihood ratio and sufficient statistics. In both examples the likelihood ratio leads to sufficient statistics, $\bar{\mathbf{x}}$ and **t**. This is generally true. Replace the parameters in (28–23) by their maximum likelihood estimators and the result follows, as it did for statistics produced by the Neyman-Pearson theorem. Briefly, if **u** is sufficient,

$$p(E; \theta) = g(u; \theta)h(E),$$

from (25–19). Also,

$$L = \frac{p_{\max}(E|\omega)}{p_{\max}(E|\Omega)} = \frac{g(u; \text{max. like. est. of } \theta \text{ in } \omega)}{g(u; \text{max. like. est. of } \theta \text{ in } \Omega)}.$$

But we have seen on page 261 that maximum likelihood estimators are functions of sufficient statistics; therefore L is a function only of sufficient statistics. More directly written,

$$L = \frac{\max_\omega g(u; \theta)h(E)}{\max_\Omega g(u; \theta)h(E)},$$

a function only of **u**. For further and valuable results see the two papers by Karlin and Rubin [98], [99].

28–12 Power functions. In the examples of this chapter, such statistics as $\bar{\mathbf{x}}$, **t**, and χ^2 have emerged as "best," from one principle or another, and though we gave no example, the statistic **F** emerges as "best" in a wide class of linear hypotheses to which we will come in Chapter 30. Recall that $R(\theta)$, the region of rejection of H_0 when H_0 is true, requires knowledge of the distribution of the appropriate statistic only for H_0 true. In each example, the principle which yielded the statistic and the critical range on it promised high power with respect to (certain) other parameter values $H_1\colon \theta \neq \theta_0$. But while we are sure that the statistics used have good properties when $\theta \neq \theta_0$, we may want to know how good. For this we need the probability distribution of the statistic for the alternatives $\theta \neq \theta_0$. In the case of $\bar{\mathbf{x}}$ in samples from normal populations, this is not difficult. On page 291 the student was asked to construct power curves for such regions as $\bar{\mathbf{x}} > c$ and $\bar{\mathbf{x}} < c$, for various alternative values of the mean $\lambda \neq \lambda_0$. The distributions of **t**, χ^2 and **F** for H_0 not true are, on the other hand, more complex. We merely indicate the best sources of tables of the probability behavior of such "noncentral" statistics. For t see Resnikoff's and Lieberman tables [165], which are remarkably complete; an example is given on (their) page 23 showing the application to the power function of the normal mean. Other good tables are by Johnson and

Welch [96]. For χ^2 see the excellent discussion by Patnaik [150] and by Sankaron [173]. For F see tables, now 30 years old, by Tang [184]; these are probably the best, and are reproduced in Mann [120]. Note also Pearson and Hartley [152], based on Tang's tables and in convenient form, with good examples.

Collateral Reading

H. D. Brunk, *An Introduction to Mathematical Statistics.* Boston, Ginn and Co., 1960.

H. A. David and C. A. Perez, "On comparing different tests of the same hypothesis," *Biometrika,* Vol. 47 (1960), pp. 297–306.

D. A. S. Fraser, *Nonparametric Methods in Statistics.* New York, John Wiley and Sons, 1957. For examples of unbiased tests, see pages 69–108.

————, *Statistics: An Introduction.* New York, John Wiley and Sons, 1958. An incisive discussion.

S. Karlin and H. Rubin, "The theory of decision procedures for distributions with monotone likelihood ratio," *Ann. Math. Stat.,* Vol. 27 (1956), pp. 272–300.

————, "Distributions possessing a monotone likelihood ratio," *Jour. Amer. Stat. Assoc.,* Vol. 51 (1956), pp. 637–643.

M. G. Kendall and A. Stuart, *The Advanced Theory of Statistics, Volume 2.* London, Charles Griffin and Company, 1961. An excellent modern reference.

E. L. Lehmann, "Some principles of the theory of testing hypotheses," *Ann. Math. Stat.,* Vol. 21 (1950), pp. 1–26. An admirable summary and discussion of some problems remaining.

————, *Testing Statistical Hypotheses.* New York, John Wiley and Sons, 1960. An advanced book and the authoritative source.

J. Neyman and E. S. Pearson, "On the use and interpretation of certain best criteria for purposes of statistical inference," *Biometrika,* Vol. 20A (1928), pp. 175–240, 263–294. Includes the likelihood ratio test.

————, "On the problem of the most efficient tests of statistical hypotheses," *Phil. Trans. Roy. Soc. London,* Series A, Vol. 231, (1933), pp. 289–337. Two of the pioneering papers.

J. Neyman and B. Tokarska, "Errors of the second kind in testing Student's hypothesis," *Jour. Amer. Stat. Assoc.,* Vol. 31 (1936), pp. 318–326.

E. S. Pearson and H. O. Hartley, "Charts of the power function for analysis of variance tests, derived from the non-central F distribution," *Biometrika,* Vol. 38 (1951), pp. 112–130.

R. M. Sundrum, "On the relation between estimating efficiency and the power of tests," *Biometrika,* Vol. 41 (1954), pp. 542–544. A statistic which has high efficiency in estimating an unknown parameter may not yield a powerful test for that parameter; good tests may better be based on inefficient estimators. For conditions when the statistic is normally distributed, see this note.

P. S. Tang, "The power function of the analysis of variance tests with tables and illustrations of their use," *Statistical Research Memoirs,* Vol. 2 (1938), pp. 126–157.

A. Wald, *Notes on the Theory of Statistical Estimation and Testing Hypotheses.* About 1941. Mimeographed notes on Wald's lectures at Columbia. Masterful and clear.

———, "Tests of statistical hypotheses concerning several parameters when the number of observations is large," *Trans. Amer. Math. Soc.*, Vol. 54 (1943), pp. 426–483. On large-sample properties of the likelihood ratio.

S. S. Wilks, *Mathematical Statistics.* New York, John Wiley and Sons, 1962.

E. B. Wilson, Jr., *An Introduction to Scientific Research.* New York, McGraw-Hill Book Co., 1952. A unique (and admirable) work.

Problems

28–1. Let $\mathbf{x}$ be normal with unknown mean and unknown variance. Show that the test of $H_0 : \sigma^2 = \sigma_0^2$ by the Neyman-Pearson theorem leads to the χ^2 distribution.

28–2. Obtain the likelihood ratio test (28–26) by direct development of $\mathbf{L}$.

CHAPTER 29

INTERVAL ESTIMATION

29–1 Fundamentals of confidence intervals. It may be argued that apart from situations in which one and only one number must be used as an estimate of a parameter as, for example, when the estimate is to be introduced as a number into an equation, even best point estimators are of mild interest. For note that if $\mathbf{E} : \mathbf{x}_1, \mathbf{x}_2, \ldots, \mathbf{x}_n$ are the variables of a random sample from a population described by the *continuous* random variable $\mathbf{x}$, then for *any* point function $\mathbf{T}$ of $\mathbf{E}$, even the "best" function, we have

$$p(\mathbf{T}|\theta_0) = 0,$$

where θ_0 is the true value of the parameter θ. It may therefore be better to seek two functions of $\mathbf{E}$, say,

$$\bar{\theta}(\mathbf{E}), \quad \underline{\theta}(\mathbf{E}), \qquad \bar{\theta}(\mathbf{E}) > \underline{\theta}(\mathbf{E}),$$

such that if in fact θ is equal to θ_0, the inequality

$$\underline{\theta}(\mathbf{E}) \leq \theta_0 \leq \bar{\theta}(\mathbf{E}),$$

preferably with $\bar{\theta}(\mathbf{E}) - \underline{\theta}(\mathbf{E})$ small (short), can be asserted with high probability,

$$p[\underline{\theta}(\mathbf{E}) \leq \theta_0 \leq \bar{\theta}(\mathbf{E})|\theta_0] = 1 - \alpha, \quad \alpha \text{ small.} \tag{29–1}$$

But θ_0, the true value of θ, is unknown to us; *any* possible value of θ could be the true value. Consequently we must require that (29–1) hold for *all* values of θ,

$$p[\underline{\theta}(\mathbf{E}) \leq \theta \leq \bar{\theta}(\mathbf{E})|\theta] = 1 - \alpha, \tag{29–2}$$

which is to be carefully read and understood. Not "the probability is $1 - \alpha$ that the true value of θ lies between two known functions $\bar{\theta}$ and $\underline{\theta}$ of an observed and therefore fixed point E". Such a probability is 1 or 0; the true value of θ which we regard as a constant either lies between such fixed quantities or it does not. The correct interpretation is: $1 - \alpha$ is the probability that the *random* interval defined by $\bar{\theta}(\mathbf{E})$ and $\underline{\theta}(\mathbf{E})$ includes the true value θ. Probabilities are always on random variables [here $\bar{\theta}(\mathbf{E})$ and $\underline{\theta}(\mathbf{E})$], not on parameters (here θ) nor on fixed points in the event space (a particular value E of $\mathbf{E}$). The probability (29–2) relates to the relative frequency with which the *random* interval $\bar{\theta}(\mathbf{E}) - \underline{\theta}(\mathbf{E})$ covers the truth;

it is a probability associated with $\mathbf{E}: \mathbf{x}_1, \mathbf{x}_2, \ldots, \mathbf{x}_n$ *before* (random variable), and not with $E: x_1, x_2, \ldots, x_n$ *after* (fixed point) the sample is drawn.

One may think of such random intervals as a way of life, and interpret (29–2) as follows: "if a large number of statements are made on any and all parameters θ each using the inequalities shown in (29–2), 100 $(1 - \alpha)$ percent of these statements will be correct." Against the background of this repeated-use interpretation of (29–2), we can often sensibly speak of a *confidence interval*

$$\underline{\theta}(E) \leq \theta \leq \bar{\theta}(E), \tag{29–3}$$

where E is a *particular* observed point in the event space, and $\bar{\theta}(E)$ and $\underline{\theta}(E)$ are, therefore, fixed numbers. We associate no probability (other than 1 or 0) with (29–3), but we do associate a *confidence coefficient* $1 - \alpha$ with the procedure which we used to get (29–3), and we interpret the confidence interval and the confidence coefficient against the background of repeated use. Applied to a *particular* problem at hand, we treat the confidence interval appropriate to that problem in an approximate $(1 - \alpha)$ probability or relative frequency sense; it is evident that unless some such relative frequency interpretation of the interval constructed from the data of a particular problem is tolerated, confidence interval theory can have little useful application.

Some points E yield $\bar{\theta}(E)$ and $\underline{\theta}(E)$ which bracket (cover) a particular value of θ, say $\theta = \theta'$; this is illustrated in the x_1, x_2, θ sample-parameter space diagram of Fig. 29–1. Some points, such as E', do not cover θ'. For

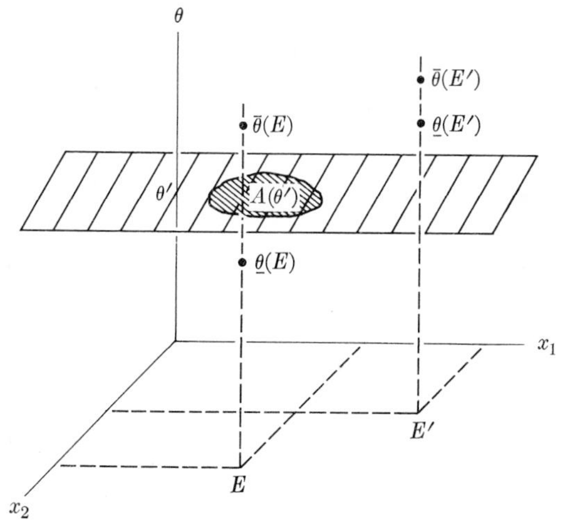

FIG. 29–1. A sample-parameter space with a region of acceptance.

all points E for which $\bar{\theta}(E)$ and $\underline{\theta}(E)$ cover θ', we can write

$$\underline{\theta}(E) \leq \theta' \leq \bar{\theta}(E),$$

and we can project such points on the plane $\theta = \theta'$, as in Fig. 29–1. In the plane $\theta = \theta'$, such points form $A(\theta')$, called the *region of acceptance* of $\theta = \theta'$. We will soon consider this region.

Summarizing the discussion to this point, we have a population described by the random variable $\mathbf{x}$ with density function $f(x;\theta)$, θ unknown. A random sample $\mathbf{E}:\mathbf{x}_1, \mathbf{x}_2, \ldots, \mathbf{x}_n$ is to be drawn; it will yield $E:x_1, x_2, \ldots, x_n$. We wish to form $\bar{\theta}(E)$ and $\underline{\theta}(E)$, with

$$\bar{\theta}(E) > \underline{\theta}(E) \qquad \text{and} \qquad p[\underline{\theta}(E) \leq \theta \leq \bar{\theta}(E)|\theta] = 1 - \alpha, \tag{29–4}$$

the latter to hold for all values of θ. Subject to certain conditions, such $\bar{\theta}(E)$ and $\underline{\theta}(E)$ are the limits of a *confidence interval* for θ with *confidence coefficient* (likelihood of correct statement) $1 - \alpha$.

Several (heavily interrelated) problems must be discussed. Among them, methods for finding *any* confidence interval (determination of $\bar{\theta}$ and $\underline{\theta}$), methods, if any, for finding the best (shortest) confidence interval, choice of $1 - \alpha$, choice of sample size n.

29–2 Relationship to testing hypotheses. We now consider the relationship between testing hypotheses and confidence intervals. In Chapter 28 we considered tests of the hypothesis $H_0:\theta = \theta_0$. As we saw there, the critical region R in the event space depended on θ_0; we can write $R(\theta_0)$. The size of the critical region $R(\theta_0)$ was α (small), and our operational rule for rejecting and "accepting" hypotheses may be stated:

$$\text{Reject } H_0:\theta = \theta_0 \text{ if } E \text{ falls in } R(\theta_0),$$

$$\text{Accept } H_0:\theta = \theta_0 \text{ if } E \text{ falls in } A(\theta_0),$$

where $A(\theta_0)$ is that part of the event or sample space not included in the critical region $R(\theta_0)$. The region $A(\theta_0)$, sometimes called the *complementary* or acceptance region, clearly also depends on θ_0. The probabilities associated with these operational rules were

$$p[\mathbf{E} \in R(\theta_0)|\theta_0] = \alpha, \qquad \text{and} \qquad p[\mathbf{E} \in A(\theta_0)|\theta_0] = 1 - \alpha, \tag{29–5}$$

for any fixed $\theta = \theta_0$.

If a region $A(\theta)$ exists for every value of θ, we write

$$p\{\mathbf{E} \in A(\theta)|\theta\} = 1 - \alpha.$$

When $\mathbf{E} \in A(\theta)$, there exists a corresponding region ω in the parameter space such that $\theta \in \omega$. As ω depends on $\mathbf{E}$, we write $\omega(\mathbf{E})$.

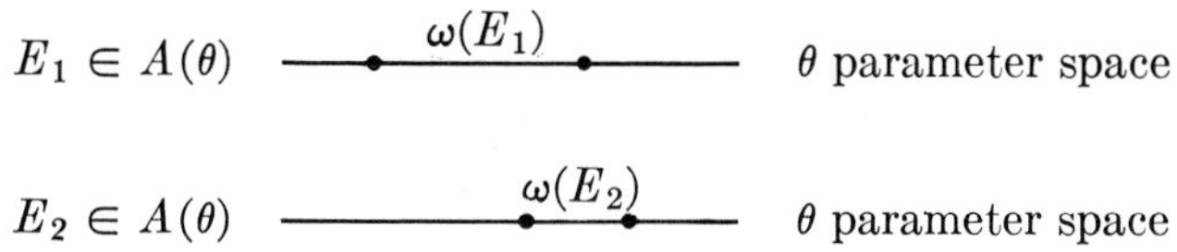

That is,

$$\mathbf{E} \in A(\theta) \Leftrightarrow \omega(\mathbf{E}) \supset \theta,$$

so we shift from probabilities on the random variable $\mathbf{E}$ to probabilities on the random interval $\omega(\mathbf{E})$:

$$p\{\mathbf{E} \in A(\theta)|\theta\} = p\{\omega(\mathbf{E}) \supset \theta|\theta\} = 1 - \alpha. \tag{29–6}$$

If the lower limit of $\omega(\mathbf{E})$ is written $\underline{\theta}(\mathbf{E})$ and the upper limit $\bar{\theta}(\mathbf{E})$, we have

$$p[\underline{\theta}(\mathbf{E}) \leq \theta \leq \bar{\theta}(\mathbf{E})|\theta] = 1 - \alpha,$$

which is (29–2). If for *every* E the set $\omega(E)$ is closed and nonempty, then $\bar{\theta}(E)$, $\underline{\theta}(E)$ form a confidence interval on θ, with confidence coefficient $1 - \alpha$; the relationship between testing hypotheses and confidence intervals is seen to be direct.

Most, perhaps all, tests, efficient or not, of a hypothesis on a parameter θ can be converted into confidence intervals on that parameter. In general, an α-size test for *each* hypothesis $\theta = \theta_0$ will produce a $(1 - \alpha)$ confidence-coefficient interval for the parameter θ. We therefore review briefly three optimal tests of statistical hypotheses introduced in Chapter 28 preliminary to writing down the (optimal) confidence interval equivalents of these tests.

Uniformly most powerful test of each $\theta = \theta_0$. We required a region R with size α equal to that of any other region R' and with power equal to or greater than R':

$$\begin{aligned} p[E \in R|\theta_0] &= p[E \in R'|\theta_0] = \alpha, \\ p[E \in R|\theta] &\geq p[E \in R'|\theta]. \end{aligned} \tag{29–7}$$

We showed that if R and a positive constant k existed such that

$$\frac{p(\mathbf{E}|\theta_0)}{p(\mathbf{E}|\theta)} \quad \begin{array}{l} \leq k \text{ inside } R, \\ \geq k \text{ outside } R, \end{array}$$

then R was most powerful with respect to θ. Such regions seldom exist.

Uniformly most powerful unbiased test of each $\theta = \theta_0$. We required a region R which was unbiased, with size α equal to that of any other *unbiased* region R', and with power equal to or greater than that of any other unbiased region R':

$$p[E \in R|\theta_0] = p[E \in R'|\theta_0] = \alpha, \qquad p[E \in R|\theta] \geq p[E \in R'|\theta]. \tag{29–8}$$

Such regions exist more often than uniformly most powerful regions.

Unbiased test with maximum power in the vicinity of $\theta = \theta_0$. We required an unbiased region R which, relative to all other *unbiased* regions R', had maximum rate of increase of power in the vicinity of θ_0:

$$\frac{\partial}{\partial\theta} p[E \in R|\theta]_{\theta=\theta_0} = 0, \qquad \frac{\partial^2}{\partial\theta^2} p[E \in R|\theta]_{\theta=\theta_0} \geq \frac{\partial^2}{\partial\theta^2} p[E \in R'|\theta]_{\theta=\theta_0}. \tag{29–9}$$

Now consider the confidence intervals which, if they exist at all, correspond to these superior tests. From (29–7), we have

$$p[E \in A|\theta_0] = p[E \in A'|\theta_0] = 1 - \alpha, \quad \text{and} \quad p[E \in A|\theta] \leq p[E \in A'|\theta], \tag{29–10}$$

or the equivalent statements

$$p\{\underline{\theta}(E) \leq \theta_0 \leq \bar{\theta}(E)|\theta_0\} = p\{\underline{\theta}'(E) \leq \theta_0 \leq \bar{\theta}'(E)|\theta_0\} = 1 - \alpha, \tag{29–11}$$

$$p\{\underline{\theta}(E) \leq \theta_0 \leq \bar{\theta}(E)|\theta\} \leq p\{\underline{\theta}'(E) \leq \theta_0 \leq \bar{\theta}'(E)|\theta\}$$

for each possible θ_0 and θ. Such optimal intervals seldom exist. Note that the confidence interval associated with the uniformly most powerful test is optimal, though not in the sense of shortest $\bar{\theta}(E) - \underline{\theta}(E)$. Rather,

$$p\{\underline{\theta}(E) \leq \theta \leq \bar{\theta}(E)|\theta_0\} \leq p\{\underline{\theta}'(E) \leq \theta \leq \bar{\theta}'(E)|\theta_0\};$$

the interval is "shortest" in the sense of being least likely to cover a false value of θ. See the admirable discussion by J. W. Pratt, *Jour. Amer. Stat. Assoc.* (1961) pp. 549–567.

A similar transposition, which the student should write out, is available for uniformly most powerful unbiased tests. Finally, the more generally available test (29–9) is equivalent to the "shortest unbiased" confidence interval,

$$\frac{\partial}{\partial\theta} p[\underline{\theta}(E) \leq \theta_0 \leq \bar{\theta}(E)|\theta]_{\theta=\theta_0} = 0,$$

and

$$\frac{\partial^2}{\partial\theta^2} p[\underline{\theta}(E) \leq \theta_0 \leq \bar{\theta}(E)|\theta]_{\theta=\theta_0} \leq \frac{\partial^2}{\partial\theta^2} p[\underline{\theta}'(E) \leq \theta_0 \leq \bar{\theta}'(E)|\theta]_{\theta=\theta_0}. \tag{29–12}$$

As our confidence intervals are essentially complements of tests of statistical hypotheses, "shortest" intervals are generally associated with "best" tests. It will therefore not be surprising that the examples shown later in this chapter utilize such statistics as $\bar{\mathbf{x}}$, $\mathbf{t}$, and χ^2 (the same statistics produced in Chapter 28 by best tests of hypotheses).

29–3 An example due to Neyman. Before considering further general aspects of confidence intervals, as well as detailed application of the theory to familiar and important distributions, let us first take up a simple but illuminating example due to Neyman [137] and used by Wald [195]. It illustrates the relationship between testing statistical hypotheses and confidence intervals, introduces points not already considered, and makes evident the superiority of one confidence interval over another.

Consider a population described by the uniform distribution,

$$f(x; \theta) = \frac{1}{\theta}, \quad 0 \leq x \leq \theta,$$

with θ unknown. A random sample of size $n = 2$ is drawn; the sample variables $\mathbf{x}_1$, $\mathbf{x}_2$ form the point $\mathbf{E}$ in the two-dimensional event or sample space. We will try four different *ad hoc* regions $R(\theta)$, that is, four tests of a simple hypothesis on θ, each of size α. We will see if these regions lead to satisfactory confidence intervals and if some intervals are better than others.

The hypothesis to be tested will be written $H : \theta = \theta$; this notation will in fact create no confusion and will facilitate the inversion to confidence intervals.

First region. The first critical region $R(\theta)$ is defined by

$$0 \leq x_1 \leq k\theta \quad \text{or} \quad 0 \leq x_2 \leq k\theta \quad \text{or} \quad x_1 > \theta \quad \text{or} \quad x_2 > \theta, \tag{29–13}$$

with $0 \leq k \leq 1$. This appears to be a reasonable region, for we reject θ if either of the sample variates is "somewhat" smaller or at all larger than θ. Inequalities (29–13) fix the complementary or acceptance region $A(\theta)$ to be

$$k\theta \leq x_1 \leq \theta \qquad \text{and} \qquad k\theta \leq x_2 \leq \theta, \tag{29–14}$$

illustrated in the sample-parameter space diagram of Fig. 29–2.

If θ is true, the complete event space is the square θ^2, while (again for θ true) $A(\theta)$ given by (29–14) is the smaller square $(\theta - k\theta)^2$. We have

$$p[E \in A(\theta)|\theta] = 1 - \alpha = \frac{(\theta - k\theta)^2}{\theta^2} = (1 - k)^2,$$

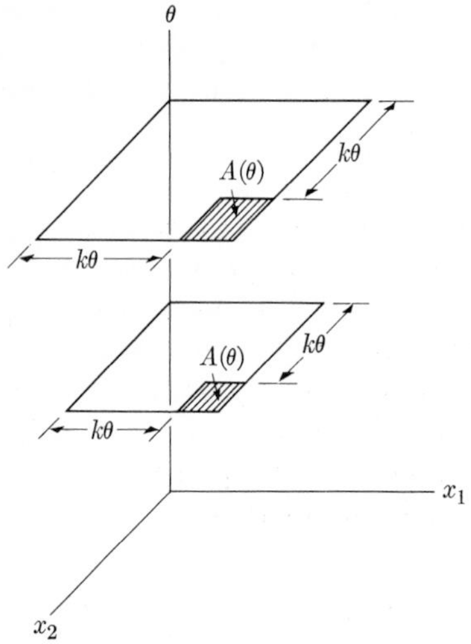

FIG. 29–2 First acceptance region $A(\theta)$ in the event space of x_1, x_2 shown as a function of θ.

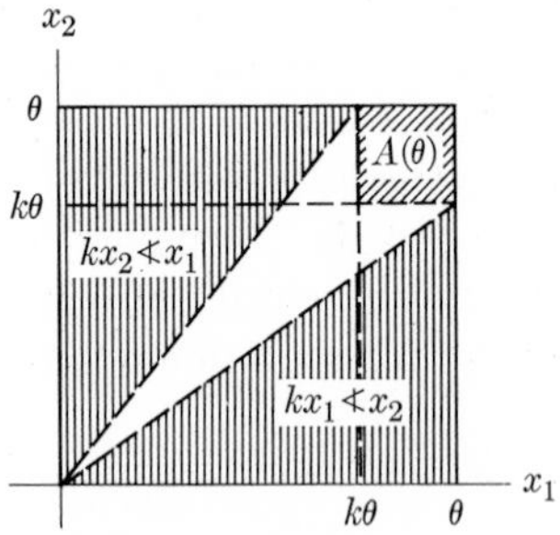

FIG. 29–3. Regions in the event space x_1, x_2 for which $kx_1 \not< x_2$ and $kx_2 \not< x_1$.

and k is fixed at

$$k = 1 - \sqrt{1 - \alpha}.$$

The inequalities (29–14) become

$$x_1 \le \theta \le \frac{x_1}{1 - \sqrt{1 - \alpha}} \qquad \text{and} \qquad x_2 \le \theta \le \frac{x_2}{1 - \sqrt{1 - \alpha}}, \tag{29–15}$$

intervals on θ, that is, the set of θ we accept for observed x_1, x_2.

Could this set be empty? That is, are there x_1, x_2 which do not cover *any* θ? The answer unfortunately is yes. Let s be the smaller of the two sample variates, l the larger. Using both inequalities of (29–15), we have

$$l \le \theta \le \frac{s}{k} \quad \text{from which} \quad kl \le s. \tag{29–16}$$

Thus if $x_2 = l$ and $x_1 = s$, the interval (29–15) requires $kx_2 \le x_1$; if $x_1 = l$ and $x_2 = s$, the interval (29–15) requires $kx_1 \le x_2$. But, as illustrated by the shaded areas in Fig. 29–3, there are many points x_1, x_2 which do not satisfy these inequalities. Such points x_1, x_2 fall in no region of acceptance, and the intervals (29–15) cannot constitute a confidence interval. Essentially, the failure is due to the fact that we have two intervals in (29–15); both intervals must hold and this is not possible here.

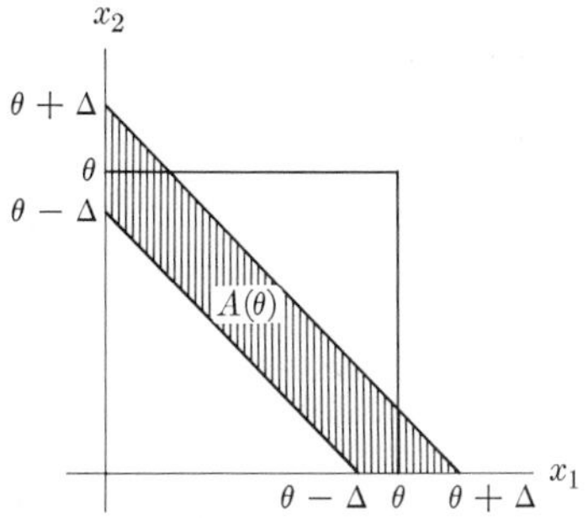

FIG. 29-4. The second region of acceptance of θ, defined by (29-18).

Second region. The second critical region $R(\theta)$ is defined by

$$|x_1 + x_2 - \theta| \geq \Delta. \tag{29-17}$$

To this corresponds a complementary or acceptance region $A(\theta)$ defined by

$$|x_1 + x_2 - \theta| \leq \Delta \qquad \text{or} \qquad \theta - \Delta \leq x_1 + x_2 \leq \theta + \Delta, \tag{29-18}$$

again an intuitively promising acceptance region, since $(\mathbf{x}_1 + \mathbf{x}_2)/2$ is an unbiased estimator of $\theta/2$ and so, therefore, is $\mathbf{x}_1 + \mathbf{x}_2$ of θ. The region, for any θ, is shown in Fig. 29-4. Note that we do not reject the hypothesis θ even if x_1 or x_2 is somewhat larger than θ. This would appear to lead to a confidence interval which in some sense must be inefficient, since θ must certainly be as large as x_1 or x_2. For θ true, the whole event space has area θ^2 while (again for θ true) $A(\theta)$ as given by (29-17) has area $\theta^2 - (\theta - \Delta)^2$. Therefore,

$$p[E \in A(\theta)|\theta] = 1 - \alpha = \frac{\theta^2 - (\theta - \Delta)^2}{\theta^2},$$

and

$$\Delta = \theta(1 - \sqrt{\alpha}).$$

Thus (29-18) becomes

$$\frac{x_1 + x_2}{2 - \sqrt{\alpha}} \leq \theta \leq \frac{x_1 + x_2}{\sqrt{\alpha}}. \tag{29-19}$$

Note that, unlike the first interval (29-15), this set of θ cannot be empty for *any* x_1, x_2.

Third region. Remove from the second complementary region the possibility of accepting the hypothesis θ when x_1 or x_2 is greater than θ. That is, add to the second critical region $R(\theta)$ as given by (29-17) the regions

$$x_1 > \theta, \qquad x_2 > \theta.$$

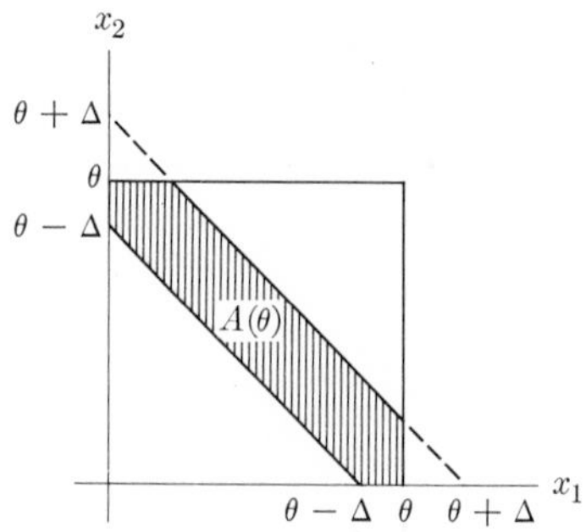

FIG. 29–5. The third region of acceptance of θ, defined by (29–20).

For θ true, the acceptance region $A(\theta)$ is not altered, and we have (29–18) as before. The confidence interval for θ becomes

$$\frac{x_1 + x_2}{2 - \sqrt{\alpha}} \leq \theta \leq \frac{x_1 + x_2}{\sqrt{\alpha}}, \qquad \theta \geq x_1, \qquad \text{and} \qquad \theta \geq x_2. \tag{29–20}$$

The situation, for all θ, is shown in Fig. 29–5.

The intervals of (29–20) may be replaced by one interval, as follows. If l is the larger of x_1, x_2, the two right-hand intervals of (29–20) may be written

$$l \leq \theta < \infty,$$

for only the larger variate matters in the right-hand intervals of (29–20). Also

$$l \leq x_1 + x_2 \leq \frac{x_1 + x_2}{\sqrt{\alpha}},$$

the lower limit l is less than the upper limit $(x_1 + x_2)/\sqrt{\alpha}$ of (29–20). So we finally have

$$u \leq \theta \leq \frac{x_1 + x_2}{\sqrt{\alpha}}, \tag{29–21}$$

where u is the larger of l and $(x_1 + x_2)/(2 - \sqrt{\alpha})$.

Fourth region. Let l be the larger of x_1, x_2 and let $0 \leq e \leq 1$. The fourth critical region $R(\theta)$ is defined by

$$l > \theta \qquad \text{or} \qquad l < e\theta. \tag{29–22}$$

Again this appears to be a reasonable critical region for we are rejecting the hypothesis θ when the larger sample variate is larger than θ or "somewhat" smaller than θ. For θ true, the acceptance region $A(\theta)$ is

$$e\theta \leq l \leq \theta; \tag{29–23}$$

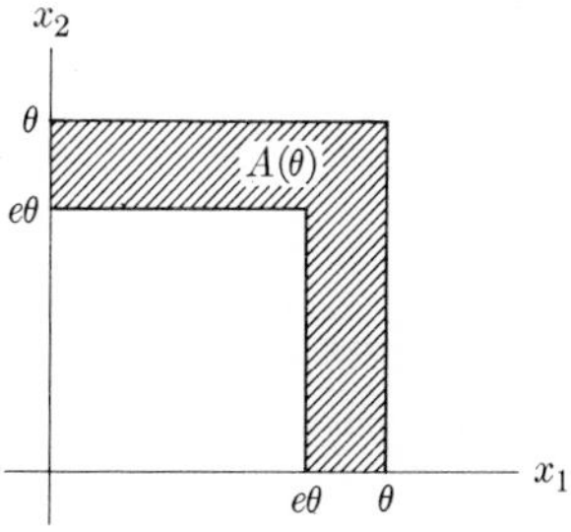

FIG. 29–6. The fourth region of acceptance of θ, defined by (29–23).

the region of acceptance is shown in Fig. 29–6;

$$p[E \in A(\theta)|\theta] = 1 - \alpha = \frac{\theta^2 - (e\theta)^2}{\theta^2}, \qquad \text{and} \qquad e = \sqrt{\alpha};$$

the confidence interval is the closed and nonempty interval

$$l \leq \theta \leq \frac{l}{\sqrt{\alpha}}.$$

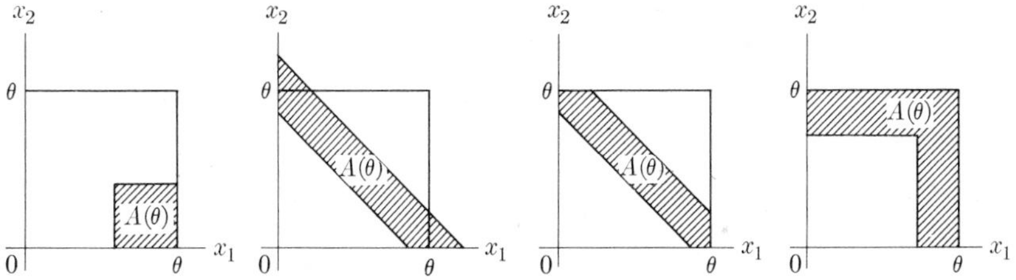

FIG. 29–7. The four regions of acceptance.

Now we turn to a comparison of the four intervals (of many more that might have been considered) which we assemble in Fig. 29–7. In the first interval, since some sample points lead to empty sets, we do not have a proper confidence interval. As the second interval is *always* wider than the third interval, we call the second interval *inadmissible*. Finally, let us judge the third and fourth intervals by comparison of their lengths $\bar{\theta}(E) - \underline{\theta}(E)$:

Length of third interval:

$$\text{smaller of } (x_1 + x_2)\left(\frac{1}{\sqrt{\alpha}} - \frac{1}{2 - \sqrt{\alpha}}\right) \quad \text{and} \quad \frac{x_1 + x_2}{\sqrt{\alpha}} - l,$$

Length of fourth interval: $\dfrac{l}{\sqrt{\alpha}} - l.$

Consider $\alpha = \frac{1}{4}$; we have

Length of third interval: smaller of $\frac{4}{3}(x_1 + x_2)$ and $2(x_1 + x_2) - l$,

Length of fourth interval: l.

But $x_1 + x_2 = l$ only if the smaller variate is zero, so

$$\tfrac{4}{3}(x_1 + x_2) > \tfrac{4}{3}l > l,$$

and the fourth interval is shorter than the third. It is readily seen to be shorter for all α. In fact, the uniformly most powerful test of θ with size α is easily shown to be

$$l > \theta, \qquad l < \theta\sqrt{\alpha},$$

and, therefore, the complementary region leading to the confidence interval

$$l \leq \theta \leq \frac{l}{\sqrt{\alpha}},$$

with confidence coefficient $1 - \alpha$, is uniformly "shortest" in the (apparently unrelated [but see Pratt, *loc. cit.*]) sense of Neyman (page 312).

Note that while the fourth interval is shortest, actual intervals depend, of course, on values of the variates. For $x_1 = 1$ and $x_2 = 1$, and $\alpha = \frac{1}{4}$, the third and fourth intervals are respectively

$$\tfrac{4}{3} \leq \theta \leq 4 \qquad \text{and} \qquad 1 \leq \theta \leq 2,$$

which overlap. For $x_1 = .1$ and $x_2 = 1.9$, and $\alpha = \frac{1}{4}$, the third and fourth intervals are respectively

$$\tfrac{4}{3} \leq \theta \leq 4 \qquad \text{and} \qquad 1.9 \leq \theta \leq 3.8,$$

one contained in the other.

29–4 Pivotal random variables. As we have seen, any random variable which provides a test of a hypothesis on θ at every possible value of θ will lead to a confidence interval on θ. We can find a confidence interval for θ if we can find a statistic $\mathbf{z}$ whose probability behavior is known for each value of θ. If a test by $\mathbf{z}$ is available, with size α, for each hypothesis $\theta = \theta_0$, the set of θ's *not rejected* by $\mathbf{z}$ form, as we have seen, a confidence set for θ with confidence coefficient $1 - \alpha$. We prefer a random variable which is based on a sufficient statistic, since such statistics lead to optimal tests of hypotheses, and therefore (generally) to optimal intervals. In particular, a *pivotal* random variable (better, a pivotal quantity) is preferred. A pivotal quantity is defined as having the following (good) properties: (a) it depends on $\mathbf{x}_1, \mathbf{x}_2, \ldots, \mathbf{x}_n$ only through a sufficient statistic; (b) it depends on the parameter on which a confidence interval is wanted and on no other, and (c) it has a fixed distribution independent of all parameters. For example, in testing a hypothesis on the mean of a normal distribution of known variance, the normally distributed sufficient statistic $\bar{\mathbf{x}}$ is the statistic of choice. It leads to the random variable

$$\mathbf{u} = \frac{\bar{\mathbf{x}} - \lambda}{\sigma/\sqrt{n}}, \tag{29–24}$$

which has a *fixed* normal distribution (mean 0, variance 1) and is pivotal for a confidence interval on the mean λ of a normal variable with *known* variance σ^2. Similarly

$$\mathbf{t} = \frac{\bar{\mathbf{x}} - \lambda}{\mathbf{s}/\sqrt{n-1}}, \tag{29–25}$$

which has a *fixed* Student's distribution with $n - 1$ degrees of freedom, is pivotal (and $\mathbf{u}$ is not) for a confidence interval on the mean λ of a normal variable with *unknown* variance σ^2.

To illustrate $\mathbf{u}$ and $\mathbf{t}$ in detail,

$$p(-1.96 \le \mathbf{u} \le 1.96) = 0.95 \quad \text{for all } n,$$

and

$$p(-2.23 \le \mathbf{t} \le 2.23) = 0.95 \quad \text{for } n - 1 = 10,$$

and since pivotal quantities can always be rearranged to show directly the confidence interval on the parameter,

$$p\left(\bar{x} - 1.96\frac{\sigma}{\sqrt{n}} \le \lambda \le \bar{x} + 1.96\frac{\sigma}{\sqrt{n}}\right) = 0.95 \quad \text{for all } n,$$

$$p\left(\bar{x} - 2.23\frac{s}{\sqrt{n-1}} \le \lambda \le \bar{x} + 2.23\frac{s}{\sqrt{n-1}}\right) = 0.95$$

$$\text{for } n - 1 = 10.$$

The argument in somewhat more general terms is as follows: if we have a random variable $\mathbf{z}$ whose distribution is known for every possible value of $\theta = \theta_0$, we can always find two numbers a and b (which generally depend on θ_0) with $b > a$ and such that the probability that $\mathbf{z}$ will lie in the ranges

$$\mathbf{z} \ge b(\theta_0) \qquad \text{or} \qquad \mathbf{z} \le a(\theta_0) \tag{29–26}$$

is α; the hypothesis $\theta = \theta_0$ is rejected when (29–26) occurs. Or,

$$p\{a(\theta) \le \mathbf{z} \le b(\theta) | \theta\} = 1 - \alpha, \tag{29–27}$$

and (29–27) may (generally) be rearranged to produce the interval

$$p\{\underline{\theta}(E) \le \theta \le \bar{\theta}(E) | \theta\} = 1 - \alpha, \tag{29–28}$$

where $\underline{\theta}(E)$ and $\bar{\theta}(E)$ depend on z but not on θ. As we have seen, the transition from (29–27) to (29–28) may occasionally fail to yield a proper confidence interval.

29–5 Geometrical interpretation. A geometrical interpretation of confidence intervals is useful. Let the density function of $\mathbf{x}$ be $f(x; \theta)$; we have a random variable $\mathbf{z}$ by which we can test the hypothesis $H_0 : \theta = \theta_0$ with size α, for every possible value of θ, the region of rejection of H_0 being the tails of the distribution of $\mathbf{z}$. That is, we can find a and b in

$$\int_a^b g(z; \theta)\, dz = 1 - \alpha \tag{29–29}$$

for every value of θ. Now a and b will generally be functions of θ. Plot $z = a(\theta)$ and $z = b(\theta)$ in the z, θ plane, as illustrated in Fig. 29–8. An observed sample E yields a particular value of z, say z'. Given z', we can read from Fig. 29–8 the values of $a(\theta)$ and $b(\theta)$ that correspond to z', and the lower and upper limits of θ corresponding to z'. Usually $a(\theta)$ and $b(\theta)$ are monotonic functions of θ, so there will be only one pair of limits $\underline{\theta}(z')$ and $\bar{\theta}(z')$ corresponding to z'.

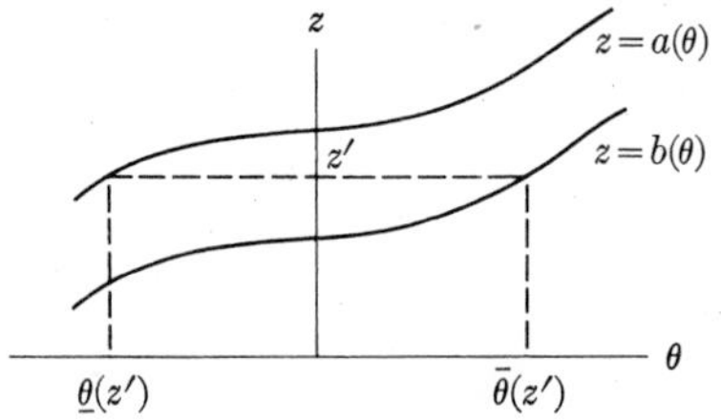

FIG. 29–8. The z, θ plane and the functions $z = a(\theta)$ and $z = b(\theta)$.

We illustrate this geometrical argument by the first of the two earlier examples. Consider a normal variable with unknown mean λ and known variance $\sigma^2 = 1$. The statistic by which we test a hypothesis on λ for every possible value of λ is $\bar{\mathbf{x}}$; we illustrate the geometry on $\bar{\mathbf{x}}$ itself, though the pivotal quantity (29–24) would have tied in more immediately with the discussion of this section.

$$\int_{a(\lambda)}^{b(\lambda)} N\left(\bar{x}; \lambda, \frac{1}{n}\right) d\bar{x} = 1 - \alpha = \text{say } 0.95, \tag{29–30}$$

with

$$\frac{a(\lambda) + b(\lambda)}{2} = \lambda,$$

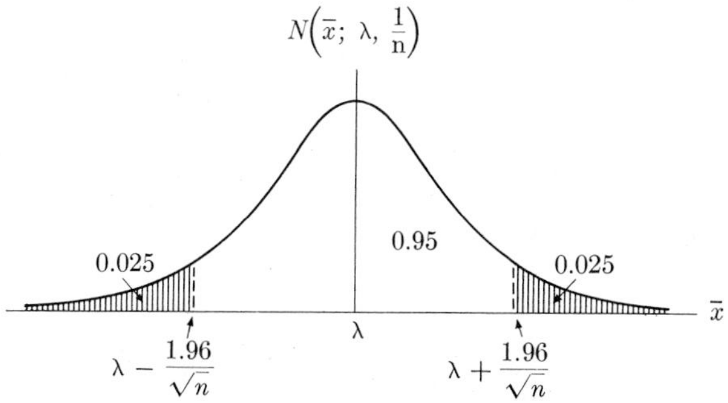

FIG. 29–9. Critical regions on $\bar{x}$ in a test of the normal mean, variance 1.

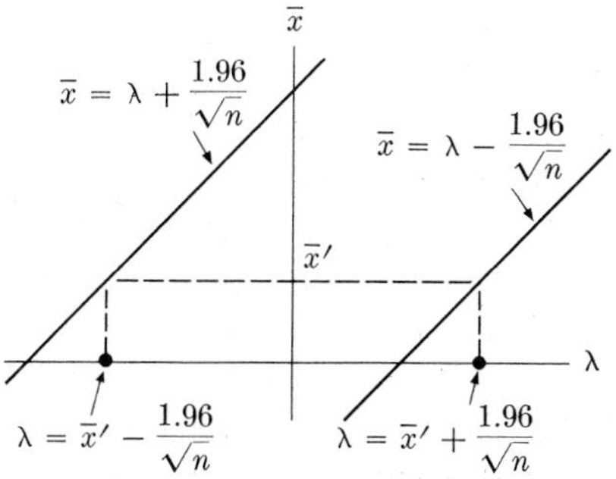

FIG. 29–10. The $\bar{x}$, λ plane and the functions $b(\lambda) = \lambda + 1.96/\sqrt{n}$ and $a(\lambda) = \lambda - 1.96/\sqrt{n}$.

where, for every λ, $\bar{\mathbf{x}}$ is normally distributed with mean λ and variance $1/n$. Integral (29–30) is illustrated in Fig. 29–9, where $b(\lambda) = \lambda + 1.96/\sqrt{n}$ and $a(\lambda) = \lambda - 1.96/\sqrt{n}$. In Fig. 29–10, these functions $a(\lambda)$ and $b(\lambda)$ are shown in the plane of $\bar{x}$, λ, and the upper and lower 0.95 confidence limits on λ, corresponding to an observed sample mean $\bar{x}'$, are seen to be $\bar{x}' + 1.96/\sqrt{n}$ and $\bar{x}' - 1.96/\sqrt{n}$.

29–6 Examples of confidence intervals. The following are examples of confidence intervals on parameters of familiar probability distributions. In each example the confidence interval is the complement of a "best" test; the interval will therefore generally be optimal in a sense corresponding to that in which the test was best.

Example 1. The mean λ of a normal distribution of known variance σ^2. On several criteria the preferred test statistic of a hypothesis on λ has been shown to be $\bar{\mathbf{x}}$, which, for any sample size n, is normally distributed with mean λ and variance σ^2/n. With alternatives on both sides of the hypothesis on λ, the likelihood ratio text leads to

$$p\left[\lambda - 1.96\frac{\sigma}{\sqrt{n}} \leq \bar{\mathbf{x}} \leq \lambda + 1.96\frac{\sigma}{\sqrt{n}}\right] = 0.95,$$

or generally,

$$p\left[\lambda - z_{\alpha/2}\frac{\sigma}{\sqrt{n}} \leq \bar{\mathbf{x}} \leq \lambda + z_{\alpha/2}\frac{\sigma}{\sqrt{n}}\right] = 1 - \alpha,$$

which may be written

$$p\left[-z_{\alpha/2} \leq \frac{\bar{\mathbf{x}} - \lambda}{\sigma/\sqrt{n}} \leq z_{\alpha/2}\right] = 1 - \alpha,$$

or to exhibit the confidence interval on λ directly,

$$p\left[\bar{x} - z_{\alpha/2}\frac{\sigma}{\sqrt{n}} \leq \lambda \leq \bar{x} + z_{\alpha/2}\frac{\sigma}{\sqrt{n}}\right] = 1 - \alpha.$$

Numerical example. Given $\sigma = 0.0021$, $\alpha = 0.10$, $n = 100$, $\bar{x} = 0.1022$; we find $z_{\alpha/2} = 1.645$, and $0.10185 \leq \lambda \leq 0.10255$.

For this example, we write out somewhat elaborately what has already been said generally. From sampling theory we know that 95 percent of all sample means formed from random samples of fixed size n will fall with $\pm 1.96\sigma(\bar{\mathbf{x}})$ of the fixed population mean λ. Or, what is the same thing, in 95 percent of samples, each of fixed size n, the random interval $\bar{\mathbf{x}} \pm 1.96\sigma(\bar{\mathbf{x}})$ will contain the population mean λ. These two statements are described by the equivalent expressions

$$p\{|\bar{\mathbf{x}} - \mu| < 1.96\sigma(\bar{\mathbf{x}})\} \qquad \text{and} \qquad p\{\bar{\mathbf{x}} - 1.96\sigma(\bar{\mathbf{x}}) \leq \lambda \leq \bar{\mathbf{x}} + 1.96\sigma(\bar{\mathbf{x}})\},$$

both of which equal 0.95. We now go further and claim that given a single observed sample of size n, the *fixed* interval $\bar{x} + 2\sigma(\bar{\mathbf{x}})$ calculated from the data of this sample contains λ, though only in the sense that 95 percent of such claims are true. The last is a confidence, not a probability, statement, but, as already indicated, it is evident that unless some $1 - \alpha$ level quasi-probability or relative frequency interpretation of the interval formed from the outcome of a single experiment is tolerated, confidence interval theory will have little useful application.

Example 2. Difference of means $E(\mathbf{x}) - E(\mathbf{y})$ of two normal variables, variances $\sigma^2(\mathbf{x})$ and $\sigma^2(\mathbf{y})$ known. The preferred test statistic $\bar{\mathbf{x}} - \bar{\mathbf{y}}$ is, for any sample sixes n_x and n_y, normally distributed with mean

$$E(\mathbf{x}) - E(\mathbf{y})$$

and variance

$$\frac{\sigma^2(\mathbf{x})}{n_x} + \frac{\sigma^2(\mathbf{y})}{n_y}.$$

For alternatives on both sides of $E(\mathbf{x}) - E(\mathbf{y})$, the likelihood ratio test leads to

$$p\left\{E(\mathbf{x}) - E(\mathbf{y}) - z_{\alpha/2}\sqrt{\frac{\sigma^2(\mathbf{x})}{n_x} + \frac{\sigma^2(\mathbf{y})}{n_y}} \le \bar{\mathbf{x}} - \bar{\mathbf{y}} \le E(\mathbf{x}) - E(\mathbf{y}) + z_{\alpha/2}\sqrt{\frac{\sigma^2(\mathbf{x})}{n_x} + \frac{\sigma^2(\mathbf{y})}{n_y}}\right\}$$

which is equal to $1 - \alpha$; or

$$p\left\{-z_{\alpha/2} \le \frac{\bar{\mathbf{x}} - \bar{\mathbf{y}} - [E(\mathbf{x}) - E(\mathbf{y})]}{\sqrt{[\sigma^2(\mathbf{x})/n_x] + [\sigma^2(\mathbf{y})/n_y]}} \le z_{\alpha/2}\right\} = 1 - \alpha.$$

Or, to exhibit the confidence interval on $E(\mathbf{x}) - E(\mathbf{y})$ directly,

$$p\left[\bar{x} - \bar{y} - z_{\alpha/2}\sqrt{\frac{\sigma^2(\mathbf{x})}{n_x} + \frac{\sigma^2(\mathbf{y})}{n_y}} \le E(\mathbf{x}) - E(\mathbf{y}) \le \bar{x} - \bar{y} + z_{\alpha/2}\sqrt{\frac{\sigma^2(\mathbf{x})}{n_x} + \frac{\sigma^2(\mathbf{y})}{n_y}}\right]$$

which is equal to $1 - \alpha$.

Numerical example. Given $\sigma(\mathbf{x}) = 8.83$, $\sigma(\mathbf{y}) = 8.81$, $n_x = 580$, $n_y = 786$, $\bar{x} = 34.45$, $\bar{y} = 28.02$, $\alpha = 0.05$; we find $z_{\alpha/2} = 1.96$ and

$$5.48 \le E(\mathbf{x}) - E(\mathbf{y}) \le 7.38.$$

Example 3. If, in the problem of finding the confidence interval on a mean of a normal variable, the variance of the normal variable is known, we use this knowledge and proceed as in Example 1. If the variance is unknown, a statistic (Student's **t**) not requiring knowledge of the variance is required, and we proceed as in Examples 5 and 6, below. Occasionally, the variance of a random variable is not an independent parameter; it may, for example, be a *function* of the mean. We give such an example now.

Consider the mean of a rectangular distribution of range θ. Let n be large. We already know that

$$E(\mathbf{x}) = \frac{\theta}{2}, \qquad \sigma^2(\mathbf{x}) = \frac{\theta^2}{12}; \tag{29–31}$$

the variance is a function of the mean. Also, for *n large*,

$$p\left[\frac{\theta}{2} - z_{\alpha/2}\frac{\sigma(\mathbf{x})}{\sqrt{n}} \le \bar{\mathbf{x}} \le \frac{\theta}{2} + z_{\alpha/2}\frac{\sigma(\mathbf{x})}{\sqrt{n}}\right] = 1 - \alpha \tag{29–32}$$

for $\bar{\mathbf{x}}$, the test statistic here (one which, as we have seen in page 241, is not optimal for n *not* large) is normally distributed in large samples from a rectangular distribution. Using (29–31) and rearranging (29–32) to exhibit the confidence interval on $\theta/2$ directly, we have

$$p\left[\frac{\bar{x}}{1 + z_{\alpha/2}/\sqrt{3n}} \leq \frac{\theta}{2} \leq \frac{\bar{x}}{1 - z_{\alpha/2}/\sqrt{3n}}\right] = 1 - \alpha.$$

Numerical example. Given $\alpha = 0.10$, $n = 20$ (hardly a large sample), $\bar{x} = 3.2$; we find $z_{\alpha/2} = 1.65$ and $2.64 \leq \theta/2 \leq 4.06$.

This example should be regarded merely as technique. The statistic $\bar{\mathbf{x}}$ does not lead to a powerful test of a hypothesis on θ; in fact, it is inadmissible. Consequently, the confidence interval based on $\bar{\mathbf{x}}$ is inadmissible (see page 318).

Example 4. To find a confidence interval on the unknown mean λ of a normal variable, variance σ^2 unknown. The likelihood ratio test of a hypothesis on the mean of a normal distribution of unknown variance leads to the pivotal quantity (29–25) which is distributed as Student's **t** with $n - 1$ degrees of freedom. We found

$$p\left[-t_{n-1,\alpha/2} \leq \frac{\bar{\mathbf{x}} - \lambda}{\mathbf{s}/\sqrt{n-1}} \leq t_{n-1,\alpha/2}\right] = 1 - \alpha,$$

or, exhibiting the confidence interval on λ directly,

$$p\left[\bar{x} - t_{n-1,\alpha/2}\frac{s}{\sqrt{n-1}} \leq \lambda \leq \bar{x} + t_{n-1,\alpha/2}\frac{s}{\sqrt{n-1}}\right] = 1 - \alpha.$$

Numerical example. Given $\alpha = 0.05$, $n = 11$, $\bar{x} = 3.92$, $s = 0.61$; we find $t_{10,0.02} = 2.228$ and $3.51 \leq \lambda \leq 4.33$. If σ^2 is in fact known, the use of $(\bar{\mathbf{x}} - \lambda)/(\mathbf{s}/\sqrt{n-1})$ (distributed as Student's **t**) will lead to a wider interval (unless s happens to be unusually small) than the superior $(\mathbf{x} - \lambda)/\sigma/\sqrt{n}$ (normally distributed), as one would expect.

Example 5. Difference of means $E(\mathbf{x}) - E(\mathbf{y})$ of two normal distributions, variances $\sigma^2(\mathbf{x})$, $\sigma^2(\mathbf{y})$ equal but unknown, n_x and n_y any values. Briefly,

$$p\left[\bar{x} - \bar{y} - t\sqrt{\frac{n_x s_x^2 + n_y s_y^2}{n_x + n_y - 2}} \leq E(\mathbf{x}) - E(\mathbf{y}) \leq \bar{x} - \bar{y} + t\sqrt{\frac{n_x s_x^2 + n_y s_y^2}{n_x + n_y - 2}}\right]$$

which is equal to $1 - \alpha$.

Numerical example. Given $\alpha = 0.05$, $n_x = 4$, $n_y = 5$, $\bar{x} = 3.255$, $\bar{y} = 3.258$, $s_x^2 = 0.000033$, $s_y^2 = 0.000092$, we find $t_{n_x - n_y - 2, \alpha/2} = 2.447$ and $0.013 \leq E(\mathbf{x}) - E(\mathbf{y}) \leq 0.041$.

Example 6. We have obtained confidence intervals by working with random variables whose probability distributions depended only on the parameter to be estimated. Occasionally such random variables cannot be found. For example, the proportion of successes $\mathbf{x}/n$ in a random sample from a binomial population is a sufficient statistic for the parameter p of the binomial distribution, but we cannot find a statistic $\mathbf{z}$ whose probability distribution depends only on $\mathbf{x}/n$ and p. That is to say, no pivotal quantity $\phi(\mathbf{z}, p)$ with fixed distribution independent of p exists. For example, *for n large,*

$$\phi = \frac{\mathbf{x} - np}{\sqrt{npq}}, \tag{29–33}$$

where $\mathbf{x}$ is the number of successes in n trials, does follow the fixed normal distribution (mean 0, variance 1); for n large ϕ is pivotal. But for n not large, (29–33) is the standardized binomial variable with probability distribution depending on p. In such cases we must be content with *a* test of every p. Such a test follows.

For any p, we can always find (approximately) values of r and s such that

$$\sum_{x=0}^{r} C_x^n p^x (1-p)^{n-x} = \frac{\alpha}{2}, \qquad \sum_{x=s}^{n} C_x^n p^x (1-p)^{n-x} = \frac{\alpha}{2}$$

are realized as close to $\alpha/2$ as is possible. That is,

$$p\{r < \mathbf{x} < s\} \cong 1 - \alpha \qquad \text{or} \qquad p\left\{\frac{r}{n} < \frac{\mathbf{x}}{n} < \frac{s}{n}\right\} \cong 1 - \alpha,$$

or, better,

$$p\left\{\frac{r(p)}{n} < \frac{\mathbf{x}}{n} < \frac{s(p)}{n}\right\} \cong 1 - \alpha, \tag{29–34}$$

for r and s surely depend on p. We have, therefore, a test of the hypothesis p which has size α for every p. Those values of p not rejected in (29–34) form a confidence set with confidence coefficient $1 - \alpha$.

Clopper and Pearson [30] have simplified the practical solution of (29–34); one of their charts (approximate for small n) is shown on page 431. For example, for $n = 100$ and $x/n = 0.60$, the 0.95 confidence limits on p are 0.45 and 0.70. See also Pachares [148]. For an alternative discussion, with tables, see Blyth and Hutchinson [18], [19].

Example 7. Confidence interval for the variance σ^2 of a normal variable. The likelihood ratio test of a hypothesis on the variance σ^2 of a normal variable leads to the test statistic $\mathbf{s}^2$, with $n\mathbf{s}^2/\sigma^2$ distributed as χ^2 with $n - 1$ degrees of freedom. The two-sided test, in which alternative values of σ^2 lie on both sides of the hypothesized value, leads to

$$p\left\{a \le \frac{n\mathbf{s}^2}{\sigma^2} \le b\right\} = 1 - \alpha,$$

which transforms into the confidence interval

$$p\left\{\frac{ns^2}{b} \leq \sigma^2 \leq \frac{ns^2}{a}\right\} = 1 - \alpha. \tag{29–35}$$

If the confidence interval (29–35), which is of length

$$\frac{ns^2}{a} - \frac{ns^2}{b},$$

is to have minimum length, then

$$\frac{1}{a} - \frac{1}{b}$$

must be minimum. As χ^2 is not symmetrically distributed (see Fig. 29–11) this requires considerable labor in the tables of χ^2. An approximate solution, good unless n is quite small, is given by equal tails (equal shaded areas). For further discussion and tables, see Tate and Klett [185].

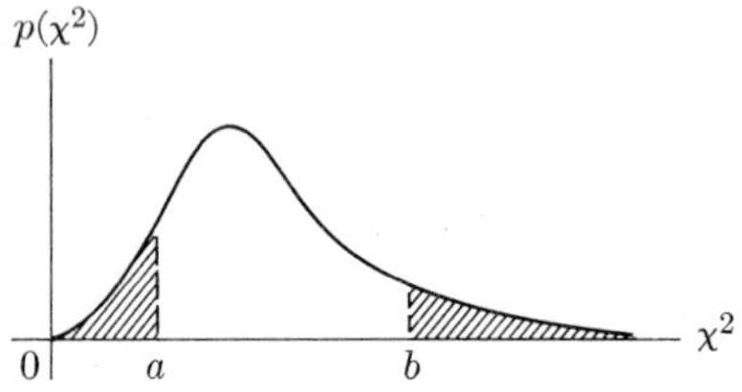

FIG. 29–11. The density function of χ^2.

29–7 One-sided confidence intervals. All confidence intervals found to this point have been two-sided; we have determined *two* limits $a(\theta)$ and $b(\theta)$. In some practical problems confidence intervals are one-sided; for example, $a(\theta)$ is often 0. In place of (29–29) we have

$$\int_0^{b(\theta)} g(z; \theta)\, dz = 1 - \alpha, \tag{29–36}$$

which can be solved for b [usually as (and it had better be) a function of θ].

Concluding remarks. Over recent years interest in estimation by confidence intervals has grown somewhat at the expense of hypothesis testing, although, of course, no less knowledge of the behavior of the random variable is required in confidence intervals than in testing hypotheses. In mathematically well-structured areas of inquiry, such as physics, modern economics, mass production, and gambling, models are easier to come by and tests of hypotheses on the parameters of such models are natural.

But in less structured areas, interval estimation appears to be the more natural course, although tests of hypotheses of independence, equality of means, etc. are, of course, possible and sometimes interesting. As we have seen, these two branches of statistical inference are closely related; the issues between them are steadily being resolved by modern decision theory.

Collateral Reading

C. A. Bennett and N. L. Franklin, *Statistical Analysis in Chemistry and the Chemical Industry.* New York, John Wiley and Sons, 1954. A good general reference. Also for confidence intervals based on order statistics, which we omit, and for confidence intervals on the correlation coefficient.

C. R. Blyth and D. W. Hutchinson, "Tables of Neyman—shortest unbiased confidence intervals for the binomial parameter," *Biometrika,* Vol. 47 (1960), pp. 381–391.

————, "Tables of Neyman—shortest unbiased confidence intervals for the Poisson parameter," *Biometrika,* Vol. 48 (1961), pp. 191–194.

C. J. Clopper and E. S. Pearson, "The use of confidence or fiducial limits illustrated in the case of the binomial," *Biometrika,* Vol. 26 (1934), pp. 404–413.

D. B. DeLury and J. H. Chung, *Confidence Limits for the Hypergeometric Distribution.* Toronto, University of Toronto Press, 1950.

D. A. S. Fraser, *Nonparametric Statistics.* New York, John Wiley and Sons, 1957.

————, *Statistics: An Introduction.* New York, John Wiley and Sons, 1958.

R. V. Hogg and A. T. Craig, *Introduction to Mathematical Statistics.* New York, Macmillan Co., 1959.

M. G. Kendall and A. Stuart, *The Advanced Theory of Statistics, Volume 2.* London, Charles Griffin and Co., 1961. An excellent reference.

E. Lehmann, *Notes on the Statistical Theory of Estimation.* Chapters I to V. Berkeley, University of California, Associated Student's Store, 1950. An outstanding work, along with Lehmann's Wiley book noted earlier.

A. Mood, *Introduction to the Theory of Statistics.* New York, McGraw-Hill Book Co., 1950. In particular pp. 227–229 for a discussion of joint estimation of two parameters.

J. Neyman, *Lectures and Conferences on Mathematical Statistics.* Washington, Graduate School, U.S. Dept. of Agriculture, 2nd ed., revised, 1952.

————, "On two different aspects of the representative method," *Jour. Roy. Stat. Soc.,* Vol. 97 (1934), pp. 558–625.

————, "Foundations of the general theory of statistical estimation." *Congrès International de Philosophie des Sciences : IV : Calcul des Probabilités.* Paris, Hermann et Cie, 1951. An admirable summary.

————, "Outline of a theory of statistical estimation based on the classical theory of probability," *Phil. Trans. Roy. Soc., London,* Vol. 236 (1937), pp. 338–380. A pathbreaking paper.

W. E. RICKER, "The concept of confidence or fiducial limits applied to the Poisson frequency," *Jour. Amer. Stat. Assoc.*, Vol. 32 (1937), pp. 349–356.

R. F. TATE and G. W. KLETT, "Optimal confidence intervals for the variance of a normal distribution," *Jour. Amer. Stat. Assoc.*, Vol. 54 (1959), pp. 674–682.

A. WALD, *Notes on the Theory of Statistical Estimation and Testing Hypotheses.* New York, Columbia University, mimeographed lecture notes. A masterful account.

S. S. WILKS, *Mathematical Statistics.* New York, John Wiley and Sons, 1962.

PROBLEMS

29–1. Find the 0.95 confidence interval for binomial p, given $n = 50$ and $x = 10$. Find the 0.95 confidence interval for binomial p, given $n = 20$, $x = 15$. If the Pearson tables of the incomplete beta-function are available to you, check the latter answer there.

29–2. If the charts prepared by DeLury and Chung [44] are available (an excerpt is shown in Fig. A–2) find the 0.95 and 0.99 confidence interval on hypergeometric A (number of successes in population of size N), given $N = 500$, $n = 100$, $x = 18$. Also the 0.90, 0.95, and 0.99 confidence intervals, given $N = 500$, $n = 50$, $x = 3$.

29–3. From the tables prepared by Ricker [167], reproduced in Table A–17, find the 0.95 and 0.99 confidence intervals on the Poisson parameter, given $x = 6$. Similarly, find the 0.95 interval for $x = 3$. If Molina's tables are available, consider if the symmetrical interval (0.025 on each side) is the shortest. For further discussion and tables, see Blyth and Hutchinson [18], [19].

Part V

An Application To Regression

In Part V some of the theory and method of earlier chapters, particularly of chapters of Part IV, is applied to the study of the important problem of the relationship between two variables.

CHAPTER 30

REGRESSION: TWO VARIABLES

30–1 The model and its analysis. We have considered several estimation and hypothesis-testing problems in connection with a single random variable which we here label **Y**.* Assuming knowledge of the form f of the density function $f(Y; \theta)$, we have formed point and interval estimators of the parameter θ, and we have tested hypotheses on θ. For example, assuming **Y** to be normally distributed, we have estimated or tested hypotheses on $\sigma^2(\mathbf{Y})$. It is evident that we have not challenged the *form* f of the density function; rather we have estimated or tested hypotheses on the unknown parameter or parameters contained in the density function.

In this final chapter we continue in a similar vein, now considering problems involving two variables **Y** and X, both for the importance of the problems themselves and to show an application of some of the results of earlier chapters.

Consider a pair of variables **Y** and X. In this chapter, values of the variable X will be *selected*† by the investigator. At each of the n selected values of X, say $X_1, X_2, \ldots, X_n$, a value of the random variable **Y** will be observed; the corresponding observed values of **Y** are $Y_1, Y_2, \ldots, Y_n$.

Since **Y** is a random variable, these observed values of **Y** will vary from one sample of size n to another, but the n selected values of X remain the same from one sample to another; **Y** has a probability distribution, X has not. Note that any statistical inferences we may draw from the data hold only for the selected set of n values of X.

This situation and extensions of it to selected values of several variables is the basis of a common model in modern experimentation. The experimenter selects values of X (or of each of several variables) and he makes determinations of **Y** at these selected values of X; his general objective is to study the behavior of **Y** in relation to X.

Our interest will center on $E(\mathbf{Y}|X)$, the conditional mean of **Y** given X, the regression function of **Y** on X. In this chapter, we confine ourselves

---

* In this chapter, capital letters are retained for the original random variables and their values; lower case letters will presently be given a special meaning.

† The argument here could be extended to a random variable **X** with respect to which our inferences are conditional; but selected X is the more common model in modern experimentation.

to regression functions which are *linear* in the regression coefficients (parameters) α and β,

$$E(\mathbf{Y}|X) = \int_{-\infty}^{\infty} Yh(Y|X)\,dY = \alpha + \beta X. \tag{30–1}$$

The situation is illustrated in Fig. 30–1.

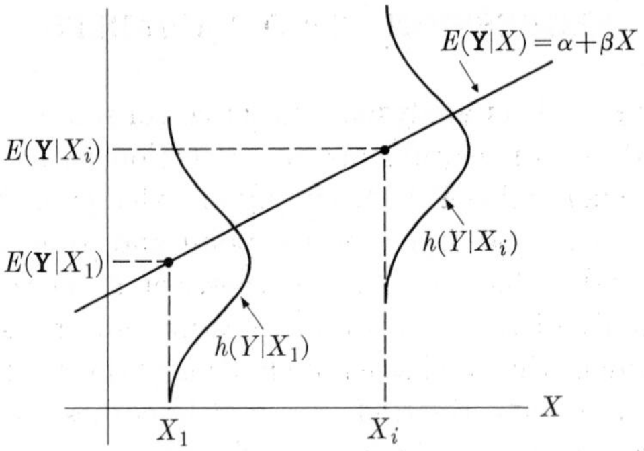

FIG. 30–1. A linear regression function of **Y** on X.

As we found in Chapter 17, if **X** and **Y** follow a bivariate normal, $E(\mathbf{Y}|X)$ *is* linear; in (30–1), however, X is not a random variable, let alone a normal random variable. But we argue, and reasonably, that over short ranges of X, almost all regression functions of **Y** on X are fairly approximated by straight lines [to this argument must also be added the mathematical tractability of (30–1)]; therefore, the regression model shown in (30–1) is of genuine practical interest.

The model by which we describe or "explain" observed values Y in terms of X will be the linear model

$$Y_i = E(\mathbf{Y}|X_i) + \epsilon_i = \alpha + \beta X_i + \epsilon_i, \quad i = 1, \ldots, n. \tag{30–2}$$

The model affirms that the observed value Y_i of **Y** at a selected X_i is "explained" by the linear regression function of **Y** on X_i plus an "error" ϵ_i which we regard as a value of a random variable $\boldsymbol{\epsilon}_i$.* The quantity ϵ_i may be regarded as that part of Y_i which is "not explained" by the linear regression function of **Y** on X_i.

As it stands, the linear model (30–2) is hardly restrictive. It "explains" the observed value Y_i of **Y** at a selected value X_i by a constant $\alpha + \beta X_i$ plus the value ϵ_i of the unrestricted random variable $\boldsymbol{\epsilon}_i$; *all* random variables **Y** satisfy so broad a model. As the argument develops, however, several restrictive conditions will be placed on $\{\boldsymbol{\epsilon}_i\}$; these are essentially

* In type, the distinction between lightface and boldface epsilon is largely one of size. Boldface epsilon is $\boldsymbol{\epsilon}$ and lightface is ϵ.

equivalent to conditions on $\mathbf{Y}$ since, for fixed X_i, the expectation $E(\mathbf{Y}|X_i)$ is a constant, and Y_i differs from ϵ_i only by that constant. Depending on the interests of the experimenter and on mathematical convenience, these conditions on $\{\epsilon_i\}$ may include: for the given selected set $X_1, X_2, \ldots, X_n$, the random variables $\epsilon_i, \epsilon_2, \ldots, \epsilon_n$ are mutually independent, normally distributed, with means 0 and *constant* variance. If *all* these conditions are placed on $\{\epsilon_i\}$, (30–2) becomes the model of a normal random variable $\mathbf{Y}$ of constant variance, with mean depending on the selected value of X.

The true linear regression function of $\mathbf{Y}$ on X described by (30–2) will be unknown, that is to say, α and β are unknown. With an eye to estimating this linear function from observed data on X and $\mathbf{Y}$, we introduce the estimator $\hat{E}(\mathbf{Y}|X)$, which we define by

$$\hat{E}(\mathbf{Y}|X_i) = a + bX_i, \tag{30–3}$$

and we connect observed values Y_i with this function by

$$Y_i = \hat{E}(\mathbf{Y}|X_i) + e_i = a + bX_i + e_i, \quad i = 1, \ldots, n. \tag{30–4}$$

The relation (30–4) is a "working model" of values of the random variable $\mathbf{Y}$; Y_i is described or "explained" by the linear estimator $a + bX_i$ of the true linear regression function $\alpha + \beta X_i$ plus an error e_i, where e_i can be regarded as that part of Y_i "not explained" by the estimator $a + bX_i$. Relationships among the quantities of (30–2) and (30–4) are illustrated in Fig. 30–2.

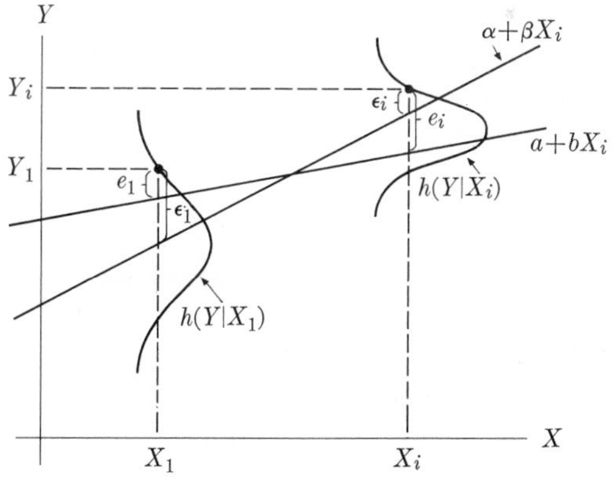

FIG. 30–2. Interrelationships of the quantities of (30–2) and (30–4).

How best to construct the linear estimator $a + bX$ from observed data on X, $\mathbf{Y}$? Of various principles for determining $a + bX$ the preferred one, both for its statistical properties and for its remarkable mathematical tractability in dealing with linear models, is *least squares*. As we have seen in Chapter 27, the least-squares linear estimator of $\alpha + \beta X$ is unbiased and has minimum variance among the class of all linear estimators.

We now consider the construction of the least-squares estimator $a + bX$. We have n observed paired observations $(X_1, Y_1), (X_2, Y_2), \ldots, (X_n, Y_n)$; the $\{X_i\}$ are selected variates, the $\{Y_i\}$ are random variates corresponding to $\{X_i\}$. We construct $a + bX$ by minimizing the sum of squares of the differences between observed values $Y_1, Y_2, \ldots, Y_n$ and the corresponding predicted values $a + bX_1, a + bX_2, \ldots, a + bX_n$. We minimize

$$\sum_{i=1}^{n} e_i^2 = (Y_i - a - bX_i)^2. \tag{30–5}$$

Differentiating (30–5) with respect to the (continuous) unknown a and b, we find

$$\begin{aligned} \sum_{i=1}^{n} Y_i &= an + b \sum_{i=1}^{n} X_i, \\ \sum_{i=1}^{n} Y_i X_i &= a \sum_{i=1}^{n} X_i + b \sum_{i=1}^{n} X_i^2, \end{aligned} \tag{30–6}$$

which may generally be solved for a and b; (30–6) involve only simple arithmetical operations on the bivariate sample data, and lead precisely to (27–43), found in Chapter 27. As the student may show, (30–6) leads to minimum and not maximum $\sum_{i=1}^{n} e_i^2$.

30–2 Algebra of the sample. We note a few useful *algebraic* properties of statistics of the bivariate sample data $\{X_i, Y_i\}$. From this point on in this chapter, all sums are over the sample $i = 1$ to $i = n$, and all quantities $(X, \epsilon, \ldots)$ inside summation signs are understood to have running index $i(X_i, \epsilon_i, \ldots)$.

From (30–6)

$$a = \frac{\sum Y}{n} - \frac{b \sum X}{n} = \overline{Y} - b\overline{X}, \tag{30–7}$$

and, noting (30–3), we have that if $X_i = \overline{X}$, $\hat{E}(\mathbf{Y}|\overline{X}) = \overline{Y}$; the sample regression line $a + bX$ goes through $(\overline{X}, \overline{Y})$.

Some of the sample algebra is better exhibited in terms of the variables

$$x_i = X_i - \overline{X}, \qquad y_i = Y_i - \overline{Y}. \tag{30–8}$$

Thus

$$e_i = Y_i - (a + bX_i) = Y_i - (\overline{Y} - b\overline{X}) - bX_i = y_i - bx_i,$$

and

$$\sum e^2 = \sum (Y - a - bX)^2 = \sum (y - bx)^2. \tag{30–9}$$

In terms of x and y, the two equations (30–6) in a and b reduce to one equation in b:

$$\sum xy = b\sum x^2, \tag{30–10}$$

the intercept of the sample regression line of y on x being 0. We get

$$\hat{E}(\mathbf{y}|x_i) = bx_i \qquad \text{and} \qquad Y_i - \hat{E}(\mathbf{Y}|X_i) = y_i - E(\mathbf{y}|x_i). \tag{30–11}$$

The sum of squares of deviations of observed values Y about the sample regression line of Y on X is

$$\sum e^2 = \sum(y - bx)^2 = \sum(y^2 - 2byx + b^2x^2) = \sum y^2 - b\sum xy,$$

while the sum of squares of deviations of observed values Y about the sample mean $\overline{Y}$ is $\sum y^2$. Using (30–10), the proportional reduction in *sample* variance attributable to or explained by (better, associated with) the sample linear regression of Y on X is

$$\frac{\sum y^2 - \sum e^2}{\sum y^2} = \frac{b\sum xy}{\sum y^2} = \left[\frac{\sum xy}{\sqrt{\sum x^2 \sum y^2}}\right]^2, \tag{30–12}$$

which by definition is r^2, the square of the *sample correlation coefficient,*

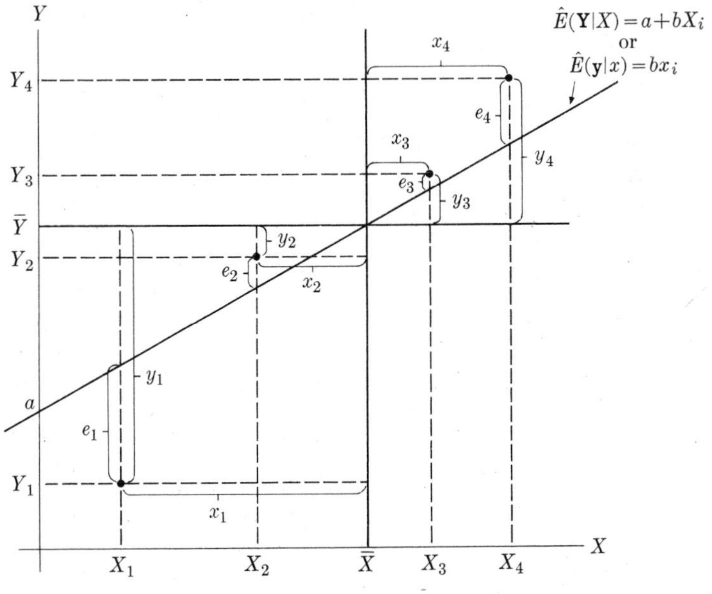

FIG. 30–3. Four observed points $(X_1, Y_1), \ldots, (X_4, Y_4)$ and the sample linear regression line of Y on X.

first introduced on page 259 in connection with the estimation of the correlation coefficient ρ of two normal variables. This result corresponds precisely to (6–21).

Some of these features of the sample regression line of Y on X are illustrated in Fig. 30–3. Note the greater variance of the observed Y_i about $\overline{Y}$, compared with the variance of Y_i about $a + bX_i$ as measured by $\sum e_i^2$. The linear estimator $a + bX_i$, which depends on X, explains better than the constant $\overline{Y}$ (which is indifferent to X) the variation in the observed Y_i.

30–3 Relationships between corresponding statistics and parameters. *Between* $\overline{\mathbf{Y}}$ *and* $E(\mathbf{Y})$. Taking expectations over the variables corresponding to (30–2), we have

$$E(\mathbf{Y}) = \alpha + \beta\overline{X} + E(\epsilon_i), \tag{30–13}$$

while summing (30–2) over the sample of n pairs of observations $(X_1, Y_1), \ldots, (X_n, Y_n)$, we have

$$\overline{Y} = \alpha + \beta\overline{X} + \bar{\epsilon}. \tag{30–14}$$

As values of X are selected, there is no distinction between population and sample for X; we will generally write $\overline{X}$ for the mean of $X_1, \ldots, X_n$. Now we introduce and retain throughout the remainder of this chapter the first (reasonable) condition on the random variables $\{\epsilon_i\}$; let $E(\epsilon_i) = 0$, $i = 1, \ldots, n$. From (30–13) and (30–14), we have

$$\overline{Y} = E(\mathbf{Y}) + \bar{\epsilon}; \tag{30–15}$$

the sample mean of Y differs from the true mean of Y by the mean of n (a finite number) of unknown true errors $\epsilon_1, \ldots, \epsilon_n$. Note that $\bar{\epsilon} = 0$ does not follow from $E(\epsilon_i) = 0$, $i = 1, \ldots, n$.

Between **b** *and* β. First we express y in terms of x by use of (30–2) and (30–14):

$$y_i = Y_i - \overline{Y} = (\alpha + \beta X_i + \epsilon_i) - (\alpha + \beta\overline{X} + \bar{\epsilon}) = \beta x_i + (\epsilon_i - \bar{\epsilon}), \tag{30–16}$$

and (30–10) becomes

$$b = \frac{\sum xy}{\sum x^2} = \frac{\sum x(\beta x + \epsilon - \bar{\epsilon})}{\sum x^2} = \beta + \frac{\sum x\epsilon}{\sum x^2}. \tag{30–17}$$

The slope b of the sample regression line differs from the true slope β by a function of n (a finite number) of unknown true errors $\epsilon_1, \ldots, \epsilon_n$. Note that if $\{x_i\}$ and $\{\epsilon_i\}$ are positively correlated in the sample, b overestimates β, etc.

Between **a** *and* α. From (30–15), (30–13), and (30–17),

$$a = \overline{Y} - b\overline{X} = \alpha + \beta\overline{X} + \bar{\epsilon} - \left(\beta + \frac{\sum x\epsilon}{\sum x^2}\right)\overline{X}$$

$$= \alpha + \bar{\epsilon} - \overline{X}\frac{\sum x\epsilon}{\sum x^2}, \qquad (30\text{–}18)$$

a relationship similar to that between b and β except that here, if $\{x_i\}$ and $\{\epsilon_i\}$ are positively correlated in the sample, a underestimates α, etc.

30–4 Expected values of principal statistics. *Means of statistics.* Recalling the condition $E(\epsilon_i) = 0$, $i = 1, \ldots, n$, imposed on $\{\epsilon_i\}$ we readily obtain from (30–15),

$$E(\overline{\mathbf{Y}}) = E(\mathbf{Y}), \qquad (30\text{–}19)$$

from (30–17),

$$E(\mathbf{b}) = \beta + E\frac{\sum x\epsilon}{\sum x^2} = \beta + \frac{1}{\sum x^2}E(x_1\epsilon_1 + \cdots + x_n\epsilon_n) = \beta, \qquad (30\text{–}20)$$

from (30–18),

$$E(\mathbf{a}) = \alpha + E(\bar{\epsilon}) - \frac{\overline{X}}{\sum x^2}E\sum x\epsilon = \alpha; \qquad (30\text{–}21)$$

$\overline{\mathbf{Y}}$, $\mathbf{b}$, and $\mathbf{a}$ are unbiased estimators of $E(\mathbf{Y})$, β, and α respectively.

Variances of statistics. We calculate the variances of the principal statistics, now making the further assumption that

$$\sigma^2(\epsilon_1) = \cdots = \sigma^2(\epsilon_n)[= \sigma^2(\epsilon)].$$

From (30–2),

$$\sigma^2(\mathbf{Y}) = E[\mathbf{Y} - E(\mathbf{Y})]^2 = \sigma^2(\epsilon);$$

from (30–15),

$$\sigma^2(\overline{\mathbf{Y}}) = E[\overline{\mathbf{Y}} - E(\overline{\mathbf{Y}})]^2 = \sigma^2(\epsilon) - \frac{\sigma^2(\epsilon)}{n}. \qquad (30\text{–}22)$$

From (30–17),

$$\sigma^2(\mathbf{b}) = \frac{1}{(\sum x^2)^2}\sigma^2(x_1\epsilon_1 + \cdots + x_n\epsilon_n)$$

$$= \frac{x_1^2\sigma^2(\epsilon) + \cdots + x_n^2\sigma^2(\epsilon)}{(\sum x^2)^2} = \frac{\sigma^2(\epsilon)}{\sum x^2}. \qquad (30\text{–}23)$$

From (30–18)

$$\sigma^2(\mathbf{a}) = \frac{\sigma^2(\epsilon)}{n} + (\overline{X})^2\frac{\sigma^2(x_1\epsilon_1 + \cdots + x_n\epsilon_n)}{(\sum x^2)^2}$$

$$= \frac{\sigma^2(\epsilon)}{n} + \frac{\overline{X}^2\sigma^2(\epsilon)}{\sum x^2}. \qquad (30\text{–}24)$$

If the experimenter is satisfied with best linear unbiased estimators of the parameters α and β [and of $E(\mathbf{Y})$], least squares analysis of his data, together with the conditions $E(\epsilon_i) = 0$, $i = 1, \ldots, n$ and $\sigma^2(\epsilon_i)$ constant for $i = 1, \ldots, n$ will produce such estimators.

Note that the variances of the sample regression coefficients **a** and **b**, involving as they do $\sum x^2$ in the denominator, can be reduced (and thereby the reliability increased) by adroit selection of the values of X. For example, the variance of **b** can be reduced (minimized) by having half of all selected values of X at the largest possible X, half at the smallest possible X, though such selection would leave us without adequate check, should we wish to check, on the validity of the assumption of linearity of the regression function.*

30–5 Confidence intervals on $\alpha + \beta X$ and on Y. The experimenter may ask for a confidence interval on the unknown linear regression function $E(\mathbf{Y}|X)$ or, asking more, on a value of **Y**. As an example, let X be human weight and Y human height. He may ask for a confidence interval on the true *mean* height $E(\mathbf{Y}|X)$ at a selected weight $X = 150$ pounds, or he may ask for a confidence interval on an *individual* weight Y at the selected weight $X = 150$ pounds. The situation is shown in Fig. 30–4. In *both* cases the best linear estimator we have is $a + bX$. For the former problem,

$$\hat{E}(\mathbf{Y}|X) = a + bX \quad \text{is an estimator of} \quad E(\mathbf{Y}|X) = \alpha + \beta X,$$

and for the latter,

$$\hat{E}(\mathbf{Y}|X) = a + bX \quad \text{is an estimator of} \quad Y = \alpha + \beta X + \epsilon.$$

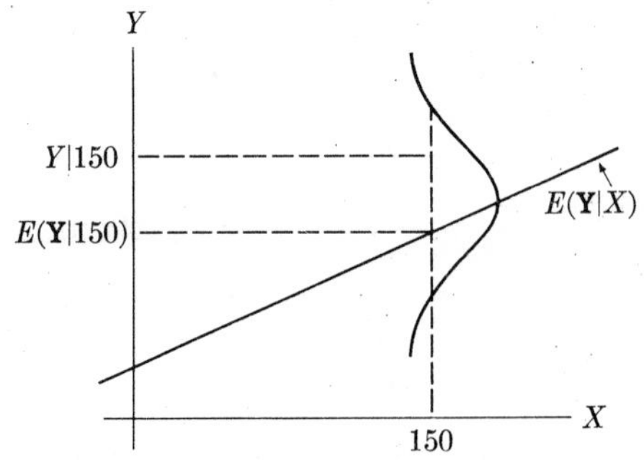

FIG. 30–4. The quantities $Y|X$ and $E(\mathbf{Y}|X)$ on which confidence intervals are to be constructed.

* For further discussion, see C. Daniel and N. Heerema [39].

Let us consider the former problem. For a confidence interval on $\alpha + \beta X$ we need a statistic which will provide a test of every hypothesis on $\alpha + \beta X$. Preferably we seek a pivotal random variable, a variable which involves the parameters α and β in the form $\alpha + \beta X$ and in no other form, involves no other parameters, and whose distribution is independent of all parameters. Unless further conditions are placed on $\{\boldsymbol{\epsilon}_i\}$ no such random variable exists; as matters stand now the distribution of $\mathbf{Y}$ which is subject only to the condition $E(\boldsymbol{\epsilon}_i) = 0$ could be said to be too loosely restricted for further analysis. If for the set $X_1, X_2, \ldots, X_n$, we also require $\boldsymbol{\epsilon}_1, \boldsymbol{\epsilon}_2, \ldots, \boldsymbol{\epsilon}_n$ to be mutually independent normal random variables (with common mean 0 and common known variance σ^2) a requirement which appears to be compatible with some experimental situations, a pivotal random variable emerges quickly. For from (30–17) and (30–18), we have

$$a + bX = \alpha + \bar{\boldsymbol{\epsilon}} - \overline{X}\frac{\sum x\boldsymbol{\epsilon}}{x^2} + \beta X + X\frac{\sum x\boldsymbol{\epsilon}}{x^2}, \tag{30–25}$$

and the estimator $\mathbf{a} + \mathbf{b}X$ is seen to be a *linear* function of independent normal random variables $\boldsymbol{\epsilon}_1, \boldsymbol{\epsilon}_2, \ldots, \boldsymbol{\epsilon}_n$ and therefore itself normal. Moreover, we know that

$$E(\mathbf{a} + \mathbf{b}X) = \alpha + \beta X, \tag{30–26}$$

and we shall presently show that

$$\sigma^2(\mathbf{a} + \mathbf{b}X) = \frac{\sigma^2(\boldsymbol{\epsilon})}{n} + \frac{x^2}{\sum x^2}\sigma^2(\boldsymbol{\epsilon}). \tag{30–27}$$

Therefore, the standardized random variable

$$\frac{(\mathbf{a} + \mathbf{b}X) - (\alpha + \beta X)}{\sigma(\mathbf{a} + \mathbf{b}X)}, \tag{30–28}$$

which has a fixed normal distribution with mean 0 and variance 1, is a pivotal quantity for testing hypotheses on $\alpha + \beta X$. Finally, by the usual inversion of pivotal quantities,

$$\begin{aligned} p\{a + bX - 1.96\sigma(\mathbf{a} + \mathbf{b}X) \le \alpha + \beta X \le a + bX + 1.96\sigma(\mathbf{a} + \mathbf{b}X)\} \\ = 0.95, \end{aligned} \tag{30–29}$$

where $\sigma(\mathbf{a} + \mathbf{b}X)$ is the positive square root of (30–27). Results similar to (30–29) may, of course, be written down for values of the confidence coefficient other than $1 - \alpha = 0.95$.

Note again that no conditions beyond $E(\boldsymbol{\epsilon}_i) = 0$ and equal $\sigma^2(\boldsymbol{\epsilon}_i)$ are necessary if the experimenter is content with best linear unbiased point estimators; it is in interval estimation that further conditions on $\{\boldsymbol{\epsilon}_i\}$ are needed.

We return to prove (30–27). Some care is needed in this evaluation, for $\mathbf{a}$ and $\mathbf{b}$ are not statistically independent, that is,

$$\sigma^2(\mathbf{a} + \mathbf{b}X) \neq \sigma^2(\mathbf{a}) + X^2\sigma^2(\mathbf{b}). \tag{30–30}$$

That $\mathbf{a}$ and $\mathbf{b}$ are not independent is readily seen. From (30–6),

$$\mathbf{a} + \mathbf{b}X = (\bar{Y} - \mathbf{b}\bar{X}) + \mathbf{b}X; \tag{30–31}$$

evidently $\mathbf{a}$ directly involves $\mathbf{b}$. We could have eliminated such dependence earlier by considering the linear regression of $\mathbf{Y}$ on $x(= X - \bar{X})$ rather than on X. Perhaps simpler now is to compute the variance of $\mathbf{a} + \mathbf{b}X$ from (30–31), since $\mathbf{b}$ and $\bar{Y}$ can be shown to be independent. Or directly from (5–10), we have

$$\sigma^2(\mathbf{a} + \mathbf{b}X) = \sigma^2(\mathbf{a}) + X^2\sigma^2(\mathbf{b}) + 2X \operatorname{cov}(\mathbf{a}, \mathbf{b}). \tag{30–32}$$

The covariance term $\operatorname{cov}(\mathbf{a}, \mathbf{b})$ is found to be

$$\begin{aligned} E(\mathbf{a} - \alpha)(\mathbf{b} - \beta) &= E(\mathbf{ab}) - \alpha\beta \\ &= E\left\{\alpha + \bar{\boldsymbol{\epsilon}} - E(X)\frac{\sum x\boldsymbol{\epsilon}}{\sum x^2}\right\}\left\{\beta + \frac{\sum x\boldsymbol{\epsilon}}{\sum x^2}\right\}. \end{aligned}$$

By any of these three methods, the result is (30–27). But $\sigma^2(\boldsymbol{\epsilon})$ on which $\sigma^2(\mathbf{a} + \mathbf{b}X)$ depends is typically unknown and itself must be estimated from the data. In our earlier study of tests of a hypothesis on the mean λ of a normal variable $\mathbf{Y}$ of unknown variance $\sigma^2(\mathbf{Y})$, the quantity

$$\frac{\sum(\mathbf{Y} - \bar{\mathbf{Y}})^2}{n - 1}, \tag{30–33}$$

calculated from n observed values of $\mathbf{Y}$, was shown to be an unbiased estimator of $\sigma^2(\mathbf{Y})$. The quantity $\sum(\mathbf{Y} - \bar{\mathbf{Y}})^2\sigma^2$ was shown to be distributed as χ^2 with $n - 1$ degrees of freedom, and

$$\frac{\bar{\mathbf{Y}} - \lambda}{\sqrt{\dfrac{\sum(\mathbf{Y} - \bar{\mathbf{Y}})^2/(n - 1)}{n}}}, \tag{30–34}$$

which is distributed as Student's $\mathbf{t}$ with $n - 1$ degrees of freedom, is a pivotal quantity for a test of the hypothesis on λ, with σ^2 unknown. The situation is similar here. We are testing a hypothesis on $\alpha + \beta X$, which has variance depending on $\sigma^2(\boldsymbol{\epsilon})$ which is unknown. The quantity

$$\frac{\sum(\mathbf{Y} - \mathbf{a} - \mathbf{b}X)^2}{n - 2}, \tag{30–35}$$

calculated from n observed *pairs* of values $(X_1, Y_1), \ldots, (X_n, Y_n)$, is, as we shall show, an unbiased estimator of $\sigma^2(\mathbf{Y})$ [which is equal to $\sigma^2(\boldsymbol{\epsilon})$]. Moreover,

$$\frac{\sum(\mathbf{Y} - \mathbf{a} - \mathbf{b}X)^2}{\sigma^2(\boldsymbol{\epsilon})}$$

will be shown to be distributed as χ^2 with $n - 2$ degrees of freedom. Finally the quantity

$$\frac{(\mathbf{a} + \mathbf{b}X) - (\alpha + \beta X)}{\sqrt{\dfrac{\sum(\mathbf{Y} - \mathbf{a} - \mathbf{b}X)^2/(n-2)}{n}}}, \tag{30–36}$$

which is distributed as Student's $\mathbf{t}$ with $n - 2$ degrees of freedom, will be a pivotal quantity for a test of the hypothesis on $\alpha + \beta X$, with $\sigma^2(\boldsymbol{\epsilon})$ unknown.

To show that (30–35) is an unbiased estimator of $\sigma^2(\boldsymbol{\epsilon})$, we rewrite the numerator of (30–35) in terms of $\boldsymbol{\epsilon}$ (for we know the mean and variance of $\boldsymbol{\epsilon}$) and the nonrandom variates X_i. Thus

$$\sum e^2 = \sum(Y - a - bX)^2 = \sum(y - bx)^2,$$

which, introducing (30–16) and (30–17), equals

$$\sum\left[(\epsilon - \bar{\epsilon}) - x\frac{\sum x\epsilon}{\sum x^2}\right]^2 = \sum(\epsilon - \bar{\epsilon})^2 - \left(\frac{\sum x\epsilon}{\sum x^2}\right)^2 \sum x^2; \tag{30–37}$$

the cross-product term involving $\bar{\epsilon}x$ vanishes since $\sum x = 0$. Now determine the expected value of the right-side of (30–37). As $\boldsymbol{\epsilon}_i$ are independent (normal) variables with means zero and variances σ^2, we know, from (25–4),

$$E\sum(\epsilon - \bar{\epsilon})^2 = (n - 1)\sigma^2(\boldsymbol{\epsilon}).$$

As for the second term of (30–37),

$$\begin{aligned} -\frac{1}{\sum x^2} E(\sum x\boldsymbol{\epsilon})^2 &= -\frac{1}{\sum x^2} E\{x_1\epsilon_1 + \cdots + x_n\epsilon_n\}^2 \\ &= -\frac{\sum x^2}{\sum x^2} E(\boldsymbol{\epsilon}^2) \\ &= -\sigma^2(\boldsymbol{\epsilon}). \end{aligned}$$

Therefore

$$E(\sum\boldsymbol{\epsilon}^2) = (n - 1)\sigma^2(\boldsymbol{\epsilon}) - \sigma^2(\boldsymbol{\epsilon}),$$

and (30–35) follows.

We now show that $\sum \mathbf{e}^2/\sigma^2(\epsilon)$ is distributed as χ^2 with $n - 2$ degrees of freedom. Again consider the two right-hand terms of (30–37). From Chapter 23,

$$\frac{\sum(\epsilon - \bar{\epsilon})^2}{\sigma^2(\epsilon)} = \frac{n\mathbf{s}^2}{\sigma^2(\epsilon)}$$

has the χ^2 distribution with $n - 1$ degrees of freedom. As for the second term of (30–37),

$$\frac{(\sum x\epsilon)^2}{\sum x^2} = \frac{(b - \beta)^2}{\sum x^2}(\sum x^2)^2 = (b - \beta)^2 \sum x^2$$

$$= (b - \beta)^2 \frac{\sigma^2(\epsilon)}{\sigma^2(\mathbf{b})} = \sigma^2(\epsilon)\left(\frac{b - \beta}{\sigma(\mathbf{b})}\right)^2,$$

again, $\sigma^2(\epsilon)$ multiplied by the square of a normal variable with zero mean and unit variance. Therefore $(\sum x\epsilon)^2/\sigma^2(\epsilon)\sum x^2$ is distributed as χ^2 with 1 degree of freedom.

Moreover, as follows from Section 23–11, the two terms of (30–37) are independently distributed, though we do not show this in detail. Accordingly, $\sum \mathbf{e}^2/\sigma^2(\epsilon)$ is distributed as χ^2 with $n - 2$ degrees of freedom.

Now recall from Chapter 23 that any random variable which is the ratio of a normal variable of mean 0 and variance 1 to the square root of a quantity which itself is a χ^2 variable divided by its (m) degrees of freedom, is distributed as Student's $\mathbf{t}$ with m degrees of freedom. Such a random variable is

$$\frac{(\mathbf{a} + \mathbf{b}X) - (\alpha + \beta X)}{\sqrt{\hat{\sigma}^2(\epsilon)/n + x^2\hat{\sigma}^2(\epsilon)/\sum x^2}},$$

where the estimator $\hat{\sigma}^2(\epsilon)$ is given by (30–35). Accordingly,

$$p\left\{t_{\alpha/2,n-2} \le \frac{\mathbf{a} + \mathbf{b}X - \alpha - \beta X}{\sqrt{\hat{\sigma}^2(\epsilon)/n + x^2\hat{\sigma}^2(\epsilon)/\sum x^2}} \le t_{\alpha/2,n-2}\right\} = 1 - \alpha, \qquad (30\text{–}38)$$

or, showing the confidence interval directly,

$$p\left\{a + bx - t_{\alpha/2,n-2}\sqrt{\frac{\hat{\sigma}^2(\epsilon)}{n} + \frac{x^2\hat{\sigma}^2(\epsilon)}{\sum x^2}}\right.$$

$$\left.\le \alpha + \beta X \le a + bx + t_{\alpha/2,n-2}\sqrt{\frac{\hat{\sigma}^2(\epsilon)}{n} + \frac{x^2\hat{\sigma}^2(\epsilon)}{\sum x^2}}\right\} = 1 - \alpha. \qquad (30\text{–}39)$$

To construct a confidence interval on a single value of $\mathbf{Y}$ rather than on $E(\mathbf{Y}|X)$, the argument is entirely similar. The difference between the estimator (which remains $a + bX$) and the quantity now to be estimated (Y) is

$$a + bX - (\alpha + \beta X + \epsilon),$$

with variance given by

$$(30\text{–}27) \text{ plus } \sigma^2(\epsilon),$$

so we must add to the term under the roots in (30–39) the estimate $\hat{\sigma}^2(\epsilon)$. The (more demanding) interval on Y is naturally wider than that on $E(\mathbf{Y}|X)$. The confidence intervals have the form shown in Fig. 30–5.

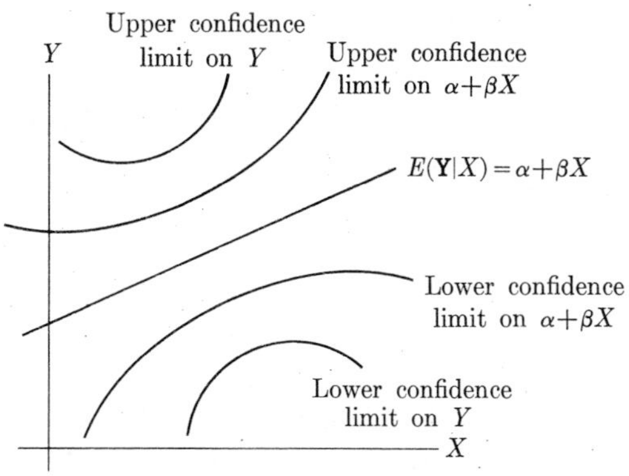

FIG. 30–5. Confidence intervals on $\alpha + \beta X$ and on Y.

30–6 Testing hypotheses on β. Finally we view the subject of this chapter from the point of view of testing hypotheses. Among other results, one link between regression and analysis of variance, the latter introduced in Chapter 23, will be seen.

First consider the hypothesis $H_0 : \beta - \beta_0$. If $\{\epsilon_i\}$ are normally and independently distributed with common variance $\sigma^2(\epsilon)$, the likelihood ratio statistic for $H_0 : \beta = \beta_0$ is $\mathbf{b}$. As we have already seen, $\mathbf{b}$ is normally distributed with variance $\sigma^2(\epsilon)/\sum x^2$ and, if H_0 is true, mean β_0. That is, if H_0 is true,

$$\frac{\mathbf{b} - \beta_0}{\sigma(\epsilon)} \sqrt{\sum x^2}$$

is normally distributed with mean 0 and variance 1. If $\sigma^2(\epsilon)$ is estimated from observed data $(x_1, y_1), \ldots, (x_n, y_n)$ by $\sum e^2/(n - 2)$, the pivotal quantity for tests of hypotheses on β is

$$\frac{(\mathbf{b} - \beta_0)}{\sqrt{\sum \mathbf{e}^2/(n-2)}} \sqrt{\sum x^2}, \tag{30–40}$$

which is distributed as Student's $\mathbf{t}$ with $n - 2$ degrees of freedom.

Now specialize H_0 to the popular hypothesis, $\beta = \beta_0 = 0$, the hypothesis that linear regression of **Y** on X contributes nothing to an understanding of the variation of **Y**. This hypothesis can, of course, be tested by (30–40), but it also fits directly into the structure of analysis of variance. Since this is the statistical structure of much of modern experimentation, we consider it now. The sum of squares not associated with linear regression $a + bX$ is

$$\sum[\mathbf{Y} - (\mathbf{a} + \mathbf{bX})]^2 = \sum \mathbf{e}^2$$

which, divided by $\sigma^2(\epsilon)$, has the χ^2 distribution with $n - 2$ degrees of freedom. From (30–12), the sum of squares associated with linear regression is

$$\mathbf{b}\sum xy = \mathbf{b}^2\sum x^2 = (\mathbf{b} - \beta + \beta)^2\sum x^2 = [(\mathbf{b} - \beta)\sqrt{\sum x^2} + \beta\sqrt{\sum x^2}]^2,$$

which, divided by $\sigma^2(\epsilon)$, evidently has, for the hypothesis $\beta = 0$ true, the χ^2 distribution with one degree of freedom.

These variance estimates, each distributed as χ^2 if H_0 is true, can be shown to be independent. Consequently, if H_0: $\beta = 0$ is true, their ratio, first divided by their respective degrees of freedom, has the F distribution with 1 and $n - 2$ degrees of freedom. Moreover, since the expected value of the mean square associated with the regression line

$$\begin{aligned} E\mathbf{b}^2\sum x^2 &= E[(\mathbf{b} - \beta)\sqrt{\sum x^2} + \beta\sqrt{\sum x^2}]^2 \\ &= E[(\mathbf{b} - \beta)^2\sum x^2 + 2(\mathbf{b} - \beta)\beta\sqrt{\sum x^2} + \beta^2\sum x^2] \\ &= \sigma^2(\epsilon) + 0 + \beta^2\sum x^2 \end{aligned}$$

is a quantity equal to or *greater* than $\sigma^2(\epsilon)$, this suggests that the test of the hypothesis $\beta = 0$ achieves maximum power if the *upper* tail of the F distribution is used as the critical region.

TABLE 30–1

ANALYSIS OF VARIANCE FOR SIMPLE LINEAR REGRESSION

	Sum of squares	Degrees of freedom	Mean square	F	Expected value of mean square
Regression	$b^2\sum x^2$	1	$b^2\sum x^2/1$	$\dfrac{b^2\sum x^2}{\dfrac{\sum y^2 - b^2\sum x^2}{n-2}}$	$\sigma^2(\epsilon) + \beta^2\sum x^2$
Remainder	$\sum y^2 - b^2\sum x^2$	$n - 2$	$(\sum y^2 - b^2\sum x^2)/(n - 2)$		$\sigma^2(\epsilon)$
Total	$\sum y^2$	$n - 1$			

Arranged in an "analysis of variance" table, these results appear as shown in Table 30–1. The parameters of central interest in modern experimental design can readily and naturally be expressed as regression coefficients of Y on one or more selected variables $X_1, X_2, \ldots$, and the analysis begun in this chapter can be extended, on an elaborate scale (see Graybill [76] or Kempthorne [100] or Scheffé [175]), to the study of hypotheses on these parameters.

Collateral Reading

R. L. Anderson and T. A. Bancroft, *Statistical Theory in Research.* New York, McGraw-Hill Book Co., 1952. One of the best elementary discussions in print; this chapter is closely patterned after their Chapter 13.

C. A. Bennett and N. L. Franklin, *Statistical Analysis in Chemistry and the Chemical Industry.* New York, John Wiley and Sons, 1954.

C. Daniel and N. Heerema, "Design of experiments for most precise slope estimation or linear extrapolation," *Jour. Amer. Stat. Assoc.*, Vol. 45 (1950), pp. 546–556.

F. A. Graybill, *An Introduction to Linear Statistical Models, Vol. 1.* New York, McGraw-Hill Book Co., 1961.

P. G. Hoel, "Confidence regions for linear regression," *Second Berkeley Symposium on Math. Statistics and Probability,* Berkeley, Univ. of Cal. Press, 1951, pp. 75–81.

A. M. Mood, *Introduction to the Theory of Statistics.* New York, McGraw-Hill Book Co., 1950.

H. Scheffé, *The Analysis of Variance.* New York, John Wiley and Sons, 1959.

C. E. Weatherburn, *A First Course in Mathematical Statistics.* Cambridge (Eng.), Cambridge University Press, 1946.

Discussion of
and
Answers to Problems

CHAPTER 1

1–1. Since S and O are disjoint,

$$p(S + O) = p(S) + p(O) = 1 + p(O), \qquad \text{since} \quad p(S) = 1.$$

$S + O$ is the set of all points belonging to S or O. But O is empty:

$$S + O = S, \qquad p(S + O) = p(S) = 1,$$
$$1 = 1 + p(O), \qquad p(O) = 0.$$

1–2. Our calendar is periodic with a period of 400 years, every fourth year of which is a leap year except those years which begin a new century but which are not divisible by 400.

In the solution of this problem, a 28-year period is useful. In 4 centuries there are 4800 months in which the 13th day falls on some day of some week. The enumeration (best by a perpetual calendar) of these 4800 days follows:

Sunday	Monday	Tuesday	Wednesday	Thursday	Friday	Saturday
687	685	685	687	684	688	684

a different (and better) solution than the one we would reach were we merely to assume equal likelihood that the 13th falls on any day of the week in a randomly chosen month [$p(\text{Friday}) = \frac{1}{7}$].

1–3. If first in line, probability of winning is $\frac{1}{52}$. If second in line, probability of winning is $\frac{51}{52} \cdot \frac{1}{51} = \frac{1}{52}$. If third in line, probability of winning is $\frac{51}{52} \cdot \frac{50}{51} \cdot \frac{1}{50} = \frac{1}{52}$, etc.

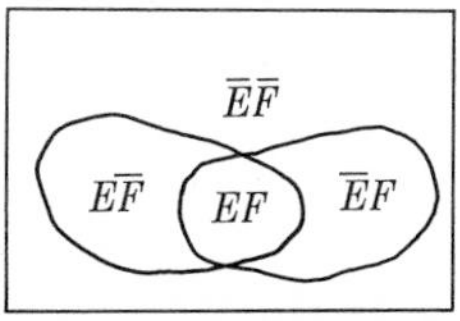

1–4. Note that

$$\bar{E}\bar{F} + E\bar{F} = \bar{F}, \qquad E\bar{F} + EF = E.$$

Thus

$$\begin{aligned} p(\bar{E}\bar{F}) + p(E\bar{F}) &= p(\bar{F}), \\ p(\bar{E}\bar{F}) &= p(\bar{F}) - p(E\bar{F}) \\ &= [1 - p(F)] - [p(E) - p(EF)] \\ &= 1 - p(F) - p(E) + p(E)p(F) \\ &= [1 - p(E)][1 - p(F)] \\ &= p(\bar{E})p(\bar{F}). \end{aligned}$$

Better, by a chain argument, one can get all possible results; begin with

$$\begin{aligned} p(E\overline{F}) &= p(E) - p(EF) \\ &= p(E) - p(E)p(F) \\ &= p(E)[1 - p(F)] \\ &= p(E)p(\overline{F}) \end{aligned}$$

and proceed to $p(\overline{E}F)$, and finally to $p(\overline{E}\overline{F})$.

1–5. To show $p(AB) \leq p(A)$:

$$p(A) = p(AB) + p(A\overline{B}).$$

But $p(A\overline{B}) \geq 0$, so $p(AB) \leq p(A)$.

To show $p(A) \leq p(A + B)$:

$$p(A + B) = p(A) + p(B) - p(AB) = p(A) + p(\overline{A}B).$$

But $p(\overline{A}B) \geq 0$, so $p(A) \leq p(A + B)$.

To show $p(A + B) \leq p(A) + p(B)$:

$$p(A + B) = p(A) + p(B) - p(AB), \qquad p(A + B) \leq p(A) + p(B).$$

1–6. $p(A + B + C)$ is the probability of A or B or C or any two of these events or any three. $p(A + B + C)$ is the probability of all possible events involving A, B, C excepting $\overline{A}\overline{B}\overline{C}$.

$$p(A + B + C) = 1 - p(\overline{A}\overline{B}\overline{C}),$$

which, if A, B, C, are mutually independent, reduces to

$$1 - p(\overline{A})p(\overline{B})p(\overline{C}).$$

Similarly

$$p(A + B + C + \cdots) = 1 - p(\overline{A})p(\overline{B})p(\overline{C}) \ldots$$

if A, B, C, ... are mutually independent. Or, formally, for 3 events,

$$\begin{aligned} p(A + B + C) &= p(A) + p(B) + p(C) - p(AB) - p(AC) - p(BC) + p(ABC) \\ &= p(A) + p(B) + p(C) - p(A)p(B) - p(A)p(C) - p(B)p(C) \\ &\quad + p(A)p(B)p(C) \\ &= p(A)[1 - p(B)] + p(B)[1 - p(C)] + p(C)[1 - p(A)] \\ &\quad + p(A)p(B)p(C) \\ &= p(A)p(\overline{B}) + p(B)p(\overline{C}) + p(C)p(\overline{A}) + p(A)p(B)p(C) \\ &= [1 - p(\overline{A})]p(\overline{B}) + [1 - p(\overline{B})]p(\overline{C}) + [1 - p(\overline{C})]p(\overline{A}) \\ &\quad + [1 - p(\overline{A})][1 - p(\overline{B})][1 - p(\overline{C})] \\ &= 1 - p(\overline{A})p(\overline{B})p(\overline{C}). \end{aligned}$$

1–7. Let A, B, C have values 0 and 1 only. Consider

$B = 0$

	$C = 0$	$C = 1$	
$A = 0$	$\frac{3}{8}$	$\frac{2}{8}$	$\frac{5}{8}$
$A = 1$	$\frac{1}{8}$	$\frac{2}{8}$	$\frac{3}{8}$
	$\frac{4}{8}$	$\frac{4}{8}$	

$B = 1$

	$C = 0$	$C = 1$	
$A = 0$	$\frac{2}{8}$	$\frac{3}{8}$	$\frac{5}{8}$
$A = 1$	$\frac{2}{8}$	$\frac{1}{8}$	$\frac{3}{8}$
	$\frac{4}{8}$	$\frac{4}{8}$	

A and C are not independent, given B. For example, if $B = 0$,

$$p(A = 0, C = 0 | B = 0) = \tfrac{3}{8},$$

while

$$p(A = 0 | B = 0) \cdot p(C = 0 | B = 0) = \tfrac{5}{8} \cdot \tfrac{1}{2} = \tfrac{5}{16} \neq \tfrac{3}{8}.$$

Now suppose $p(B = 0) = p(B = 1) = \frac{1}{2}$. Then

$$p(A = 0, C = 0) = \tfrac{1}{2} \cdot \tfrac{3}{8} + \tfrac{1}{2} \cdot \tfrac{2}{8} = \tfrac{5}{16},$$
$$p(A = 0, C = 1) = \tfrac{1}{2} \cdot \tfrac{2}{8} + \tfrac{1}{2} \cdot \tfrac{3}{8} = \tfrac{5}{16},$$
$$p(A = 1, C = 0) = \tfrac{1}{2} \cdot \tfrac{1}{8} + \tfrac{1}{2} \cdot \tfrac{2}{8} = \tfrac{3}{16},$$
$$p(A = 1, C = 1) = \tfrac{1}{2} \cdot \tfrac{2}{8} + \tfrac{1}{2} \cdot \tfrac{1}{8} = \tfrac{3}{16},$$

	$C = 0$	$C = 1$	
$A = 0$	$\frac{5}{16}$	$\frac{5}{16}$	$\frac{10}{16}$
$A = 1$	$\frac{3}{16}$	$\frac{3}{16}$	$\frac{6}{16}$
	$\frac{8}{16}$	$\frac{8}{16}$	

and the joint distribution of A and C are seen to be independent. The Bernstein example on page 10 will also suffice:

$$p(AC|B) = \frac{p(ABC)}{p(B)} = \frac{p(E)}{p(B)} = \frac{\frac{1}{4}}{\frac{1}{2}} = \frac{1}{2},$$

$$p(A|B) = \frac{p(AB)}{p(B)} = \frac{p(E)}{p(B)} = \frac{\frac{1}{4}}{\frac{1}{2}} = \frac{1}{2},$$

$$p(C|B) = \frac{p(BC)}{p(B)} = \frac{p(E)}{p(B)} = \frac{\frac{1}{4}}{\frac{1}{2}} = \frac{1}{2}.$$

1–8. (a) A to win: $(\frac{1}{2})^1 + (\frac{1}{2})^4 + (\frac{1}{2})^7 + \cdots,$
since A gets a second opportunity only if A, B, C fail in the first try (probability $\frac{1}{2} \cdot \frac{1}{2} \cdot \frac{1}{2}$), and his chance to win (if he gets a second chance) is $\frac{1}{2}$. Similarly for later opportunities.

(b) B to win: $(\frac{1}{2})^2 + (\frac{1}{2})^5 + (\frac{1}{2})^8 + \cdots.$

(c) C to win: $(\frac{1}{2})^3 + (\frac{1}{2})^6 + (\frac{1}{2})^9 + \cdots.$

To sum (a) write

$$\varphi = a + a^4 + a^7 + \cdots + a^k,$$
$$a^3\varphi = a^4 + a^7 + \cdots + a^{k+3},$$
$$\varphi = \frac{a^{k+3} - a}{a^3 - 1} \quad \text{which, for } a < 1 \text{ and } k \to \infty, \quad = \frac{a}{1 - a^3}.$$

For $a = \frac{1}{2}$, $\varphi = \frac{4}{7}$ [and similarly for (b) and (c)]; the respective probabilities of winning are

$$p(A) = \tfrac{4}{7}, \qquad p(B) = \tfrac{2}{7}, \qquad p(C) = \tfrac{1}{7}.$$

One can also argue that repetition of the pattern after the first round will not disturb the relative prospects of A, B, and C. So, normalizing $\frac{1}{2}, \frac{1}{4}, \frac{1}{8}$, which add to $\frac{7}{8}$, we would have

$$p(A) = \tfrac{1}{2}/\tfrac{7}{8} = \tfrac{4}{7},$$
$$p(B) = \tfrac{1}{4}/\tfrac{7}{8} = \tfrac{2}{7},$$
$$p(C) = \tfrac{1}{8}/\tfrac{7}{8} = \tfrac{1}{7},$$

as before.

1–9. *Without bus*: Probability of failure to open = pr(both A and B remain closed or both C and D remain closed or both events occur)

$$= p^2 + p^2 - p^4 = p^2(2 - p^2)$$
$$= 0.0199 \quad \text{for } p = 0.1.$$

Probability of failure to close = pr(either A or B or both remain open *and* either C or D or both remain open)

$$= (p + p - p^2)(p + p - p^2) = p^2(2 - p)^2$$
$$= 0.0361 \quad \text{for } p = 0.1.$$

With bus: Probability of failure to open = pr(either A or C or both remain closed *and* either B or D or both remain closed)

$$= (p + p - p^2)(p + p - p^2) = p^2(2 - p)^2$$
$$= 0.0361 \quad \text{for } p = 0.1.$$

Probability of failure to close = pr(both A and C remain open or both B and D remain open or both events occur)

$$= p^2 + p^2 - p^4) = p^2(2 - p^2)$$
$$= 0.0199 \quad \text{for } p = 0.1.$$

Without the bus the (limiting) value of the ratio of the probability of closing to opening failure is 2. The bus merely reverses the ratio. If both errors are

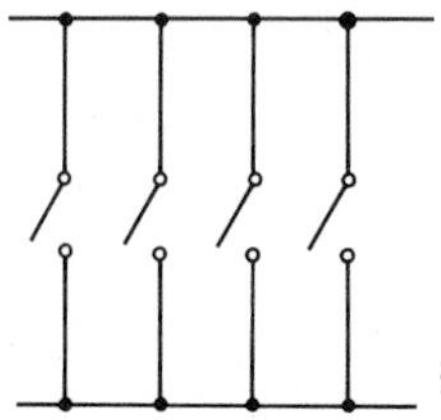

FIGURE A

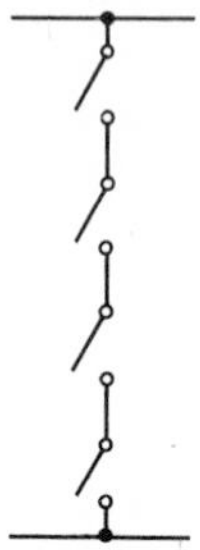

FIGURE B

equally harmful, installation of a bus is not indicated. The probability of failure over a cycle (closing plus opening) is the same, with or without a bus.

The situation is different if one error is more harmful than another. If one is more concerned with closing errors, n parallel switches may be indicated, as in Fig. A. Here the failure to close is p^n (small), but the probability of failure to open is maximized. If one is more concerned with failure to open, n series switches may be indicated, as in Fig. B. Here the failure to open is p^n (small), but the probability of failure to close is maximized. Combinations are common.

CHAPTER 3

3–1.

$$\begin{aligned}\int_{-\infty}^{\infty} (x - A)^2 f(x)\, dx &= \int_{-\infty}^{\infty} (x - \lambda + \lambda - A)^2 f(x)\, dx, \quad \text{where } \lambda \text{ is } E(\mathbf{x}),\\ &= \int_{-\infty}^{\infty} [(x - \lambda) + (\lambda - x)]^2 f(x)\, dx\\ &= \int_{-\infty}^{\infty} (x - \lambda)^2 f(x)\, dx \text{ (nonnegative)} + (\lambda - A)^2,\end{aligned}$$

which is *minimum* if $A = \lambda$. The problem can also be solved by differentiation of the original integral with respect to A.

3–2.

$$\int_{-\infty}^{\infty} |x - A| f(x)\, dx = \int_{-\infty}^{A} (A - x) f(x)\, dx + \int_{A}^{\infty} (x - A) f(x)\, dx,$$

which, after suitable addition and subtraction of terms, reduces to

$$2A \int_{-\infty}^{A} f(x)\, dx - 2\int_{-\infty}^{A} x f(x)\, dx - A + E(\mathbf{x}),$$

which we call ϕ.

$$\frac{\partial \phi}{\partial A} = 2Af(A) + 2\int_{-\infty}^{A} f(x)\, dx - 2Af(A) - 1 = 0,$$

from which

$$\int_{-\infty}^{A} f(x)\, dx = \tfrac{1}{2},$$

and A is the median of $\mathbf{x}$.

3–3.

$$\mathbf{u} = \frac{\mathbf{x} - E(\mathbf{x})}{\sigma(\mathbf{x})}$$

$$E(\mathbf{u}) = \frac{1}{\sigma(\mathbf{x})} E[\mathbf{x} - E(\mathbf{x})] = \frac{1}{\sigma(\mathbf{x})} [E(\mathbf{x}) - E(\mathbf{x})] = 0$$

$$\sigma^2(\mathbf{u}) = \frac{1}{\sigma^2(\mathbf{x})} E[\mathbf{x} - E(\mathbf{x})]^2 = \frac{\sigma^2(\mathbf{x})}{\sigma^2(\mathbf{x})} = 1.$$

3–4. Write $E(\mathbf{x}) = \lambda$,

$$\lambda = \int_0^\infty xf(x)\,dx = \int_0^c xf(x)\,dx + \int_c^\infty xf(x)\,dx.$$

But

$$\int_c^\infty xf(x)\,dx \geq \int_c^\infty cf(x)\,dx,$$

so

$$\lambda - \int_0^c xf(x)\,dx \geq c\int_c^\infty f(x)\,dx$$

$$\geq cp(\mathbf{x} \geq c),$$

and

$$\lambda \geq cp(\mathbf{x} \geq c),$$

and the theorem follows. It is most useful for $c > \lambda$.

The function $p\{|\mathbf{x} - \lambda| \geq k\sigma\}$ may be replaced by $p\{(\mathbf{x} - \lambda)^2 \geq k^2\sigma^2\}$, for $(x - \lambda)^2$ is a nonnegative random variable. By the Markov inequality,

$$p\{(\mathbf{x} - \lambda)^2 \geq k^2\sigma^2\} \leq E(\mathbf{x} - \lambda)^2/k^2\sigma^2,$$

that is, $\leq 1/k^2$.

CHAPTER 4

4–1. In axiom (a)

$$F(b, d) - F(a, d) - F(b, c) + F(a, c) \geq 0;$$

let $a \to -\infty$. Using axiom (b) [that $F(-\infty, y) = 0$], axiom (a) becomes

$$[F(b, d) - F(b, c)] \geq 0, \qquad \text{or} \qquad F(b, d) \geq F(b, c) \quad \text{for } d > c,$$

and $F(x, y)$ is monotonic nondecreasing in $\mathbf{y}$. The argument is similar for $\mathbf{x}$.

4–2.

$$
\begin{aligned}
f(x, y) &= g(x)h(y|x) \\
f(0, 0) &= g(0)h(0|0) = \tfrac{2}{7} \cdot \tfrac{1}{6} = \tfrac{2}{42} \\
f(1, 0) &= g(1)h(0|1) = \tfrac{5}{7} \cdot \tfrac{2}{6} = \tfrac{10}{42} \\
f(0, 1) &= g(0)h(1|0) = \tfrac{2}{7} \cdot \tfrac{5}{6} = \tfrac{10}{42} \\
f(1, 1) &= g(1)h(1|1) = \tfrac{5}{7} \cdot \tfrac{4}{6} = \tfrac{20}{42} \\
g(0) &= f(0, 0) + f(0, 1) = \tfrac{2}{42} + \tfrac{10}{42} = \tfrac{12}{42} \quad \text{(as earlier)} \\
g(1) &= f(1, 0) + f(1, 1) = \tfrac{10}{42} + \tfrac{20}{42} = \tfrac{30}{42} \quad \text{(as earlier)} \\
h(0) &= f(0, 0) + f(1, 0) = \tfrac{2}{42} + \tfrac{10}{42} = \tfrac{12}{42} \\
h(1) &= f(0, 1) + f(1, 1) = \tfrac{10}{42} + \tfrac{20}{42} = \tfrac{30}{42} \\
h(0|0) + h(1|0) &= \tfrac{1}{6} + \tfrac{5}{6} = 1 \quad (x \text{ fixed at } 0) \\
h(0|1) + h(1|1) &= \tfrac{2}{6} + \tfrac{4}{6} = 1 \quad (x \text{ fixed at } 1)
\end{aligned}
$$

Here we have calculated the joint probabilities $f(x, y)$ from the conditional probabilities $h(y|x)$ by piecemeal consideration of what is left in the lot after one unit is drawn. $f(x, y)$ can and will be obtained *directly*, in problem 12–3, where the example will be viewed as one of sampling (drawing 2 objects) without replacement from a finite population (7 objects).

4–3. We can define independence of $\mathbf{x}$ and $\mathbf{y}$ as follows: for *any* sets of real numbers S_x and S_y, the random variables $\mathbf{x}$, $\mathbf{y}$ are independent if the joint event $x \in S_x$, $y \in S_y$ satisfies

$$p\{x \in S_x, y \in S_y\} = p\{x \in S_x\} p\{y \in S_y\}.$$

Consider two functions of the random variables $\mathbf{x}$, $\mathbf{y}$:

$$u = u(x), \qquad v = v(y).$$

For any set S_x of real numbers, let s_x be the set of real numbers x such that $u(x) \in S_x$. Then the event $u \in S_x$ occurs only if $x \in s_x$. Apply a similar argument to any set S_y of real numbers and to $v(y)$. Finally,

$$\begin{aligned} p\{u \in S_x, v \in S_y\} &= p\{x \in s_x, y = s_y\} \\ &= p\{x \in s_x\} p\{y \in s_y\} = p\{u \in S_x\} p\{v \in S_y\}. \end{aligned}$$

CHAPTER 5

5–1.

$$\begin{aligned} M_{\mathbf{x}+\mathbf{y}}(\theta) &= \int_{-\infty}^{\infty} \int_{-\infty}^{\infty} e^{\theta(x+y)} f(x, y)\, dx\, dy \\ &= \int_{-\infty}^{\infty} e^{\theta x} g(x)\, dx \int_{-\infty}^{\infty} e^{\theta y} h(y)\, dy = M_{\mathbf{x}}(\theta) M_{\mathbf{y}}(\theta). \end{aligned}$$

5–2. If $\rho(\mathbf{u}, \mathbf{v}) = 0$, we have $\mu_{11}(\mathbf{u}, \mathbf{v}) = 0$;

$$\begin{aligned} \mu_{11} &= E[\mathbf{u} - E(\mathbf{u})][\mathbf{v} - E(\mathbf{v})] \\ &= E[a\{\mathbf{x} - E(\mathbf{x})\} + b\{\mathbf{y} - E(\mathbf{y})\}][c\{\mathbf{x} - E(\mathbf{x})\} + d\{\mathbf{y} - E(\mathbf{y})\}] \\ &= ac\sigma^2(\mathbf{x}) + (ad + bc)\rho(\mathbf{x}, \mathbf{y})\sigma(\mathbf{x})\sigma(\mathbf{y}) + bd\sigma^2(\mathbf{y}). \end{aligned}$$

If this is to vanish for all $\sigma^2(\mathbf{x})$, $\sigma^2(\mathbf{y})$, $\rho(\mathbf{x}, \mathbf{y})$, three of the four constants a, b, c, d would have to vanish ($\mathbf{u}$ *or* $\mathbf{v}$ must be identically zero). If $\mathbf{x}$ and $\mathbf{y}$ are independent, $b = 0$ and $c = 0$ or $a = 0$, $d = 0$ will suffice.

5–3. $E(a\mathbf{x} - \mathbf{y})^2 = [a^2 E\mathbf{x}^2 - 2aE\mathbf{xy} + E\mathbf{y}^2] \geq 0.$

Let $a = E\mathbf{xy}/E\mathbf{x}^2$. We have

$$\left[\frac{(E\mathbf{xy})^2}{(E\mathbf{x}^2)^2} E\mathbf{x}^2 - 2\frac{E\mathbf{xy}}{E\mathbf{x}^2} E\mathbf{xy} + E\mathbf{y}^2\right] \geq 0,$$

$$[(E\mathbf{xy})^2 - 2(E\mathbf{xy})^2 + E\mathbf{x}^2 E\mathbf{y}^2] \geq 0,$$

$$(E\mathbf{xy})^2 \leq E\mathbf{x}^2 E\mathbf{y}^2,$$

or by considering

$$E(a\mathbf{x} + b\mathbf{y})^2 = a^2 E\mathbf{x}^2 + 2abE\mathbf{xy} + b^2 E\mathbf{y}^2$$

as a positive semidefinite quadratic form in the variables a and b.

5–4.

$$\rho = \frac{E[\mathbf{x} - E(\mathbf{x})][\mathbf{y} - E(\mathbf{y})]}{\sigma(\mathbf{x})\sigma(\mathbf{y})} = \frac{E(\mathbf{xy}) - E(\mathbf{x})E(\mathbf{y})}{\sigma(\mathbf{x})\sigma(\mathbf{y})}$$

$$= \frac{E(\mathbf{x})E(\mathbf{y}) - E(\mathbf{x})E(\mathbf{y})}{\sigma(\mathbf{x})\sigma(\mathbf{y})} = 0.$$

If $\rho(\mathbf{x}, \mathbf{y}) = 0$, the random variables $\mathbf{x}$ and $\mathbf{y}$ need not be independent. We have

$$\rho(\mathbf{x}, \mathbf{y}) = \frac{E(\mathbf{xy}) - E(\mathbf{x})E(\mathbf{y})}{\sigma(\mathbf{x})\sigma(\mathbf{y})},$$

so, for $\rho(\mathbf{x}, \mathbf{y}) = 0$,

$$E(\mathbf{xy}) = E(\mathbf{x})E(\mathbf{y}),$$

that is,

$$\int_{-\infty}^{\infty}\int_{-\infty}^{\infty} xyf(x, y)\,dx\,dy = \int_{-\infty}^{\infty} xg(x)\,dx \int_{-\infty}^{\infty} yh(y)\,dy = \int_{-\infty}^{\infty}\int_{-\infty}^{\infty} xyg(x)h(y)\,dx\,dy,$$

or

$$\int_{-\infty}^{\infty}\int_{-\infty}^{\infty} xy[f(x, y) - g(x)h(y)]\,dx\,dy = 0,$$

which does not imply that the integrand is identically zero.

5–5.

$$\mathbf{u} = a + b\mathbf{x}, \qquad \mathbf{v} = c + d\mathbf{y}$$

$$\rho(\mathbf{u}, \mathbf{v}) = \frac{E[\mathbf{u} - E(\mathbf{u})][\mathbf{u} - E(\mathbf{v})]}{\sigma(\mathbf{u})\sigma(\mathbf{v})}$$

$$= \frac{E[a + b\mathbf{x} - a - bE(\mathbf{x})][c + d\mathbf{y} - c - dE(\mathbf{y})]}{b\sigma(\mathbf{x})\,d\sigma(\mathbf{y})}$$

$$= \frac{bdE[\mathbf{x} - E(\mathbf{x})][\mathbf{y} - E(\mathbf{y})]}{bd\sigma(\mathbf{x})\sigma(\mathbf{y})}$$

$$= \rho(\mathbf{x}, \mathbf{y}).$$

5–6.

$$\sigma^2(\mathbf{xy}) = \int_{-\infty}^{\infty}\int_{-\infty}^{\infty} (xy - E\mathbf{xy})^2 f(x, y)\,dx\,dy$$

$$= \int_{-\infty}^{\infty}\int_{-\infty}^{\infty} (xy - E\mathbf{x}E\mathbf{y})^2 g(x)h(y)\,dx\,dy$$

$$= \int_{-\infty}^{\infty}\int_{-\infty}^{\infty} (x^2y^2 - 2xyE\mathbf{x}E\mathbf{y} + E^2\mathbf{x}E^2\mathbf{y})g(x)h(y)\,dx\,dy$$

$$= \int_{-\infty}^{\infty} x^2 g(x)\,dx \int_{-\infty}^{\infty} y^2 h(y)\,dy$$

$$- 2E\mathbf{x}E\mathbf{y}\int_{-\infty}^{\infty} xg(x)\,dx \int_{-\infty}^{\infty} yh(y)\,dy + E^2\mathbf{x}E^2\mathbf{y}\int_{-\infty}^{\infty} g(x)\,dx \int_{-\infty}^{\infty} h(y)\,dy.$$

Using (3–9),

$$= [\sigma^2(\mathbf{x}) + E^2\mathbf{x}][\sigma^2(\mathbf{y}) + E^2\mathbf{y}] - E^2\mathbf{x}E^2\mathbf{y}$$
$$= \sigma^2(\mathbf{x})\sigma^2(\mathbf{y}) + \sigma^2(\mathbf{x})E^2\mathbf{y} + \sigma^2(\mathbf{y})E^2\mathbf{x}.$$

CHAPTER 6

6–1. Most directly, expanding (6–20), we get

$$\rho^2\sigma^2(\mathbf{y}) = \sigma^2(\mathbf{y}) - E(\mathbf{y}^2 + \alpha^2 + \beta^2\mathbf{x}^2 - 2\mathbf{y}\alpha - 2\mathbf{y}\beta\mathbf{x} + 2\alpha\beta\mathbf{x}).$$

Substituting (6–11) for α and β, and recalling that

$$E(\mathbf{x}^2) = \sigma^2(\mathbf{x}) + E^2(\mathbf{x}),$$

we are led to

$$\rho^2\sigma^2(\mathbf{y}) = 2\rho\,\frac{\sigma(\mathbf{y})}{\sigma(\mathbf{x})}\,[E(\mathbf{xy}) - E(\mathbf{x})E(\mathbf{y})] - \rho^2\sigma^2(\mathbf{y}),$$

and the result follows.

CHAPTER 7

7–1.

$$\mathbf{L} = \sum_{i=1}^{n} a_i\mathbf{x}_i, \qquad \mathbf{M} = \sum_{i=1}^{n} b_i\mathbf{x}_i$$

$$\rho(\mathbf{L}, \mathbf{M}) = 0 = \frac{E[\mathbf{L} - E(\mathbf{L})][\mathbf{M} - E(\mathbf{M})]}{\sigma(\mathbf{L})\sigma(\mathbf{M})}.$$

Writing $\lambda_i = E(\mathbf{x}_i)$, we have

$$0 = E[\textstyle\sum a_i\mathbf{x}_i - \sum a_i\lambda_i][\sum b_i\mathbf{x}_i - \sum b_i\lambda_i]$$
$$= E[\textstyle\sum a_i(\mathbf{x}_i - \lambda_i)][\sum b_i(\mathbf{x}_i - \lambda_i)]$$
$$= E[a_1(\mathbf{x}_1 - \lambda_1) + \cdots + a_n(\mathbf{x}_n - \lambda_n)][b_1(\mathbf{x}_1 - \lambda_1) + \cdots + b_n(\mathbf{x}_n - \lambda_n)]$$
$$= (a_1b_1 + \cdots + a_nb_n)\sigma^2(\mathbf{x}) + \text{cross-product terms of the form}$$
$$E[a_1(\mathbf{x}_1 - \lambda_1)b_1(\mathbf{x}_2 - \lambda_2)],$$

all of which cross-product terms vanish by virtue of the mutual independence of $\mathbf{x}_1, \ldots, \mathbf{x}_n$. Since $\sigma^2(\mathbf{x}) > 0$, we have

$$\sum_{i=1}^{n} a_ib_i = 0.$$

7–2.

$$\tfrac{1}{2}\cdot 1 + \tfrac{1}{4}\cdot 2 + \tfrac{1}{8}\cdot 4 + \cdots = \tfrac{1}{2} + \tfrac{1}{2} + \tfrac{1}{2} + \cdots;$$

the gain is not finite and the entry fee is infinite, but the payer of the entry fee will have to wait through a large number of trials for the gain to be realized. Few would pay the "large" entry fee; evidently, limited as we are by finite resources, money losses and gains are valued at something different from their expected values.

7–3.

$$ax^2 + 2bx + c = 0, \qquad x = \frac{-b \pm \sqrt{b^2 - ac}}{a},$$

which is imaginary when $ac > b^2$.

$$\text{prob (real)} = 1 - \text{prob (imaginary)}.$$

We have

$$\int_0^1 \int_0^{\sqrt{ac}} \int_0^1 f(a)f(b)f(c)\, da\, db\, dc = \tfrac{4}{9},$$

$$1 - \tfrac{4}{9} = \tfrac{5}{9}.$$

CHAPTER 8

8–1. We have $\mathbf{t}$, $0 \leq t < \infty$, density $f(t)$, and $\mathbf{z} = \mathbf{t} - T_0$.

$$f(t)\, dt \to f\{t(z)\} \frac{dt}{dz}\, dz,$$

and

$$g(z)\, dz = f(z + T_0)\, dz, \quad -T_0 \leq z < \infty.$$

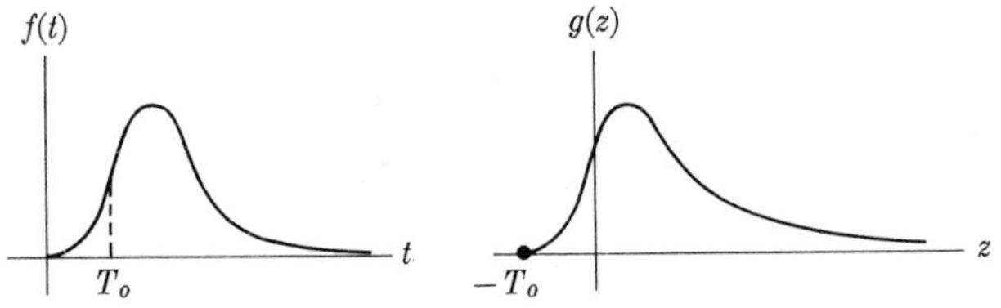

We want $g(z)$ relative to $t \not< k$. For $k \leq t < \infty$, $0 \leq z < \infty$, and $g(z)$ must be normalized (area adjusted to 1) by division by

$$\int_0^\infty f(z + T_0)\, dz.$$

8–2. Simplest, show

$$\begin{vmatrix} \dfrac{\partial x}{\partial u} & \dfrac{\partial x}{\partial v} \\ \dfrac{\partial y}{\partial u} & \dfrac{\partial y}{\partial v} \end{vmatrix} \cdot \begin{vmatrix} \dfrac{\partial u}{\partial x} & \dfrac{\partial u}{\partial y} \\ \dfrac{\partial v}{\partial x} & \dfrac{\partial v}{\partial y} \end{vmatrix} = 1.$$

Multiply out; then consider the partial derivatives of

$$x = x(u, v), \qquad y = y(u, v).$$

The rest of the argument will be algebraic.

8–3.

$$\mathbf{u} = \frac{\mathbf{x}}{\mathbf{y}}, \quad \mathbf{v} = \mathbf{x}, \qquad f(x, y) = g(x)h(y)$$

$$J = \begin{vmatrix} \dfrac{\partial x}{\partial u} & \dfrac{\partial x}{\partial v} \\ \dfrac{\partial y}{\partial u} & \dfrac{\partial y}{\partial v} \end{vmatrix} = \begin{vmatrix} 0 & 1 \\ -\dfrac{v}{u^2} & \dfrac{1}{u} \end{vmatrix} = \frac{v}{u^2}$$

$$\phi(u) = \int_v g(v)h\left(\frac{v}{u}\right) \cdot \frac{v}{u^2}\, dv. \tag{a}$$

Or

$$\mathbf{u} = \frac{\mathbf{x}}{\mathbf{y}}, \quad \mathbf{v} = \mathbf{y}, \qquad f(x, y) = g(x)h(y)$$

$$J = \begin{vmatrix} \dfrac{\partial x}{\partial u} & \dfrac{\partial x}{\partial v} \\ \dfrac{\partial y}{\partial u} & \dfrac{\partial y}{\partial v} \end{vmatrix} = \begin{vmatrix} v & u \\ 0 & 1 \end{vmatrix} = v,$$

$\phi(u) = \int_v g(uv)h(v) \cdot v\, dv$, which is identical with (a).

CHAPTER 9

9–1.

$$C_m^n = \frac{n!}{m!(n - m)!} = C_{n-m}^n.$$

For each different combination of m things taken from n things, a different combination of $n - m$ things remains. There are, therefore, *at least* as many combinations of $n - m$ things as of n things. Similarly for each different combination of $n - m$ things. That is,

$$C_{n-m}^n \geq C_m^n, \qquad C_m^n \geq C_{n-m}^n,$$

and the result follows.

9–2.

$$\begin{aligned} C_m^{n-1} + C_{m-1}^{n-1} &= \frac{(n-1)!}{m!(n-1-m)!} \frac{(n-1)!}{(m-1)!(n-m)!} \\ &= \frac{(n-1)!}{(m-1)!(n-m-1)!}\left[\frac{1}{m} + \frac{1}{n-m}\right] \\ &= \frac{(n-1)!}{(m-1)!(n-m-1)!} \frac{n}{m(n-m)} \\ &= \frac{n!}{m!(n-m)!} = C_m^n. \end{aligned}$$

The result also follows immediately from the identity

$$(n - 1)![n - m] + (n - 1)![m] = (n - 1)![n] = n!;$$

divide through by $m!(n - m)!$.

9–4. We want

$$C^n_{n-m}P^{n-m}_{n-m}(n_1, n_2, \ldots, n_k)$$
$$= \frac{n!}{(n - m)!m!} \frac{(n - m)!}{n_1!n_2! \ldots n_k!}$$
$$= \frac{n!}{m!n_1!n_2! \ldots n_k!}.$$

9–5.

$$C^N_0 + C^N_1 + C^N_2 + \cdots + C^N_N = \sum_{i=0}^{N} C^N_i.$$

But

$$(a + b)^N = \sum_{i=0}^{i=N} C^N_i a^i b^{N-i}.$$

Putting $a = b = 1$, we have

$$2^N = \sum_{i=0}^{i=N} C^N_i.$$

If the subset of no elements is excluded, the answer is $2^N - 1$.

9–6. The number of different ways of drawing 1, 2, . . . , N coins from N coins is, from problem 9–5,

$$C^N_1 + C^N_2 + \cdots + C^N_N = (1 + 1)^N - 1 = 2^N - 1.$$

Let N be even (the less favorable case to the proposition); the number of even outcomes is

$$A = C^N_2 + C^N_4 + \cdots + C^N_N.$$

But

$$(1 + 1)^N = C^N_0 + C^N_1 + \cdots + C^N_N,$$

from which

$$(1 + 1)^N + (1 - 1)^N = 2C^N_0 + 2A,$$

from which

$$A = \frac{2^N - 2}{2} = 2^{N-1} - 1.$$

Similarly, letting B be the number of subsets containing an odd number of outcomes,

$$(1 + 1)^N - (1 - 1)^N = 2B,$$

from which

$$B = \frac{2^N - 0}{2} = 2^{N-1}.$$

$$p(\text{even}) = \frac{2^{N-1} - 1}{2^N - 1}, \qquad p(\text{odd}) = \frac{2^{N-1}}{2^N - 1}, \qquad p(\text{even}) < p(\text{odd}).$$

CHAPTER 10

10–1.

$$f(x) = C_x^n p^x (1 - p)^{n-x} = \frac{n!}{x!(n - x)!} p^x (1 - p)^{n-x}$$

$$f(x + 1) = C_{x+1}^n p^{x+1} (1 - p)^{n-x-1} = \frac{n!}{(x + 1)!(n - x - 1)!} p^x (1 - p)^{n-x-1},$$

which reduces to

$$f(x + 1) = \frac{n - x}{x + 1} \frac{p}{1 - p} f(x).$$

10–2.

$$M_{\mathbf{x}}(\theta) = (q + pe^{\theta})^n$$

$$M_{\mathbf{x}+\mathbf{y}}(\theta) = (q_1 + p_1 e^{\theta})^{n_1} (q_2 + p_2 e^{\theta})^{n_2}.$$

For

$$p_1 = p_2 = p$$

$$M_{\mathbf{x}+\mathbf{y}}(\theta) = (q + pe^{\theta})^{n_1} (q + pe^{\theta})^{n_2} = (q + pe^{\theta})^{n_1+n_2};$$

$\mathbf{x} + \mathbf{y}$ follows a binomial distribution with parameters p and $n_1 + n_2$.

10–3. From (10–1) we have

$$\frac{f(x + 1)}{f(x)} = \frac{n - x}{x + 1} \frac{p}{1 - p}.$$

Under what conditions is $f(x + 1)/f(x) > 1$?

$$(n - x)p > (x + 1)(1 - p)$$

$$np > x + 1 - p$$

$$x < np - q.$$

The binomial frequency increases as long as

$$x < p(n + 1) - 1 \text{ [or } x + 1 < (n + 1)p],$$

then decreases.

10–4.

p_i = prob of success in ith trial

$\mathbf{x}_i$ = number of successes in ith trial (1 or 0)

$\mathbf{x}$ = number of successes in n independent trials

$$E(\mathbf{x}) = E(\mathbf{x}_1) + \cdots + E(\mathbf{x}_n) = p_1 + \cdots + p_n$$

or, writing

$$p = \frac{p_1 + \cdots + p_n}{n},$$

$$E(\mathbf{x}) = np$$

$$\begin{aligned}
\sigma^2(\mathbf{x}) &= E(\mathbf{x}_1 - p_1)^2 + \cdots + E(\mathbf{x}_n - p_n)^2 \\
&= p_1q_1 + \cdots + p_nq_n \\
&= \sum p_iq_i \\
&= \sum p_i - \sum p_i^2 \\
&= np - \sum (p_i - p + p)^2 \\
&= np - \sum (p_i - p)^2 - \sum p^2 \\
&= np - np^2 - \sum (p_i - p)^2 \\
&= np(1 - p) - \sum (p_i - p)^2 \\
&= npq - \sum (p_i - p)^2.
\end{aligned}$$

10–5.

$$x = 0 \text{ with probability } 1 - p$$
$$x = 1 \text{ with probability } p$$

r is the number of n observations which are large as x_1.

$$\begin{aligned}
p(\mathbf{r} = n) &= \text{prob } (\mathbf{x}_1 = 0) \\
&\quad + \text{prob } (\mathbf{x}_1 = 1) \text{ prob (remaining } n - 1 \text{ observations} = 1) \\
&= (1 - p) + p(p^{n-1}) = 1 - p + p^n,
\end{aligned}$$

or generally

$$\begin{aligned}
p(\mathbf{r}) &= \text{prob } (\mathbf{x}_1 = 0) \\
&\quad + \text{prob } (\mathbf{x}_1 = 1) \text{ prob } (\mathbf{r} - 1 \text{ of remaining } n - 1 \text{ observations} = 1) \\
&= (1 - p) + pC_{r-1}^{n-1}p^{r-1}(1 - p)^{n-r} \\
&= 1 - p + C_{r-1}^{n-1}p^r(1 - p)^{n-r}.
\end{aligned}$$

10–6. The probability of 3 or 4 small motors failing (and thereby, of the rocket failing) is $p^4 + 4p^3q$. The probability of 2 large motors failing (and thereby, of the rocket failing) is p^2. We have

$$p^4 + 4p^3q = p^2,$$

from which

$$p = \tfrac{1}{3}.$$

CHAPTER 11

11–1. The moment generating function of a Poisson random variable with parameter A is

$$e^{A(e^\theta - 1)}.$$

For independent Poisson random variables $\mathbf{x}$ and $\mathbf{y}$ with parameters A and B, we have

$$M_{\mathbf{x}+\mathbf{y}}(\theta) = e^{A(e^\theta - 1)} \cdot e^{B(e^\theta - 1)} = e^{(A+B)(e^\theta - 1)},$$

the moment generating function of a Poisson random variable with parameter $A + B$. $\mathbf{x} - \mathbf{y}$ has range $-\infty$ to ∞ which is sufficient to exclude the possibility of a Poisson distribution. Or

$$M_{\mathbf{x}-\mathbf{y}}(\theta) = M_{\mathbf{x}}(\theta)M_{\mathbf{y}}(-\theta) = e^{A(e^\theta - 1)}e^{B(e^{-\theta} - 1)},$$

which cannot be reduced to Poisson form.

11–2.

$$g(x|y) = e^{-y}y^x/x!, \qquad h(y) = e^{-y}, \quad y \geq 0.$$

$$g(x) = \int_0^\infty f(x, y)\,dy = \int_0^\infty g(x\,|\,y)h(y)\,dy = \int_0^\infty \frac{e^{-y}y^x}{x!}e^{-y}\,dy$$

$$= \frac{1}{x!}\int_0^\infty e^{-u}\left(\frac{u}{2}\right)^x \frac{du}{2}, \quad \text{with } u = 2y;$$

$$= \frac{1}{x!2^{x+1}}\int_0^\infty e^{-u}u^{-x}\,dx = \frac{\Gamma(x+1)}{x!2^{x\;1}},$$

which, since x is integral, equals $1/2^{x+1}$. Note that $\sum g(x) = 1$.

CHAPTER 12

12–1. To show that

$$\sum_{x=0}^{n} C_x^A C_{n-x}^B = C_n^{A+B},$$

write

$$(1 + t)^{A+B} = (1 + t)^A(1 + t)^B,$$

an identity in t (it must hold for any value of t). Therefore, the coefficient of t^n on the left side must equal the coefficient of t^n on the right side, that is,

$$C_n^{A+B} = \sum C_u^A C_v^B, \quad \text{the sum over } u + v = n,$$

$$C_n^{A+B} = C_0^A C_n^B + C_1^A C_{n-1}^B + \cdots + C_n^A C_0^B,$$

and the answer follows.

12–2. To find the value of N which maximizes the hypergeometric probability, for given A, n, x, we study

$$\frac{C_x^A C_{n-x}^{N-A}}{C_n^N} \div \frac{C_x^A C_{n-x}^{N-1-A}}{C_n^{N-1}} \quad \begin{matrix} > 1 \\ < 1, \end{matrix}$$

which reduces to

$$\frac{(N - n)(N - A)}{(N - A - n + x)} \quad \begin{matrix} > 1 \\ < 1. \end{matrix}$$

The probability increases if

$$N < \frac{nA}{x},$$

and decreases if

$$N > \frac{nA}{x}.$$

12–3.

$$p(0) = \frac{C_0^5 C_2^2}{C_2^7} = \frac{1}{21},$$

$$p(1) = \frac{C_1^5 C_1^2}{C_2^7} = \frac{10}{21} \quad [=p(1, 0) + p(0, 1)],$$

$$p(2) = \frac{C_2^5 C_0^2}{C_2^7} = \frac{10}{21}.$$

12–4. Perhaps the simplest demonstration is via a recursion formula (Wadsworth-Bryan [192]). First, obtain for the hypergeometric random variable **x** the recursion formula

$$f(x + 1) = \frac{(n - x)(A - x)}{(x + 1)(N - A - n + 1 + x)} f(x), \tag{a}$$

as well as the immediate result, from (12–5),

$$f(0) = \frac{C_n^{N-A}}{C_n^N}, \tag{b}$$

for n small compared with A (and therefore with respect to $N \geq A$), we divide (a) by N. Now write $p = A/N$ and $q = (N - A)/N$, and the binomial recursion formula, problem 10–3, follows. Moreover, under the stated conditions, $f(0)$ in (b) approaches the binomial q^n. The conditions for induction are satisfied; all terms of the hypergeometric density function approach the corresponding terms of the binomial for N, A large, n small compared with A, N.

Alternatively, expand the factorial terms involving N in

$$f(x) = \frac{C_x^{pN} C_{n-x}^{qN}}{C_n^N};$$

let $N \to \infty$ and argue that, in the polynomials of N, the highest term will dominate.

CHAPTER 14

14–1. We have, from Chapter 7,

$$g(x_2, x_3, \ldots, x_h|x_1) = \frac{f(x_1, x_2, \ldots, x_h)}{h(x_1)}, \tag{a}$$

where $h(x_1)$ is the marginal distribution of $\mathbf{x}_1$ and $g(x_2, x_3, \ldots, x_h|x_1)$ is the joint conditional distribution of $\mathbf{x}_2, \mathbf{x}_3, \ldots, \mathbf{x}_h$, given x_1. We have

$$f(x_1, x_2, \ldots, x_h) = \frac{n!}{x_1!x_2!\ldots x_h!} p_1^{x_1}p_2^{x_2}\ldots p_h^{x_h}. \tag{b}$$

Hold x_1 (for example) fixed and write $m = n - x_1$; normalize the remaining probabilities $p_2, p_3, \ldots, p_h$ by

$$p_{2*} = \frac{p_2}{1-p_1}, \quad p_{3*} = \frac{p_3}{1-p_1}, \ldots, \quad p_{h*} = \frac{p_h}{1-p_1}.$$

We then have

$$p_{2*} + p_{3*} + \cdots + p_{h*} = 1,$$

as well as

$$\mathbf{x}_2 + \mathbf{x}_3 + \cdots + \mathbf{x}_h = \mathbf{m}.$$

Now write (b) in the form

$$\frac{n!}{x_1!x_2!\ldots x_h!} p_1^{x_1}p_2^{x_2}\ldots p_h^{x_h}$$
$$= \left[\frac{n!}{x_1!(n-x_1)!} p_1^{x_1}(1-p_1)^{n-x_1}\right]\cdot\left[\frac{m!}{x_2!\ldots x_h!} p_{2*}^{x_2}\ldots p_{h*}^{x_h}\right].$$

The second term on the right is now a *bona fide* joint density function, the joint conditional density function of $\mathbf{x}_2, \ldots, \mathbf{x}_h$ given x_1; it is multinomial with parameters $p_{2*}, \ldots, p_{h*}, m$. The first term on the right is the marginal distribution of $\mathbf{x}_1$; it is binomial with parameters p_1, n.

CHAPTER 16

16–1. The expression

$$\frac{\partial^2 f}{\partial x^2} = 1 + \left(\frac{x-\lambda}{\sigma}\right)^2 = 0$$

defines the point of inflection. We find

$$x - \lambda = \pm\sigma.$$

16–2.

	Normal	Bienamé-Chebychev
Outside $\pm 1\sigma$	0.3174	≤ 0.5000
Outside $\pm 2\sigma$	0.0456	≤ 0.2500
Outside $\pm 3\sigma$	0.0027	≤ 0.1111
Outside $\pm 4\sigma$	0.0001	≤ 0.0625

16–3. From

$$M_{\mathbf{x}-c}(\theta) = e^{\theta^2/4b}, \qquad c = \lambda, \qquad \frac{1}{2b} = \sigma^2$$

we have, for a normal variable $\mathbf{x}$ of mean λ and variance σ^2

$$M_{\mathbf{x}}(\theta) = e^{\theta\lambda} e^{\theta^2\sigma^2/2},$$

$$M_{a+b\mathbf{x}}(\theta) = e^{\theta a} M_{\mathbf{x}}(\theta b) = e^{\theta a} e^{\theta b\lambda} e^{\theta^2 b^2 \sigma^2/2} = e^{\theta(a+b\lambda)} e^{(\theta^2/2)(b^2\sigma^2)}.$$

Comparing $M_{a+b\mathbf{x}}(\theta)$ and $M_{\mathbf{x}}(\theta)$, we see that $a + b\mathbf{x}$ is also normally distributed, with mean $a + b\lambda$ and variance $b^2\sigma^2$. More generally, for the linear sum of independent random variables $\mathbf{x}_1, \ldots, \mathbf{x}_n$,

$$\mathbf{y} = a_1\mathbf{x}_1 + \cdots + a_n\mathbf{x}_n,$$

we have, for each $\mathbf{x}_i$

$$M_{a_i\mathbf{x}_i}(\theta) = \int_{-\infty}^{\infty} e^{a_i x_i \theta} f(x_i)\, dx_i = M_{\mathbf{x}_i}(a_i\theta),$$

and from (7–8),

$$M_{\mathbf{y}}(\theta) = M_{\mathbf{x}_1}(a_1\theta) \ldots M_{\mathbf{x}_n}(a_n\theta).$$

But the moment generating function of a normal random variable $\mathbf{x}_i$ of mean λ_i and variance σ_i^2 is, repeating (23–2),

$$M_{\mathbf{x}_i}(\theta) = e^{\theta\lambda_i} e^{\theta^2\sigma_i^2/2}$$

from which

$$M_{a_i\mathbf{x}_i}(\theta) = e^{a_i\theta\lambda_i} e^{a_i^2\theta^2\sigma_i^2/2},$$

and

$$M_{\mathbf{y}}(\theta) = e^{\theta\Sigma a_i\lambda_i} e^{(\theta^2/2)\Sigma a_i^2\sigma_i^2};$$

$\mathbf{y}$ is normal with mean $a_1\lambda_1 + \cdots + a_n\lambda_n$ and variance $a_1^2\sigma_1^2 + \cdots + a_n^2\sigma_n^2$.

16–4.

$$E|\mathbf{x}| = \int_{-\infty}^{\infty} |x| N(0, \sigma^2)\, dx$$

$$= 2\int_0^{\infty} x \frac{1}{\sigma\sqrt{2\pi}} e^{-(1/2)(x/\sigma)^2} dx.$$

Let

$$-\frac{1}{2}\left(\frac{x}{\sigma}\right)^2 = u, \qquad x\, dx = -\sigma^2\, du;$$

then

$$E|\mathbf{x}| = \frac{2\sigma}{\sqrt{2\pi}} \int_0^{\infty} e^{-u}\, du = \sigma\sqrt{\frac{2}{\pi}},$$

a positive quantity. In Chapter 25, we will need the related quantity

$$E\left\{\frac{\sum_{i=1}^{n} |\mathbf{x}_i - \lambda|}{n}\right\},$$

where $\mathbf{x}_1, \ldots, \mathbf{x}_n$ are independent normal random variables of common mean λ and common variance σ^2. We get

$$E\left\{\frac{\sum_{i=1}^{n} |\mathbf{x}_i - \lambda|}{n}\right\} = \int_{-\infty}^{\infty} |x - \lambda| \frac{1}{\sigma\sqrt{2\pi}} e^{-(1/2)[(x-\lambda)/\sigma]^2} \, dx$$

which, following the earlier argument, integrates out to $\sigma\sqrt{2/\pi}$.

16–5. For $\mathbf{x}$ normal,

$$M_{(\mathbf{x}-\lambda)/\sigma}(\theta) = e^{\theta^2/2} = M_{(\mathbf{x}-\lambda)}\left(\frac{\theta}{\sigma}\right),$$

that is, regarding θ/σ as generator, $e^{\theta^2/2}$ is the m.g.f. of $\mathbf{x} - \lambda$. Now

$$e^{\theta^2/2} = 1 + \frac{\theta^2}{2} + \frac{\theta^4}{2^2}\frac{1}{2!} + \frac{\theta^6}{2^3}\frac{1}{3!} + \cdots + \frac{\theta^k}{2^{k/2}}\frac{1}{(k/2)!} + \cdots + \frac{\theta^{2k}}{2^k k!} + \cdots. \quad \text{(a)}$$

But

$$\left(\frac{\theta}{\sigma}\right)^{2k} \frac{1}{(2k)!}$$

is the multiplier of μ_{2k}. Rewriting the last term of (a)

$$\left(\frac{\theta}{\sigma}\right)^{2k} \frac{1}{(2k)!} \frac{\sigma^{2k}(2k)!}{2^k k!},$$

we have

$$\mu_{2k} = \frac{(2k)!\mu_2^k}{2^k k!}$$

CHAPTER 17

17–1. Since

$$f(x, y)\, dx\, dy = \frac{1}{2\pi} e^{-(1/2)(x^2+y^2)} \, dx\, dy$$

$$r^2 = x^2 + y^2,$$

$$r\, dr\, d\theta = dx\, dy,$$

the joint probability element becomes

$$f(r, \theta)\, dr\, d\theta = \frac{1}{2\pi} e^{-(1/2)r^2} r\, dr\, d\theta.$$

We have

$$\frac{1}{2\pi}\int_0^{2\pi} d\theta \int_0^R re^{-(1/2)r^2}\, dr = 0.95.$$

Let $-\frac{1}{2}r^2 = u$; the integration yields

$$e^{(1/2)R^2} = 20, \qquad R = \sqrt{2 \log_e 20} = 2.45.$$

17–2. We let $\sigma^2 = 1$;

$$f(x, y)\, dx\, dy = g(x) dx h(y) dy = \frac{1}{2\pi} e^{-(x^2+y^2)/2}\, dx\, dy$$

$$r = \sqrt{x^2 + y^2}, \qquad dx\, dy = r\, dr\, d\theta,$$

$$J = \begin{vmatrix} \frac{\partial x}{\partial r} & \frac{\partial y}{\partial r} \\ \frac{\partial x}{\partial \theta} & \frac{\partial y}{\partial \theta} \end{vmatrix} = \begin{vmatrix} \cos\theta & \sin\theta \\ -r\sin\theta & r\cos\theta \end{vmatrix} = r$$

$$f(x, y)\, dx\, dy \to \phi(r, \theta)\, dr\, d\theta \to \frac{1}{2\pi} e^{-r^2/2} r\, dr\, d\theta$$

$$\psi(r)\, dr = \frac{1}{2\pi} \int_0^{2\pi} e^{-r^2/2} r\, dr\, d\theta = re^{-r^2/2}\, dr,$$

which is not normal. The distribution function has the form

$$\int_0^r g(r)\, dr = 1 - e^{-r^2/2}.$$

Note that

$$\int_0^\infty re^{-r^2/2}\, dr = 1,$$

as must be true; also

$$E(\mathbf{r}) = \sqrt{\frac{\pi}{2}},$$

$$\sigma^2(\mathbf{r}) = \frac{4 - \pi}{2}.$$

17–3.

$$J^{-1} = \begin{vmatrix} \frac{\partial u}{\partial x} & \frac{\partial u}{\partial y} \\ \frac{\partial v}{\partial x} & \frac{\partial v}{\partial y} \end{vmatrix} = \begin{vmatrix} \frac{1}{\sigma_{\mathbf{x}}} & \frac{\rho_{\mathbf{xy}}}{\sigma_{\mathbf{y}}} \\ 0 & \frac{\sqrt{1 - \rho_{\mathbf{xy}}^2}}{\sigma_{\mathbf{x}}} \end{vmatrix},$$

from which

$$J = \frac{\sigma_{\mathbf{x}}\sigma_{\mathbf{y}}}{\sqrt{1 - \rho_{\mathbf{xy}}^2}}.$$

Also

$$\mathbf{u}^2 + \mathbf{v}^2 = \frac{\mathbf{x}^2}{\sigma_{\mathbf{x}}^2} - \frac{2\rho_{\mathbf{xy}}}{\sigma_{\mathbf{x}}\sigma_{\mathbf{y}}} + \frac{\mathbf{y}^2}{\sigma_{\mathbf{y}}^2},$$

which permits factoring:

$$f(u, v) = g(u)h(v);$$

u and **v** are independently normal.

17–4. Area A = Area B. If in

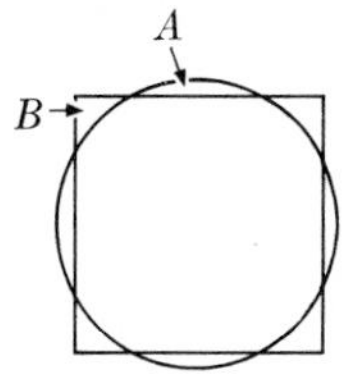

$$f(x, y) = \frac{1}{\sigma^2 2\pi} e^{-(1/2\sigma^2)(x^2+y^2)}, \qquad \text{(a)}$$

$$x_1^2 + y_1^2 > x_2^2 + y_2^2, \qquad f(x_1, y_1) < f(x_2, y_2).$$

This is readily seen by the change of variable

$$x = r\cos\theta, \qquad y = r\sin\theta, \qquad x^2 + y^2 = r^2, \qquad dx\,dy = r\,dr\,d\theta;$$

(a) becomes

$$\frac{1}{\sigma^2 2\pi} e^{-(1/2)(r/\sigma)^2} r\,dr\,d\theta,$$

which is evidently a monotonic decreasing function of increasing r. But area A is inside the circle ($x^2 + y^2 \le r^2$) while B is outside the circle ($x^2 + y^2 > r^2$). Thus the probability is higher for the circle.

CHAPTER 20

20–1.

$$\mathbf{x} = \mathbf{x}_1 + \mathbf{x}_2 + \mathbf{x}_3,$$

$$\int_0^x u\,du = \frac{x^2}{2}, \qquad 0 \le x \le 1,$$

$$\int_{x-1}^1 u\,du + \int_1^x (2 - u)\,du = \tfrac{1}{2}(-2x^2 + 6x - 3), \qquad 1 \le x \le 2,$$

$$\int_{x-1}^2 (2 - u)\,du = \tfrac{1}{2}(x - 3)^2, \qquad 2 \le x \le 3.$$

A graph of the function $h_3(x)$, $x = x_1 + x_2 + x_3$ is shown below:

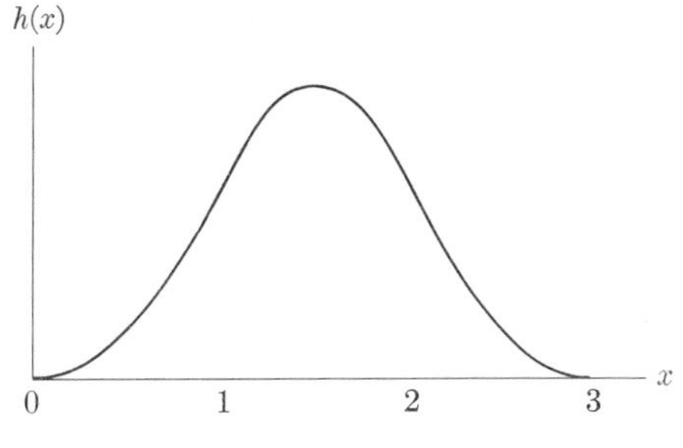

$0 \le x \le 1,$	$1 \le x \le 2,$	$2 \le x \le 3$
$\frac{1}{2}x^2$	$\frac{1}{2}(-2x^2 + 6x - 3)$	$\frac{1}{2}(x - 3)^2$

x	$h(x)$	x	$h(x)$	x	$h(x)$
0.0	0.000	1.0	0.50	2.0	0.500
0.1	0.005	1.1	0.59	2.1	0.405
0.2	0.020	1.2	0.67	2.2	0.320
0.3	0.045	1.3	0.70	2.3	0.245
0.4	0.080	1.4	0.74	2.4	0.180
0.5	0.125	1.5	0.75	2.5	0.125
0.6	0.180	1.6	0.74	2.6	0.080
0.7	0.245	1.7	0.70	2.7	0.045
0.8	0.320	1.8	0.67	2.8	0.020
0.9	0.405	1.9	0.59	2.9	0.005
1.0	0.500	2.0	0.50	3.0	0.000

20–2. For $\mathbf{u} = \mathbf{x} + \mathbf{y}$, we have

$$f(u) = \sum_y g(u - y)h(y),$$

which, for binomial $\mathbf{x}$, $\mathbf{y}$ becomes

$$f(u) = \sum_y C^n_{u-y} p^{u-y} q^{n-u+y} C^n_y p^y q^{n-y} = p^u q^{2n-u} \sum_y C^n_y C^n_{u-y}.$$

As for the limits on y,

$$0 \leq x \leq n, \quad 0 \leq u - y \leq n, \quad -n \leq y - u \leq 0, \quad u - n \leq y \leq u.$$

Repeating problem 12–1,

$$\sum_{x=0}^{x=n} C^A_x C^{N-A}_{n-x} = C^N_n,$$

we readily obtain

$$\sum_{y=u-n}^{y=u} C^n_y C^n_{u-y} = C^{2n}_u,$$

and

$$f(u) = C^{2n}_u p^u q^{2n-u},$$

a binomial density function.

20–3. Generally, for $\mathbf{u} = \mathbf{x} + \mathbf{y}$

$$h(u) = \int_y f(u - y)g(y)\, dy.$$

For $\mathbf{x}$, $\mathbf{y}$ independent Cauchy variables,

$$h(u) = \int_{-\infty}^{\infty} \frac{1}{\pi} \frac{1}{1 + (u - y)^2} \cdot \frac{1}{\pi} \frac{1}{1 + y^2}\, dy;$$

the integration on y is perhaps best performed via the replacement (use partial fractions) of the integrand by $1/\pi^2$ multiplied by

$$\frac{1}{u^2(u^2+4)}\left[\frac{u^2}{1+y^2}+\frac{2uy}{1+y^2}+\frac{u^2}{1+(y-u)^2}-\frac{2u(y-u)}{1+(y-u)^2}\right],$$

the terms in square brackets integrating to

$$u \log (y^2+1), \quad -u \ln [1+(u-y)^2], \quad u^2 \tan^{-1} y, \quad u^2 \tan^{-1}(y-u);$$

finally, for $v = u/2$,

$$h(v) = \frac{1}{\pi}\frac{1}{1+v^2}, \quad -\infty < v < \infty.$$

The mean of two independent random variables, each of which follows a Cauchy distribution, itself follows the same Cauchy distribution.

20–4. The expression

$$\begin{aligned} h(u) &= \int_y f(u-y)g(y)\,dy \\ &= \int_{-\infty}^{\infty} \frac{1}{\sigma_{\mathbf{x}}\sqrt{2\pi}} \exp\left[-\frac{1}{2}\left\{\frac{u-y-\lambda_{\mathbf{x}}}{\sigma_{\mathbf{x}}}\right\}^2\right] \cdot \frac{1}{\sigma_{\mathbf{y}}\sqrt{2\pi}} \exp\left[-\frac{1}{2}\left\{\frac{y-\lambda_{\mathbf{y}}}{\sigma_{\mathbf{y}}}\right\}^2\right] dy \end{aligned}$$

integrates out to $\mathbf{u}$ normal with mean $\lambda_{\mathbf{x}} + \lambda_{\mathbf{y}}$ and variance $\sigma_{\mathbf{x}}^2 + \sigma_{\mathbf{y}}^2$. Direct expansion and simplification of the exponent, or expression of the integral in polar coordinates or, immediately, use of a normal function identity (Parzen [149] page 191) will give the answer.

CHAPTER 22

22–1. Following the argument for the distribution of the maximum, we have

$$n\left[\int_{x_1}^{\infty} f(x)\,dx\right]^{n-1} f(x_1)\,dx_1,$$

a function of the random variable $\mathbf{x}_1$ and the sample size n; also, the distribution function,

$$G(x_1) = 1 - [1 - F(x_1)]^n.$$

22–2. The joint density function of $\mathbf{r} = \mathbf{x}_n - \mathbf{x}_1$ and $\mathbf{s} = \mathbf{x}_1$ was given in (22–8). Letting

$$f(x) = \theta e^{-\theta x}, \quad x \geq 0,$$

we get

$$g(r, s) = n(n-1)\left[\int_s^{s+r} \theta e^{-\theta x}\, dx\right]^{n-2} \theta e^{-\theta(r+s)}\theta e^{-\theta s}\, dr\, ds,$$

which reduces to

$$n(n-1)\,\theta^2 e^{-\theta r}(1 - e^{-\theta r})^{n-2} e^{-\theta n s}\, dr\, ds.$$

Integrating on s (from 0 to ∞), we have

$$h(r) = (n - 1)\theta e^{-\theta r}(1 - e^{\theta r})^{n-2}.$$

22–3. The variable $\mathbf{x}$ is normal with mean 0 and variance σ^2; $\mathbf{x}_{\max} = \mathbf{z}$. Write $f(x)$ for the normal density function, mean 0, variance σ^2, and $F(x)$ for the normal distribution function, same moments.

$$\phi(z)\, dz = n\left[\int_{-\infty}^{z} f(x)\, dx\right]^{n-1} f(z)\, dz.$$

For $n = 2$,

$$= 2\left[\int_{-\infty}^{z} f(x)\, dx\right] f(z)\, dz = 2F(z)f(z)\, dz,$$

$$E(\mathbf{z}) = 2\int_{-\infty}^{\infty} zF(z)f(z)\, dz.$$

Let $u = F(z)$, $dv = 2zf(z)$ and the integration yields

$$E(\mathbf{z}) = \frac{\sigma}{\sqrt{\pi}}.$$

If $\mathbf{x}$ has mean λ,

$$E(\mathbf{z}) = \lambda + \frac{\sigma}{\sqrt{\pi}}.$$

22–4. The relationship is natural. Write $p - \theta = v$ and $k = n - 2$; the binomial yields

$$\sum_{j=0}^{n-2} C_j^n p^j q^{n-j} = \int_p^1 n(n - 1)v^{n-2}(1 - v)\, dv.$$

The correspondence to Wilks' tolerance theorem is evident. Define p as the probability that an element of the population lies within the sample range (equivalent to β of the tolerance theorem). The left-hand side is the probability that as many as $n - 2$ of the observations in the sample of size n lie within the sample range, given p.

22–5. Ordering the independent random variables $\mathbf{x}_1, \ldots, \mathbf{x}_n$ orders the (independent) random variables $\mathbf{F}(\mathbf{x}_1), \ldots, \mathbf{F}(\mathbf{x}_n)$, where

$$F(x_1) = \int_{-\infty}^{x_1} f(x)\, dx, \ldots, F(x_n) = \int_{-\infty}^{x_n} f(x)\, dx.$$

We know that $\mathbf{F}(\mathbf{x}_1), \ldots, \mathbf{F}(\mathbf{x}_n)$ are uniformly distributed (8–10). The density function of the range $\mathbf{r} = \mathbf{F}(\mathbf{x}_n) - \mathbf{F}(\mathbf{x}_1)$ of uniform random variables is

$$n(n-1)r^{n-2}(1-r).$$

But the random variable $\mathbf{v}$ of Wilks' tolerance theorem has density

$$n(n-1)v^{n-2}(1-v),$$

and

$$v = \int_{x_1}^{x_n} f(x)\,dx = \int_{-\infty}^{x_n} f(x)\,dx - \int_{-\infty}^{x_1} f(x)\,dx = F(x_n) - F(x_1) = r.$$

CHAPTER 23

23–1.

$$\mathbf{M}_{\mathbf{x}}(\theta) = (1-2\theta)^{-n/2}, \quad \text{writing } \mathbf{x} \text{ for } \chi^2.$$

$$E(\mathbf{x}) = \left.\frac{\partial \mathbf{M}_{\mathbf{x}}(\theta)}{\partial \theta}\right|_{\theta=0} = -\frac{n}{2}(1-2\theta)^{-(n/2)-1}(-2)\Big|_{\theta=0} = n$$

$$E(\mathbf{x}^2) = \left.\frac{\partial^2 \mathbf{M}_{\mathbf{x}}(\theta)}{\partial \theta^2}\right|_{\theta=0} = \left[\frac{\partial}{\partial\theta} n(1-2\theta)^{-(n/2)-1}\right]_{\theta=0} = 2n\left(\frac{n}{2}+1\right)$$

$$\sigma^2(\mathbf{x}) = n^2 + 2n - n^2 = 2n$$

23–2. By direct integration of

$$\int_0^{\infty} Ff(F)\,dF,$$

or by

$$E(\mathbf{F}_{a,b}) = E\left(\frac{\chi_a^2}{a} \cdot \frac{b}{\chi_b^2}\right)$$

$$= E\left(\frac{\chi_a^2}{a}\right) E\left(\frac{b}{\chi_b^2}\right), \quad \text{since } \chi_a^2 \text{ and } \chi_b^2 \text{ are independent.}$$

From problem 23–1, $E(\chi_a^2) = a$. Also, writing $\mathbf{x} = \chi_b^2$, we have

$$E\left(\frac{1}{\chi_b^2}\right) = \frac{1}{2^{b/2}\Gamma(b/2)} \int_0^{\infty} \frac{1}{x} x^{(b/2)-1} e^{-x/2}\,dx$$

$$= \frac{(b/2-2)!}{2(b/2-1)!} = \frac{1}{b-2},$$

and $E(\mathbf{F}_{a,b}) = b/(b-2)$ follows.

23–3.

$$f(t) = \frac{1}{\sqrt{m\pi}} \cdot \frac{\Gamma(m+1)/2}{\Gamma(m/2)} \cdot \left(1 + \frac{t^2}{m}\right)^{-(m+1)/2}$$

Let $m = 1; f(t)$ reduces to

$$\frac{\Gamma(1)}{\sqrt{\pi}\,\Gamma(\frac{1}{2})} \cdot \frac{1}{1+t^2} = \frac{1}{\pi(1+t^2)}.$$

23–4.

$$\mathbf{F}_{a,b} = \frac{\chi_a^2}{a} \Big/ \frac{\chi_b^2}{b}$$

$$\mathbf{t} = \frac{\mathbf{x}}{\sqrt{\chi_m^2/m}}, \quad \text{where } \mathbf{x} \text{ is } N(0, 1) \text{ and } \chi^2 \text{ has m d.f.}$$

$$\mathbf{F}_{1,n-1} = \frac{\chi_1^2}{1} \Big/ \frac{\chi_{n-1}^2}{n-1} = \frac{\mathbf{x}^2}{\chi_{n-1}^2/(n-1)} = \mathbf{t}_{n-1}^2.$$

The problem can also be solved by direct examination of the distributions of $\mathbf{t}^2$ (obtained by change of variable from the distribution of $\mathbf{t}$) and $\mathbf{F}$. We have

$$f(t)\,dt = \frac{1}{\sqrt{(n-1)\pi}} \frac{[(n-2)/2]!}{[(n-3)/2]!} \left(\frac{t^2}{n-1} + 1\right)^{-n/2} dt.$$

Writing

$$u = t^2, \qquad \frac{dt}{du} = \frac{1}{2u^{1/2}};$$

the distribution of $\mathbf{u}$ is

$$\frac{1}{\sqrt{(n-1)\pi}} \frac{[(n-2)/2]!}{[(n-3)/2]!} \left(\frac{u}{n-1} + 1\right)^{-n/2} \frac{1}{2u^{1/2}}\,du,$$

which (multiplied by 2—why?) is precisely the distribution of $\mathbf{F}_{1,n-1}$.

CHAPTER 24

24–1. In (24–7) put

$$n_1 = n, \qquad n_2 = 1, \qquad m_1 = b, \qquad m_2 = 1.$$

$$z(m_2|n_2, m_1, n_1) = \frac{\int_0^1 p^{b+1}(1-p)^{n-b}\,dp}{\int_0^1 p^b(1-p)^{n-b}\,dp} = \frac{B(b+2, n-b+1)}{B(b+1, n-b+1)}$$

$$= \frac{\Gamma(b+2)\Gamma(n-b+1)}{\Gamma(n+3)} \frac{\Gamma(n+2)}{\Gamma(b+1)\Gamma(n-b+1)} = \frac{b+1}{n+2}.$$

Our knowledge of these situations suggests that the "trials" are not independent. After infancy, the older a man becomes, the lower his probability of survival. Once Fermat's last theorem is disproved, the probability of it being disproved again is probably almost one.

24–2. Let A be the proportion defective in a lot, $p(A)$ the relative frequency (proportion) of lots containing proportion defective A. Let B be the observed number of defectives ($B = 6$). We seek $p(A|B)$. From Bayes' theorem,

$$p(A|B) = \frac{p(B|A)p(A)}{\sum_A p(B|A)p(A)}.$$

From the hypergeometric distribution,

$$p(B|A) = \frac{C_k^M C_{n-k}^{N-M}}{C_n^N},$$

where M = number defective in lot ($1200A$), k = number defective in sample (6), N = total number in lot (1200), n = total number in sample (50). We have

$$p(B|A) = \frac{C_6^{1200A} C_{50-6}^{1200-1200A}}{C_{50}^{1200}},$$

and we find

$$p(A|B) = \frac{p(A)C_6^{1200A} C_{44}^{1200-1200A}}{\sum_A p(A)C_6^{1200A} C_{44}^{1200-1200A}}.$$

Computations for six values of A (given in detail by Fry [72], along with valuable discussion) lead to:

A	$p(A)$	$p(A\|B)$
0.00	0.78	0.001
0.01	0.17	0.004
0.02	0.034	0.070
0.03	0.009	0.170
0.04	0.005	0.374
0.05	0.002	0.382
0.06	0.000	0.000

It appears very likely that manufacturing standards have not been maintained.

CHAPTER 25

25–1.

$$\begin{aligned} E\{c_1\mathbf{T}_1 + \cdots + c_k\mathbf{T}_k\} &= c_1E(\mathbf{T}_1) + \cdots + c_kE(\mathbf{T}_k) \\ &= \theta(c_1 + \cdots + c_k) \\ &= \theta, \quad \text{if } c_1 + \cdots + c_k = 1 \end{aligned}$$

$$\sigma^2(\mathbf{t}) = w^2\sigma^2(\mathbf{u}) + (1 - w)^2\sigma^2(\mathbf{v}) \quad \text{minimum}$$

$$\frac{d\sigma^2(\mathbf{t})}{dw} = 2w\sigma^2(\mathbf{u}) - 2(1 - w)\sigma^2(\mathbf{v}) = 0,$$

$$w[\sigma^2(\mathbf{u}) + \sigma^2(\mathbf{v})] = \sigma^2(\mathbf{v}),$$

$$w = \frac{\sigma^2(\mathbf{v})}{\sigma^2(\mathbf{u}) + \sigma^2(\mathbf{v})}, \qquad 1 - w = \frac{\sigma^2(\mathbf{u})}{\sigma^2(\mathbf{u}) + \sigma^2(\mathbf{v})},$$

25–2.

$$f(x) = \frac{1}{\pi}\frac{1}{1 + (x - \theta)^2}$$

The density function $p(\tilde{x})$ of the median $\tilde{\mathbf{x}}$ in random samples of size $2k + 1$ from a Cauchy population is

$$p(\tilde{x}) = \frac{(2k+1)!}{k!k!}\left\{\int_{-\infty}^{\tilde{x}} \frac{1}{\pi}\frac{dx}{1 + (x - \theta)^2}\right\}^k \left\{\int_{\tilde{x}}^{\infty} \frac{1}{\pi}\frac{dx}{1 + (x - \theta)^2}\right\}^k f(\tilde{x}),$$

where

$$f(\tilde{x}) = \frac{1}{\pi}\frac{1}{1 + (\tilde{x} - \theta)^2}.$$

To evaluate the integrals, let $y = x - \theta$ and recall that

$$\int \frac{dy}{1 + y^2} = \tan^{-1} y.$$

We are led to

$$E\{\tilde{\mathbf{x}} - \theta\} = \frac{(2k+1)!}{k!k!\pi}\int_{-\infty}^{\infty}\left\{\frac{1}{4} - \left[\frac{1}{\pi}\tan^{-1}(\tilde{x} - \theta)\right]^2\right\}^k \frac{(\tilde{x} - \theta)}{1 + (\tilde{x} - \theta)^2}\, d\tilde{x}.$$

The expression in braces is of the form $(\frac{1}{4} + t)^n$. Expansion in series leads to

$$E(\tilde{\mathbf{x}} - \theta) = \frac{(2k+1)!}{k!k!\pi}\int_{-\infty}^{\infty}\sum_{i=1}^{k}\frac{k!}{i!(k-i)!}\left(\frac{1}{4}\right)^{k-i} \times \left\{-\left[\frac{1}{\pi}\tan^{-1}(x - \theta)\right]^2\right\}^i \frac{(\tilde{x} - \theta)\, dx}{1 + (\tilde{x} - \theta)^2},$$

which, by change of variable $\tilde{x} - \theta = \tan y$, finally reduces to 0.

25–3.

$$\mathbf{r} \text{ is } N(R, \sigma^2), \qquad E\mathbf{r} = R$$

$$E[\mathbf{r} - R]^2 = \sigma^2 \qquad \text{or} \qquad E\mathbf{r}^2 = \sigma^2 + R^2.$$

We want $E\{\phi(\mathbf{r}_1, \mathbf{r}_2, \ldots, \mathbf{r}_n)\} = \pi R^2$. Consider

$$\phi(r_1, r_2, \ldots, r_n) = r_1^2 + r_2^2 + \cdots + r_n^2,$$

$$E(\mathbf{r}_1^2 + \mathbf{r}_2^2 + \cdots + \mathbf{r}_n^2) = nE\mathbf{r}^2 = n(\sigma^2 + R^2) = n\sigma^2 + nR^2,$$

or

$$\frac{\pi}{n} E(\mathbf{r}_1^2 + \mathbf{r}_2^2 + \cdots + \mathbf{r}_n^2) - \pi\sigma^2 = \pi R^2.$$

Therefore,

$$\frac{\pi}{n} (r_1^2 + r_2^2 + \cdots + r_n^2) - \pi\sigma^2$$

is an unbiased estimator of πR^2. If σ^2 is unknown, an unbiased estimator of πR^2 is

$$\frac{\pi}{n} (r_1^2 + r_2^2 + \cdots + r_n^2) - \pi \frac{ns^2}{n-1},$$

where s^2 is the sample variance.

25–4. Write $\mathbf{x}$ for χ^2. The density function of $\mathbf{x}$ is

$$f(x) = \frac{1}{2\Gamma(n-1)/2} \left(\frac{x}{2}\right)^{(n-3)/2} e^{-(1/2)x}.$$

The change of variable from $\mathbf{x}$ to $\mathbf{s}$ yields

$$g(s) = \frac{(n/2\sigma^2)^{(n-3)/2}}{\Gamma(n-1)/2} s^{n-3} e^{ns^2/2\sigma^2} \frac{ns}{\sigma^2}$$

and, noting that

$$\Gamma\left(\frac{m}{2}\right) = 2\int_0^\infty y^{m-1} e^{-y^2}\, dy,$$

we find

$$E(\mathbf{s}) = \sqrt{\frac{2\sigma^2}{n}} \cdot \frac{\Gamma(n/2)}{\Gamma(n-1)/2},$$

and preferably left in this form, since in the form

$$\sqrt{\frac{2\sigma^2}{n}} \frac{[(n-2)/2]!}{[(n-3)/2]!}$$

$n/2$ and $(n-1)/2$ are integers simultaneously, and this is impossible.

25–5. The expression

$$E\{\mathbf{v}(\mathbf{x})\} = \frac{1}{2\theta}\int_{-\theta}^{\theta} v(x)\, dx = \int_{-\theta}^{\theta} v(x)\, dx = 0$$

is satisfied (for example) by $v(x) = x \neq 0$.

CHAPTER 26

26–1.

$$f(x; \lambda) = \frac{e^{-\lambda}\lambda^x}{x!} \tag{a}$$

$$L = \sum_{i=1}^{n} \log f(x_i, \lambda) = \log e^{-n\lambda} + \log \lambda^{\Sigma_{i=1}^n x_i} - \log \sum_{i=1}^{n} x_i!$$

$$\frac{\partial L}{\partial \lambda} = -n + \frac{1}{\lambda}\sum_{i=1}^{n} x_i = 0, \text{ and the answer follows.}$$

$$\frac{1}{\sigma^2(\mathbf{T})} = -nE\left\{\frac{\partial^2 \log f(x, \lambda)}{\partial \lambda^2}\right\},$$

$$\frac{1}{\sigma^2(\mathbf{T})} = -n\sum_{x=0}^{\infty} \frac{x_i}{\lambda^2}\frac{e^{-\lambda}\lambda^x}{x!} = \frac{n}{\lambda},$$

$$\sigma^2(\mathbf{T}) = \frac{\lambda}{n}.$$

26–2.

$$f(x;\lambda) = \frac{1}{\sigma\sqrt{2\pi}}\, e^{-(1/2)(x-\lambda)^2}$$

$$p(E|\lambda) = \frac{1}{\sigma^n(\sqrt{2\pi})^n}\exp\left[-\frac{1}{2\sigma^2}\sum_{i=1}^{n}(x_i-\lambda)^2\right]$$

$$\log p(E|\lambda) = \log \text{constant} - \frac{1}{2\sigma^2}\sum_{i=1}^{n}(x_i-\lambda)^2$$

$$\frac{\partial \log p(E|\lambda)}{\partial \lambda} = 0 + \frac{\sum_{i=1}^{n}(x_i-\lambda)}{\sigma^2} = 0$$

$$\sum_{i=1}^{n}(x_i-\lambda) = 0 \text{ with unique solution } \mathbf{T} = \sum_{i=1}^{n} x_i/n.$$

26–3.

$$f(x;\theta) = \theta e^{-\theta x}$$

$$p(E|\theta) = \theta^n e^{-\theta\Sigma_i^n x_i}, \qquad \log p(E|\theta) = n\log\theta - \theta\sum_{i=1}^{n} x_i$$

$$\frac{\partial \log p(E|\theta)}{\partial \theta} = \frac{n}{\theta} - \sum_{i=1}^{n} x_i = 0;$$

the maximum likelihood estimator of θ is the reciprocal of the sample mean.

26–4. The maximum likelihood estimator of N is nA/a, where n is the sample size, A is the number of the successes in the population of size N, and a is the number of successes in the sample:

$$E\left(\frac{nA}{\mathbf{a}}\right) = nAE\left(\frac{1}{\mathbf{a}}\right).$$

But among the possible values of $\mathbf{a}$ is 0. All other values of $\mathbf{a}$ lead to positive $E(1/\mathbf{a})$. The expectation is therefore ∞.

26–5. We have

$$-\frac{1}{n\sigma_{\mathbf{T}}^2} = -\frac{2}{\pi}\int_{-\infty}^{\infty}\frac{1-y^2}{(1+y^2)^3}\,dy.$$

Let $y = \tan u$

$$= -\frac{2}{\pi}\int_{-\pi/2}^{+\pi/2}\frac{1-\tan^2 u}{\sec^4 u}\,du$$

$$= -\frac{2}{\pi}\left[2\int_{-\pi/2}^{+\pi/2}\cos^4 u\,du - \int_{-\pi/2}^{+\pi/2}\cos^2 u\,du\right]$$

$$= -\frac{4}{\pi}\left[\frac{3u}{8}+\frac{3\sin 2u}{16}+\frac{\cos^3 u\sin u}{4}\right]_{-\pi/2}^{\pi/2}+\frac{2}{\pi}\left[\frac{u}{2}+\frac{\sin 2u}{4}\right]_{-\pi/2}^{\pi/2} = -\tfrac{1}{2},$$

$$\sigma_{\mathbf{T}}^2 = \frac{2}{n}.$$

CHAPTER 28

28–1.

$$\frac{p_0(E)}{p_1(E)} = (\sigma_1^n/\sigma_0^n)e^{-(1/2)ns^2\left\{\frac{1}{\sigma_0^2}-\frac{1}{\sigma_1^2}\right\}} \quad \begin{matrix}\leq k \text{ inside } R,\\ > k \text{ outside } R,\end{matrix}$$

which leads to

$$s^2\{\sigma_0^2 - \sigma_1^2\} \quad \begin{matrix}\leq \text{ inside } R,\\ > \text{ outside } R,\end{matrix}$$

from which the region R is defined by

$$s^2 \leq k_2 \quad \text{if } \sigma_1^2 < \sigma_0^2$$

$$s^2 > k_2 \quad \text{if } \sigma_1^2 > \sigma_0^2.$$

No uniformly most powerful test exists. As found in (23–21), $\mathbf{s}^2$ follows a χ^2 distribution.

28–2. On change of variables we have, generally,

$$f(\bar{x})\,d\bar{x} = f[\bar{x}(L)]\,\frac{d\bar{x}}{dL}\,dL.$$

As for $f(\bar{x})\,d\bar{x}$, if H_0 is true, $\bar{\mathbf{x}}$ is normally distributed with mean λ_0 and variance $1/n$,

$$f(\bar{x}) = \sqrt{\frac{n}{2\pi}}\,e^{-(n/2)(\bar{x}-\lambda_0)^2},$$

so

$$f[\bar{x}(L)] = \sqrt{\frac{n}{2\pi}}\,L.$$

From (28–25)

$$\frac{dL}{d\bar{x}} = -\frac{1}{nL(\bar{x} - \lambda_0)},$$

which gives us

$$g(L)\,dL = \frac{1}{\sqrt{2\pi}}\sqrt{-2\log_e L}\,dL,$$

and we now need only solve for $L_{0.05}$ in

$$\int_0^{L_{0.05}} \frac{1}{\sqrt{2\pi}}\sqrt{-2\log_e L}\,dL = 0.05,$$

with the solution here valid for any n.

Write $-2\log L = u^2$, $dL = -ue^{-(1/2)u^2}\,du$.

$$\int_0^{L_c} \frac{1}{\sqrt{2\pi}\sqrt{-2\log L}}\,dL = \int_{-\infty}^{-\sqrt{-2\log L_c}} \frac{1}{\sqrt{2\pi}} e^{-(1/2)u^2}\,du + \int_{+\sqrt{-2\log L_c}}^{\infty} \frac{1}{\sqrt{2\pi}} e^{-(1/2)u^2}\,du,$$

since the L_c level for L corresponds to two equal extreme regions for u whose limits are found from

$$0 \le L \le 1, \quad -\infty < \log L \le 0, \quad 0 \le -2\log L < \infty,$$
$$0 \le u^2 < \infty, \quad -\infty < u < \infty.$$

From normal areas we have

$$\sqrt{-2\log L_c} = 1.96, \qquad L_c = 0.147.$$

This value of L_c could, of course, now be employed to calculate limits on the likelihood ratio statistic $\bar{x}$. At $L = L_c = 0.147$,

$$L = e^{-(n/2)(\bar{x}-\lambda_0)^2}$$

reduces to the familiar

$$\pm 1.96 = \frac{\bar{x} - \lambda_0}{1/\sqrt{n}}.$$

The region of rejection of H_0 is (28–26).

CHAPTER 29

29–1. *Binomial*

$$n = 50, \qquad x = 10$$

$$0.10 \le p \le 0.34, \qquad \text{conf. coeff. } 0.95,$$

Also $$n = 20, \qquad x = 15$$

$$0.515 \leq p \leq 0.92, \quad \text{conf. coeff. } 0.95.$$

As we found in (10–11), the binomial sum can be expressed in terms of the incomplete beta-function. For the last example, from Pearson's tables [157],

$$0.523 < p < 0.920.$$

29–2. *Hypergeometric* (from the DeLury-Chung charts)

N	n	x	Lower limit of A	Upper limit of A	Confidence coefficient
500	100	18	62	131	0.95
500	100	18	54	138	0.99
500	50	3	14	73	0.90
500	50	3	11	81	0.95
500	50	3	7	99	0.99

29–3. *Poisson*

$$x = 6$$

$$2.2 \leq \lambda \leq 13.1, \quad \text{conf. coeff. } 0.95$$

$$1.5 \leq \lambda \leq 15.6, \quad \text{conf. coeff. } 0.99$$

For $x = 3$, $1 - \alpha = 0.95$, for symmetrical bounds,

$$\sum_{x=3}^{\infty} \frac{e^{-\lambda_2}\lambda_2^x}{x!} = 0.975, \qquad \sum_{x=3}^{\infty} \frac{e^{-\lambda_1}\lambda_1^x}{x!} = 0.025,$$

from which (from Molina's tables)

$$p\{0.06 \leq \lambda \leq 7.23\} = 0.95$$

with length $7.23 - 0.06 = 7.17$. Is this shortest? Several approximate nonsymmetrical bounds with $1 - \alpha = 0.95$ are shown below (from Molina's tables):

$$0.01,\ 0.96; \quad \lambda_1 = 0.4, \quad \lambda_2 = 6.6, \quad \lambda_2 - \lambda_1 = 6.2$$

$$0.04,\ 0.99; \quad \lambda_1 = 0.7, \quad \lambda_2 = 8.4, \quad \lambda_2 - \lambda_1 = 7.7$$

$$0.00,\ 0.95; \quad \lambda_1 = 0.0, \quad \lambda_2 = 6.3, \quad \lambda_2 - \lambda_1 = 6.3$$

$$0.005,\ 0.955; \quad \lambda_1 = 0.33, \quad \lambda_2 = 6.43, \quad \lambda_2 - \lambda_1 = 6.10$$

Comparison of binomial, Poisson, and hypergeometric probabilities, for $N = 500$, $n = 50$, $\sum x_i = 5$:

	Lower bound	Upper bound
Binomial p	0.04	0.22
Hypergeometric p	0.045	0.215
Poisson λ/n	0.032	0.234

LIST OF SELECTED TEXTBOOKS

We list a few of the many good textbooks in probability and/or statistics. All are fairly general in coverage, and the level of difficulty is generally either elementary or intermediate.

R. L. ANDERSON and T. A. BANCROFT, *Statistical Theory in Research.* New York, McGraw-Hill Book Co., 1952. A general text. Part II on least squares and experimental design is one of the best discussions at this level in print.

N. ARLEY and K. R. BUCH, *Introduction to the Theory of Probability and Statistics.* New York, John Wiley and Sons, 1952. A text on probability with a nonpedestrian applied spirit throughout.

H. D. BRUNK, *An Introduction to Mathematical Statistics.* Boston, Ginn and Co., 1960. Surely among the best statistical texts in the market. Includes a modern introduction to probability and random variables.

H. CRAMÉR, *Mathematical Methods of Statistics.* Princeton, Princeton University Press, 1946. An advanced book in probability and statistics. One of the principal book sources of reliable proofs of properties of statistical estimators.

H. CRAMÉR, *The Elements of Probability Theory and Some of Its Applications.* New York, John Wiley and Sons, 1955. An elementary text with the quality that a first-class research man can give such a work. Carefully written, good problems, and a valuable historical sense.

D. DUGUÉ, *Traité de Statistique Théorique et Appliquée.* Paris, Masson et Cie, 1958. A first-class text in statistical theory at an intermediate level by a French master who has made important contributions to statistical theory.

W. FELLER, *An Introduction to Probability Theory and Its Applications, Vol. I.*, 2nd ed., New York, John Wiley and Sons, 1957. A masterpiece. Incomplete, not always systematic, but always spirited, profound, and provocative. Unifying theory original with Feller is introduced almost casually. A student who can understand all of this and who can solve all the (remarkable) problems has come a long way.

D. A. S. FRASER, *Statistics: An Introduction.* New York, John Wiley and Sons, 1958. An excellent modern textbook. Reliable and incisive discussion.

P. G. HOEL, *Introduction to Mathematical Statistics.* 3rd ed., New York, John Wiley and Sons, 1962. A successful and good elementary book in statistical theory and application; one of the first of modern vintage to appear.

R. V. HOGG and A. T. CRAIG, *Introduction to Mathematical Statistics.* New York, Macmillan Co., 1959. A strong modern text with reliable discussion of modern developments and good problems. The proofs here are a long cut above those of most intermediate texts.

M. G. KENDALL and A. STUART, *The Advanced Theory of Statistics,* Vol. 1 (1958, Vol. 2 (1961), [and Vol. 3, 1964]. London, Charles Griffin and Co. A valuable reference work in statistical theory. Whatever the faults of the earlier edition, there are few teachers or students who have not learned much from this first extensive compendium of intermediate to advanced statistical theory. The current edition is indeed impressive, particularly Volume 2.

A. M. Mood, *Introduction to the Theory of Statistics*. New York, McGraw-Hill Book Co., 1950. A thorough, first-class text, for years the alternative choice to Hoel. A second edition will soon appear.

M. E. Munroe, *Theory of Probability*. New York, McGraw-Hill Book Co., 1951. An offbeat book with interesting material not found elsewhere. The author is peculiarly sensitive to the difficulties serious beginners face in studying probability theory.

E. Parzen, *Modern Probability Theory and Its Applications*. New York, John Wiley and Sons, 1960. A systematic and beautiful book, with careful attention to many topics Feller gracefully omits.

L. Schmetterer, *Einführung in die Mathematische Statistik*. Vienna, Springer Press, 1956. A first-class text, of high mathematical rigor, with intermediate-level discussion of probability, statistical theory, and statistical inference.

G. P. Wadsworth and J. G. Bryan, *Introduction to Probability and Random Variables*. New York, McGraw-Hill Book Co., 1960. A fine and different book, with uncommon detailed examples, and with continuous emphasis of important mathematical detail.

B. L. van der Waerden, *Mathematische Statistik*. Berlin, Springer, 1957. A sharp concise text which includes many modern results, all abetted by the impeccable publishing standards of Springer.

S. S. Wilks, *Mathematical Statistics*. New York, John Wiley and Sons, 1962. Earlier version: Princeton, Princeton University Press, 1943. The 1943 volume is a dry, careful, intermediate work on statistical theory. Tightly written, solid and valuable. The admirable 1962 volume is concise, sharply modernized, and more advanced. By an American pioneer, distinguished for his statistical research, editing, and teaching.

I would like to mention Professor Paul Rider's elementary but pioneering *An Introduction to Modern Statistical Methods* (John Wiley and Sons) published twenty-three years ago, the first printed American text with a modern point of view. Rider's text played an important role in the modernization of elementary courses in statistics in this country.

The student may wish to buy one or two books. If his interests are principally in probability, Feller or Parzen will be the choice; if in statistics at the level of this text, Brunk or Fraser; if in statistics at a more advanced level, Wilks' 1962 book or Kendall-Stuart. If his interests are in both areas and if he is well prepared, Cramér's *Mathematical Methods*.

LIST OF SPECIALIZED BOOKS

I would like to mention several books which are either too advanced or too specialized for inclusion in the working list above, but which a student may well want to examine. They will give him an idea of some areas of probability and statistical theory (some beyond the level of difficulty of the present text and some not) which are of much current interest.

M. S. Bartlett, *An Introduction to Stochastic Processes.* Cambridge (Eng.), Cambridge University Press, 1955.

D. Blackwell and M. Girshick, *Theory of Games and Statistical Decisions.* New York, John Wiley and Sons, 1954.

H. Chernoff and L. E. Moses, *Elementary Decision Theory.* New York, John Wiley and Sons, 1959.

K. L. Chung, *Markov Chains.* Berlin, Springer-Verlag, 1960.

D. A. S. Fraser, *Nonparametric Methods in Statistics.* New York, John Wiley and Sons, 1957. An area in which minimal assumptions are made on distribution functions. See also the survey by S. S. Wilks, "Non-parametric statistical inference" (with 51 references), in *Probability and Statistics,* the Harold Cramér volume, Almqvist and Wicksell, Stockholm (New York, John Wiley and Sons).

B. W. Gnedenko, *Lehrbuch der Wahrscheinlichkeitsrechnung.* Berlin, Akademie-Verlag, 1957.

F. A. Graybill, *An Introduction to Linear Statistical Models, Vol. I.* New York, McGraw-Hill Book Co., 1961.

E. L. Lehmann, *Notes on the Statistical Theory of Estimation,* Chapters I–V. Berkeley, University of California Associated Student's Store, 1950.

E. L. Lehmann, *Testing Statistical Hypotheses.* New York, John Wiley and Sons, 1959.

M. Loève, *Probability Theory,* 2nd ed. New York, D. Van Nostrand, Inc., 1960.

E. Parzen, *Stochastic Processes.* San Francisco, Holden-Day, Inc., 1962.

H. Richter, *Wahrscheinlichkeitsrechnung.* Berlin, Springer-Verlag, 1956.

H. Scheffé, *The Analysis of Variance.* New York, John Wiley and Sons, 1959.

A. Wald, *Sequential Analysis.* New York, John Wiley and Sons, 1947.

A. Wald, *Statistical Decision Functions.* New York, John Wiley and Sons, 1950.

L. Weiss, *Statistical Decision Theory.* New York, McGraw-Hill Book Co., 1961.

LIST OF BOOKS ON APPLICATIONS

The following is a short list of books which deal admirably with some of the most successful areas of application of probability and statistical theory.

Biometrics

C. R. Rao, *Advanced Statistical Methods in Biometric Research.* New York, John Wiley and Sons, 1952. Admirable, also, on modern statistical theory.

Chemical

C. A. Bennett and N. L. Franklin, *Statistical Analysis in Chemistry and the Chemical Industry.* New York, John Wiley and Sons, 1954.

Communication

W. B. Davenport, Jr., and W. L. Root, *An Introduction to the Theory of Random Signals and Noise.* New York, McGraw-Hill Book Co., 1958.

J. H. Laning, Jr., and R. H. Battin, *Random Processes in Automatic Control.* New York, McGraw-Hill Book Co., 1956.

Y. W. Lee, *Statistical Theory of Communication.* New York, John Wiley and Sons, 1960.

Engineering

K. A. Brownlee, *Statistical Theory and Methodology in Science and Engineering.* New York, John Wiley and Sons, 1960.

A. Hald, *Statistical Theory with Engineering Applications.* New York, John Wiley and Sons, 1952.

Experimentation, generally

R. A. Fisher, *Statistical Methods for Research Workers.* 13th ed., Edinburgh, Oliver and Boyd, 1958.

R. A. Fisher, *Design of Experiments.* 7th ed., Edinburgh, Oliver and Boyd, 1960.

O. Kempthorne, *The Design and Analysis of Experiments.* New York, John Wiley and Sons, 1952.

E. B. Wilson, Jr., *An Introduction to Scientific Research.* New York, McGraw-Hill Book Co., 1952. A provocative work; the only one of its kind.

Genetics

O. Kempthorne, *An Introduction to Genetic Statistics.* New York, John Wiley and Sons, 1957.

Learning theory

R. R. Bush and F. Mosteller, *Stochastic Models for Learning.* New York, John Wiley and Sons, 1955.

Physics

R. B. Lindsay, *Introduction to Physical Statistics*. New York, John Wiley and Sons, 1941. Probability heavily permeates modern physical theory and sharp advances have been made since the date of this book; but many of these make severe mathematical demands on students.

Sampling human populations

W. E. Deming, *Some Theory of Sampling*. New York, John Wiley and Sons, 1950.

M. H. Hansen, W. N. Hurwitz, and W. G. Madow, *Sample Survey Methods and Theory, Volumes I and II*. New York, John Wiley and Sons, 1953.

NAMES MENTIONED IN TEXT (WITHOUT REFERENCES)

A. C. Aitken
A. Bhattacharyya
T. Bayes
D. Bernoulli
J. Bernoulli
S. N. Bernstein
J. Bienaymé
B. H. Camp
A. L. Cauchy
C. V. L. Charlier
P. L. Chebychev
H. Cramér
J. deR. D'Alembert
G. Darmois
W. E. Deming
D. Dugué
A. K. Erlang
P. S. Fermat
R. A. Fisher
S. Fox
M. Fréchet
T. C. Fry
F. Galton
C. F. Gauss
J. P. Gram
F. R. Helmert
C. Hermite
J. Herschel
N. R. Jørgensen
M. Kendall
J. Kiefer
J. L. Lagrange
P. S. Laplace
A. M. Legendre
P. Lévy
W. Lexis
A. A. Markov
J. C. Maxwell
M. B. Meidell
A. de Moivre
E. C. Molina
P. R. de Montmart
J. Neyman
B. Pascal
E. S. Pearson
K. Pearson
G. A. A. Plana
S. D. Poisson
G. Pólya
M. Prendergast
J. W. S. Rayleigh
G. F. B. Riemann
H. A. Schwarz
H. Silverstone
R. M. Solow
B. Taylor
L. H. C. Tippett
J. Venn
A. Wald
S. S. Wilks

REFERENCES

1. A. C. Aitken, "On least squares and linear combination of observations," *Proc. Roy. Soc., Edinburgh,* Vol. 55 (1935), pp. 42–48.

2. R. L. Anderson and T. A. Bancroft, *Statistical Theory in Research.* New York, McGraw-Hill Book Co., 1952.

3. F. J. Anscombe, "Sampling theory of the negative binomial and logarithmic series distributions," *Biometrika,* Vol. 37 (1950), pp. 358–382.

4. N. Arley, *On the Theory of Stochastic Processes and Their Application to the Theory of Cosmic Radiation.* Copenhagen, G.E.C. Gads Forlag, 1943.

5. N. Arley and K. R. Buch, *Introduction to the Theory of Probability and Statistics.* New York, John Wiley and Sons, 1950.

6. R. R. Bahadur, "Examples of inconsistency of maximum likelihood estimates," *Sankhyā,* Vol. 20 (1958), pp. 207–210.

7. N. J. Bailey, "On estimating the size of mobile populations from recapture data," *Biometrika,* Vol. 38 (1951), pp. 293–306.

8. M. S. Bartlett, *An Introduction to Stochastic Processes.* Cambridge (Eng.), Cambridge University Press, 1955.

9. D. E. Barton, "A class of distributions for which the maximum likelihood estimate is unbiased and of minimum variance for all sample sizes," *Biometrika,* Vol. 43 (1956), pp. 200–202.

10. G. E. Bates and J. Neyman, "Contributions to the theory of accident proneness," *University of California Publications in Statistics,* Vol. I (1951), pp. 215–234, pp. 255–276. Berkeley, Univ. of Cal. Press, 1952.

11. T. Bayes, "An essay toward solving a problem in the doctrine of chances," *Phil. Trans. Roy. Soc.,* Vol. 53 (1763), pp. 370–418. Reproduced (with a second paper) by the Graduate School, United States Dept. of Agriculture, with an editorial preface and remarks by W. E. Deming and comments by E. C. Molina.

12. R. Beale, *Generation of the Pascal Distribution Tables on the 704 Electronic Computer.* M.I.T. term paper, 1959.

13. C. A. Bennett and N. L. Franklin, *Statistical Analysis in Chemistry and the Chemical Industry.* New York, John Wiley and Sons, 1954.

14. A. T. Bharucha-Reid, *Elements of the Theory of Markov Processes and Their Applications.* New York, McGraw-Hill Book Co., 1960.

15. D. Blackwell, "Conditional expectation and unbiased sequential estimation," *Ann. Math. Stat.,* Vol. 18 (1947), pp. 105–110.

16. D. Blackwell and M. A. Girshick, *Theory of Games and Statistical Decisions.* New York, John Wiley and Sons, 1954.

17. G. Blanch, *Tables of the Bivariate Normal Distribution Function and Related Functions.* National Bureau of Standards, Applied Math. Series 50, 1959.

18. C. R. Blyth and D. W. Hutchinson, "Tables of Neyman—shortest unbiased confidence intervals for the binomial parameter," *Biometrika,* Vol. 47 (1960), pp. 381–391.

19. C. R. Blyth, "Tables of Neyman—shortest unbiased confidence intervals for the Poisson parameter," *Biometrika*, Vol. 48 (1961), pp. 191–194.

20. E. Borel, *Mécanique Statistique Classique*. Paris, Gauthier-Villars et Cie, 1925.

21. B. H. Brown, "Problem," *Amer. Math. Monthly*, Vol. 40 (1933), p. 607.

22. K. A. Brownlee, *Statistical Theory and Methodology in Science and Engineering*. New York, John Wiley and Sons, 1960.

23. H. D. Brunk, *An Introduction to Mathematical Statistics*. Boston, Ginn & Co., 1960.

24. R. R. Bush and F. Mosteller, *Stochastic Models for Learning*. New York, John Wiley and Sons, 1955.

25. D. G. Chapman, "Some properties of the hypergeometric distribution with application to zoological sample censuses," *University of California Publications in Statistics*, Vol. I (1951), pp. 131–160; Berkeley, Univ. of Cal. Press, 1952.

26. C. V. L. Charlier, *Application de la Théorie des Probabilités à l'Astronomie*. Paris, Gauthier-Villars et Cie, 1931.

27. H. Chernoff and L. E. Moses, *Elementary Decision Theory*. New York, John Wiley and Sons, 1959.

28. J. T. Chu and H. Hotelling, "The moments of the sample median," *Ann. Math. Stat.*, Vol. 26 (1955), pp. 593–606.

29. K. L. Chung, *Markov Chains*. Berlin, Springer-Verlag, 1960.

30. C. J. Clopper and E. S. Pearson, "The use of confidence or fiducial limits illustrated in the case of the binomial," *Biometrika*, Vol. 26 (1934), pp. 404–413.

31. J. L. Coolidge, *An Introduction to Mathematical Probability*. Oxford, Oxford University Press, 1925.

32. R. T. Cox, *The Algebra of Probable Inference*. Baltimore, The Johns Hopkins Press, 1961.

33. A. T. Craig, "On the mathematics of the representative method of sampling," *Ann. Math. Stat.*, Vol. 10 (1939), pp. 26–34.

34. C. C. Craig, "On frequency distributions of the quotient and of the product of two statistical variables," *Amer. Math. Monthly*, Vol. 49 (1942), pp. 24–32.

35. H. Cramér, "A contribution to the theory of statistical estimation," *Skandinavisk Actuarietidskrift*, Vol. 29 (1946), pp. 85–94.

36. ———, *Mathematical Methods of Statistics*. Princeton, Princeton University Press, 1950.

37. ———, *The Elements of Probability Theory and Some of Its Applications*. New York, John Wiley and Sons, 1955.

38. J. H. Curtiss, "On the distribution of the quotient of two chance variables," *Ann. Math. Stat.*, Vol. 12 (1941), pp. 409–421.

39. C. Daniel and N. Heerema, "Design of experiments for most precise slope estimation or linear extrapolation," *Jour. Amer. Stat. Assoc.*, Vol. 45 (1950), pp. 546–556.

40. W. B. Davenport, Jr., and W. L. Root, *An Introduction to the Theory of Random Signals and Noise*. New York, McGraw-Hill Book Co., 1958.

41. F. N. David, *Probability Theory for Statistical Methods.* Cambridge (Eng.), Cambridge University Press, 1949.

42. F. N. David and J. Neyman, "Extension of the Markoff theorem on least squares," *Statistical Research Memoirs II* (1938), pp. 105–116.

43. H. A. David and C. A. Perez, "On comparing different tests of the same hypothesis," *Biometrika,* Vol. 47 (1960), pp. 297–306.

44. D. B. DeLury and J. H. Chung, *Confidence Limits for the Hypergeometric Distribution.* Toronto, University of Toronto Press, 1950.

45. W. E. Deming, *Some Theory of Sampling.* New York, John Wiley and Sons, 1950.

46. Department of Commerce, "Tables of the cumulative binomial probabilities," *Ordnance Corps Pamphlet* PB111389. Washington, 1952.

47. L. G. Dion, *An Approximate Solution of Wilks' Tolerance Limit Equation.* M.I.T. thesis (S.B.), 1951.

48. J. L. Doob, "Probability and statistics," *Trans. Amer. Math. Soc.,* Vol. 36 (1937), pp. 759–775.

49. R. Dorfman, "The detection of defective members of large populations," *Ann. Math. Stat.,* Vol. 14 (1943), pp. 436–440.

50. D. Dugué, *Traité de Statistique Théorique et Appliquée.* Paris, Masson et Cie., 1958.

51. D. B. Duncan, *Statistical Theory II.* Mimeographed notes. Chapel Hill, University of North Carolina.

52. J. Durbin and M. G. Kendall, "The geometry of estimation," *Biometrika,* Vol. 38 (1951), pp. 150–158.

53. F. Y. Edgeworth, "The law of error," *Trans. Cambridge Phil. Soc.,* Vol. 120 (1904), pp. 36–65, pp. 113–141, and an appendix bound (only) with the reprints.

54. F. Eggenberger and G. Pólya, "Über die Statistik verketteter Vorgänge," *Zeitschrift für Angew. Mathematik und Mechanik,* Vol. 1 (1923), pp. 279–289.

55. E. T. Federighi, "Extended tables of the percentage points of Student's t distribution," *Jour. Amer. Stat. Assoc.,* Vol. 54 (1959), pp. 683–688.

56. W. Feller, "Note on the law of large numbers and fair games," *Ann. Math. Stat.,* Vol. 16 (1945), pp. 301–304.

57. ———, *An Introduction to Probability Theory and Its Applications.* 2nd ed. New York, John Wiley and Sons, 1957.

58. A. Fisher, *The Mathematical Theory of Probabilities and Its Application to Frequency Curves and Statistical Methods.* 2nd ed. New York, Macmillan Co., 1922.

59. R. A. Fisher, "On the mathematical foundations of theoretical statistics," *Phil. Trans. Roy. Soc. London,* Series A, Vol. 222 (1922), pp. 309–368.

60. ———, "On a distribution yielding the error functions of several well known statistics," *Proc. Internat. Math. Congress,* Vol. II, Toronto (1924), pp. 805–813.

61. ———, "Theory of statistical estimation," *Proc. Camb. Phil. Soc.,* Vol. 22 (1925), pp. 702–725.

62. ———, "Applications of Student's distribution," *Metron.* Vol. 5 (1925), pp. 90–104.

63. ———, *Statistical Theory of Estimation.* Calcutta, University of Calcutta, 1938.
64. ———, *Statistical Methods for Research Workers.* 13th ed. Edinburgh, Oliver and Boyd, 1958.
65. ———, *Contributions to Mathematical Statistics.* New York, John Wiley and Sons, 1950.
66. ———, *Design of Experiments,* 7th ed. Edinburgh, Oliver and Boyd, 1960.
67. R. A. Fisher and L. H. C. Tippett, "Limiting forms of the frequency-distribution of the largest or smallest member of a sample," *Proc. Camb. Phil. Soc.,* Vol. 24 (1928), pp. 180–190.
68. A. R. Forsythe, *A Treatise on Differential Equations.* 6th ed. London, Macmillan Co., 1933.
69. D. A. S. Fraser, *Nonparametric Methods in Statistics.* New York, John Wiley and Sons, 1957.
70. ———, *Statistics: An Introduction.* New York, John Wiley and Sons, 1958.
71. M. Fréchet, "Sur l'extension de certaines évaluations statistique au cas de petits échantillons," *Revue Inst. Intern. Statistique,* Vol. 11 (1943), pp. 182–205.
72. T. C. Fry, *Probability and Its Engineering Uses.* New York, D. Van Nostrand Co., 1928.
73. R. P. Gillespie, *Integration.* London, Oliver and Boyd, 1951.
74. B. W. Gnedenko, *Lehrbuch der Wahrscheinlichkeitsrechnung.* Berlin, Akademie-Verlag, 1957.
75. S. Goldberg, *Probability, An Introduction.* Englewood Cliffs (N.J.), Prentice Hall, 1960.
76. F. A. Graybill, *An Introduction to Linear Statistical Models,* Vol. I. New York, McGraw-Hill Book Co., 1961.
77. M. Greenwood, "Accident proneness," *Biometrika,* Vol. 37 (1950), pp. 24–29.
78. M. Greenwood and G. U. Yule, "An enquiry into the nature of frequency distributions representative of multiple happenings . . . ," *Jour. Roy. Stat. Soc.,* Vol. 83 (1920), pp. 255–279.
79. E. J. Gumbel, *Statistics of Extremes.* New York, Columbia University Press, 1958.
80. ———, "Bivariate exponential distributions," *Jour. Amer. Stat. Assoc.* Vol. 55 (1960), pp. 698–707.
81. A. Hald, *Statistical Theory with Engineering Applications.* New York, John Wiley and Sons, 1952.
82. P. R. Halmos, "The theory of unbiased estimation," *Ann. Math. Stat.,* Vol. 17 (1946), pp. 34–43.
83. P. R. Halmos and L. J. Savage, "Applications of the Radon-Nikodym theorem to the theory of sufficient statistics," *Ann. Math. Stat.,* Vol. 20 (1949), pp. 225–241.
84. M. H. Hansen, W. N. Hurwitz, and W. G. Madow, *Sample Survey Methods and Theory,* Volumes I and II. New York, John Wiley and Sons, 1953.
85. Harvard University, Computation Laboratory, *Tables of the Error Func-*

tion and Its First Twenty Derivatives, Vol. 23. Cambridge (Mass.), Harvard University Press, 1952.

86. ———, *Tables of the Cumulative Binomial Probability Distribution*, Vol. 35. Cambridge (Mass.), Harvard University Press, 1955.

87. K. HAYASHI, *Fünfstellige Funktionentafeln*. Berlin, Springer-Verlag, 1930.

88. P. G. HOEL, "Confidence regions for linear regression," *Proc. Second Berkeley Symposium on Math. Statistics and Probability*. Berkeley, Univ. of Cal. Press, 1951, pp. 75–81.

89. ———, *Introduction to Mathematical Statistics*. 3rd ed. New York, John Wiley and Sons, 1962.

90. L. HOGBEN, *Chance and Choice by Cardpack and Chessboard*, Volumes I and II. New York, Chanticleer Press, 1950, 1955.

91. R. V. HOGG and A. T. CRAIG, *Introduction to Mathematical Statistics*. New York, Macmillan Co., 1959.

92. P. L. HSU, "On the best unbiased quadratic estimate of the variance," *Statistical Research Memoirs*, Vol. II (1938), pp. 91–104.

93. H. JEFFREYS, *Theory of Probability*. 3rd ed. New York, Oxford University Press, 1961.

94. N. L. JOHNSON, "Uniqueness of a result in the theory of accident proneness," *Biometrika*, Vol. 44 (1957), pp. 530–531.

95. N. L. JOHNSON and H. TETLEY, *Statistics*, Vol. II. Cambridge (Eng.), Cambridge University Press, 1950.

96. N. L. JOHNSON and B. L. WELCH, "Applications of the non-central t distribution," *Biometrika*, Vol. 31 (1940), pp. 362–389.

97. A. R. KAMAT, "Some more estimates of circular probable error," *Jour. Amer. Stat. Assoc.*, Vol. 57 (1962), pp. 191–195.

98. S. KARLIN and H. RUBIN, "The theory of decision procedures for distributions with monotone likelihood ratio," *Ann. Math. Stat.*, Vol. 27 (1956), pp. 272–300.

99. ———, "Distributions possessing a monotone likelihood ratio," *Jour. Amer. Stat. Soc.*, Vol. 51 (1956), pp. 637–643.

100. O. KEMPTHORNE, *The Design and Analysis of Experiments*. New York, John Wiley and Sons, 1952.

101. ———, *An Introduction to Genetic Statistics*. New York, John Wiley and Sons, 1957.

102. M. G. KENDALL, "Regression, structure and functional relationship, part I," *Biometrika*, Vol. 38 (1951), pp. 11–25.

103. ———, *Exercises in Theoretical Statistics*. New York, Hafner Publishing Co., 1954.

104. M. C. KENDALL and A. STUART, *The Advanced Theory of Statistics*, Volumes 1 and 2. London, Charles Griffin & Co., 1958, 1961. Volume 3 to appear.

105. J. KIEFER, "On Minimum Variance Estimators," *Ann. Math. Stat.*, Vol. 23 (1952), pp. 627–629.

106. T. KITIGAWA, *Tables of the Poisson Distribution*. Tokyo, Baifûkon Book Publishing Co., 1952.

107. W. KNEALE, *Probability and Induction*. Oxford, Clarendon Press, 1949.

108. A. A. KOLMOGOROV, *Foundations of the Theory of Probability*. New York, Chelsea Publishing Co., 1954. Original German edition in 1933.

109. H. O. Lancaster, "Zero correlation and independence," *The Australian Jour. of Stat.*, Vol. 1 (1959), pp. 53–56.

110. J. H. Laning, Jr. and R. H. Battin, *Random Processes in Automatic Control.* New York, McGraw-Hill Book Co., 1956.

111. Y. W. Lee, *Statistical Theory of Communication.* New York, John Wiley and Sons, 1960.

112. E. Lehmann, *Notes on the Theory of Estimation, Chapters I–V.* Berkeley, University of California Associated Students Store, 1950.

113. ———, "Some principles of the theory of testing hypotheses," *Ann. Math. Stat.*, Vol. 21 (1950), pp. 1–26.

114. ———, *Testing Statistical Hypotheses.* New York, John Wiley and Sons, 1960.

115. H. Levy and L. Roth. *Elements of Probability.* Oxford, Clarendon Press, 1936.

116. G. J. Lieberman and D. B. Owen, *Tables of the Hypergeometric Distribution.* Stanford, Stanford University Press, 1961.

117. R. B. Lindsay, *Introduction to Physical Statistics.* New York, John Wiley and Sons, 1941.

118. M. Loève, *Probability Theory.* 2nd ed. New York, D. Van Nostrand Co., 1960.

119. O. Lundberg, *On Random Processes and Their Application to Sickness and Accident Statistics.* Uppsala, Almqvist and Wiksells, 1940.

120. H. B. Mann, *Analysis and Design of Experiments.* New York, Dover Publications, 1949.

121. A. A. Markov, *Wahrscheinlichkeitsrechnung.* Leipzig, Teubner, 1912.

122. M. Merrington, "Tables of percentage points of the t distribution," *Biometrika*, Vol. 32 (1941–1942), p. 300.

123. M. Merrington and C. M. Thompson, "Tables of percentage points of the inverted beta (F) distribution," *Biometrika*, Vol. 33 (1943), pp. 73–88.

124. R. von Mises, *Wahrscheinlichkeitsrechnung.* Leipzig and Vienna, Fr. Deuticke, 1931.

125. ———, *Probability, Statistics and Truth.* Second revised English edition, New York, Macmillan Co., 1957.

126. ———, "On the correct use of Bayes' formula," *Ann. Math. Stat.*, Vol. 13 (1942), pp. 156–165.

127. ———, *Notes on Mathematical Theory of Probability and Statistics*, Spec. Pub. No. 1. Cambridge (Mass.), Harvard University Graduate School of Engineering, 1946.

128. E. C. Molina, *Poisson's Exponential Binomial Limit.* New York, D. Van Nostrand Co., 1942.

129. A. M. Mood, *Introduction to the Theory of Statistics.* New York, McGraw-Hill Book Co., 1950.

130. M. E. Munroe, *Theory of Probability.* New York, McGraw-Hill Book Co., 1951.

131. National Bureau of Standards, *Tables of the Binomial Probability Distribution.* Applied Mathematics Series 6. Washington, 1950.

132. ———, *Tables of Normal Probability Functions.* Applied Mathematics Series 23. Washington, 1953.

133. ———, *Tables of the Error Function and Its Derivatives.* Applied Mathematics Series 41. Washington, 1954.

134. E. F. Neild, *Hypergeometric Distribution, Volumes I–IX.* M.I.T. thesis (S.B.), 1960.

135. J. Neyman, "On two different aspects of the representative method," *Jour. Roy. Stat. Soc.*, Vol. 97 (1934), pp. 558–625.

136. ———, "Su un teorema concernente le cosiddette statistiche sufficienti," *Giorn. Ist. Italiano d. Attuari*, Vol. 6 (1935), pp. 320–334.

137. ———, "Outline of a theory of statistical estimation based on the classical theory of probability," *Phil. Trans. Roy. Soc. London*, Vol. A 236 (1937), pp. 338–380.

138. ———, "On statistics the distribution of which is independent of the parameters involved in the original probability law of the observed variables," *Statistical Research Memoirs*, Vol 2 (1938), pp. 58–59.

139. ———, "Contribution to the theory of sampling human populations," *Jour. Amer. Stat. Assoc.*, Vol. 33 (1938), pp. 101–116.

140. ———, *First Course in Probability and Statistics.* New York, Henry Holt and Co., 1950.

141. ———, "Foundations of the general theory of statistical estimation," *Congrès international de philosophie des Sciences: IV: Calcul des Probabilités.* Paris, Hermann et Cie, 1951.

142. ———, *Lectures and Conferences on Mathematical Statistics and Probability.* 2nd ed. rev. Washington: Graduate School, United States Dept. of Agriculture, 1952.

143. J. Neyman and E. S. Pearson, "On the use and interpretation of certain best criteria for purposes of statistical inference," *Biometrika*, Vol. 20A (1928), pp. 175–240; pp. 263–294.

144. ———, "On the problem of the most efficient tests of statistical hypotheses," *Phil. Trans. Roy. Soc.*, Vol. A 231 (1933), pp. 289–337.

145. J. Neyman and E. L. Scott, "Consistent estimates bases on partially consistent observations," *Econometrica*, Vol. 16 (1948), pp. 1–32.

146. J. Neyman and B. Tokarska, "Errors of the second kind in testing Student's hypothesis," *Jour. Amer. Stat. Assoc.*, Vol. 31 (1936), pp. 318–326.

147. D. B. Owen, *The Bivariate Normal Probability Distribution.* Washington, Dept. of Commerce, 1957.

148. J. Pachares, "Tables of confidence limits for the binomial distribution," *Jour. Amer. Stat. Assoc.*, Vol. 55 (1960), pp. 521–533.

149. E. Parzen, *Modern Probability Theory and Its Applications.* New York, John Wiley and Sons, 1960.

150. P. B. Patnaik, "The non-central χ^2 and F-distributions and their applications," *Biometrika*, Vol. 36 (1949), pp. 202–232.

151. E. S. Pearson and H. O. Hartley, "The probability integral of the range in samples of n observations from a normal population," *Biometrika*, Vol. 32 (1941–1942), pp. 301–310.

152. ———, *Biometrika Tables for Statisticians.* Cambridge (Eng.), Cambridge University Press, 1954.

153. E. S. PEARSON and J. WISHART, eds., *"Student's" Collected Papers.* Cambridge (Eng.), Cambridge University Press, 1942.

154. K. PEARSON, "On a criterion that a given system of deviations from the probable in the case of a correlated system of variables is such that it can be reasonably supposed to have arisen in random sampling," *Phil. Mag.*, 5th Series, Vol. 50 (1900), pp. 157–175.

155. ———, "Moments of the hypergeometrical series," *Biometrika,* Vol. 16 (1924), pp. 157–162.

156. ———, *Tables for Statisticians and Biometricians, Part II.* Biometric Laboratory, University College, London, 1931.

157. ———, *Tables of the Incomplete Beta-function.* London, Biometrika Office, 1934.

158. J. PIERCE, *Tables of the Multinomial Distribution.* Cambridge (Mass.), M.I.T. thesis (M.S.), 1957.

159. R. L. PLACKETT, "A historical note on the method of least squares," *Biometrika,* Vol. 36 (1949), pp. 458–460.

160. H. C. PLUMMER, *Probability and Frequency.* London, Macmillan Co., 1940.

161. J. N. PUROHIT and R. M. PHATARFORD, "Use of the reciprocal of a normal variate in textile statistics," *Calcutta Stat. Bull.*, Vol. 6 (1955), pp. 99–101.

162. (THE) RAND CORPORATION: *A Million Random Digits with 100,000 Normal Deviates.* Glencoe (Ill.), The Free Press, 1955.

163. C. R. RAO, "Information and accuracy attainable in the estimation of statistical parameters," *Bull. Calcutta Math. Soc.*, Vol. 37 (1945), pp. 81–91.

164. ———, *Advanced Statistical Methods in Biometric Research.* New York, John Wiley and Sons, 1952.

165. G. J. RESNIKOFF and G. J. LIEBERMAN, *Tables of the Non-central t-Distribution.* Stanford, Stanford University Press, 1957.

166. H. RICHTER, *Wahrscheinlichkeitsrechnung.* Berlin, Springer-Verlag, 1956.

167. W. E. RICKER, "The concept of confidence or fiducial limits applied to the Poisson frequency," *Jour. Amer. Stat. Assoc.*, Vol. 32 (1937), pp. 349–356.

168. P. RIDER, *An Introduction to Modern Statistical Methods.* New York, John Wiley and Sons, 1939.

169. H. L. RIETZ, "Mathematical Statistics," *Carus Mathematical Monograph No. 3. Chicago, Math. Assoc. of America.* Open Court Publishing Co., 1927.

170. J. RIORDAN, *An Introduction to Combinatorial Analysis.* New York, John Wiley and Sons, 1958.

171. "Table of binomial coefficients," *Royal Society Mathematical Tables, Vol. 3.* Cambridge (Eng.), Cambridge University Press, 1954.

172. P. A. SAMUELSON, "Prices of Factors and Goods in General Equilibrium," *The Review of Economic Studies,* Vol. 21 (1953–1954), pp. 1–20.

173. M. SANKARAN, "On the non-central chi-square distribution," *Biometrika* Vol. 46 (1959), pp. 235–237.

174. A. E. SARHAN, "Estimation of the mean and standard deviation by order statistics," *Ann. Math. Stat.*, Vol. 25 (1954), pp. 317–328.

175. H. SCHEFFÉ, *The Analysis of Variance.* New York, John Wiley and Sons, 1959.

176. L. Schmetterer, *Einführung in die Mathematische Statistik.* Vienna, Springer-Verlag, 1956.

177. W. F. Sheppard, "The Probability Integral," *British Assoc. Math. Tables,* Vol. 7. Cambridge (Eng.), Cambridge University Press, 1939.

178. S. Siegel, *Nonparametric Statistics for the Behavioral Sciences.* New York, McGraw-Hill Book Co., 1956.

179. J. G. Smith and A. J. Duncan, *Sampling Statistics and Applications,* Vol. II. New York, McGraw-Hill Book Co., 1945.

180. "Student," "The probable error of a mean," *Biometrika,* Vol. 6 (1908), pp. 1–25.

181. "Studies in the history of probability and statistics, IX," Bayes' essay towards solving a problem in the doctrine of chance. *Biometrika,* Vol. 45 (1958), pp. 293–315.

182. R. M. Sundrum, "On the relation between estimating efficiency and the power of tests," *Biometrika,* Vol. 41 (1954), pp. 542–544.

183. G. Taguchi, "Tables of the 5 percent and 1 percent points for the Pólya-Eggenberger distribution function," *Reports of Statistical Application Research, Union of Japanese Scientists and Engineers,* Vol. 2 (1952–1953), pp. 27–32.

184. P. C. Tang, "The power function of the analysis of variance tests with tables and illustrations of their use," *Statistical Research Memoirs,* Vol. 2 (1938), pp. 126–157.

185. R. F. Tate and G. W. Klett, "Optimal confidence intervals for the variance of a normal distribution," *Jour. Amer. Stat. Assoc.,* Vol. 54 (1959), pp. 674–682.

186. C. M. Thompson, "Tables of percentage points of the χ^2 distribution," *Biometrika,* Vol. 32 (1941), pp. 188–189.

187. I. Todhunter, *A History of the Mathematical Theory of Probability from the Time of Pascal to That of Laplace.* London, Macmillan Co., 1865. Reprint, N. Y. Chelsea Pub. Co., 1949.

188. J. Topping, *Errors of Observation and Their Treatment.* Rev. ed. London, Chapman and Hall, 1957.

189. J. Tukey, "Questions and Answers," *The American Statistician,* Vol. 3 (Oct. 1949), p. 12.

190. ———, "A smooth invertibility theorem," *Ann. Math. Stat.,* Vol. 29 (1958), pp. 581–584.

191. J. V. Uspensky, *Introduction to Modern Probability.* New York, McGraw-Hill Book Co., 1937.

192. G. P. Wadsworth and J. G. Bryan, *Introduction to Probability and Random Variables.* New York, McGraw-Hill Book Co., 1960.

193. B. L. van der Waerden, *Mathematische Statistik.* Berlin, Springer-Verlag, 1957.

194. A. Wald, "Tests of statistical hypotheses concerning several parameters when the number of observations is large," *Trans. Amer. Math. Soc.,* Vol. 54 (1943), pp. 426–483.

195. ———, *Notes on the Theory of Statistical Estimation and Testing Hypotheses.* New York, Columbia University. Mimeographed, 1941.

196. ———, *Sequential Analysis*. New York, John Wiley and Sons, 1947.
197. ———, "Note on the consistency of the maximum likelihood estimate," *Ann. Math. Stat.*, Vol. 20 (1949), pp. 595–601.
198. ———, *Statistical Decision Functions*. New York, John Wiley and Sons, 1950.
199. L. WEISS, *Statistical Decision Theory*. New York, McGraw-Hill Book Co., 1961.
200. E. T. WHITTAKER and G. ROBINSON, *The Calculus of Observations*. 3rd ed. London, Blackie and Sons, 1940.
201. L. WHITTAKER, "Tables of the Poisson Distribution," *Tables for Statisticians and Biometricians, Part I*. 3rd ed. Cambridge (Eng.), Cambridge University Press, 1930.
202. W. A. WHITWORTH, *DCC Exercises*. New York, Stechert (Hafner Publishing Co.), 1945.
203. ———, *Choice and Chance*. 5th ed. New York, Hafner Publishing Co., 1951.
204. S. S. WILKS, *Lectures on the Theory of Statistical Inference*. Ann Arbor, Edwards Bros., 1937.
205. ———, "The large-sample distribution of the likelihood ratio for testing composite hypotheses," *Ann. Math. Stat.*, Vol. 9 (1938), pp. 60–62.
206. ———, *Mathematical Statistics*. Princeton, Princeton University Press, 1943. Same title, New York, John Wiley and Sons, 1962.
207. ———, "Order statistics," *Bull. Amer. Math. Soc.*, Vol. 54 (1948), pp. 6–50.
208. ———, "Non-parametric statistical inference," in *Probability and Statistics*, the Harald Cramér volume. Almqvist and Wiksell, Stockholm (New York, John Wiley and Sons).
209. E. BRIGHT WILSON, *An Introduction to Scientific Research*. New York, McGraw-Hill Book Co., 1952.
210. P. M. WOODWARD, *Probability and Information Theory, with Applications to Radar*. London, Pergamon Press, 1953.
211. G. U. YULE, "On the distribution of deaths with age when causes of death act cumulatively, and similar frequency distributions," *Jour. Roy. Stat. Soc.*, Vol. 73 (1910), pp. 26–38.

APPENDIX

On the following pages, seventeen tables and two figures are shown. I am grateful to the authors and their publishers for permission to reproduce these tables here. Acknowledgment of the source is indicated on each table or figure; to these names must be added E. S. Pearson. In the interest of simplicity, I have confined the acknowledgment to the essentials; these brief statements hardly indicate the extent of my obligation.

A valuable listing of tables is found in J. Arthur Greenwood and H. O. Hartley, *A Guide to Tables in Mathematical Statistics.* Princeton, Princeton University Press, 1962. Also, a recent and admirable collection of tables has been formed by D. W. Owen, *Handbook of Statistical Tables,* Reading (Mass.), Addison-Wesley Publishing Co., 1962.

TABLE A-1*

FACTORIALS

$n!$

n	$n!$					
1						1
2						2
3						6
4						24
5						120
6						720
7						5040
8						40320
9					3	62880
10					36	28800
11					399	16800
12					4790	01600
13					62270	20800
14				8	71782	91200
15				130	76743	68000
16				2092	27898	88000
17				35568	74280	96000
18			6	40237	37057	28000
19			121	64510	04088	32000
20			2432	90200	81766	40000
21			51090	94217	17094	40000
22		11	24000	72777	76076	80000
23		258	52016	73888	49766	40000
24		6204	48401	73323	94393	60000
25	1	55112	10043	33098	59840	00000

* K. Hayashi, *Fünfstellige Funktionentafeln*. Springer-Verlag, 1930.

Table A–2*

Number of Permutations

$$P_m^n$$

n \ m	1	2	3	4	5	6	7	8
1	1							
2	2	2						
3	3	6	6					
4	4	12	24	24				
5	5	20	60	120	120			
6	6	30	120	360	720	720		
7	7	42	210	840	2520	5040	5040	
8	8	56	336	1680	6720	20160	40320	40320
9	9	72	504	3024	15120	60480	1 81440	3 62880
10	10	90	720	5040	30240	1 51200	6 04800	18 14400
11	11	110	990	7920	55440	3 32640	16 63200	66 52800
12	12	132	1320	11880	95040	6 65280	39 91680	199 58400
13	13	156	1716	17160	1 54440	12 35520	86 48640	518 91840
14	14	182	2184	24024	2 40240	21 62160	172 97280	1210 80960
15	15	210	2730	32760	3 60360	36 03600	324 32400	2594 59200

n \ m	9	10	11	12	13	14	15
8							
9	3 62880						
10	36 28800	36 28800					
11	199 58400	399 16800	399 16800				
12	798 33600	2395 00800	4790 01600	4790 01600			
13	2594 59200	10378 36800	31135 10400	62270 20800	62270 20800		
14	7264 85760	36324 28800	1 45297 15200	4 35891 45600	8 71782 91200	8 71782 91200	
15	18162 14400	1 08972 86400	5 44864 32000	21 79457 28000	65 38371 84000	130 76743 68000	130 76743 68000

* K. Hayashi, *Fünfstellige Funktionentafeln.* Springer-Verlag, 1930.

TABLE A–3*

NUMBER OF COMBINATIONS

C_m^n

n \ m	2	3	4	5	6	7	8	9	10	11
1										
2	1									
3	3	1								
4	6	4	1							
5	10	10	5	1						
6	15	20	15	6	1					
7	21	35	35	21	7	1				
8	28	56	70	56	28	8	1			
9	36	84	126	126	84	36	9	1		
10	45	120	210	252	210	120	45	10	1	
11	55	165	330	462	462	330	165	55	11	1
12	66	220	495	792	924	792	495	220	66	12
13	78	286	715	1287	1716	1716	1287	715	286	78
14	91	364	1001	2002	3003	3432	3003	2002	1001	364
15	105	455	1365	3003	5005	6435	6435	5005	3003	1365
16	120	560	1820	4368	8008	11440	12870	11440	8008	4368
17	136	680	2380	6188	12376	19448	24310	24310	19448	12376
18	153	816	3060	8568	18564	31824	43758	48620	43758	31824
19	171	969	3876	11628	27132	50388	75582	92378	92378	75582
20	190	1140	4845	15504	38760	77520	1 25970	1 67960	1 84756	1 67960
21	210	1330	5985	20349	54264	1 16280	2 03490	2 93930	3 52716	3 52716
22	231	1540	7315	26334	74613	1 70544	3 19770	4 97420	6 46646	7 05432
23	253	1771	8855	33649	1 00947	2 45157	4 90314	8 17190	11 44066	13 52078
24	276	2024	10626	42504	1 34596	3 46104	7 35471	13 07504	19 61256	24 96144

25	300	2300	12650	53130	1 77100	4 80700	10 81575	20 42975	32 68760	44 57400
26	325	2600	14950	65780	2 30230	6 57800	15 62275	31 24550	53 11735	77 26160
27	351	2925	17550	80730	2 96010	8 88030	22 20075	46 86825	84 36285	130 37895
28	378	3276	20475	98280	3 76740	11 84040	31 08105	69 06900	131 23110	214 74180
29	406	3654	23751	1 18755	4 75020	15 60780	42 92145	100 15005	200 30010	345 97290
30	435	4060	27405	1 42506	5 93775	20 35800	58 52925	143 07150	300 45015	546 27300
31	465	4495	31465	1 69911	7 36281	26 29575	78 88725	201 60075	443 52165	846 72315
32	496	4960	35960	2 01376	9 06192	33 65856	105 18300	280 48800	645 12240	1290 24480
33	528	5456	40920	2 37336	11 07568	42 72048	138 84156	385 67100	925 61040	1935 36720
34	561	5984	46376	2 78256	13 44904	53 79616	181 56204	524 51256	1311 28140	2860 97760
35	595	6545	52360	3 24632	16 23160	67 24520	235 35820	706 07460	1835 79396	4172 25900
36	630	7140	58905	3 76992	19 47792	83 47680	302 60340	941 43280	2541 86856	6008 05296
37	666	7770	66045	4 35897	23 24784	102 95472	386 08020	1244 03620	3483 30136	8549 92152
38	703	8436	73815	5 01942	27 60681	126 20256	489 03492	1630 11640	4727 33756	12033 22288
39	741	9139	82251	5 75757	32 62623	153 80937	615 23748	2119 15132	6357 45396	16760 56044
40	780	9880	91390	6 58008	38 38380	186 43560	769 04685	2734 38880	8476 60528	23118 01440
41	820	10660	1 01270	7 49398	44 96388	224 81940	955 48245	3503 43565	11210 99408	31594 61968
42	861	11480	1 11930	8 50668	52 45786	269 78328	1180 30185	4458 91810	14714 42973	42805 61376
43	903	12341	1 23410	9 62598	60 96454	322 24114	1450 08513	5639 21995	19173 34783	57520 04349
44	946	13244	1 35751	10 86008	70 59052	383 20568	1772 32627	7089 30508	24812 56778	76693 39132
45	990	14190	1 48995	12 21759	81 45060	453 79620	2155 53195	8861 63135	31901 87286	1 01505 95910
46	1035	15180	1 63185	13 70754	93 66819	535 24680	2609 32815	11017 16330	40763 50421	1 33407 83196
47	1081	16215	1 78365	15 33939	107 37573	628 91499	3144 57495	13626 49145	51780 66751	1 74171 33617
48	1128	17296	1 94580	17 12304	122 71512	736 29072	3773 48994	16771 06640	65407 15896	2 25952 00368
49	1176	18424	2 11876	19 06884	139 83816	859 00584	4509 78066	20544 55634	82178 22536	2 91359 16264
50	1225	19600	2 30300	21 18760	158 90700	998 84400	5368 78650	25054 33700	1 02722 78170	3 73537 38800

* *Royal Society Math. Tables, Vol. 3,* Cambridge University Press, 1954.

Table A–4*

Binomial Density Function

$$C_x^n p^x q^{n-x}$$

$n = 10$, $x = 9$	$n = 10$, $x = 8$	$n = 10$, $x = 7$	$n = 10$, $x = 6$	p
0.0000000	0.0000000	0.0000000	0.0000000	0.01
.0000000	.0000000	.0000000	.0000000	.02
.0000000	.0000000	.0000000	.0000001	.03
.0000000	.0000000	.0000000	.0000007	.04
.0000000	.0000000	.0000001	.0000027	.05
.0000000	.0000000	.0000003	.0000076	.06
.0000000	.0000000	.0000008	.0000185	.07
.0000000	.0000001	.0000019	.0000395	.08
.0000000	.0000002	.0000043	.0000765	.09
.0000000	.0000004	.0000087	.0001378	.10
.0000000	.0000008	.0000165	.0002334	.11
.0000000	.0000015	.0000293	.0003761	.12
.0000001	.0000028	.0000496	.0005807	.13
.0000002	.0000049	.0000805	.0008649	.14
.0000003	.0000084	.0001259	.0012486	.15
.0000006	.0000136	.0001909	.0017542	.16
.0000010	.0000216	.0002816	.0024056	.17
.0000017	.0000333	.0004051	.0032293	.18
.0000026	.0000501	.0005701	.0042528	.19
.0000041	.0000737	.0007865	.0055050	.20
.0000062	.0001063	.0010656	.0070152	.21
.0000094	.0001502	.0014205	.0088132	.22
.0000139	.0002089	.0018653	.0109282	.23
.0000201	.0002861	.0024160	.0133888	.24
.0000286	.0003862	.0030899	.0162220	.25
.0000402	.0004946	.0039256	.0194530	.26
.0000556	.0006773	.0048831	.0231043	.27
.0000761	.0008814	.0060434	.0271955	.28
.0001030	.0011348	.0074087	.0317425	.29
.0001378	.0014467	.0090017	.0367569	.30
.0001824	.0018273	.0108458	.0422460	.31
.0002392	.0022879	.0129645	.0482120	.32
.0003110	.0028410	.0153816	.0546516	.33
.0004008	.0035005	.0181203	.0615557	.34
.0005123	.0042814	.0212030	.0689098	.35
.0006499	.0051999	.0246512	.0766928	.36
.0008187	.0062735	.0284849	.0848774	.37
.0010243	.0075208	.0327222	.0934302	.38
.0012732	.0089617	.0373786	.1023119	.39
.0015728	.0106169	.0424673	.1114767	.40
.0019316	.0125080	.0479981	.1208733	.41
.0023587	.0146576	.0539772	.1304449	.42
.0028648	.0170887	.0604067	.1401295	.43
.0034615	.0198248	.0672844	.1498606	.44
.0041617	.0228896	.0746031	.1595678	.45
.0049798	.0263065	.0823507	.1691770	.46
.0059314	.0300986	.0905095	.1786117	.47
.0070335	.0342885	.0990559	.1877933	.48
.0083049	.0388975	.1079604	.1966421	.49
.0097656	.0439453	.1171875	.2050781	.50

* *National Bureau of Standards, Applied Math. Series No. 6,* 1950.

TABLE A-4

$n = 10$ $x = 5$	$n = 10$ $x = 4$	$n = 10$ $x = 3$	$n = 10$ $x = 2$	$n = 10$ $x = 1$	$n = 10$ $x = 0$	p
0.0000000	0.0000020	0.0001118	0.0041524	0.0913517	0.9043821	0.01
.0000007	.0000298	.0008334	.0153137	.1667496	.8170728	.02
.0000053	.0001417	.0026179	.0317416	.2280693	.7374241	.03
.0000211	.0004208	.0057711	.0519401	.2770136	.6648326	.04
.0000609	.0009648	.0104751	.0746348	.3151247	.5987369	.05
.0001438	.0018776	.0168085	.0987502	.3437969	.5386151	.06
.0002946	.0032622	.0247660	.1233878	.3642878	.4839823	.07
.0005442	.0052156	.0342741	.1478071	.3777290	.4343885	.08
.0009286	.0078242	.0452062	.1714071	.3851368	.3894161	.09
.0014880	.0111603	.0573956	.1937103	.3874205	.3486784	.10
.0022663	.0152802	.0706463	.2143473	.3853920	.3118172	.11
.0033092	.0202227	.0847430	.2330432	.3797740	.2785010	.12
.0046635	.0260081	.0994594	.2496050	.3712074	.2484234	.13
.0063758	.0326379	.1145656	.2639102	.3602584	.2213016	.14
.0084909	.0400957	.1298337	.2758967	.3474254	.1968744	.15
.0110508	.0483476	.1450428	.2855531	.3331452	.1749012	.16
.0140940	.0573434	.1599833	.2929107	.3177984	.1551604	.17
.0176535	.0670182	.1744599	.2980358	.3017152	.1374480	.18
.0217568	.0772936	.1882943	.3010231	.2851798	.1215767	.19
.0264241	.0880804	.2013266	.3019899	.2684354	.1073742	.20
.0316689	.0992794	.2134168	.3010703	.2516883	.0946828	.21
.0374961	.1107842	.2244458	.2984109	.2351116	.0833578	.22
.0439029	.1224828	.2343149	.2941670	.2188489	.0732668	.23
.0508774	.1342597	.2429462	.2884987	.2030175	.0642889	.24
.0583992	.1459980	.2502823	.2815676	.1877117	.0563135	.25
.0664394	.1575807	.2562851	.2735350	.1730051	.0492399	.26
.0749607	.1688930	.2609351	.2645593	.1589532	.0429763	.27
.0839177	.1798234	.2642305	.2547936	.1455963	.0374391	.28
.0932571	.1902661	.2661851	.2443854	.1329607	.0325524	.29
.1029193	.2001210	.2668279	.2334745	.1210608	.0282475	.30
.1128378	.2092958	.2662012	.2221921	.1099015	.0244619	.31
.1229405	.2177072	.2643586	.2106609	.0994787	.0211392	.32
.1331509	.2252806	.2613646	.1989935	.0897815	.0182284	.33
.1433887	.2319522	.2572916	.1872931	.0807931	.0156834	.34
.1535704	.2376685	.2522196	.1756530	.0724917	.0134627	.35
.1636111	.2423869	.2462343	.1641563	.0648518	.0115292	.36
.1734251	.2460761	.2394254	.1528764	.0578451	.0098493	.37
.1829267	.2487160	.2318857	.1418774	.0514409	.0083930	.38
.1920317	.2502976	.2237092	.1312141	.0456072	.0071334	.39
.2006581	.2508227	.2149908	.1209324	.0403108	.0060466	.40
.2087276	.2503034	.2058245	.1110698	.0355183	.0051112	.41
.2161657	.2487623	.1963021	.1016565	.0311962	.0043080	.42
.2229036	.2462307	.1865137	.0927146	.0273113	.0036203	.43
.2288780	.2427494	.1765450	.0842601	.0238311	.0030331	.44
.2340327	.2383667	.1664782	.0763026	.0207241	.0025330	.45
.2383189	.2331381	.1563907	.0688460	.0179598	.0021083	.46
.2416958	.2271255	.1463545	.0618892	.0155089	.0017489	.47
.2441312	.2203963	.1364359	.0554270	.0133435	.0014456	.48
.2456019	.2130221	.1266954	.0494500	.0114374	.0011904	.49
.2460938	.2050781	.1171875	.0439453	.0097656	.0009766	.50

Table A–5*

Binomial Distribution Function

$$1 - F(x - 1) = \sum_{r=x}^{r=n} C_r^n p^r q^{n-r}$$

$n = 10$ $x = 10$	$n = 10$ $x = 9$	$n = 10$ $x = 8$	$n = 10$ $x = 7$	p
0.0000000	0.0000000	0.0000000	0.0000000	0.01
.0000000	.0000000	.0000000	.0000000	.02
.0000000	.0000000	.0000000	.0000000	.03
.0000000	.0000000	.0000000	.0000000	.04
.0000000	.0000000	.0000000	.0000001	.05
.0000000	.0000000	.0000000	.0000003	.06
.0000000	.0000000	.0000000	.0000008	.07
.0000000	.0000000	.0000001	.0000020	.08
.0000000	.0000000	.0000002	.0000045	.09
.0000000	.0000000	.0000004	.0000091	.10
.0000000	.0000000	.0000008	.0000173	.11
.0000000	.0000000	.0000015	.0000308	.12
.0000000	.0000001	.0000029	.0000525	.13
.0000000	.0000002	.0000051	.0000856	.14
.0000000	.0000003	.0000087	.0001346	.15
.0000000	.0000006	.0000142	.0002051	.16
.0000000	.0000010	.0000226	.0003042	.17
.0000000	.0000017	.0000350	.0004401	.18
.0000001	.0000027	.0000528	.0006229	.19
.0000001	.0000042	.0000779	.0008644	.20
.0000002	.0000064	.0001127	.0011783	.21
.0000003	.0000097	.0001599	.0015804	.22
.0000004	.0000143	.0002232	.0020885	.23
.0000006	.0000207	.0003068	.0027228	.24
.0000010	.0000296	.0004158	.0035057	.25
.0000014	.0000416	.0005362	.0044618	.26
.0000021	.0000577	.0007350	.0056181	.27
.0000030	.0000791	.0009605	.0070039	.28
.0000042	.0001072	.0012420	.0086507	.29
.0000059	.0001437	.0015904	.0105921	.30
.0000082	.0001906	.0020179	.0128637	.31
.0000113	.0002505	.0025384	.0155029	.32
.0000153	.0003263	.0031673	.0185489	.33
.0000206	.0004214	.0039219	.0220422	.34
.0000276	.0005399	.0048213	.0260243	.35
.0000366	.0006865	.0058864	.0305376	.36
.0000481	.0008668	.0071403	.0356252	.37
.0000628	.0010871	.0086079	.0413301	.38
.0000814	.0013546	.0103163	.0476949	.39
.0001049	.0016777	.0122946	.0547619	.40
.0001342	.0020658	.0145738	.0625719	.41
.0001708	.0025295	.0171871	.0711643	.42
.0002161	.0030809	.0201696	.0805763	.43
.0002720	.0037335	.0235583	.0908427	.44
.0003405	.0045022	.0273918	.1019949	.45
.0004242	.0054040	.0317105	.1140612	.46
.0005260	.0064574	.0365560	.1270655	.47
.0006493	.0076828	.0419713	.1410272	.48
.0007979	.0091028	.0480003	.1559607	.49
.0009766	.0107422	.0546875	.1718750	.50

* *National Bureau of Standards, Applied Math. Series No. 6,* 1950.

TABLE A–5

$n = 10$ $x = 6$	$n = 10$ $x = 5$	$n = 10$ $x = 4$	$n = 10$ $x = 3$	$n = 10$ $x = 2$	$n = 10$ $x = 1$	p
0.0000000	0.0000000	0.0000020	0.0001138	0.0042662	0.0956179	0.01
.0000000	.0000007	.0000305	.0008639	.0161776	.1829272	.02
.0000001	.0000054	.0001471	.0027650	.0345066	.2625759	.03
.0000007	.0000218	.0004426	.0062137	.0581538	.3351674	.04
.0000028	.0000637	.0010285	.0115036	.0861384	.4012631	.05
.0000079	.0001517	.0020293	.0188378	.1175880	.4613849	.06
.0000193	.0003139	.0035761	.0283421	.1517299	.5160177	.07
.0000415	.0005857	.0058013	.0400754	.1878825	.5656115	.08
.0000810	.0010096	.0088338	.0540400	.2254471	.6105839	.09
.0001469	.0016349	.0127952	.0701908	.2639011	.6513216	.10
.0002507	.0025170	.0177972	.0884435	.3027908	.6881828	.11
.0004069	.0037161	.0239388	.1086818	.3417250	.7214990	.12
.0006332	.0052967	.0313048	.1307642	.3803692	.7515766	.13
.0009505	.0073263	.0399642	.1545298	.4184400	.7786984	.14
.0013832	.0098741	.0499698	.1798035	.4557002	.8031256	.15
.0019593	.0130101	.0613577	.2064005	.4919536	.8250988	.16
.0027098	.0168038	.0741472	.2341305	.5270412	.8448396	.17
.0036694	.0213229	.0883411	.2628010	.5608368	.8625520	.18
.0048757	.0266325	.1039261	.2922204	.5932435	.8784233	.19
.0063694	.0327935	.1208739	.3222005	.6241904	.8926258	.20
.0081935	.0398624	.1391418	.3525586	.6536289	.9053172	.21
.0103936	.0478897	.1586739	.3831197	.6815306	.9166422	.22
.0130167	.0569196	.1794024	.4137173	.7078843	.9267332	.23
.0161116	.0669890	.2012487	.4441949	.7326936	.9357111	.24
.0197277	.0781269	.2241249	.4744072	.7559748	.9436865	.25
.0239148	.0903542	.2479349	.5042200	.7777550	.9507601	.26
.0287224	.1036831	.2725761	.5335112	.7980705	.9570237	.27
.0341994	.1181171	.2979405	.5621710	.8169646	.9625609	.28
.0403932	.1336503	.3239164	.5901015	.8344869	.9674476	.29
.0473490	.1502683	.3503893	.6172172	.8506917	.9717525	.30
.0551097	.1679475	.3772433	.6434445	.8656366	.9755381	.31
.0637149	.1866554	.4043626	.6687212	.8793821	.9788608	.32
.0732005	.2063514	.4316320	.6929966	.8919901	.9817716	.33
.0835979	.2269866	.4589388	.7162304	.9035235	.9843166	.34
.0949341	.2485045	.4861730	.7383926	.9140456	.9865373	.35
.1072304	.2708415	.5132284	.7594627	.9236190	.9884708	.36
.1205026	.2939277	.5400038	.7794292	.9323056	.9901507	.37
.1347603	.3176870	.5664030	.7982887	.9401661	.9916070	.38
.1500068	.3420385	.5923361	.8160453	.9472594	.9928666	.39
.1662386	.3668967	.6177194	.8327102	.9536426	.9939534	.40
.1834452	.3921728	.6424762	.8483007	.9593705	.9948888	.41
.2016092	.4177749	.6665372	.8628393	.9644958	.9956920	.42
.2207058	.4436094	.6898401	.8763538	.9690684	.9963797	.43
.2407033	.4695813	.7123307	.8888757	.9731358	.9969669	.44
.2615627	.4955954	.7339621	.9004403	.9767429	.9974670	.45
.2832382	.5215571	.7546952	.9110859	.9799319	.9978917	.46
.3056772	.5473730	.7744985	.9208530	.9827422	.9982511	.47
.3288205	.5729517	.7933480	.9297839	.9852109	.9985544	.48
.3526028	.5982047	.8112268	.9379222	.9873722	.9988096	.49
.3769531	.6230469	.8281250	.9453125	.9892578	.9990234	.50

Table A–6*

Poisson Density Function

$$e^{-a}a^x/x!$$

x	$a = 0.2$	$a = 0.3$	$a = 0.4$	$a = 0.5$	$a = 0.6$
0	0.8187308	0.7408182	0.6703200	0.606531	0.548812
1	.1637462	.2222455	.2681280	.303265	.329287
2	.0163746	.0333368	.0536256	.075816	.098786
3	.0010916	.0033337	.0071501	.012636	.019757
4	.0000546	.0002500	.0007150	.001580	.002964
5	.0000022	.0000150	.0000572	.000158	.000356
6	.0000001	.0000008	.0000038	.000013	.000036
7			.0000002	.0000001	.0000003
x	$a = 0.7$	$a = 0.8$	$a = 0.9$	$a = 1.0$	$a = 1.2$
0	0.496585	0.449329	0.406570	0.367879	0.301194
1	.347610	.359463	.365913	.367879	.361433
2	.121663	.143785	.164661	.183940	.216860
3	.028388	.038343	.049398	.061313	.086744
4	.004968	.007669	.011115	.015328	.026023
5	.000696	.001227	.002001	.003066	.006246
6	.000081	.000164	.000300	.000511	.001249
7	.000008	.000019	.000039	.000073	.000214
8	.000001	.000002	.000004	.000009	.000004
9				.000001	.000001
x	$a = 1.4$	$a = 1.6$	$a = 1.8$	$a = 2.0$	
0	0.246597	0.201897	0.165299	0.135335	
1	.345236	.323034	.297538	.270671	
2	.241665	.258428	.267784	.270671	
3	.112777	.137828	.160671	.180447	
4	.039472	.055131	.072302	.090224	
5	.011052	.017642	.026029	.036089	
6	.002579	.004705	.007809	.012030	
7	.000516	.001075	.002008	.003437	
8	.000090	.000215	.000452	.000859	
9	.000014	.000038	.000090	.000191	
10	.000002	.000006	.000016	.000038	
11		.000001	.000003	.000007	
12				.000001	

* E. C. Molina, *Poisson's Exponential Binomial Limit.* D. Van Nostrand, Inc., 1947.

TABLE A–6

x	$a = 2.5$	$a = 3.0$	$a = 3.5$	$a = 4.0$	$a = 4.5$	$a = 5.0$
0	0.082085	0.049787	0.030197	0.018316	0.011109	0.006738
1	.205212	.149361	.105691	.073263	.049990	.033690
2	.256516	.224042	.184959	.146525	.112479	.084224
3	.213763	.224042	.215785	.195367	.168718	.140374
4	.133602	.168031	.188812	.195367	.189808	.175467
5	.066801	.100819	.132169	.156293	.170827	.175467
6	.027834	.050409	.077098	.104196	.128120	.146223
7	.009941	.021604	.038549	.059540	.082363	.104445
8	.003106	.008102	.016865	.029770	.046329	.065278
9	.000863	.002701	.006559	.013231	.023165	.036266
10	.000216	.000810	.002296	.005292	.010424	.018133
11	.000049	.000221	.000730	.001925	.004264	.008242
12	.000010	.000055	.000213	.000642	.001599	.003434
13	.000002	.000013	.000057	.000197	.000554	.001321
14		.000003	.000014	.000056	.000178	.000472
15		.000001	.000003	.000015	.000053	.000157
16			.000001	.000004	.000015	.000049
17				.000001	.000004	.000014
18					.000001	.000004
19						.000001

TABLE A-7*

POISSON DISTRIBUTION FUNCTION

$$1 - F(x - 1) = \sum_{r=x}^{r=\infty} \frac{e^{-a} a^{r}}{r!}$$

x	$a = 0.2$	$a = 0.3$	$a = 0.4$	$a = 0.5$	$a = 0.6$
0	1.0000000	1.0000000	1.0000000	1.0000000	1.0000000
1	.1812692	.2591818	.3296800	.393469	.451188
2	.0175231	.0369363	.0615519	.090204	.121901
3	.0011485	.0035995	.0079263	.014388	.023115
4	.0000568	.0002658	.0007763	.001752	.003358
5	.0000023	.0000158	.0000612	.000172	.000394
6	.0000001	.0000008	.0000040	.000014	.000039
7			.0000002	.000001	.000003

x	$a = 0.7$	$a = 0.8$	$a = 0.9$	$a = 1.0$	$a = 1.2$
0	1.0000000	1.0000000	1.0000000	1.0000000	1.0000000
1	.503415	.550671	.593430	.632121	.698806
2	.155805	.191208	.227518	.264241	.337373
3	.034142	.047423	.062857	.080301	.120513
4	.005753	.009080	.013459	.018988	.033769
5	.000786	.001411	.002344	.003660	.007746
6	.000090	.000184	.000343	.000594	.001500
7	.000009	.000021	.000043	.000083	.000251
8	.000001	.000002	.000005	.000010	.000037
9				.000001	.000005
10					.000001

x	$a = 1.4$	$a = 1.6$	$a = 1.8$		
0	1.000000	1.000000	1.000000		
1	.753403	.798103	.834701		
2	.408167	.475069	.537163		
3	.166502	.216642	.269379		
4	.053725	.078813	.108708		
5	.014253	.023682	.036407		
6	.003201	.006040	.010378		
7	.000622	.001336	.002569		
8	.000107	.00260	.000562		
9	.000016	.000045	.000110		
10	.000002	.000007	.000019		
11		.000001	.000003		

* E. C. Molina, *Poisson's Exponential Binomial Limit*. D. Van Nostrand, Inc., 1947.

TABLE A–7

x	$a = 2.5$	$a = 3.0$	$a = 3.5$	$a = 4.0$	$a = 4.5$	$a = 5.0$
0	1.000000	1.000000	1.000000	1.000000	1.000000	1.000000
1	.917915	.950213	.969803	.981684	.988891	.993262
2	.712703	.800852	.864112	.908422	.938901	.959572
3	.456187	.576810	.679153	.761897	.826422	.875348
4	.242424	.352768	.463367	.566530	.657704	.734974
5	.108822	.184737	.274555	.371163	.467896	.559507
6	.042021	.083918	.142386	.214870	.297070	.384039
7	.014187	.033509	.065288	.110674	.168949	.237817
8	.004247	.011905	.026739	.051134	.086586	.133372
9	.001140	.003803	.009874	.021363	.040257	.068094
10	.000277	.001102	.003315	.008132	.017093	.031828
11	.000062	.000292	.001019	.002840	.006669	.013695
12	.000013	.000071	.000289	.000915	.002404	.005453
13	.000002	.000016	.000076	.000274	.000805	.002019
14		.000003	.000019	.000076	.000252	.000698
15		.000001	.000004	.000020	.000074	.000226
16			.000001	.000005	.000020	.000069
17				.000001	.000005	.000020
18					.000001	.000005
19						.000001

1.000000
.080301
.919699

.876663

1.00000

A = 1
λ = 0
λ = 3

632121
264241
080301

Table A-8*

Hypergeometric Density and Distribution Functions

$$f(x) = C_x^k C_{n-x}^{N-k} / C_n^N, \quad F(x) = \sum_{r=0}^{r=x} C_r^k C_{n-r}^{N-k} / C_n^N$$

N	n	k	x	$F(x)$	$f(x)$	N	n	k	x	$F(x)$	$f(x)$
2	1	1	0	0.500000	0.500000	6	2	2	2	1.000000	0.066667
2	1	1	1	1.000000	0.500000	6	3	1	0	0.500000	0.500000
3	1	1	0	0.666667	0.666667	6	3	1	1	1.000000	0.500000
3	1	1	1	1.000000	0.333333	6	3	2	0	0.200000	0.200000
3	2	1	0	0.333333	0.333333	6	3	2	1	0.800000	0.600000
3	2	1	1	1.000000	0.666667	6	3	2	2	1.000000	0.200000
3	2	2	1	0.666667	0.666667	6	3	3	0	0.050000	0.050000
3	2	2	2	1.000000	0.333333	6	3	3	1	0.500000	0.450000
4	1	1	0	0.750000	0.750000	6	3	3	2	0.950000	0.450000
4	1	1	1	1.000000	0.250000	6	3	3	3	1.000000	0.050000
4	2	1	0	0.500000	0.500000	6	4	1	0	0.333333	0.333333
4	2	1	1	1.000000	0.500000	6	4	1	1	1.000000	0.666667
4	2	2	0	0.166667	0.166667	6	4	2	0	0.066667	0.066667
4	2	2	1	0.833333	0.666667	6	4	2	1	0.600000	0.533333
4	2	2	2	1.000000	0.166667	6	4	2	2	1.000000	0.400000
4	3	1	0	0.250000	0.250000	6	4	3	1	0.200000	0.200000
4	3	1	1	1.000000	0.750000	6	4	3	2	0.800000	0.600000
4	3	2	1	0.500000	0.500000	6	4	3	3	1.000000	0.200000
4	3	2	2	1.000000	0.500000	6	4	4	2	0.400000	0.400000
4	3	3	2	0.750000	0.750000	6	4	4	3	0.933333	0.533333
4	3	3	3	1.000000	0.250000	6	4	4	4	1.000000	0.066667
5	1	1	0	0.800000	0.800000	6	5	1	0	0.166667	0.166667
5	1	1	1	1.000000	0.200000	6	5	1	1	1.000000	0.833333
5	2	1	0	0.600000	0.600000	6	5	2	1	0.333333	0.333333
5	2	1	1	1.000000	0.400000	6	5	2	2	1.000000	0.666667
5	2	2	0	0.300000	0.300000	6	5	3	2	0.500000	0.500000
5	2	2	1	0.900000	0.600000	6	5	3	3	1.000000	0.500000
5	2	2	2	1.000000	0.100000	6	5	4	3	0.666667	0.666667
5	3	1	0	0.400000	0.400000	6	5	4	4	1.000000	0.333333
5	3	1	1	1.000000	0.600000	6	5	5	4	0.833333	0.833333
5	3	2	0	0.100000	0.100000	6	5	5	5	1.000000	0.166667
5	3	2	1	0.700000	0.600000	7	1	1	0	0.857143	0.857143
5	3	2	2	1.000000	0.300000	7	1	1	1	1.000000	0.142857
5	3	3	1	0.300000	0.300000	7	2	1	0	0.714286	0.714286
5	3	3	2	0.900000	0.600000	7	2	1	1	1.000000	0.285714
5	3	3	3	1.000000	0.100000	7	2	2	0	0.476190	0.476190
5	4	1	0	0.200000	0.200000	7	2	2	1	0.952381	0.476190
5	4	1	1	1.000000	0.800000	7	2	2	2	1.000000	0.047619
5	4	2	1	0.400000	0.400000	7	3	1	0	0.571429	0.571429
5	4	2	2	0.000000	0.600000	7	3	1	1	1.000000	0.428571
5	4	3	2	0.600000	0.600000	7	3	2	0	0.285714	0.285714
5	4	3	3	1.000000	0.400000	7	3	2	1	0.857143	0.571429
5	4	4	3	0.800000	0.800000	7	3	2	2	1.000000	0.142857
5	4	4	4	1.000000	0.200000	7	3	3	0	0.114286	0.114286
6	1	1	0	0.833333	0.833333	7	3	3	1	0.628571	0.514286
6	1	1	1	1.000000	0.166667	7	3	3	2	0.971428	0.342857
6	2	1	0	0.666667	0.666667	7	3	3	3	1.000000	0.028571
6	2	1	1	1.000000	0.333333	7	4	1	0	0.428571	0.428571
6	2	2	0	0.400000	0.400000	7	4	1	1	1.000000	0.571429
6	2	2	1	0.933333	0.533333	7	4	2	0	0.142857	0.142857

* G. J. Lieberman and D. B. Owen, *Tables of the Hypergeometric Probability Distribution*. Stanford University Press, 1961.

TABLE A–8

N	n	k	x	$F(x)$	$f(x)$
7	4	2	1	0.714286	0.571429
7	4	2	2	1.000000	0.285714
7	4	3	0	0.028571	0.028571
7	4	3	1	0.371429	0.342857
7	4	3	2	0.885714	0.514286
7	4	3	3	1.000000	0.114286
7	4	4	1	0.114286	0.114286
7	4	4	2	0.628571	0.514286
7	4	4	3	0.971428	0.342857
7	4	4	4	1.000000	0.028571
7	5	1	0	0.285714	0.285714
7	5	1	1	1.000000	0.714286
7	5	2	0	0.047619	0.047619
7	5	2	1	0.523809	0.476190
7	5	2	2	1.000000	0.476190
7	5	3	1	0.142857	0.142857
7	5	3	2	0.714286	0.571429
7	5	3	3	1.000000	0.285714
7	5	4	2	0.285714	0.285714
7	5	4	3	0.857143	0.571429
7	5	4	4	1.000000	0.142857
7	5	5	3	0.476190	0.476190
7	5	5	4	0.952381	0.476190
7	5	5	5	1.000000	0.047619
7	6	1	0	0.142857	0.142857
7	6	1	1	1.000000	0.857143
7	6	2	1	0.285714	0.285714
7	6	2	2	1.000000	0.714286
7	6	3	2	0.428571	0.428571
7	6	3	3	1.000000	0.571429
7	6	4	3	0.571429	0.571429
7	6	4	4	1.000000	0.428571
7	6	5	4	0.714286	0.714286
7	6	5	5	1.000000	0.285714
7	6	6	5	0.857143	0.857143
7	6	6	6	1.000000	0.142857
8	1	1	0	0.875000	0.875000
8	1	1	1	1.000000	0.125000
8	2	1	0	0.750000	0.750000
8	2	1	1	1.000000	0.250000
8	2	2	0	0.535714	0.535714
8	2	2	1	0.964286	0.428571
8	2	2	2	1.000000	0.035714
8	3	1	0	0.625000	0.625000
8	3	1	1	1.000000	0.375000
8	3	2	0	0.357143	0.357143
8	3	2	1	0.892857	0.535714
8	3	2	2	1.000000	0.107143
8	3	3	0	0.178571	0.178571
8	3	3	1	0.714286	0.535714
8	3	3	2	0.982143	0.267857
8	3	3	3	1.000000	0.017857
8	4	1	0	0.500000	0.500000
8	4	1	1	1.000000	0.500000
8	4	2	0	0.214286	0.214286
8	4	2	1	0.785714	0.571429
8	4	2	2	1.000000	0.214286
8	4	3	0	0.071429	0.071429
8	4	3	1	0.500000	0.428571
8	4	3	2	0.928571	0.428571
8	4	3	3	1.000000	0.071429
8	4	4	0	0.014286	0.014286
8	4	4	1	0.242857	0.228571
8	4	4	2	0.757143	0.514286
8	4	4	3	0.985714	0.228571
8	4	4	4	1.000000	0.014286
8	5	1	0	0.375000	0.375000
8	5	1	1	1.000000	0.625000
8	5	2	0	0.107143	0.107143
8	5	2	1	0.642857	0.535714
8	5	2	2	1.000000	0.357143
8	5	3	0	0.017857	0.017857
8	5	3	1	0.285714	0.267857
8	5	3	2	0.821429	0.535714
8	5	3	3	1.000000	0.178571
8	5	4	1	0.071429	0.071429
8	5	4	2	0.500000	0.428571
8	5	4	3	0.928571	0.428571
8	5	4	4	1.000000	0.071429
8	5	5	2	0.178571	0.178571
8	5	5	3	0.714286	0.535714
8	5	5	4	0.982143	0.267857
8	5	5	5	1.000000	0.017857
8	6	1	0	0.250000	0.250000
8	6	1	1	1.000000	0.750000
8	6	2	0	0.035714	0.035714
8	6	2	1	0.464286	0.428571
8	6	2	2	1.000000	0.535714
8	6	3	1	0.107143	0.107143
8	6	3	2	0.642857	0.535714
8	6	3	3	1.000000	0.357143
8	6	4	2	0.214286	0.214286
8	6	4	3	0.785714	0.571429
8	6	4	4	1.000000	0.214286
8	6	5	3	0.357143	0.357143
8	6	5	4	0.892857	0.535714
8	6	5	5	1.000000	0.107143
8	6	6	4	0.535714	0.535714
8	6	6	5	0.964286	0.428571
8	6	6	6	1.000000	0.035714

(Continued)

Table A–8

N	n	k	x	$F(x)$	$f(x)$	N	n	k	x	$F(x)$	$f(x)$
8	7	1	0	0.125000	0.125000	9	5	3	1	0.404762	0.357143
8	7	1	1	1.000000	0.875000	9	5	3	2	0.880952	0.476190
8	7	2	1	0.250000	0.250000	9	5	3	3	1.000000	0.119048
8	7	2	2	1.000000	0.750000	9	5	4	0	0.007936	0.007936
8	7	3	2	0.375000	0.375000	9	5	4	1	0.166667	0.158730
8	7	3	3	1.000000	0.625000	9	5	4	2	0.642857	0.476190
8	7	4	3	0.500000	0.500000	9	5	4	3	0.960317	0.317460
8	7	4	4	1.000000	0.500000	9	5	4	4	1.000000	0.039683
8	7	5	4	0.625000	0.625000	9	5	5	1	0.039683	0.039683
8	7	5	5	1.000000	0.375000	9	5	5	2	0.357143	0.317460
8	7	6	5	0.750000	0.750000	9	5	5	3	0.833333	0.476190
8	7	6	6	1.000000	0.250000	9	5	5	4	0.992063	0.158730
8	7	7	6	0.875000	0.875000	9	5	5	5	1.000000	0.007936
8	7	7	7	1.000000	0.125000	9	6	1	0	0.333333	0.333333
9	1	1	0	0.888889	0.888889	9	6	1	1	1.000000	0.666667
9	1	1	1	1.000000	0.111111	9	6	2	0	0.083333	0.083333
9	2	1	0	0.777778	0.777778	9	6	2	1	0.583333	0.500000
9	2	1	1	1.000000	0.222222	9	6	2	2	1.000000	0.416667
9	2	2	0	0.583333	0.583333	9	6	3	0	0.011905	0.011905
9	2	2	1	0.972222	0.388889	9	6	3	1	0.226190	0.214286
9	2	2	2	1.000000	0.027778	9	6	3	2	0.761905	0.535714
9	3	1	0	0.666667	0.666667	9	6	3	3	1.000000	0.238095
9	3	1	1	1.000000	0.333333	9	6	4	1	0.047619	0.047619
9	3	2	0	0.416667	0.416667	9	6	4	2	0.404762	0.357143
9	3	2	1	0.916667	0.500000	9	6	4	3	0.880952	0.476190
9	3	2	2	1.000000	0.083333	9	6	4	4	1.000000	0.119048
9	3	3	0	0.238095	0.238095	9	6	5	2	0.119048	0.119048
9	3	3	1	0.773809	0.535714	9	6	5	3	0.595238	0.476190
9	3	3	2	0.988095	0.214286	9	6	5	4	0.952381	0.357143
9	3	3	3	1.000000	0.011905	9	6	5	5	1.000000	0.047619
9	4	1	0	0.555556	0.555556	9	6	6	3	0.238095	0.238095
9	4	1	1	1.000000	0.444444	9	6	6	4	0.773809	0.535714
9	4	2	0	0.277778	0.277778	9	6	6	5	0.988095	0.214286
9	4	2	1	0.833333	0.555556	9	6	6	6	1.000000	0.011905
9	4	2	2	1.000000	0.166667	9	7	1	0	0.222222	0.222222
9	4	3	0	0.119048	0.119048	9	7	1	1	1.000000	0.777778
9	4	3	1	0.595238	0.476190	9	7	2	0	0.027778	0.027778
9	4	3	2	0.952381	0.357143	9	7	2	1	0.416667	0.388889
9	4	3	3	1.000000	0.047619	9	7	2	2	1.000000	0.583333
9	4	4	0	0.039683	0.039683	9	7	3	1	0.083333	0.083333
9	4	4	1	0.357143	0.317460	9	7	3	2	0.583333	0.500000
9	4	4	2	0.833333	0.476190	9	7	3	3	1.000000	0.416667
9	4	4	3	0.992063	0.158730	9	7	4	2	0.166667	0.166667
9	4	4	4	1.000000	0.007936	9	7	4	3	0.722222	0.555556
9	5	1	0	0.444444	0.444444	9	7	4	4	1.000000	0.277778
9	5	1	1	1.000000	0.555556	9	7	5	3	0.277778	0.277778
9	5	2	0	0.166667	0.166667	9	7	5	4	0.833333	0.555556
9	5	2	1	0.722222	0.555556	9	7	5	5	1.000000	0.166667
9	5	2	2	1.000000	0.277778	9	7	6	4	0.416667	0.416667
9	5	3	0	0.047619	0.047619	9	7	6	5	0.916667	0.500000

TABLE A–8

N	n	k	x	$F(x)$	$f(x)$	N	n	k	x	$F(x)$	$f(x)$
9	7	6	6	1.000000	0.083333	10	5	1	0	0.500000	0.500000
9	7	7	5	0.583333	0.583333	10	5	1	1	1.000000	0.500000
9	7	7	6	0.972222	0.388889	10	5	2	0	0.222222	0.222222
9	7	7	7	1.000000	0.027778	10	5	2	1	0.777778	0.555556
9	8	1	0	0.111111	0.111111	10	5	2	2	1.000000	0.222222
9	8	1	1	1.000000	0.888889	10	5	3	0	0.083333	0.083333
9	8	2	1	0.222222	0.222222	10	5	3	1	0.500000	0.416667
9	8	2	2	1.000000	0.777778	10	5	3	2	0.916667	0.416667
9	8	3	2	0.333333	0.333333	10	5	3	3	1.000000	0.083333
9	8	3	3	1.000000	0.666667	10	5	4	0	0.023810	0.023810
9	8	4	3	0.444444	0.444444	10	5	4	1	0.261905	0.238095
9	8	4	4	1.000000	0.555556	10	5	4	2	0.738095	0.476190
9	8	5	4	0.555556	0.555556	10	5	4	3	0.976190	0.238095
9	8	5	5	1.000000	0.444444	10	5	4	4	1.000000	0.023810
9	8	6	5	0.666667	0.666667	10	5	5	0	0.003968	0.003968
9	8	6	6	1.000000	0.333333	10	5	5	1	0.103175	0.099206
9	8	7	6	0.777778	0.777778	10	5	5	2	0.500000	0.396825
9	8	7	7	1.000000	0.222222	10	5	5	3	0.896825	0.396825
9	8	8	7	0.888889	0.888889	10	5	5	4	0.996032	0.099206
9	8	8	8	1.000000	0.111111	10	5	5	5	1.000000	0.003968
10	1	1	0	0.900000	0.900000	10	6	1	0	0.400000	0.400000
10	1	1	1	1.000000	0.100000	10	6	1	1	1.000000	0.600000
10	2	1	0	0.800000	0.800000	10	6	2	0	0.133333	0.133333
10	2	1	1	1.000000	0.200000	10	6	2	1	0.666667	0.533333
10	2	2	0	0.622222	0.622222	10	6	2	2	1.000000	0.333333
10	2	2	1	0.977778	0.355556	10	6	3	0	0.033333	0.033333
10	2	2	2	1.000000	0.022222	10	6	3	1	0.333333	0.300000
10	3	1	0	0.700000	0.700000	10	6	3	2	0.833333	0.500000
10	3	1	1	1.000000	0.300000	10	6	3	3	1.000000	0.166667
10	3	2	0	0.466667	0.466667	10	6	4	0	0.004762	0.004762
10	3	2	1	0.933333	0.466667	10	6	4	1	0.119048	0.114286
10	3	2	2	1.000000	0.066667	10	6	4	2	0.547619	0.428571
10	3	3	0	0.291667	0.291667	10	6	4	3	0.928571	0.380952
10	3	3	1	0.816667	0.525000	10	6	4	4	1.000000	0.071429
10	3	3	2	0.991667	0.175000	10	6	5	1	0.023810	0.023810
10	3	3	3	1.000000	0.008333	10	6	5	2	0.261905	0.238095
10	4	1	0	0.600000	0.600000	10	6	5	3	0.738095	0.476190
10	4	1	1	1.000000	0.400000	10	6	5	4	0.976190	0.238095
10	4	2	0	0.333333	0.333333	10	6	5	5	1.000000	0.023810
10	4	2	1	0.866667	0.533333	10	6	6	2	0.071429	0.071429
10	4	2	2	1.000000	0.133333	10	6	6	3	0.452381	0.380952
10	4	3	0	0.166667	0.166667	10	6	6	4	0.880952	0.428571
10	4	3	1	0.666667	0.500000	10	6	6	5	0.995238	0.114286
10	4	3	2	0.966667	0.300000	10	6	6	6	1.000000	0.004762
10	4	3	3	1.000000	0.033333	10	7	1	0	0.300000	0.300000
10	4	4	0	0.071429	0.071429	10	7	1	1	1.000000	0.700000
10	4	4	1	0.452381	0.380952	10	7	2	0	0.066667	0.066667
10	4	4	2	0.880952	0.428571	10	7	2	1	0.533333	0.466667
10	4	4	3	0.995238	0.114286	10	7	2	2	1.000000	0.466667
10	4	4	4	1.000000	0.004762	10	7	3	0	0.008333	0.008333

Table A–9*

Negative Binomial Density and Distribution Functions

$$f(n) = C_{r-1}^{n-1} p^r q^{n-r}, \qquad F(n) = \sum_{k=r}^{n} C_{r-1}^{k-1} p^r q^{k-r}$$

$p = 0.900,\ r = 1$		
n	$f(n)$	$F(n)$
1	0.90000	0.9000
2	0.09000	0.9900

$p = 0.900,\ r = 2$		
n	$f(n)$	$F(n)$
2	0.81000	0.8100
3	0.16200	0.9720
4	0.02430	0.9963

$p = 0.900,\ r = 3$		
n	$f(n)$	$F(n)$
3	0.72900	0.7290
4	0.21870	0.9477
5	0.04374	0.9914

$p = 0.900,\ r = 4$		
n	$f(n)$	$F(n)$
4	0.65610	0.6561
5	0.26244	0.9185
6	0.06561	0.9841
7	0.01312	0.9973

$p = 0.900,\ r = 5$		
n	$f(n)$	$F(n)$
5	0.59049	0.5905
6	0.29524	0.8857
7	0.08857	0.9743
8	0.02067	0.9950

$p = 0.900,\ r = 6$		
n	$f(n)$	$F(n)$
6	0.53144	0.5314
7	0.31886	0.8503
8	0.11160	0.9619
9	0.02976	0.9917

* R. Beale, M.I.T., 1959.

TABLE A–12

w \ n	2	3	4	5	6	7	8	9	10
2.50	0.9229	0.8195	0.7110	0.6075	0.5132	0.4300	0.3579	0.2964	0.2443
2.55	.9286	.8315	.7282	.6283	.5364	.4541	.3820	.3198	.2665
2.60	.9340	.8429	.7448	.6487	.5592	.4782	.4064	.3437	.2894
2.65	.9390	.8537	.7607	.6685	.5816	.5022	.4309	.3680	.3130
2.70	.9438	.8640	.7759	.6877	.6036	.5259	.4555	.3927	.3372
2.75	0.9482	0.8737	0.7905	0.7063	0.6252	0.5494	0.4801	0.4175	0.3617
2.80	.9523	.8828	.8045	.7242	.6461	.5725	.5044	.4425	.3867
2.85	.9561	.8915	.8177	.7415	.6665	.5952	.5286	.4675	.4119
2.90	.9597	.8996	.8304	.7580	.6863	.6174	.5525	.4923	.4372
2.95	.9630	.9073	.8424	.7739	.7055	.6390	.5760	.5171	.4625
3.00	0.9661	0.9145	0.8537	0.7891	0.7239	0.6601	0.5991	0.5415	0.4878
3.05	.9690	.9212	.8645	.8036	.7416	.6806	.6216	.5656	.5129
3.10	.9716	.9275	.8746	.8174	.7587	.7003	.6436	.5892	.5378
3.15	.9741	.9334	.8842	.8305	.7750	.7194	.6649	.6124	.5623
3.20	.9763	.9388	.8931	.8429	.7905	.7377	.6856	.6350	.5864
3.25	0.9784	0.9439	0.9016	0.8546	0.8053	0.7553	0.7055	0.6569	0.6099
3.30	.9804	.9487	.9095	.8657	.8194	.7721	.7248	.6782	.6329
3.35	.9822	.9531	.9168	.8761	.8327	.7881	.7432	.6988	.6553
3.40	.9838	.9572	.9237	.8859	.8454	.8034	.7609	.7186	.6769
3.45	.9853	.9609	.9302	.8951	.8573	.8179	.7778	.7376	.6978
3.50	0.9867	0.9644	0.9361	0.9037	0.8685	0.8316	0.7939	0.7558	0.7180
3.55	.9879	.9677	.9417	.9117	.8790	.8446	.8091	.7732	.7373
3.60	.9891	.9706	.9468	.9192	.8889	.8568	.8236	.7898	.7558
3.65	.9901	.9734	.9516	.9261	.8981	.8683	.8372	.8055	.7735
3.70	.9911	.9759	.9559	.9326	.9067	.8790	.8501	.8204	.7902
3.75	0.9920	0.9782	0.9600	0.9386	0.9148	0.8891	0.8622	0.8345	0.8062
3.80	.9928	.9803	.9637	.9441	.9222	.8985	.8736	.8477	.8212
3.85	.9935	.9822	.9672	.9493	.9291	.9073	.8842	.8602	.8355
3.90	.9942	.9839	.9703	.9540	.9355	.9155	.8941	.8718	.8488
3.95	.9948	.9856	.9732	.9583	.9415	.9230	.9034	.8827	.8614
4.00	0.9953	0.9870	0.9758	0.9623	0.9469	0.9300	0.9120	0.8929	0.8731
4.05	.9958	.9883	.9782	.9660	.9519	.9365	.9199	.9024	.8841
4.10	.9963	.9895	.9804	.9693	.9566	.9425	.9273	.9112	.8943
4.15	.9967	.9906	.9824	.9724	.9608	.9480	.9341	.9193	.9038
4.20	.9970	.9916	.9842	.9752	.9647	.9530	.9404	.9269	.9126
4.25	0.9974	0.9925	0.9859	0.9777	0.9682	0.9576	0.9461	0.9338	0.9208
4.30	.9976	.9933	.9874	.9800	.9715	.9619	.9514	.9402	.9283
4.35	.9979	.9941	.9887	.9821	.9744	.9657	.9562	.9460	.9352
4.40	.9981	.9947	.9899	.9840	.9771	.9692	.9607	.9514	.9416
4.45	.9984	.9953	.9910	.9857	.9795	.9724	.9647	.9563	.9474
4.50	0.9985	0.9958	0.9920	0.9873	0.9817	0.9754	0.9684	0.9608	0.9527
4.55	.9987	.9963	.9929	.9887	.9837	.9780	.9717	.9649	.9575
4.60	.9989	.9967	.9937	.9899	.9855	.9804	.9747	.9686	.9620
4.65	.9990	.9971	.9944	.9911	.9871	.9825	.9775	.9719	.9660
4.70	.9991	.9974	.9951	.9921	.9885	.9845	.9799	.9750	.9696
4.75	0.9992	0.9977	0.9956	0.9930	0.9898	0.9862	0.9822	0.9777	0.9729
4.80	.9993	.9980	.9962	.9938	.9910	.9878	.9842	.9802	.9759
4.85	.9994	.9983	.9966	.9945	.9920	.9892	.9860	.9824	.9786
4.90	.9995	.9985	.9970	.9952	.9930	.9904	.9876	.9844	.9810
4.95	.9995	.9987	.9974	.9958	.9938	.9916	.9890	.9862	.9832
5.00	0.9996	0.9988	0.9977	0.9963	0.9946	0.9926	0.9903	0.9878	0.9851

TABLE A–13*

WILKS' TOLERANCE THEOREM

Values of n satisfying $n\beta^{n-1} + (n - 1)\beta^n = 1 - \alpha$

$1-\alpha$ \ β	0.99	0.98	0.95	0.90	0.80	0.70	0.50	0.30	0.20	0.10	0.05	0.02	0.01
0.01	661	330	130	64	31	20	11	7	5	4	3	3	3
0.02	581	290	115	56	27	17	9	7	5	4	3	3	2
0.05	473	236	93	46	22	14	8	5	4	3	3	2	2
0.10	388	194	77	38	18	12	7	4	4	3	2	2	2
0.20	299	149	59	29	14	9	5	4	3	2	2	2	2
0.30	244	122	49	24	12	8	5	3	3	2	2	2	2
0.50	168	84	34	17	9	6	3	3	2	2	2	2	2
0.70	110	55	22	11	6	4	3	2	2	2	2	2	2
0.80	83	42	17	9	5	3	2	2	2	2	2	2	2
0.90	54	27	11	6	3	3	2	2	2	2	2	2	2
0.95	36	19	8	4	3	2	2	2	2	2	2	2	2
0.98	22	12	5	3	2	2	2	2	2	2	2	2	2
0.99	16	8	4	2	2	2	2	2	2	2	2	2	2

* L. G. Dion, M.I.T., 1951.

TABLE A–14*

CHI SQUARE

Values of χ^2_α satisfying $\int_{\chi^2_\alpha}^{\infty} f(\chi^2; m)\, d\chi^2 = \alpha$

$m \backslash \alpha$	0.995	0.990	0.975	0.950	0.900	0.750
1	$392704 \cdot 10^{-10}$	$157088 \cdot 10^{-9}$	$982069 \cdot 10^{-9}$	$393214 \cdot 10^{-8}$	0.0157908	0.1015308
2	0.0100251	0.0201007	0.0506356	0.102587	0.210720	0.575364
3	0.0717212	0.114832	0.215795	0.351846	0.584375	1.212534
4	0.206990	0.297110	0.484419	0.710721	1.063623	1.92255
5	0.411740	0.554300	0.831211	1.145476	1.61031	2.67460
6	0.675727	0.872085	1.237347	1.63539	2.20413	3.45460
7	0.989265	1.239043	1.68987	2.16735	2.83311	4.25485
8	1.344419	1.646482	2.17973	2.73264	3.48954	5.07064
9	1.734926	2.087912	2.70039	3.32511	4.16816	5.89883
10	2.15585	2.55821	3.24697	3.94030	4.86518	6.73720
11	2.60321	3.05347	3.81575	4.57481	5.57779	7.58412
12	3.07382	3.57056	4.40379	5.22603	6.30380	8.43842
13	3.56503	4.10691	5.00874	5.89186	7.04150	9.29906
14	4.07468	4.66043	5.62872	6.57063	7.78953	10.1653
15	4.60094	5.22935	6.26214	7.26094	8.54675	11.0365
16	5.14224	5.81221	6.90766	7.96164	9.31223	11.9122
17	5.69724	6.40776	7.56418	8.67176	10.0852	12.7919
18	6.26481	7.01491	8.23075	9.39046	10.8649	13.6753
19	6.84398	7.63273	8.90655	10.1170	11.6509	14.5620
20	7.43386	8.26040	9.59083	10.8508	12.4426	15.4518
21	8.03366	8.89720	10.28293	11.5913	13.2396	16.3444
22	8.64272	9.54249	10.9823	12.3380	14.0415	17.2396
23	9.26042	10.19567	11.6885	13.0905	14.8479	18.1373
24	9.88623	10.8564	12.4011	13.8484	15.6587	19.0372
25	10.5197	11.5240	13.1197	14.6114	16.4734	19.9393
26	11.1603	12.1981	13.8439	15.3791	17.2919	20.8434
27	11.8076	12.8786	14.5733	16.1513	18.1138	21.7494
28	12.4613	13.5648	15.3079	16.9279	18.9392	22.6572
29	13.1211	14.2565	16.0471	17.7083	19.7677	23.5666
30	13.7867	14.9535	16.7908	18.4926	20.5992	24.4776
40	20.7065	22.1643	24.4331	26.5093	29.0505	33.6603
50	27.9907	29.7067	32.3574	34.7642	37.6886	42.9421
60	35.5346	37.4848	40.4817	43.1879	46.4589	52.2938
70	43.2752	45.4418	48.7576	51.7393	55.3290	61.6983
80	51.1720	53.5400	57.1532	60.3915	64.2778	71.1445
90	59.1963	61.7541	65.6466	69.1260	73.2912	80.6247
100	67.3276	70.0648	74.2219	77.9295	82.3581	90.1332

* Catherine M. Thompson, *Biometrika,* Vol. 32 (1941), pp. 188–189.

(Continued)

Table A–14

$m \backslash \alpha$	0.500	0.250	0.100	0.050	0.025	0.010	0.005
1	0.454937	1.32330	2.70554	3.84146	5.02389	6.63490	7.87944
2	1.38629	2.77259	4.60517	5.99147	7.37776	9.21034	10.5966
3	2.36597	4.10835	6.25139	7.81473	9.34840	11.3449	12.8381
4	3.35670	5.38527	7.77944	9.48773	11.1433	13.2767	14.8602
5	4.35146	6.62568	9.23635	11.0705	12.8325	15.0863	16.7496
6	5.34812	7.84080	10.6446	12.5916	14.4494	16.8119	18.5476
7	6.34581	9.03715	12.0170	14.0671	16.0128	18.4753	20.2777
8	7.34412	10.2188	13.3616	15.5073	17.5346	20.0902	21.9550
9	8.34283	11.3887	14.6837	16.9190	19.0228	21.6660	23.5893
10	9.34182	12.5489	15.9871	18.3070	20.4831	23.2093	25.1882
11	10.3410	13.7007	17.2750	19.6751	21.9200	24.7250	26.7569
12	11.3403	14.8454	18.5494	21.0261	23.3367	26.2170	28.2995
13	12.3398	15.9839	19.8119	22.3621	24.7356	27.6883	29.8194
14	13.3393	17.1170	21.0642	23.6848	26.1190	29.1413	31.3193
15	14.3389	18.2451	22.3072	24.9958	27.4884	30.5779	32.8013
16	15.3385	19.3688	23.5418	26.2962	28.8454	31.9999	34.2672
17	16.3381	20.4887	24.7690	27.5871	30.1910	33.4087	35.7185
18	17.3379	21.6049	25.9894	28.8693	31.5264	34.8053	37.1564
19	18.3376	22.7178	27.2036	30.1435	32.8523	36.1908	38.5822
20	19.3374	23.8277	28.4120	31.4104	34.1696	37.5662	39.9968
21	20.3372	24.9348	29.6151	32.6705	35.4789	38.9321	41.4010
22	21.3370	26.0393	30.8133	33.9244	36.7807	40.2894	42.7956
23	22.3369	27.1413	32.0069	35.1725	38.0757	41.6384	44.1813
24	23.3367	28.2412	33.1963	36.4151	39.3641	42.9798	45.5585
25	24.3366	29.3389	34.3816	37.6525	40.6465	44.3141	46.9278
26	25.3364	30.4345	35.5631	38.8852	41.9232	45.6417	48.2899
27	26.3363	31.5284	36.7412	40.1133	43.1944	46.9630	49.6449
28	27.3363	32.6205	37.9159	41.3372	44.4607	48.2782	50.9933
29	28.3362	33.7109	39.0875	42.5569	45.7222	49.5879	52.3356
30	29.3360	34.7998	40.2560	43.7729	46.9792	50.8922	53.6720
40	39.3354	45.6160	51.8050	55.7585	59.3417	63.6907	66.7659
50	49.3349	56.3336	63.1671	67.5048	71.4202	76.1539	79.4900
60	59.3347	66.9814	74.3970	79.0819	83.2976	88.3794	91.9517
70	69.3344	77.5766	85.5271	90.5312	95.0231	100.425	104.215
80	79.3343	88.1303	96.5782	101.879	106.629	112.329	116.321
90	89.3342	98.6499	107.565	113.145	118.136	124.116	128.299
100	99.3341	109.141	118.498	124.342	129.561	135.807	140.169

Table A–15*

Student's t

Values of t_α satisfying $\int_{-t_\alpha}^{+t_\alpha} f(t; m)\, dt = 1 - \frac{\alpha}{2}$

m \ α	0.50	0.25	0.10	0.05	0.025	0.01	0.005
1	1.00000	2.4142	6.3138	12.706	25.452	63.657	127.32
2	0.81650	1.6036	2.9200	4.3027	6.2053	9.9248	14.089
3	0.76489	1.4226	2.3534	3.1825	4.1765	5.8409	7.4533
4	0.74070	1.3444	2.1318	2.7764	3.4954	4.6041	5.5976
5	0.72669	1.3009	2.0150	2.5706	3.1634	4.0321	4.7733
6	0.71756	1.2733	1.9432	2.4469	2.9687	3.7074	4.3168
7	0.71114	1.2543	1.8946	2.3646	2.8412	3.4995	4.0293
8	0.70639	1.2403	1.8595	2.3060	2.7515	3.3554	3.8325
9	0.70272	1.2297	1.8331	2.2622	2.6850	3.2498	3.6897
10	0.69981	1.2213	1.8125	2.2281	2.6338	3.1693	3.5814
11	0.69745	1.2145	1.7959	2.2010	2.5931	3.1058	3.4966
12	0.69548	1.2089	1.7823	2.1788	2.5600	3.0545	3.4284
13	0.69384	1.2041	1.7709	2.1604	2.5326	3.0123	3.3725
14	0.69242	1.2001	1.7613	2.1448	2.5096	2.9768	3.3257
15	0.69120	1.1967	1.7530	2.1315	2.4899	2.9467	3.2860
16	0.69013	1.1937	1.7459	2.1199	2.4729	2.9208	3.2520
17	0.68919	1.1910	1.7396	2.1098	2.4581	2.8982	3.2225
18	0.68837	1.1887	1.7341	2.1009	2.4450	2.8784	3.1966
19	0.68763	1.1866	1.7291	2.0930	2.4334	2.8609	3.1737
20	0.68696	1.1848	1.7247	2.0860	2.4231	2.8453	3.1534
21	0.68635	1.1831	1.7207	2.0796	2.4138	2.8314	3.1352
22	0.69580	1.1816	1.7171	2.0739	2.4055	2.8188	3.1188
23	0.68531	1.1802	1.7139	2.0687	2.3979	2.8073	3.1040
24	0.68485	1.1789	1.7109	2.0639	2.3910	2.7969	3.0905
25	0.68443	1.1777	1.7081	2.0595	2.3846	2.7874	3.0782
26	0.68405	1.1766	1.7056	2.0555	2.3788	2.7787	3.0669
27	0.68370	1.1757	1.7033	2.0518	2.3734	2.7707	3.0565
28	0.68335	1.1748	1.7011	2.0484	2.3685	2.7633	3.0469
29	0.68304	1.1739	1.6991	2.0452	2.3638	2.7564	3.0380
30	0.68276	1.1731	1.6973	2.0423	2.3596	2.7500	3.0298
40	0.68066	1.1673	1.6839	2.0211	2.3289	2.7045	2.9712
60	0.67862	1.1616	1.6707	2.0003	2.2991	2.6603	2.9146
120	0.67656	1.1559	1.6577	1.9799	2.2699	2.6174	2.8599
∞	0.67449	1.1503	1.6449	1.9600	2.2414	2.5758	2.8070

TABLE A-16*

F

Values of F_α satisfying $\int_0^{F_\alpha} f(F; a, b)\, dF = \alpha$

$\alpha = 0.95$

b \ a	1	2	3	4	5	6	7	8	9
1	161.45	199.50	215.71	224.58	230.16	233.99	236.77	238.88	240.54
2	18.513	19.000	19.164	19.247	19.296	19.330	19.353	19.371	19.385
3	10.128	9.5521	9.2766	9.1172	9.0135	8.9406	8.8868	8.8452	8.8123
4	7.7086	6.9443	6.5914	6.3883	6.2560	6.1631	6.0942	6.0410	5.9988
5	6.6079	5.7861	5.4095	5.1922	5.0503	4.9503	4.8759	4.8183	4.7725
6	5.9874	5.1433	4.7571	4.5337	4.3874	4.2839	4.2066	4.1468	4.0990
7	5.5914	4.7374	4.3468	4.1203	3.9715	3.8660	3.7870	3.7257	3.6767
8	5.3177	4.4590	4.0662	3.8378	3.6875	3.5806	3.5005	3.4381	3.3881
9	5.1174	4.2565	3.8626	3.6331	3.4817	3.3738	3.2927	3.2296	3.1789
10	4.9646	4.1028	3.7083	3.4780	3.3258	3.2172	3.1355	3.0717	3.0204
11	4.8443	3.9823	3.5874	3.3567	3.2039	3.0946	3.0123	2.9480	2.8962
12	4.7472	3.8853	3.4903	3.2592	3.1059	2.9961	2.9134	2.8486	2.7964
13	4.6672	3.8056	3.4105	3.1791	3.0254	2.9153	2.8321	2.7669	2.7144
14	4.6001	3.7389	3.3439	3.1122	2.9582	2.8477	2.7642	2.6987	2.6458
15	4.5431	3.6823	3.2874	3.0556	2.9013	2.7905	2.7066	2.6408	2.5876
16	4.4940	3.6337	3.2389	3.0069	2.8524	2.7413	2.6572	2.5911	2.5377
17	4.4513	3.5915	3.1968	2.9647	2.8100	2.6987	2.6143	2.5480	2.4943
18	4.4139	3.5546	3.1599	2.9277	2.7729	2.6613	2.5767	2.5102	2.4563
19	4.3808	3.5219	3.1274	2.8951	2.7401	2.6283	2.5435	2.4768	2.4227
20	4.3513	3.4928	3.0984	2.8661	2.7109	2.5990	2.5140	2.4471	2.3928
21	4.3248	3.4668	3.0725	2.8401	2.6848	2.5727	2.4876	2.4205	2.3661
22	4.3009	3.4434	3.0491	2.8167	2.6613	2.5491	2.4638	2.3965	2.3419
23	4.2793	3.4221	3.0280	2.7955	2.6400	2.5277	2.4422	2.3748	2.3201
24	4.2597	3.4028	3.0088	2.7763	2.6207	2.5082	2.4226	2.3551	2.3002
25	4.2417	3.3852	2.9912	2.7587	2.6030	2.4904	2.4047	2.3371	2.2821
26	4.2252	3.3690	2.9751	2.7426	2.5868	2.4741	2.3883	2.3205	2.2655
27	4.2100	3.3541	2.9604	2.7278	2.5719	2.4591	2.3732	2.3053	2.2501
28	4.1960	3.3404	2.9467	2.7141	2.5581	2.4453	2.3593	2.2913	2.2360
29	4.1830	3.3277	2.9340	2.7014	2.5454	2.4324	2.3463	2.2782	2.2229
30	4.1709	3.3158	2.9223	2.6896	2.5336	2.4205	2.3343	2.2662	2.2107
40	4.0848	3.2317	2.8387	2.6060	2.4495	2.3359	2.2490	2.1802	2.1240
60	4.0012	3.1504	2.7581	2.5252	2.3683	2.2540	2.1665	2.0970	2.0401
120	3.9201	3.0718	2.6802	2.4472	2.2900	2.1750	2.0867	2.0164	1.9588
∞	3.8415	2.9957	2.6049	2.3719	2.2141	2.0986	2.0096	1.9384	1.8799

TABLE A–16

$\alpha = 0.95$

b \ a	10	12	15	20	24	30	40	60	120	∞
1	241.88	243.91	245.95	248.01	249.05	250.09	251.14	252.20	253.25	254.32
2	19.396	19.413	19.429	19.446	19.454	19.462	19.471	19.479	19.487	19.496
3	8.7855	8.7446	8.7029	8.6602	8.6385	8.6166	8.5944	8.5720	8.5494	8.5265
4	5.9644	5.9117	5.8578	5.8025	5.7744	5.7459	5.7170	5.6878	5.6581	5.6281
5	4.7351	4.6777	4.6188	4.5581	4.5272	4.4957	4.4638	4.4314	4.3984	4.3650
6	4.0600	3.9999	3.9381	3.8742	3.8415	3.8082	3.7743	3.7398	3.7047	3.6688
7	3.6365	3.5747	3.5108	3.4445	3.4105	3.3758	3.3404	3.3043	3.2674	3.2298
8	3.3472	3.2840	3.2184	3.1503	3.1152	3.0794	3.0428	3.0053	2.9669	2.9276
9	3.1373	3.0729	3.0061	2.9365	2.9005	2.8637	2.8259	2.7872	2.7475	2.7067
10	2.9782	2.9130	2.8450	2.7740	2.7372	2.6996	2.6609	2.6211	2.5801	2.5379
11	2.8536	2.7876	2.7186	2.6464	2.6090	2.5705	2.5309	2.4901	2.4480	2.4045
12	2.7534	2.6866	2.6169	2.5436	2.5055	2.4663	2.4259	2.3842	2.3410	2.2962
13	2.6710	2.6037	2.5331	2.4589	2.4202	2.3803	2.3392	2.2966	2.2524	2.2064
14	2.6021	2.5342	2.4630	2.3879	2.3487	2.3082	2.2664	2.2230	2.1778	2.1307
15	2.5437	2.4753	2.4035	2.3275	2.2878	2.2468	2.2043	2.1601	2.1141	2.0658
16	2.4935	2.4247	2.3522	2.2756	2.2354	2.1938	2.1507	2.1058	2.0589	2.0096
17	2.4499	2.3807	2.3077	2.2304	2.1898	2.1477	2.1040	2.0584	2.0107	1.9604
18	2.4117	2.3421	2.2686	2.1906	2.1497	2.1071	2.0629	2.0166	1.9681	1.9168
19	2.3779	2.3080	2.2341	2.1555	2.1141	2.0712	2.0264	1.9796	1.9302	1.8780
20	2.3479	2.2776	2.2033	2.1242	2.0825	2.0391	1.9938	1.9464	1.8963	1.8432
21	2.3210	2.2504	2.1757	2.0960	2.0540	2.0102	1.9645	1.9165	1.8657	1.8117
22	2.2967	2.2258	2.1508	2.0707	2.0283	1.9842	1.9380	1.8895	1.8380	1.7831
23	2.2747	2.2036	2.1282	2.0476	2.0050	1.9605	1.9139	1.8649	1.8128	1.7570
24	2.2547	2.1834	2.1077	2.0267	1.9838	1.9390	1.8920	1.8424	1.7897	1.7331
25	2.2365	2.1649	2.0889	2.0075	1.9643	1.9192	1.8718	1.8217	1.7684	1.7110
26	2.2197	2.1479	2.0716	1.9898	1.9464	1.9010	1.8533	1.8027	1.7488	1.6906
27	2.2043	2.1323	2.0558	1.9736	1.9299	1.8842	1.8361	1.7851	1.7307	1.6717
28	2.1900	2.1179	2.0411	1.9586	1.9147	1.8687	1.8203	1.7689	1.7138	1.6541
29	2.1768	2.1045	2.0275	1.9446	1.9005	1.8543	1.8055	1.7537	1.6981	1.6377
30	2.1646	2.0921	2.0148	1.9317	1.8874	1.8409	1.7918	1.7396	1.6835	1.6223
40	2.0772	2.0035	1.9245	1.8389	1.7929	1.7444	1.6928	1.6373	1.5766	1.5089
60	1.9926	1.9174	1.8364	1.7480	1.7001	1.6491	1.5943	1.5343	1.4673	1.3893
120	1.9105	1.8337	1.7505	1.6587	1.6084	1.5543	1.4952	1.4290	1.3519	1.2539
∞	1.8307	1.7522	1.6664	1.5705	1.5173	1.4591	1.3940	1.3180	1.2214	1.0000

(Continued)

TABLE A–16

$\alpha = 0.99$

b \ a	1	2	3	4	5	6	7	8	9
1	4052.2	4999.5	5403.3	5624.6	5763.7	5859.0	5928.3	5981.6	6022.5
2	98.503	99.000	99.166	99.249	99.299	99.332	99.356	99.374	99.388
3	34.116	30.817	29.457	28.710	28.237	27.911	27.672	27.489	27.345
4	21.198	18.000	16.694	15.977	15.522	15.207	14.976	14.799	14.659
5	16.258	13.274	12.060	11.392	10.967	10.672	10.456	10.289	10.158
6	13.745	10.925	9.7795	9.1483	8.7459	8.4661	8.2600	8.1016	7.9761
7	12.246	9.5466	8.4513	7.8467	7.4604	7.1914	6.9928	6.8401	6.7188
8	11.259	8.6491	7.5910	7.0060	6.6318	6.3707	6.1776	6.0289	5.9106
9	10.561	8.0215	6.9919	6.4221	6.0569	5.8018	5.6129	5.4671	5.3511
10	10.044	7.5594	6.5523	5.9943	5.6363	5.3858	5.2001	5.0567	4.9424
11	9.6460	7.2057	6.2167	5.6683	5.3160	5.0692	4.8861	4.7445	4.6315
12	9.3302	6.9266	5.9526	5.4119	5.0643	4.8206	4.6395	4.4994	4.3875
13	9.0738	6.7010	5.7394	5.2053	4.8616	4.6204	4.4410	4.3021	4.1911
14	8.8616	6.5149	5.5639	5.0354	4.6950	4.4558	4.2779	4.1399	4.0297
15	8.6831	6.3589	5.4170	4.8932	4.5556	4.3183	4.1415	4.0045	3.8948
16	8.5310	6.2262	5.2922	4.7726	4.4374	4.2016	4.0259	3.8896	3.7804
17	8.3997	6.1121	5.1850	4.6690	4.3359	4.1015	3.9267	3.7910	3.6822
18	8.2854	6.0129	5.0919	4.5790	4.2479	4.0146	3.8406	3.7054	3.5971
19	8.1850	5.9259	5.0103	4.5003	4.1708	3.9386	3.7653	3.6305	3.5225
20	8.0960	5.8489	4.9382	4.4307	4.1027	3.8714	3.6987	3.5644	3.4567
21	8.0166	5.7804	4.8740	4.3688	4.0421	3.8117	3.6396	3.5056	3.3981
22	7.9454	5.7190	4.8166	4.3134	3.9880	3.7583	3.5867	3.4530	3.3458
23	7.8811	5.6637	4.7649	4.2635	3.9392	3.7102	3.5390	3.4057	3.2986
24	7.8229	5.6136	4.7181	4.2184	3.8951	3.6667	3.4959	3.3629	3.2560
25	7.7698	5.5680	4.6755	4.1774	3.8550	3.6272	3.4568	3.3239	3.2172
26	7.7213	5.5263	4.6366	4.1400	3.8183	3.5911	3.4210	3.2884	3.1818
27	7.6767	5.4881	4.6009	4.1056	3.7848	3.5580	3.3882	3.2558	3.1494
28	7.6356	5.4529	4.5681	4.0740	3.7539	3.5276	3.3581	3.2259	3.1195
29	7.5976	5.4205	4.5378	4.0449	3.7254	3.4995	3.3302	3.1982	3.0920
30	7.5625	5.3904	4.5097	4.0179	3.6990	3.4735	3.3045	3.1726	3.0665
40	7.3141	5.1785	4.3126	3.8283	3.5138	3.2910	3.1238	2.9930	2.8876
60	7.0771	4.9774	4.1259	3.6491	3.3389	3.1187	2.9530	2.8233	2.7185
120	6.8510	4.7865	3.9493	3.4796	3.1735	2.9559	2.7918	2.6629	2.5586
∞	6.6349	4.6052	3.7816	3.3192	3.0173	2.8020	2.6393	2.5113	2.4073

TABLE A–16

$\alpha = 0.99$

$b \backslash a$	10	12	15	20	24	30	40	60	120	∞
1	6055.8	6106.3	6157.3	6208.7	6234.6	6260.7	6286.8	6313.0	6339.4	6366.0
2	99.399	99.416	99.432	99.449	99.458	99.466	99.474	99.483	99.491	99.501
3	27.229	27.052	26.872	26.690	26.598	26.505	26.411	26.316	26.221	26.125
4	14.546	14.374	14.198	14.020	13.929	13.838	13.745	13.652	13.558	13.463
5	10.051	9.8883	9.7222	9.5527	9.4665	9.3793	9.2912	9.2020	9.1118	9.0204
6	7.8741	7.7183	7.5590	7.3958	7.3127	7.2285	7.1432	7.0568	6.9690	6.8801
7	6.6201	6.4691	6.3143	6.1554	6.0743	5.9921	5.9084	5.8236	5.7372	5.6495
8	5.8143	5.6668	5.5151	5.3591	5.2793	5.1981	5.1156	5.0316	4.9460	4.8588
9	5.2565	5.1114	4.9621	4.8080	4.7290	4.6486	4.5667	4.4831	4.3978	4.3105
10	4.8492	4.7059	4.5582	4.4054	4.3269	4.2469	4.1653	4.0819	3.9965	3.9090
11	4.5393	4.3974	4.2509	4.0990	4.0209	3.9411	3.8596	3.7761	3.6904	3.6025
12	4.2961	4.1553	4.0096	3.8584	3.7805	3.7008	3.6192	3.5355	3.4494	3.3608
13	4.1003	3.9603	3.8154	3.6646	3.5868	3.5070	3.4253	3.3413	3.2548	3.1654
14	3.9394	3.8001	3.6557	3.5052	3.4274	3.3476	3.2656	3.1813	3.0942	3.0040
15	3.8049	3.6662	3.5222	3.3719	3.2940	3.2141	3.1319	3.0471	2.9595	2.8684
16	3.6909	3.5527	3.4089	3.2588	3.1808	3.1007	3.0182	2.9330	2.8447	2.7528
17	3.5931	3.4552	3.3117	3.1615	3.0835	3.0032	2.9205	2.8348	2.7459	2.6530
18	3.5082	3.3706	3.2273	3.0771	2.9990	2.9185	2.8354	2.7493	2.6597	2.5660
19	3.4338	3.2965	3.1533	3.0031	2.9249	2.8442	2.7608	2.6742	2.5839	2.4893
20	3.3682	3.2311	3.0880	2.9377	2.8594	2.7785	2.6947	2.6077	2.5168	2.4212
21	3.3098	3.1729	3.0299	2.8796	2.8011	2.7200	2.6359	2.5484	2.4568	2.3603
22	3.2576	3.1209	2.9780	2.8274	2.7488	2.6675	2.5831	2.4951	2.4029	2.3055
23	3.2106	3.0740	2.9311	2.7805	2.7017	2.6202	2.5355	2.4471	2.3542	2.2559
24	3.1681	3.0316	2.8887	2.7380	2.6591	2.5773	2.4923	2.4035	2.3099	2.2107
25	3.1294	2.9931	2.8502	2.6993	2.6203	2.5383	2.4530	2.3637	2.2695	2.1694
26	3.0941	2.9579	2.8150	2.6640	2.5848	2.5026	2.4170	2.3273	2.2325	2.1315
27	3.0618	2.9256	2.7827	2.6316	2.5522	2.4699	2.3840	2.2938	2.1984	2.0965
28	3.0320	2.8959	2.7530	2.6017	2.5223	2.4397	2.3535	2.2629	2.1670	2.0642
29	3.0045	2.8685	2.7256	2.5742	2.4946	2.4118	2.3253	2.2344	2.1378	2.0342
30	2.9791	2.8431	2.7002	2.5487	2.4689	2.3860	2.2992	2.2079	2.1107	2.0062
40	2.8005	2.6648	2.5216	2.3689	2.2880	2.2034	2.1142	2.0194	1.9172	1.8047
60	2.6318	2.4961	2.3523	2.1978	2.1154	2.0285	1.9360	1.8363	1.7263	1.6006
120	2.4721	2.3363	2.1915	2.0346	1.9500	1.8600	1.7628	1.6557	1.5330	1.3805
∞	2.3209	2.1848	2.0385	1.8783	1.7908	1.9664	1.5923	1.4730	1.3246	1.0000

Table A–17*

Confidence Limits for Poisson Parameter

x	Confidence coefficient				x	Confidence coefficient			
	0.99		0.95			0.99		0.95	
	Lower limit	Upper limit	Lower limit	Upper limit		Lower limit	Upper limit	Lower limit	Upper limit
0	0.0	5.3	0.0	3.7					
1	0.0	7.4	0.1	5.6	26	14.7	42.2	17.0	38.0
2	0.1	9.3	0.2	7.2	27	15.4	43.5	17.8	39.2
3	0.3	11.0	0.6	8.8	28	16.2	44.8	18.6	40.4
4	0.6	12.6	1.0	10.2	29	17.0	46.0	19.4	41.6
5	1.0	14.1	1.6	11.7	30	17.7	47.2	20.2	42.8
6	1.5	15.6	2.2	13.1	31	18.5	48.4	21.0	44.0
7	2.0	17.1	2.8	14.4	32	19.3	49.6	21.8	45.1
8	2.5	18.5	3.4	15.8	33	20.0	50.8	22.7	46.3
9	3.1	20.0	4.0	17.1	34	20.8	52.1	23.5	47.5
10	3.7	21.3	4.7	18.4	35	21.6	53.3	24.3	48.7
11	4.3	22.6	5.4	19.7	36	22.4	54.5	25.1	49.8
12	4.9	24.0	6.2	21.0	37	23.2	55.7	26.0	51.0
13	5.5	25.4	6.9	22.3	38	24.0	56.9	26.8	52.2
14	6.2	26.7	7.7	23.5	39	24.8	58.1	27.7	53.3
15	6.8	28.1	8.4	24.8	40	25.6	59.3	28.6	54.5
16	7.5	29.4	9.2	26.0	41	26.4	60.5	29.4	55.6
17	8.2	30.7	9.9	27.2	42	27.2	61.7	30.3	56.8
18	8.9	32.0	10.7	28.4	43	28.0	62.9	31.1	57.9
19	9.6	33.3	11.5	29.6	44	28.8	64.1	32.0	59.0
20	10.3	34.6	12.2	30.8	45	29.6	65.3	32.8	60.2
21	11.0	35.9	13.0	32.0	46	30.4	66.5	33.6	61.3
22	11.8	37.2	13.8	33.2	47	31.2	67.7	34.5	62.5
23	12.5	38.4	14.6	34.4	48	32.0	68.9	35.3	63.6
24	13.2	39.7	15.4	35.6	49	32.8	70.1	36.1	64.8
25	14.0	41.0	16.2	36.8	50	33.6	71.3	37.0	65.9

* W. E. Ricker, *Jour. Amer. Stat. Assoc.*, Vol. 32 (1937), pp. 349–356.

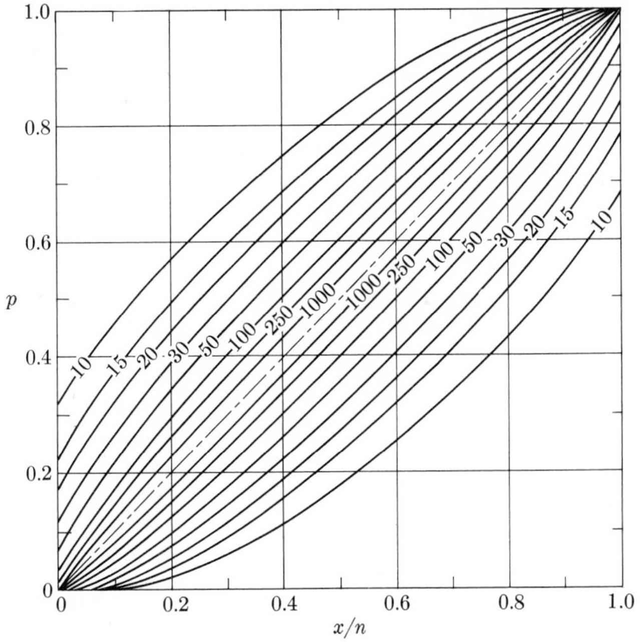

FIG. A–1. Confidence limits for binomial p. Confidence limit $= 0.95$. [From C. J. Clopper and E. S. Pearson, *Biometrika,* Vol. 26 (1934), pp. 404–413.]

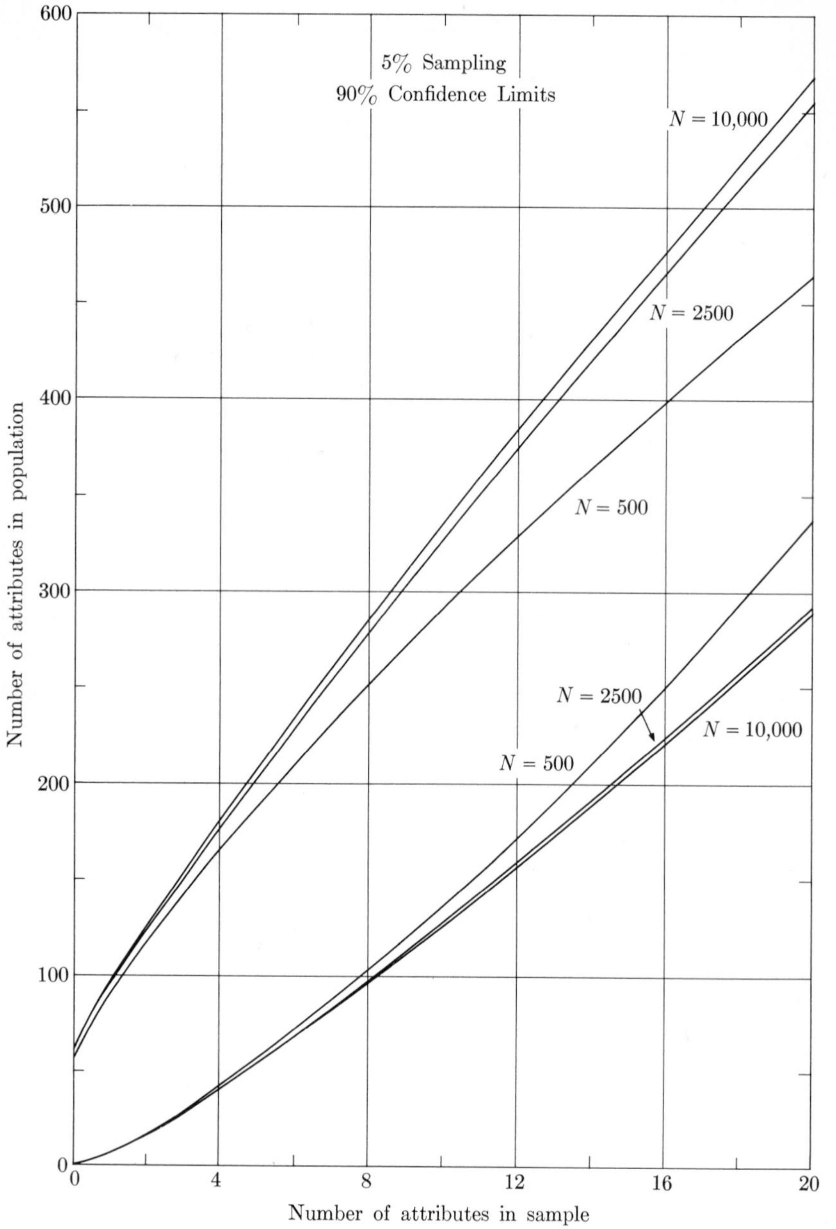

FIG. A–2. Confidence limits for hypergeometric p. [From J. H. Chung and D. B. DeLury, *Confidence Limits for the Hypergeometric Distribution*. Ontario Research Foundation. University of Toronto Press, 1950.]

Name Index

This index of names includes those noted in the preface (pages i–xiv) and in the main body of the text (pages 3–381). Those appearing in the lists following the text are not included.

Names appearing in connection with theorems (Wilks' tolerance theorem, the Neyman-Pearson theorem, ...) are generally not included here.

Classical writers (Pascal, Gauss, ...) are, with a few exceptions, not included.

NAME INDEX

Subject Index

SUBJECT INDEX

ABCDE69876543

NOTES

$$\binom{n}{x} + \binom{n}{x-1} = \binom{n+1}{x} \qquad \binom{n-1}{x} + \binom{n-1}{x-1} = \binom{n}{x} \qquad \binom{n-1}{r-1} = \binom{n-1}{n-r}$$

$$n^{[x]} = {}^{n}P_{x} = n(n-1)\ldots(n-x+1).$$

Mean $\quad E(x) = \int_{-\infty}^{\infty} x\, f(x)$

VARIANCE $\quad = E(x^2) - E^2(x)$

COV. $\quad \mathrm{Cov}(x, y) = E\left[(x - E(x))(y - E(y))\right]$
$\quad = E(xy) - E(x)E(y)$

$$M_x(\theta) = \int_{-\infty}^{\infty} e^{\theta x} f(x)\,dx$$

$$M_{x-\mu}(\theta) = e^{-\mu\theta} M_x(\theta)$$

Poisson sum of 2 dist. $\quad \dfrac{e^{-(A+B)}(A+B)^x}{x!}$

Cauchy

$$f(x)\,dx = \frac{1}{\pi} \cdot \frac{1}{1+(x-\theta)^2}\,dx$$

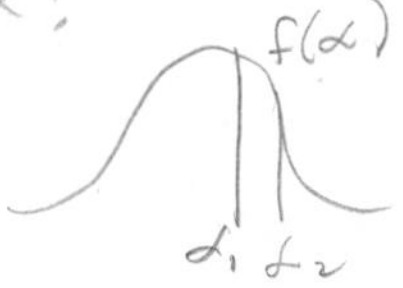

$$\int_{-\pi/2}^{\pi/2} \frac{1}{\pi}\,dv = 1 \qquad 0 \text{ elsewhere.}$$

Normal $\quad f(x) = \dfrac{1}{\sigma\sqrt{2\pi}}\, e^{\frac{(x-\lambda)^2}{2\sigma^2}} \qquad E(x) = c \qquad V(x) = \dfrac{1}{2B}$

$$F(x) = \int \frac{1}{\sigma\sqrt{2\pi}}\, e^{-\frac{1}{2}\left(\frac{t-\lambda}{\sigma}\right)^2} dt$$

Notation $\quad N(65, 10)$
Mean $= 65$
$\sigma^2 = 10$

estimate μ by

$$\frac{P(X \text{ between } \mu_1) + P(X \text{ betw } \mu_2)}{2}$$